Control of Distribution System Water Quality Using Inhibitors

Awwa
Research
Foundation

About the Awwa Research Foundation

The Awwa Research Foundation (AwwaRF) is a member-supported, international, nonprofit organization that sponsors research to enable water utilities, public health agencies, and other professionals to provide safe and affordable drinking water to consumers.

The Foundation's mission is to advance the science of water to improve the quality of life. To achieve this mission, the Foundation sponsors studies on all aspects of drinking water, including supply and resources, treatment, monitoring and analysis, distribution, management, and health effects. Funding for research is provided primarily by subscription payments from approximately 1,000 utilities, consulting firms, and manufacturers in North America and abroad. Additional funding comes from collaborative partnerships with other national and international organizations, allowing for resources to be leveraged, expertise to be shared, and broad-based knowledge to be developed and disseminated. Government funding serves as a third source of research dollars.

From its headquarters in Denver, Colorado, the Foundation's staff directs and supports the efforts of more than 800 volunteers who serve on the board of trustees and various committees. These volunteers represent many facets of the water industry, and contribute their expertise to select and monitor research studies that benefit the entire drinking water community.

The results of research are disseminated through a number of channels, including reports, the Web site, conferences, and periodicals.

For subscribers, the Foundation serves as a cooperative program in which water suppliers unite to pool their resources. By applying Foundation research findings, these water suppliers can save substantial costs and stay on the leading edge of drinking water science and technology. Since its inception, AwwaRF has supplied the water community with more than $300 million in applied research.

More information about the Foundation and how to become a subscriber is available on the Web at **www.awwarf.org**.

Control of Distribution System Water Quality Using Inhibitors

Prepared by:
J.S. Taylor, J.D. Dietz, A.A. Randall, C.D. Norris, A. Alshehri, J. Arévalo, X. Guan, P. Lintereur, D. MacNevin, E. Stone, R. Vaidya, B. Zhao, S. Glatthorn, and A. Shekhar
University of Central Florida
4000 Central Florida Blvd., Orlando, FL 32816

Sponsored by:
Awwa Research Foundation
6666 West Quincy Avenue, Denver, CO 80235-3098

and

Tampa Bay Water
2535 Landmark Drive, Clearwater, FL 33761

Published by:

Awwa Research Foundation

American Water Works Association

IWA Publishing

DISCLAIMER

This study was jointly funded by the Awwa Research Foundation (AwwaRF) and Tampa Bay Water (TBW). AwwaRF and TBW assume no responsibility for the content of the research study reported in this publication or for the opinions or statements of fact expressed in the report. The mention of trade names for commercial products does not represent or imply the approval or endorsement of AwwaRF or TBW. This report is presented solely for informational purposes.

CONTENTS

TABLES

FIGURES

FOREWORD

The Awwa Research Foundation (AwwaRF) is a nonprofit corporation that is dedicated to the implementation of a research effort to help utilities respond to regulatory requirements and traditional high-priority concerns of the industry. The research agenda is developed through a process of consultation with subscribers and drinking water professionals. Under the umbrella of a Strategic Research Plan, the Research Advisory Council prioritizes the suggested projects based upon current and future needs, applicability, and past work; the recommendations are forwarded to the Board of Trustees for final selection. The foundation also sponsors research projects through an unsolicited proposal process; the Collaborative Research, Research Applications, and Tailored Collaboration programs; and various joint research efforts with organizations such as the U.S. Environmental Protection Agency, the U.S. Bureau of Reclamation, and the Association of California Water Agencies.

This publication is a result of one of these sponsored studies, and it is hoped that its findings will be applied in communities throughout the world. The following report serves not only as a means of communicating the results of the water industry's centralized research program but also as a tool to enlist the further support of the nonmember utilities and individuals.

Projects are managed closely from their inception to the final report by the foundation's staff and large cadre of volunteers who willingly contribute their time and expertise. The foundation serves a planning and management function and awards contracts to other institutions such as water utilities, universities, and engineering firms. The funding for this research effort comes primarily from the Subscription Program, through which water utilities subscribe to the research program and make an annual payment proportionate to the volume of water they deliver and consultants and manufacturers subscribe based on their annual billings. The program offers a cost-effective and fair method for funding research in the public interest.

A broad spectrum of water supply issues is addressed by the foundation's research agenda: resources, treatment and operations, distribution and storage, water quality and analysis, toxicology, economics, and management. The ultimate purpose of the coordinated effort is to assist water suppliers to provide the highest possible quality of water economically and reliably. The true benefits are realized when the results are implemented at the utility level. The foundation's trustees are pleased to offer this publication as a contribution toward that end.

David E. Rager
Chair, Board of Trustees
Awwa Research Foundation

Robert C. Renner, P.E.
Executive Director
Awwa Research Foundation

ACKNOWLEDGMENTS

The University of Central Florida (UCF) and UCF project team wish to express their sincere gratitude to Tampa Bay Water (TBW) and TBW's member governments: Hillsborough County, Fla.; Pasco County, Fla.; Pinellas County, Fla.; City of New Port Richey, Fla.; City of St. Petersburg, Fla.; and City of Tampa, Fla.; and to Awwa Research Foundation (AwwaRF) for their support and funding for this project.

Special recognition and gratitude is given to the following individuals:

- Christine Owen (water quality assurance officer for TBW and also the tailored collaboration projects project coordinator) for her endless support and coordination of all UCF-project-related activities.
- Robert M. Powell (Pinellas County, Fla., director of laboratory) and Pick Tally (Pinellas County, Fla., director of utilities) for their support during the proposal and conduct of the project.
- Roy Martinez (AwwaRF project officer) for his continuous support and assistance in proposing, conducting, reporting, and performing all activities associated with the project.
- Jonathan Clement, Ph.D., P.E. (Black & Veatch Engineers); Koby Cohen (vice president of quality assurance for Suburban Water Systems); and Bruce Johnson (assistant director of utilities, Tucson, Ariz.) for their technical review and contributions as AwwaRF Project Advisory Committee members.
- TBW staff at the Cypress Creek Water Treatment Facility, who continuously supported this project through design, construction, permitting, and operation for all phases of the project.
- Finally, and most importantly, this report is dedicated to Dennis A. Marshall (Pinellas County, Fla., operations support), who passed away on June 24, 2007. Denny was a Pinellas County Utilities employee who worked from the initiation to the end of the TBW projects doing a multitude of tasks. He could do anything involving the construction, maintenance, and operation of the field facility, and was also responsible for transportation and delivery of water for the pilot distribution systems. This project could not have been done without him, his colleague Oz Wiesner, and the support of Pinellas County. Denny was not only very competent, he was unselfish and a joy to work with. He was a very good man and will be sincerely missed.

EXECUTIVE SUMMARY

INTRODUCTION

Drinking water utilities in Florida and similar environments have historically relied on groundwater, but overreliance on the source has led to the pursuit of other multiple sources. Tampa Bay Water (TBW) and the University of Central Florida sought, through the tailored collaboration project with Awwa Research Foundation described below, understanding of the issues involved when multiple-source waters are blended, particularly the impacts to distribution system water quality.

TBW manages drinking water resources for six member governments (MGs) on the west coast of central Florida: the cities of New Port Richey, St. Petersburg, and Tampa; and Hillsborough, Pasco, and Pinellas counties. This region, like most of Florida, has historically utilized groundwater for its drinking water supply; however, adverse environmental impacts have mandated that alternatives (nongroundwater sources such as desalinated seawater and treated surface water) be used for future supply. TBW manages two alternative supplies: a 66-mgd (2.5×10^5 m^3/day) surface water treatment facility, which utilizes enhanced coagulation, ozone, and biological filtration; and a 25-mgd (9.5×10^4 m^3/day) desalination plant. Finished water from these facilities is blended with finished groundwater and distributed to the member governments through a large regional transmission system, which runs through the three-county area. Blends of the various resources depend on the delivery location and supply availability. Although the regional water supply contract defines "acceptable water quality" (more stringently than the Safe Drinking Water Act requires), the delivered water quality varies by source, and, prior to this study, the system effects of corrosion inhibitors on the regional distribution of varying water quality were unknown.

Hence, the capabilities of corrosion inhibitors to offset the adverse effects of changing finished water quality on distribution system water quality was investigated over a 3-year period and utilized an 8,500-ft^2 (800-m^2) research facility that was constructed for this work. The field facilities included 18 pilot distribution systems (PDSs) built from actual polyvinyl chloride (PVC)-lined cement, cast-iron, and galvanized steel distribution system pipe taken from the existing MG distribution systems; a Pb–Cu (lead–copper) corrosion facility; and cradles that housed coupons for investigation of biological films and chemical scales. A trailer was fitted with equipment for continuous monitoring of electronoise (EN) measurements for corrosion control. Effluent water quality results following a 2-day PDS hydraulic residence time (HRT) was compared among 14 PDSs that had received varying doses of orthophosphate (OP), zinc orthophosphate (ZOP), blended ortho-polyphosphate (BOP), silicate, or pH-only inhibition over 1 year. The effect of inhibitors relative to pH control was investigated for (1) control of iron (Fe), copper, and lead release; (2) EN parameters and correlation of EN parameters to metal release; (3) pipe surface characterization; (4) biological characterization; (5) chloramine residual dissipation; and (6) surface roughness of unlined cast iron (UCI), galvanized steel (GS), lined cast iron (LCI), PVC, and Cu and Pb coupons.

The work found that

- Phosphorus- and silicon-based inhibitors significantly reduced lead and copper release and somewhat reduced iron release relative to pH control;

- EN could be used to describe the release of iron, copper, or lead with the same degree of confidence as water quality;
- Addition of phosphorus- or silicon-based inhibitors resulted in the formation of phosphorus- or silicon-based solids on the corrosion scales that corresponded to the reduced release of iron, copper, and lead in the presence of these inhibitors;
- Maintenance of chloramine residual was independent of phosphorus- or silicon-based inhibitors but was dependent on pipe geometry, hydraulics, material, water quality, and dose;
- Roughness of the surface of UCI, LCI, GS, PVC, Cu, and Pb was not affected by inhibitor type or dose; and
- The addition of phosphorus- or silicon-based inhibitors resulted in higher biological activity as indicated by bulk heterotrophic plate counts (HPCs) and biofilm HPC (BFHPC) densities relative to pH-only inhibition of drinking water in PDSs.

METAL RELEASE

Conflicting Objectives for Iron, Copper, and Lead Control

Prior research has identified conflicting objectives for control of iron and copper. Operation with a blend that is enriched in groundwater, with an associated elevated alkalinity, would be desirable to mitigate red water problems associated with release of iron from UCI and GS pipe materials. Unfortunately, the elevated alkalinity also creates compliance problems for copper (and to a lesser extent, lead). The current study is intended to evaluate corrosion control strategies (pH elevation and inhibitor addition) to determine whether the feasible operating domain for alkalinity (and thus the blend composition) could be expanded without violation of secondary standards for iron (or apparent color) or copper.

Empirical Models

Empirical models were developed for total and dissolved metal concentration, as shown in Equations ES.1 through ES.3. The resultant models for iron concentration are specific for the laminar hydraulic conditions (2-day HRT) and geometry of the hybrid PDS. Procedures for adjustment for alternate hydraulic conditions, pipe geometry, and water chemistry were reported in a previous study (Taylor et al. 2005). The copper models are general for a distribution system that employs copper pipe materials within a residence. The lead models are specific for the surface area of lead coupon incorporated into the experimental system; calibration of the lead prediction models to account for differing lead exposures in plumbing materials may be necessary. Specific models were developed by analysis of variance to determine the significance of water quality parameters and nonlinear least-squares parameter estimation, as summarized below for total metal species:

$$\begin{aligned} TFe = {} & (0.495BOP \times TP^{-0.104} + 0.593OP \times TP^{0.047} \\ & + 0.470ZOP \times TP^{-0.073} \times Zn^{-0.180} + 0.122Si \times SiO_2^{0.528} + 0.661pH_s) \\ & \times alk^{-0.457} \times Cl^{0.345} \times Fe_{inf}^{0.136} \end{aligned} \quad (ES.1)$$

$$TCu = (6.13BOP \times TP^{-0.174} + 5.76OP \times TP^{-0.328} + 5.24ZOP \times TP^{-0.205} \times Zn^{-0.069} + 20.9Si \times SiO_2^{-0.214} + 16.7pH_s) \times pH^{-3.48} \times alk^{0.569} \times Cl^{0.394} \quad \text{(ES.2)}$$

$$TPb = (6.64 \times 10^{-6}BOP \times TP^{-0.283} + 6.94 \times 10^{-6}OP \times TP^{-1.47} + 2.08 \times 10^{-7}ZOP \times TP^{-0.892} \times Zn^{-0.824} + 3.80 \times 10^{-4}Si \times SiO_2^{-1.16} + 4.14 \times 10^{-5}pH_s) \times pH^{-2.75} \times Cl^{1.98} \times SO_4^{-1.54} \times temp^{2.94} \quad \text{(ES.3)}$$

where TFe = total iron, mg/L
BOP = blended ortho-polyphosphate inhibitor (0, 1)
OP = orthophosphate inhibitor (0, 1)
ZOP = zinc orthophosphate inhibitor (0, 1)
Si = silicon inhibitor (0, 1)
pH_s = pH control (0, 1)
TP = total phosphorus, mg/L
Zn = zinc, mg/L
SiO_2 = silica, mg/L as SiO_2
alk = alkalinity, mg/L as $CaCO_3$
Cl = chloride, mg/L
Fe_{inf} = influent iron, mg/L
TCu = total copper, mg/L
pH = $-\log \{H^+\}$
TPb = total lead, mg/L
SO_4 = sulfate, mg/L
temp = temperature, °C

Remedial Measures for Metal Release

Alternative control measures have been identified, including

- Elevation of pH above pH_s.
- Addition of an inhibitor: BOP, OP, ZOP, or Si.
- Adjustment of alkalinity.
- Control of HRT by design or operation (flushing).
- Specification and/or replacement of undesired pipe materials (UCI or GS).

Control of Iron

Previous work has demonstrated that for laminar flow conditions, the iron flux is nearly constant for a constant water quality. Under these conditions, the elevation of iron concentration would be proportional to the HRT. The current study demonstrated that for all corrosion control measures and all blends that were studied, statistically significant increases in iron concentration were experienced for a 2-day HRT in the hybrid PDS. These findings suggest that compliance with iron standards may not be possible unless suitable hydraulic conditions are maintained in the distribution system. Avoidance of dead zones and associated elevated HRTs in UCI and GS pipe sections would be critical to prevent red water episodes.

Alteration of the water chemistry has been documented to achieve modest mitigation of iron release in comparison to operation at pH_s. Alkalinity supplementation of the surface water and reverse osmosis groundwater sources to achieve a minimum alkalinity in the resultant blends in excess of 100 mg/L as $CaCO_3$ (calcium carbonate) would be prudent. Alkalinity supplementation by addition of lime and carbon dioxide may be preferable to addition of sodium salts (e.g., sodium bicarbonate). The data indicated that pH elevation was as effective as inhibitor addition. For the inhibitors that were examined, the silicon (Si) product achieved the greatest reduction in iron concentrations. The performance of the ZOP inhibitor was inferior to the other corrosion control measures, including pH elevation.

Elevation of pH to address metal release may not be an attractive option for source water blends that contain significant levels of calcium. These blends would typically be enriched in groundwater. For such waters with a comparatively low value of pH_s, operation at an elevated pH could produce excessive calcium carbonate scale formation.

The occurrence of elevated iron concentrations is limited to exposure to UCI and galvanized pipe materials. Replacement of these pipes with alternative materials of construction (lined iron or PVC) would also realize benefits in control of iron concentrations.

Control of Copper

The experimental results and the resultant empirical model for copper concentration demonstrate a pronounced benefit associated with all of the corrosion control strategies that were investigated. Operation at pH_s in the absence of inhibitor addition was consistently reported to result in a violation of the copper action level. Although operation at an elevated pH would be successful for compliance with the copper standard, the extreme pH conditions that would be required ($pH > 8.0$) may produce unacceptable scaling conditions. All of the inhibitors were shown to produce compliance with the relevant copper standard for all water quality blends that were investigated. All of the phosphate-based inhibitors (BOP, OP, and ZOP) produced similar results for a comparable inhibitor total phosphorus dose. Even though the performance with the minimum dose of the Si inhibitor (3 mg/L SiO_2) was inferior to performance with the minimum dose of the phosphate-based inhibitors (0.5 mg/L TP), an increase in the Si dose could be practiced to achieve comparable performance. The simulated performance of the various inhibitors, at the minimum doses examined in the study and at pH_s, is presented in Figure ES.1. The variation in alkalinity in these simulations reflects a variation in the percentage composition of groundwater and surface water in a hypothetical product blend. These results are significant, as the simulations with inhibitors indicate that it would be possible to achieve compliance with copper standards for elevated alkalinity conditions without reliance on an elevated pH.

Control of Lead

The system response to implementation of lead corrosion control measures (pH elevation or inhibitor addition) was similar to the experience for control of copper discussed previously. Elevation of pH could achieve compliance with lead standards, but operation at a positive Langelier saturation index may produce unacceptable calcium carbonate scale formation. The use of an inhibitor achieved control of lead concentration for all of the water quality blends employed in the study for pH conditions that would not promote deposition of calcium carbonate scale. All of the inhibitors were effective, although the ZOP and BOP inhibitors provided somewhat better results

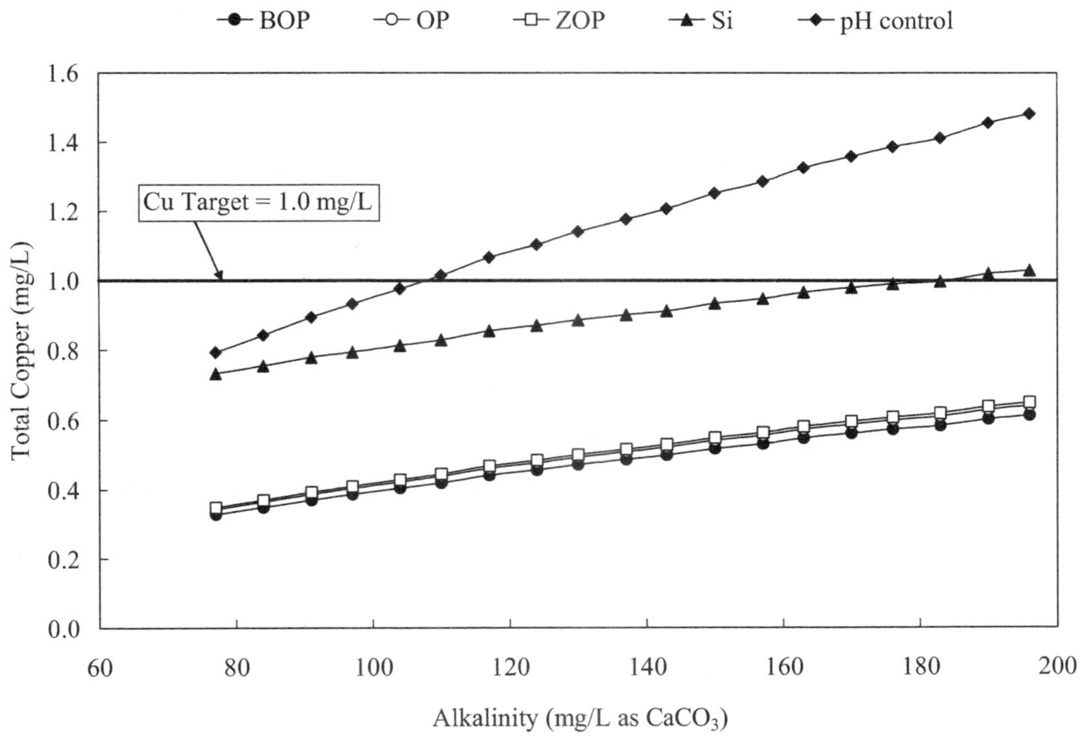

Figure ES.1 Copper release as a function of alkalinity

at the minimum inhibitor dose than either the OP or Si products. Increases in the dose of OP or Si would improve results to comparable conditions obtained with ZOP or BOP.

ELECTRONOISE

Electronoise is a very rapid technique for monitoring corrosion and is specific for any metal. EN monitoring for Fe, Cu, and Pb/Sn (lead/tin) corrosion using Fe, Cu, and Pb/Sn electrodes was conducted in all PDSs during the year of operation. Several variables were collected during EN monitoring; however, EN models describing metal release were developed for Fe, Cu, and Pb release with linear polarization resistance corrosion rate (LPRCR), harmonic distortion analysis corrosion rate (HMCR), electrochemical noise corrosion rate (ECNCR), and pitting factor (PF). The statistical significance of the EN models and the water quality models developed in this work was equivalent.

EN Models for Metal Release

Regression modeling using EN parameters to predict metal release was done for both total and dissolved iron, copper, and lead. The models used dummy variables for inhibitor type and dose to describe nonlinear relationships between EN parameters and metal release relative to inhibitor, inhibitor dose, and temperature.

The general form of EN model for metal release used dummy variables to represent inhibitor type and dose as shown in Equation ES.4.

$$\begin{aligned} Me = {} & (LPRCR_{Me}^{a1} \times HMCR_{Me}^{a2} \times ECNCR_{Me}^{a3} \times PF_{Me}^{a4} \times EN_{Me}^{ai}) \\ & \times (a \times BOP \times TP^{f} + b \times OP \times TP^{g} + c \times ZOP \times TP^{h} \times Zn^{i} + d \times Si \times SiO_2^{j} \\ & + e \times pH_s \times temp^{o} \end{aligned} \qquad (ES.4)$$

where
Me = Fe, Cu, or Pb
a1 = $LPRCR_{Me}$ exponent
a2 = $HMCR_{Me}$ exponent
a3 = $ECNCR_{Me}$ exponent
a4 = PF_{Me} exponent
EN_{Me}^{ai} = one of the terms associated with exponents a5, a6, a7, a8, a9, a10, a11, or a12
a5 = B_{Me} exponent
a6 = Ba_{Me} exponent
a7 = Bc_{Me} exponent
a8 = $PSKW_{Me}$ (skew of the potential) exponent
a9 = $PKRT_{Me}$ (kurtosis of the potential) exponent
a10 = $ISKW_{Me}$ (skew of the current) exponent
a11 = $IKRT_{Me}$ (kurtosis of the current) exponent
a12 = $RSOL_{Me}$ (solution resistance) exponent
BOP = blended ortho-polyphosphate inhibitor dummy variable (0, 1)
OP = orthophosphate inhibitor dummy variable (0, 1)
ZOP = zinc orthophosphate inhibitor dummy variable (0, 1)
Si = silicon inhibitor dummy variable (0, 1)
pH_s = pH control dummy variable (0, 1)
TP = total phosphorus, mg/L
Zn = zinc, mg/L
SiO_2 = silica, mg/L as SiO_2
temp = temperature, °C
a = BOP dummy variable coefficient
b = OP dummy variable coefficient
c = ZOP dummy variable coefficient
d = silica dummy variable coefficient
e = pH control dummy variable coefficient
f = TP exponent associated with BOP dummy variable
g = TP exponent associated with OP dummy variable
h = TP exponent associated with ZOP dummy variable
i = Zn exponent associated with ZOP dummy variable
j = SiO_2 exponent associated with silica dummy variable
o = temperature exponent

Iron Release

Iron was released by both general and localized corrosion. However, localized corrosion was more important for particulate iron, which comprised most of the total iron release in this study. EN empirical models were developed to predict total and dissolved iron (DFe) release in the hybrid PDS and are presented in Equations ES.5 and ES.6.

$$\begin{aligned} TFe = {} & (PF_{Fe}{}^{0.0694}) \times (0.240 \times BOP \times TP^{-0.071} + 0.277 \times OP \times TP^{0.047} \\ & + 0.239 \times ZOP \times TP^{-0.088} \times Zn^{-0.157} + 0.071 \times Si \times SiO_2{}^{0.456} \\ & + 0.334 \times pH_s) \times temp^{-0.144} \end{aligned} \qquad (ES.5)$$

$$\begin{aligned} DFe = {} & (HMCR_{Fe}{}^{0.486}) \times (246.71 \times BOP \times TP^{-0.128} + 157.63 \times OP \times TP^{0.275} \\ & + 342.92 \times ZOP \times TP^{-0.534} \times Zn^{0.109} + 74.79 \times Si \times SiO_2{}^{0.188} \\ & + 172.21 \times pH_s) \times temp^{-3.22} \end{aligned} \qquad (ES.6)$$

Copper Release

Copper corrosion was also dependent on general and localized corrosion. However, general corrosion was more significant for copper release, which was mostly in the dissolved form. EN models were developed to predict total and dissolved copper release (TCu and DCu) and are presented in Equations ES.7 and ES.8. The model predicts that copper release is highest for pH control and all inhibitors reduced total copper release, which is consistent with the data.

$$\begin{aligned} TCu = {} & (LPRCR_{Cu}{}^{0.028} \times ECNCR_{Cu}{}^{0.031}) \times (0.510 \times BOP \times TP^{-0.116} \\ & + 0.476 \times OP \times TP^{-0.281} + 0.471 \times ZOP \times TP^{-0.232} \times Zn^{-0.048} \\ & + 1.960 \times Si \times SiO_2{}^{-0.324} + 1.505 \times pH_s) \times temp^{-0.063} \end{aligned} \qquad (ES.7)$$

$$\begin{aligned} DCu = {} & (LPRCR_{Cu}{}^{0.0259} \times ECNCR_{Cu}{}^{0.027}) \times (0.406 \times BOP \times TP^{-0.107} \\ & + 0.372 \times OP \times TP^{-0.293} + 0.386 \times ZOP \times TP^{-0.242} \times Zn^{-0.021} \\ & + 1.528 \times Si \times SiO_2{}^{-0.320} + 1.181 \times pH_s) \times temp^{-0.028} \end{aligned} \qquad (ES.8)$$

Lead Release

Lead corrosion was also dependent on general and localized corrosion. Released lead was generally in the particulate form. EN models were developed to predict total and dissolved lead release (TPb and DPb) and are presented in Equations ES.9 and ES.10. The model predicts that lead release is highest for the pH_s control and all inhibitors reduced total lead release, which is consistent with the data. Lead release was contrary to copper release for the effect of temperature, and lead release increased with temperature.

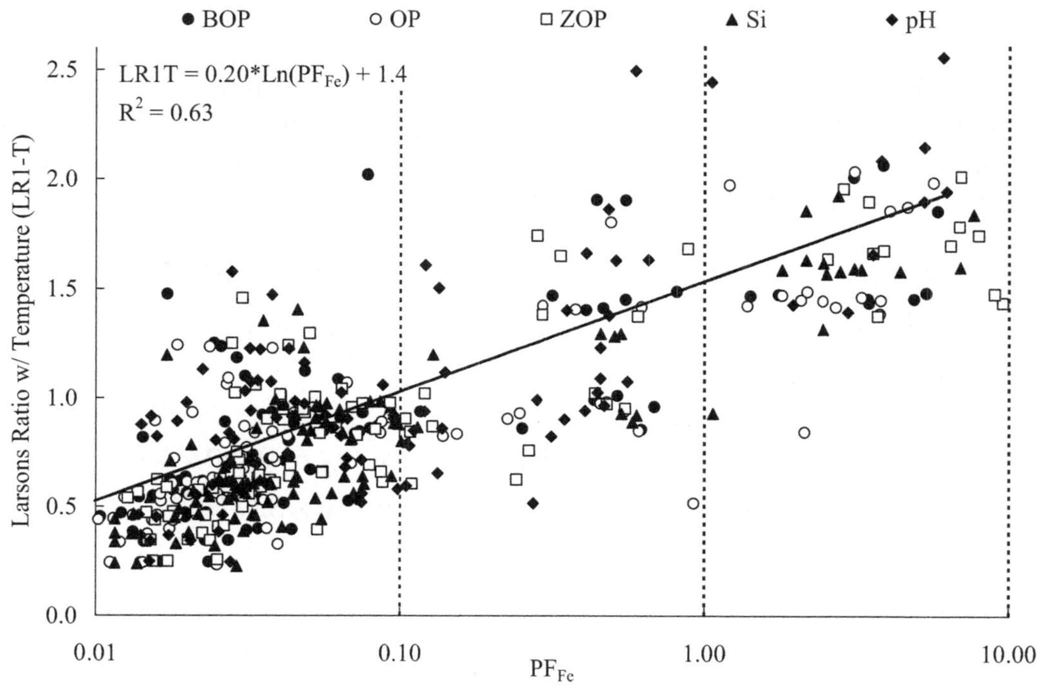

Figure ES.2 LR1-T versus PF_{Fe}

$$TPb = LPRCR_{Pb}^{0.217} \times HMCR_{Pb}^{0.337} \times PF_{Pb}^{-0.347} \times (1.58 \times 10^{-8} \times BOP \times TP^{-0.805} + 1.53 \times 10^{-9} \times OP \times TP^{-4.188} + 5.49 \times 10^{-8} \times ZOP \times TP^{-0.784} \times Zn^{0.735} + 5.56 \times 10^{-8} \times Si \times SiO_2^{-0.326} + 1.83 \times 10^{-7} \times pH_s) \times temp^{3.67} \quad (ES.9)$$

$$DPb = LPRCR_{Pb}^{0.189} \times HMCR_{Pb}^{0.281} \times PF_{Pb}^{-0.297} \times (1.09 \times 10^{-7} \times BOP \times TP^{-0.115} + 2.78 \times 10^{-8} \times OP \times TP^{2.550} + 3.72 \times 10^{-7} \times ZOP \times TP^{-0.690} \times Zn^{0.580} + 1.92 \times 10^{-7} \times Si \times SiO_2^{-0.088} + 8.3 \times 10^{-7} \times pH_s) \times temp^{2.96} \quad (ES.10)$$

EN Water Quality Correlations

PF was the only EN parameter that consistently correlated with water quality (alkalinity, chlorides, sulfate, and temperature) for iron, copper, and lead, which is shown in Figure ES.2 for the correlation with the modified form of Larson's ratio for temperature (LR1-T).

IDENTIFICATION AND MODELING OF CORROSION SCALES

The effects of inhibition on corrosion scale formation and surface roughness of pipe materials common to drinking water distribution systems with and without exposure to phosphorus- and silicon-based inhibitors were investigated using X-ray photoelectron spectroscopy (XPS) and profilometry. Scales and surface roughness were analyzed on new iron, copper, and lead coupons and used UCI, LCI, PVC, and GS coupons.

Iron

Variation in water quality by phase did not affect the composition of iron surface scales; however, inhibitor type did affect scale composition. Fe_2O_3 (ferric oxide), Fe_3O_4 (magnetite), and FeOOH (hydrated ferric oxide) were identified as major corrosion products on the unlined iron pipe coupons regardless of inhibitor; however, $FePO_4$ (ferric phosphate) and silicate compounds were identified on all iron surfaces exposed to phosphate-based and Si inhibition, respectively.

Thermodynamic modeling indicated that siderite ($FeCO_3$) was descriptive of iron release. Assuming $FeCO_3$ was the controlling solid phase for iron release in the presence or absence of inhibitors provided a thermodynamic model that slightly overpredicted actual iron release, even though only soluble iron release was considered. Consideration of $FePO_4$ as the controlling solid phase provided a model that grossly underpredicted actual iron release in the presence of phosphorus inhibition. $FeSiO_3$ was considered as the controlling solid phase in the presence of Si inhibition, but it underpredicted iron release and was less accurate than the $FeCO_3$ model. Consequently, assumption of phosphate- or silica-based controlling solid phases allowed development of models that reflected the observed trends of iron release in the presence of phosphate or silicate inhibition but were less accurate than the $FeCO_3$ model. The field data, XPS identification of scale compounds, and modeling suggests that the addition of inhibitors resulted in the formation of inhibitor-specific solid compounds that competed with iron corrosion products (i.e., Fe_2O_3, Fe_3O_4, and FeOOH) and retarded the release of iron.

Galvanized Steel

Several thermodynamic models were developed to describe dissolved iron and zinc release from GS under various inhibiting conditions. The most descriptive models for iron release were developed assuming $FeCO_3$ as the controlling solid phase in the presence of pH only or phosphate-based inhibitors. In both cases, the developed thermodynamic models slightly overpredicted actual iron release, even though only soluble iron release was predicted. Thermodynamic modeling of $FeSiO_3$ as the controlling solid phase under the Si inhibitor treatment resulted in underprediction of actual dissolved iron release. The most descriptive models for zinc release were developed by assuming zinc carbonate ($ZnCO_3$) as the controlling solid phase in the presence of pH only or phosphate inhibitors. The developed models overpredicted actual zinc concentrations even though only soluble zinc release was predicted, which indicates that the impact of phosphate-based inhibitor on zinc release is insignificant. Thermodynamic modeling of $ZnSiO_3$ as the controlling solid phase under the Si inhibitor treatment resulted in underprediction of actual dissolved zinc release and suggests that the addition of Si inhibitor results in the formation of a solid film that inhibits zinc and iron release.

Fe_2O_3, Fe_3O_4, and FeOOH were identified by XPS as major corrosion products on the GS coupons exposed to all inhibitor treatments. $FePO_4$ was only detected on the coupons exposed to phosphate inhibitors. Therefore, addition of corrosion inhibitor resulted in the formation of solid phosphate scale ($FePO_4$), which could be a barrier to metal release from GS. Silicate compounds were identified on all GS coupons exposed to Si inhibitor. Zinc deposits were present in the GS scale for most observations. The lack of zinc phosphate ($Zn_3(PO_4)_2$) scale in the presence of ZOP indicates that $Zn_3(PO_4)_2$ precipitation was not a factor in inhibiting iron release. Variation of phase water quality did not affect the distribution of solid iron forms in the GS films, which suggests that variations in water quality in this work would not adversely affect iron release.

Copper

Thermodynamic models were developed describing copper release assuming different solids that might be expected to form in copper pipes receiving no inhibitor, phosphate, or silicate inhibitors. The most descriptive models were developed assuming tenorite (CuO) and cupric hydroxide ($Cu(OH)_2$) as the controlling solid phases. The primary ligands responsible for increases in total copper were all related to alkalinity, including $CuHCO_3^-$, $CuCO_3^0$, $Cu(CO_3)_2^{2-}$, $Cu(OH)CO_3^-$, and $Cu(OH)_2CO_3^{2-}$, and clearly predicted increasing copper release with increasing alkalinity. These models do not account for suspended solids release of copper and underpredicted copper release for pH control.

Models for dissolved copper release were developed assuming $Cu(OH)_2$, CuO, and $Cu_3(PO_4)_2{\cdot}2H_2O$ each as controlling solid phases in the presence of phosphate inhibitors. The $Cu_3(PO_4)_2{\cdot}2H_2O$ model predicted decreased copper release, which was the most descriptive of actual data.

Thermodynamic models were developed assuming $Cu(OH)_2$, CuO, and cupric silicate hydride ($CuSiO_3{\cdot}H_2O$) as controlling solid phases. The $CuSiO_3{\cdot}H_2O$ model provided the best prediction of copper release and only slightly overpredicted dissolved copper release.

Survey and high-resolution XPS spectra were collected to identify the corrosion products of copper and determine other chemical compounds present. Because of nonuniformity in the surface composition of copper scales, failure to detect a compound by XPS in one scan does not imply the compound was not in the scale.

Cu_2O, (cuprous oxide) CuO, $Cu(OH)_2$, and Cu(II) salts were detected as the predominant copper corrosion compounds in the scale formed on copper coupons in the presence of inhibitors. The addition of corrosion inhibitors did not affect the relative distribution of copper corrosion products. Phosphate was detected on many coupons, but the form of phosphate was not determined. $Cu_2(PO_4)_3$ (cupric phosphate) was not in the XPS library but could have been the form of phosphorus in the copper scale. The thermodynamic model for cupric phosphate dihydride, $Cu_3(PO_4)_2{\cdot}2H_2O$, predicted dissolved copper release similar to the levels of copper release observed in the phosphate lines.

XPS analysis indicated the presence of a polymeric silica film in the surface scale, which may have been $CuSiO_3$ (cupric silicate), which was also not in the XPS library. The thermodynamic model for $CuSiO_3{\cdot}H_2O$ predicted dissolved copper release similar to the levels of copper release observed in the silicate lines.

Lead

Models for lead release were developed assuming hydrocerussite ($Pb_3(OH)_2(CO_3)_2$) and cerussite ($PbCO_3$) as controlling solid phases in the absence of inhibitors and were equally descriptive. $Pb_3(OH)_2(CO_3)_2$, $PbCO_3$, and lead(II) oxide (PbO) were the predominant lead corrosion products found by XPS. Phosphates were detected for most of the lead/tin coupons exposed to phosphate-based inhibitors. There was no XPS library for hydroxypyromorphite ($Pb_5(PO_4)_3OH$); however, $Pb_5(PO_4)_3OH$ was assumed as the controlling solid phase in the presence of phosphorus inhibition as lead release was best described by models assuming $Pb_5(PO_4)_3OH$ as the controlling solid phase.

Silica was detected in the scale of half of the coupons in the presence of Si inhibition. The scale on all lead/tin coupons exposed to Si inhibition appeared denser than other scales formed in the presence of phosphorus or pH inhibition. Although silica was found on other lead/tin coupons exposed to other treatments, the advantage of Si addition seems to be in the promotion of a more dense amorphous silica layer.

SURFACE ROUGHNESS

Surface roughness of new Pb and Cu coupons and GS, UCI, LCI, and PVC coupons taken from used water distribution system pipes was measured before and after exposure to inhibition. The changes in surface roughness were analyzed with respect to inhibitor type and dose, water quality, and initial and final roughness. The change in surface roughness of all materials was independent of inhibitor type or dose but was dependent on the initial surface roughness.

The change in roughness of UCI and GS coupons was correlated to alkalinity and temperature; however, roughness change was generally independent of water quality. As shown in Table ES.1, the decreasing order of surface roughness was UCI > GS ≈ LCI > PVC for the used material and Cu > Pb for the new materials. Although there was significant variability in roughness measurements, as shown by the standard deviations (σ) in Table ES.1, measurement of mean roughness of replicate coupons after inhibition exposure showed that experimental precision was within 20%.

Table ES.1
Roughness summary by material

Roughness*	UCI_{old}	GS_{old}	LCI_{old}	PVC_{old}	Cu_{new}	Pb_{new}
R_i (μm)	86.1	25.5	25.4	1.5	3.3	1.7
σ_i (μm)	21.7	7.4	10.4	0.5	0.3	0.7
R_f (μm)	93.3	26.3	26.7	1.7	3.9	2.1
σ_f (μm)	18.7	7.6	9.0	0.5	0.7	1.2
ΔR (μm)	7.9	0.3	1.5	0.2	0.6	0.4
σ_Δ (μm)	20.9	6.5	6.0	0.2	0.6	0.8

*where R_i = initial roughness
σ_i = standard deviation initial roughness
R_f = final roughness
σ_f = standard deviation final roughness
ΔR = delta roughness
σ_Δ = standard deviation delta roughness

RESIDUAL

The chloramine residual at the influent point of the PDS was not significantly different for any of the PDSs and averaged 5.0 mg/L as Cl_2 (chlorine) during the project. The corrosion inhibitors type or dose used had no effect on chloramine decay. The PDS conducting the pH_s+0.3 control water (PDS14) had the same chloramine demand as most of the PDSs using inhibitors, with the exception of PDS12 (12 mg/L SiO_2), which had a higher pH. The control water adjusted to pH_s (PDS13) had greater chloramine consumption than all the other PDSs.

Chloramine residual decayed in the PDS at different rates depending on pipe material, pipe geometry, water quality, and flow velocity. Chloramine residual decayed faster in UCI and GS pipes and was slower in the PVC and LCI pipes. The order of the decay rate was LCI < PVC << UCI < GS. The rate of chloramine decay also increased with temperature and total organic carbon, and decreased with increasing pH. Chloramine was observed to decline faster during summer than in winter.

Data obtained from the internal sampling ports of the hybrid PDSs were used to develop a first-order chloramine decay model by nonlinear regression based on the model obtained from a previous study, as shown in Equation ES.11. The model contains a bulk decay rate constant (K_B), a pH effect constant (K_H), a wall decay rate constant (W), and a temperature correction factor (θ). The model was developed by maintaining the bulk decay rate constant and calculating a common θ and K_H for all the pipe material as well as an individual W for each pipe material.

$$Cl_{2(t)} = Cl_{2(0)} * \exp\left[\left(K_B * UV_{254} + K_H * [H] + \frac{W_{Pipe}}{D_{Pipe}}\right) * \theta^{(T-20)} * time\right] \qquad \text{(ES.11)}$$

where W_{pipe} = –0.025 (PVC), –0.014 (LCI), –0.103 (UCI), –0.051 (GS), in inches per hour
$\theta = 1.11$
$K_B = -0.02$, cm/hr
$K_H = -106404$, L/mol-hr
D_{Pipe} = pipe diameter, in.
UV_{254} = UV absorbance at 254 nm, cm^{-1}

LCI and PVC have the lowest W at 0.014 in./hr and 0.025 in./hr, respectively, given that those materials are fairly inert, do not promote bacterial growth, and in general have the lowest chloramine demand of the materials used in the study. UCI and GS have W that are four and nine times greater than W for LCI. UCI has a W of 0.103 in./hr and GS has a W of 0.051 in./hr. The higher wall demand of the LCI and GS pipe surfaces are due in part to the greater chemical reactivity of iron and zinc relative to LCI and PVC. The diameter of the pipe affects the rate of chloramine decay because the effect of W is proportional to the ratio of volume to wetted surface area. The relative effect of W decreases when the pipe diameter increases. This effect is more noticeable for unlined metallic pipes because their W is more elevated.

The rate of chloramine decay was found to increase with the increase in temperature. The rate of decay would double for each 7°C increment of the water temperature, which corresponds to a temperature correction factor, θ, of 1.11. The pH of the water was the only corrosion control measure to have an effect on the rate of chloramine decay in the PDSs. The decay of chloramine residual was found to be slower when the pH of the water increased. Similar to organics, the effect of pH is more noticeable in LCI and PVC pipes than in UCI and GS pipes.

The chloramine decay model was subsequently modified to include the Reynolds number (Re) effect (flow velocity), as shown in Equation ES.12. Flow velocity experiments in the single material lines showed that UCI and GS wall decay rates increase with the Reynolds number. PVC and LCI did not show variation in W with the Reynolds number. UCI and GS showed a region of rapid increase of W under laminar conditions, followed by a more gradual increase under transitional and fully turbulent flow.

$$Cl_{2(t)} = Cl_{2(0)} * \exp\left[-\left(K_B * UV_{254} + K_H * [H^+] + \frac{B * Re + C}{D}\right) * \theta^{(T-20)} * time\right] \quad (ES.12)$$

where

		Pipe Material					
				UCI		GS	
	Units	PVC	LCI	Re < 500	Re > 500	Re < 500	Re > 500
B	in./hr	0	0	0.00044	0.00003	0.0004	0.00009
C	in./hr	0.025	0.014	0.076	0.27	0.083	0.12
K_B	cm/hr	0.02	0.02	0.02	0.02	0.02	0.02
K_H	L/mol-hr	106404	106404	106404	106404	106404	106404
θ		1.11	1.11	1.11	1.11	1.11	1.11
$[H^+]$	H^+ molar concentration in moles/L						

The chloramine decay model was verified using data that was not used during model development. The verification showed a good agreement between the predicted and actual chloramine residual with an average absolute error of 0.6 mg/L as Cl_2.

BIOLOGICAL CHARACTERIZATION

Bulk water and biofilm bacterial densities (measured as HPC) increased with the addition of phosphate- or silica-based inhibitors to finished waters relative to pH-only inhibition, as shown in Table ES.2. Statistically significant models for bulk water and biofilm density were developed, as shown in and Equations ES.13 and ES.14. As shown in Figure ES.3, bulk HPCs varied with inhibitor, and increased with increasing temperature and decreasing residual. As shown in Figure ES.4, BFHPCs also increased with increasing temperature and decreasing residual, but varied with material. The effect of inhibitor type on HPCs is illustrated in Figure ES.3, using Equation ES.13 for BFHPCs that were in and out of the actual residual range. The HPC model predicts significant HPC growth as residual decreases, especially as residual approaches zero. The effect of material on BFHPCs is illustrated in Figure ES.4, using Equation ES.14 for BFHPCs that were in and out of the actual residual range. The BFHPC model also predicts significant HPC growth in the biofilm as the residual approaches zero. The actual residual data ranges differ because the residuals for the effluent HPCs were measured in the PDS effluent and the effluent residuals for the BFHPCs were measured in the cradles, which

Table ES.2
Effluent HPC and BFHPC average (cfu/mL) and standard deviation summary for different inhibitors and materials

		BOP	OP	ZOP	Si	pH
HPC	Effluent	3.98E+03	2.96E+03	2.66E+03	2.63E+03	9.21E+02
BFHPC	PVC	1.54E+05	4.63E+04	5.02E+04	3.14E+04	1.26E+04
	UCI	4.67E+05	4.94E+05	7.52E+05	5.54E+05	1.99E+05
	LCI	2.26E+05	3.73E+05	2.70E+05	3.62E+05	1.19E+05
	GS	9.41E+05	1.01E+06	2.30E+05	1.35E+06	2.01E+06

were constructed entirely of PVC. However, the models correctly portray the importance of residual for controlling HPC proliferation.

$$\log HPC_{eff} = a \times temp^{b} \times TCl_{2-NH_2Cl}^{c} \qquad \text{(ES.13)}$$

where

	Inhibitor				
Parameter	BOP	OP	ZOP	Si	pH
a	0.65	1.63	0.74	0.90	0.66
b	0.56	0.26	0.50	0.48	0.46
c	–0.19	–0.21	–0.20	–0.35	–0.07

NH_2Cl = monochloramine

$$\log BFHPC = a \times temp^{b} \times TCl_{2-NH_2Cl}^{c} \qquad \text{(ES.14)}$$

where

	Pipe Material			
Parameter	PVC	LCI	UCI	GS
a	0.30	1.46	2.68	2.95
b	1.31	0.71	0.25	0.20
c	–0.94	–0.62	–0.71	–0.00

Bulk water and biofilm bacterial densities (measured as HPC) increased with phosphate- or silica-based inhibitor addition. Average PDS effluent HPCs increased 163% (0.42 log) when phosphorus-based inhibitors were added and 99% (0.31 log) when silicon-based inhibitors were added. However, HPCs were much more sensitive to residual than to the presence of inhibitor.

Average increases in biofilm density were 0.34 log for P-based inhibitors, and 0.36 log for Si-based inhibitors, which is practically equivalent. Biofilm densities are much higher than bulk densities of HPC. Pipe material affected biofilm density. For the less dense biofilms of PVC and LCI, a change in residual from 3.2 to 4.7 mg/L can lower the biofilm densities by 1.7 or 1.3 log, respectively, which implies that for these pipe materials the additional biofilm resulting from

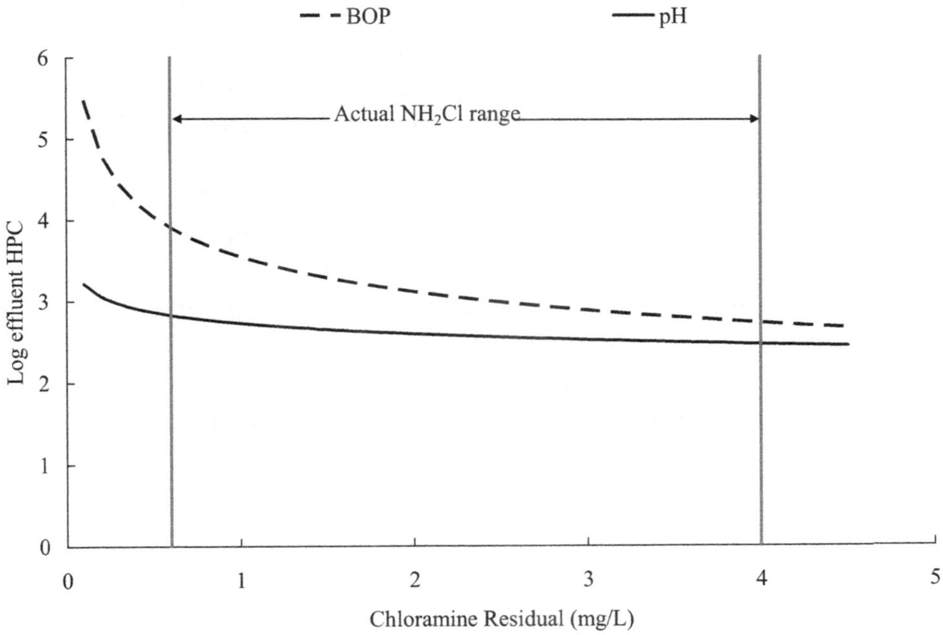

Figure ES.3 Predicted PDS log effluent HPC versus chloramine residual at average (21.2°C) temperature

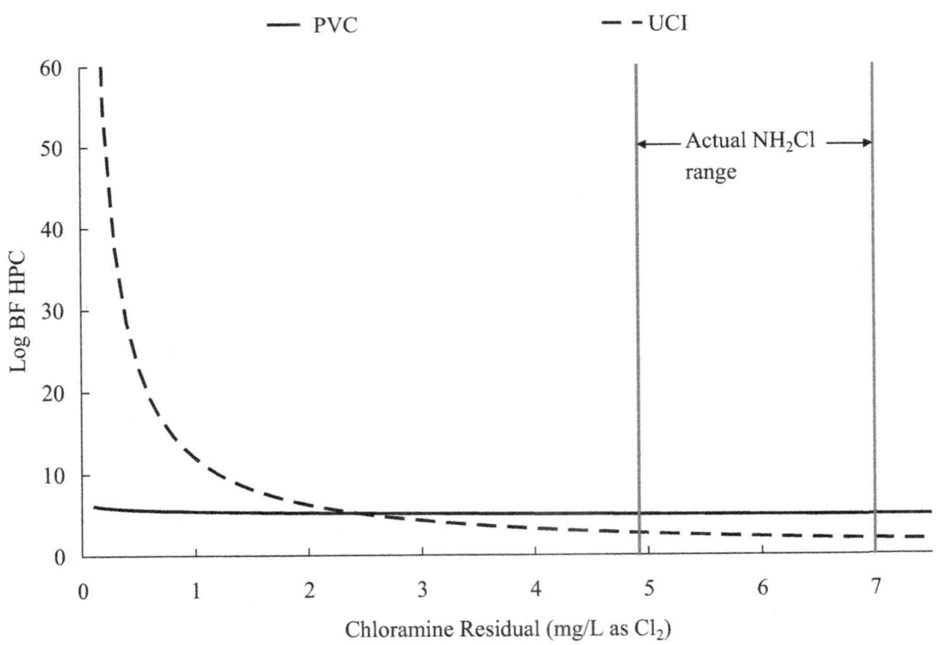

Figure ES.4 Predicted PDS log BFHPC versus chloramine residual at average (16.8°C) temperature

inhibitor addition can be managed by maintaining residual. PVC and LCI have significantly lower biofilm densities, which were typically adequate to offset HPC increases due to inhibitor addition relative to UCI and GS. Sensitivity analysis of BFHPC models showed that the denser biofilms on the unlined metals are far less sensitive to residual than the less dense biofilms on PVC and LCI. Biofilm and bulk HPC levels increased as temperature increased.

The biological data from the study reinforces the importance of residual maintenance in controlling HPC and biofilm levels. Biofilm coliforms were extremely rare in this study and were only found on one occasion. Bulk coliforms were never found in a previous study and were not sought in this study. Assimilable organic carbon (AOC) was apparently stable during the study. The AOC data and supplemental experiments, however, showed that while adding carbon (as acetate) did stimulate growth, so did the addition of phosphate salts, P-based inhibitors, and Si-based inhibitors. AOC field and laboratory data imply that carbon may not be the only limiting nutrient.

CHAPTER 1
INTRODUCTION

This document is the final report for the second of two tailored collaboration projects (TCPs) funded by Tampa Bay Water (TBW) and Awwa Research Foundation. The TCP was conducted by the University of Central Florida (UCF) and had the objective of investigating the capability of corrosion inhibitors on the reduction of the adverse effects of changing water quality in distribution systems for the member governments (MGs) of TBW, which are Hillsborough, Pasco, and Pinellas counties and the cities of New Port Richey, St. Petersburg, and Tampa.

PURPOSE

As with many metropolitan areas, the Tampa Bay area is facing increasing consumer demand and more stringent environmental and drinking water regulations. TBW is responsible for the management of drinking water resources in the Tampa Bay area. Groundwater, surface water, and brackish water resources are regionally available. Groundwater from the Floridan aquifer has been used exclusively for drinking water supply by all member governments except the City of Tampa, which predominately uses surface water. Groundwater use has been cited for adverse environmental impacts; consequently, TBW has been directed by the Southwest Water Management District to reduce groundwater use. In order to meet drinking water demands, TBW has developed regional surface water and desalinated supplies, which will offset the reductions in groundwater use and provide for growth in the region.

UCF has conducted research regarding the effect of variable finished water quality on distribution system water quality since 2000. From the previous TCP reported in *Effects of Blending on Distribution System Water Quality*, and conducted by Taylor et al. (2005), several observations have been noted on scale disruptions associated with water quality change, and these are summarized as follows:

- Apparent color, particulate iron, and turbidity are released from cast iron and galvanized pipe when the distribution system is exposed to water that, in comparison to historical water quality conditions, has lower levels of alkalinity and higher levels of sulfates and chlorides.
- Control of copper corrosion in the available groundwater requires an active corrosion control strategy to meet regulatory requirements for copper because of the high alkalinity of the groundwater.
- Nitrification can occur when the chloramine residual is lost and occurs primarily if not totally in the biofilm on the surface of the distribution system pipes.
- Biofilm growth increases directly with surface roughness of pipe and is greater for galvanized steel and unlined cast iron pipes.

The work demonstrated that control of the scale or film in the existing distribution systems is a primary factor for maintaining acceptable distribution system water quality. The finished water quality received by each member government will vary because of the use of groundwater, surface water, and seawater sources, and the associated treatment processes. Not all finished water quality parameters have the same effects on distribution system water quality.

The most significant controlling parameters found to date within the variation of water quality expected in the MG distribution systems are alkalinity, sulfates, and chlorides. Changing distribution system water quality cannot be eliminated, and the study showed that some aspects of change in water quality can adversely impact the distribution system scale or film and associated distribution system water quality.

One potential and feasible method of significantly reducing or eliminating adverse water quality impacts from the disruption of distribution system scale or film is to replace the controlling scale or film with a film that will not be disrupted when exposed to the expected changing water quality. This is possible with surface active agents such as corrosion inhibitors. Corrosion inhibitors offer an opportunity for scale control because they bond directly with the elemental metal or scale on the pipe surface, forming a barrier between the interior pipe surface and the bulk water. This action will reduce surface area for biofilm growth; reduce residual demand; and control the release of iron particles, apparent color, and turbidity from the pipe wall, thus improving distribution system water quality in general. Additionally, the new film will not be dependent on alkalinity, sulfates, chlorides, or any other of the water quality parameters that will vary seasonally in the regional finished water. The film will only be dependent on the surface active agent that is added to finished water. The advantage of using a surface active agent such as a corrosion inhibitor is that the concentration of the inhibitor is controlled by application and ideally is not affected by the changing water quality.

There is, however, little knowledge on the effect of variable water quality on the capability of corrosion inhibitors to function in a distribution system. Although it is unreasonable to assume a surface active film can offset any change in distribution system water quality, it is entirely reasonable to expect an inhibitor-generated surface active film to greatly reduce the adverse effects on distribution system water quality. Maintaining acceptable water quality may be achieved through the use of corrosion inhibitors. But proper selection of corrosion inhibitors could be critical in maintaining acceptable water quality in the distribution systems. Therefore, this project is a continuation of the previous TCP, which was expanded to investigate the reduction of the adverse effects of blending finished waters from varying sources in drinking water distribution systems using inhibitors.

LITERATURE

Existing literature on the effect of blending water quality and its impact on metal release and subsequent treatment with inhibitors is presented in chapter 2. The application of inhibitors such as orthophosphates, silicates, and by pH control for corrosion control of blended waters are discussed, as are electronoise (EN) measurements for corrosion control, and surface and biological characteristics of the pipe material.

METHODS

As described in chapter 3, 18 pilot distribution systems (PDSs) and 2 pilot treatment systems that were used in the implementation of this project were designed and built by UCF and TBW-MG personnel during the previous project, reported in *Effects of Blending on Distribution System Water Quality* (Taylor et al. 2005). The two finished waters produced at the project site were a conventionally treated groundwater (GW) and a desalinated groundwater (by reverse osmosis, or RO). The 18 PDSs consisted of polyvinyl chloride (PVC), galvanized steel (GS), unlined cast-iron (UCI), or lined cast-iron (LCI) pipe taken from the existing MG distribution systems. Four of the distribution systems consisted of each of the single-pipe materials. Fourteen distribution systems consisted of the four different pipe materials and were named hybrid lines. The 14 hybrid PDSs were used for the inhibitor study while the other four, the single-material lines, were only used to study the effect of the Reynolds number (Re) on chloramine decay. Twelve thousand gallons (45.4 m^3) of water per week were used during the project, 7,000 gal (26.5 m^3) were produced from the two pilot processes, and 5,000 gal (18.9 m^3) of surface water (SW) were hauled from the TBW regional surface water treatment plant. The three types of water were blended at predetermined ratios in an individual tank that served as the feed to 13 of the PDSs. The other PDS received the same blend; however, its pH was adjusted to pH_s. Three UCF faculty, 10 UCF graduate students, and two MG field personnel were involved in the operation and maintenance of the field facility.

To conduct the inhibitor study, four corrosion inhibitors were selected and added to the PDS system: blended ortho-polyphosphate (BOP), orthophosphate (OP), zinc orthophosphate (ZOP), and silicon (Si). Each phosphate-based inhibitor was studied at three doses: 0.5, 1, and 2 mg/L as P (phosphorus) for all the phosphate-type inhibitors. The silicate inhibitor was applied as 3, 6, and 12 mg/L as SiO_2 (silicon dioxide). A delivery system was constructed that included a common blend tank for each inhibitor and separate pumps for each PDS. Physical, chemical, and biological changes resulting from the blended water quality and the addition of inhibitors were monitored through the PDSs.

QUALITY ASSURANCE

Quality assurance plans were developed for the UCF laboratory and field lab, and followed through the course of the study. The quality assurance plan is described in chapter 4. Field and laboratory procedures for operation and analysis were developed and approved by the Project Advisory Committee before any operation and analyses were conducted. Extensive analyses were done in the field and the laboratory. Separate field operation logs were kept for each process, the PDSs, and the corrosion shed. Separate field and laboratory logs were kept for each analysis and duly recorded by work order number to maintain a paper trail for all field and laboratory activities. Physical and aesthetic water parameters measured were temperature, turbidity, and color. Chemical water quality parameters included nonpurgeable dissolved organic carbon (NPDOC), chlorine demand and residual, alkalinity, pH, total and calcium hardness, Langelier saturation index (LSI), sulfates, total dissolved solids (TDS), iron, sodium, chlorides, aluminum, silica, lead, and copper. Nitrogen and phosphorus forms were also measured to enable nutrient balances across the PDSs. Biological parameters included assimilable organic carbon (AOC), heterotrophic plate counts (HPCs), total coliform (TC), and attached cell growth.

INFORMATION GAINED

The work found that

1. Phosphorus- and silicon-based inhibitors significantly reduced lead and copper release and somewhat reduced iron release relative to pH control;
2. EN could be used to describe the release of iron, copper, or lead with the same degree of confidence as water quality;

3. Addition of phosphorus- or silicon-based inhibitors resulted in the formation of phosphorus- or silicon-based solids on the corrosion scales that corresponded to the reduced release of iron, copper, and lead in the presence of these inhibitors;
4. Maintenance of chloramine residual was independent of phosphorus- or silicon-based inhibitors but was dependent on pipe geometry, hydraulics, material, water quality, and dose;
5. Roughness of the surfaces of UCI, LCI, GS, PVC, copper, and lead were not affected by inhibitor type or dose; and
6. The addition of phosphorus- or silicon-based inhibitors resulted in higher biological activity, as indicated by bulk HPCs and biofilm.

CHAPTER 2
LITERATURE

WATER BLENDING

Numerous studies have indicated that blending different source waters or changing source waters can alter the equilibrium conditions in pipelines and cause sloughing of calcium carbonate and rust deposits that were stable and compatible to the existing source water. In 1992, the City of Tucson, Ariz., changed its source water from historical groundwater to surface water. This introduction of new source water led to the removal of aerobic deposits on the interior of the pipes and caused the anaerobic corrosion deposits to be exposed, which resulted in excessive taste, odor, and color complaints (Malcolm Pirnie, Inc. 1997).

Corrosion is a process that deteriorates metals via an oxidation–reduction chemical reaction based on thermodynamic equilibrium (Taylor et al. 2001). Corrosion of iron pipes in a distribution system can cause three distinct problems. First, pipe mass is lost through oxidation to soluble iron species or iron-bearing scale. Second, the scale can accumulate as large tubercles that increase head loss and decrease water capacity. Finally, the release of soluble or particulate iron corrosion by-products to the water leads to consumer complaints of "red water" at the tap (McNeill and Edwards 2001).

Corrosion can also lead to unacceptable levels of copper and lead being introduced into drinking water, which in turn can have deleterious effects on human health. Increasing awareness of the presence and potential health impacts of corrosion products in drinking water has led the U.S. Environmental Protection Agency (USEPA) to mandate action levels of 0.015 mg/L for lead and 1.3 mg/L for copper (Pontius 1991). Canadian drinking water guideline levels are 0.01 mg/L maximum acceptable concentration for lead on a flushed sample and 1.0 mg/L aesthetic objective for copper (Health and Welfare Canada 1996). Corrosion can also lead to premature failure of plumbing and piping systems and can cause problems such as metallic tasting water and staining of plumbing fixtures, clothing, and human hair (Churchill et al. 2000).

A study at the Paris Water Utility found that blending of nanofiltered and surface finished waters caused carbonate imbalance in the water, which resulted in excessive lead and copper release. A 1-year research project has been conducted to select a model to calculate the suitable dosage of soda ash (sodium carbonate) and contact time to obtain a stable pH. The Hallopeau–Dubin model, which uses flow rate, temperature, conductivity, calcium hardness, alkalinity, and pH as input parameters, led to very satisfactory results and could be generalized to other water mixtures (Plottu-Pecheux et al. 2001).

Imran et al. (2006) found that varying water quality to minimize iron release could lead to increased copper and lead release within the distribution system. It was found that increasing alkalinity was beneficial in reducing the release of iron products from pipes but increased the corrosion of copper and lead pipes. Increasing sulfates was found to reduce the release of lead but increased the release of iron. These conflicting water quality requirements for lead, copper, and iron release require evaluation of the trade-offs between water quality and the corrosion response of all metals subject to corrosion in the distribution system.

METAL RELEASE MODELING

Taylor et al. (2005) developed statistical linear and nonlinear relationships that described the metal release in terms of source water quality. Equations 2.1–2.3 show the copper, lead, and iron release models that were developed to analyze different blends of source waters and identify a feasible range of blends that satisfy the water quality required for control of copper, lead, and color release.

Copper Release Model

A study was conducted to evaluate the performance and stability of an artificially synthesized cuprous oxide (Cu_2O) film and naturally formed heterogeneous corrosion scales under the influence of a number of synthetic and real waters using electrochemical impedance spectroscopy. According to Palit and Pehkonenb (2000), the quantitative prediction of copper levels in drinking water was found to rely on the solubility and physical properties of cupric oxide, hydroxide, and basic carbonate solids, which comprise most scales in drinking water supplies. Because of considerable uncertainty in the solubility constants of all these minerals, the tendency for prolonged existence of thermodynamically metastable phases like cupric hydroxide ($Cu(OH)_{2(s)}$), and the possible slow rate of formation of malachite ($Cu_2(OH)_2CO_{3(s)}$), quantification becomes difficult.

The model shown in Equation 2.1 was developed by Xiao in 2004. The nonlinear copper release model indicates that alkalinity, temperature, sulfate, silica, and pH significantly affect total copper release.

$$Cu = 0.28 + temp^{0.72} \times alk^{0.73} \times pH^{-2.86} \times (SO_4^{2-})^{0.10} \times SiO_2^{-0.22} \qquad (2.1)$$

where
Cu = total copper release, mg/L
temp = temperature, °C
alk = alkalinity, mg/L as $CaCO_3$ (calcium carbonate)
SO_4^{2-} = sulfate, mg/L
SiO_2 = silica, mg/L as SiO_2

Both the experimental data and empirical model showed that increasing alkalinity will produce more total copper release than any other water quality parameters. Copper concentration increased with increasing temperature and sulfate. Increasing pH and silica were observed to decrease copper concentration for the conditions of this investigation (Xiao 2004).

Lead Release Model

A variety of regression functions combining 11 major parameters (e.g., alkalinity, pH, chloride, sulfate, dissolved oxygen [DO], temperature, calcium, SiO_2, conductivity, ultraviolet absorbance at 254 nm [UV_{254}], and chloramine) were examined and evaluated to develop the lead release model. Temperature, alkalinity, pH, chlorides, and sulfate were found to be statistically significant in the lead release model, as shown in Equation 2.2 (Tang et al. 2006).

$$Pb = 1.027^{(T-25)} (alk)^{0.677}(pH)^{-2.726}(Cl)^{1.462}(SO_4^{2-})^{-0.228} \quad (2.2)$$

where Pb = total lead release, mg/L
alk = alkalinity, mg/L as $CaCO_3$
T = temperature, °C
Cl = chloride, mg/L
SO_4^{2-} = sulfate, mg/L

The statistical model indicates that total lead release will increase as pH and/or sulfate increases and decrease as alkalinity and/or chloride increases.

Iron Release Model

Effluent values for apparent color correlate well with total iron concentrations (Cullen 2002). Because of high coefficient of correlation ($R^2 = 0.82$) between apparent color and total iron concentrations, effluent values for apparent color can be used as a substitute measurement for total iron (Fe) (Equation 2.3).

$$Fe = 0.0132 \times \text{apparent color} \quad (2.3)$$

Chlorides, sodium, sulfate, DO, temperature, hydraulic residence time (HRT), and alkalinity were statistically significant in a color release (iron) model (Imran et al. 2005a), which is shown in Equation 2.4.

$$\Delta C = \frac{(Cl^-)^{0.485}(Na^+)^{0.561}(SO_4^{2-})^{0.118}(DO)^{0.967}(T)^{0.813}(HRT)^{0.836}}{10^{1.321}(alk)^{0.912}} \quad (2.4)$$

where ΔC = increase in effluent values for apparent color, cobalt–platinum units
Cl^-, Na, SO_4^{2-} = chlorides, sodium, and sulfates, respectively, mg/L
DO = dissolved oxygen content, mg/L
T = temperature, °C
HRT = hydraulic residence time, days
alk = alkalinity, mg/L as $CaCO_3$

CORROSION INHIBITORS

Orthophosphates, blended orthophosphates, silicates, and pH control are some of the corrosion control techniques commonly used in drinking water treatment. Only linear polyphosphates are used in the water industry, and the term refers to condensed inorganic phosphate compounds ranging from a simple diphosphate to a long-chain linear polyphosphate. BOPs are a mixture of orthophosphates and polyphosphates. The ratio of ortho-to-poly in BOPs can vary because of reversion. Reversion refers to the transformation of polyphosphate to orthophosphate by hydrolysis over time.

Film Study and Surface Characterization

The use of phosphate inhibitors forms a layer on the surface of pipes, which protects against corrosion (Benjamin et al. 1990). The surface speciation of orthophosphate ions on goethite has been studied as a function of pH, time, total phosphate concentration, and ionic medium. Phosphate anions have a high affinity for corroded surface metals, which leads to the formation of a stable phosphate metal complex on the pipe surface and limits further corrosion (Persson, Nilsson, and Sjoberg 1996).

Moriarty (1990) examined surface corrosion in a cooling water system when BOP was used as an inhibitor. The study showed two types of protective film formed on the pipe surface: an inner thin monolayer film, which formed on the pipe surface from adsorption of orthophosphate, and an outer thicker layer composed of a weakly soluble compound, generally porous, which was formed by the polyphosphate. The inner layer reduces iron dissolution while the outer, protective layer serves as a physical barrier to diffusion and electron transfer.

ZOP offers control of corrosion at both the cathodic and anodic sites. Different PO_4/Zn (phosphate/zinc) ratios were evaluated in the investigation of the impact of various corrosion control strategies (Volk et al. 2000). The reduction of zinc did not affect the effectiveness of the corrosion inhibitor. Zinc molecules are supposed to react with carbonates at the cathode and form deposits, whereas phosphates block the anodic sites. The corrosion behavior of pure iron in polyphosphate solutions has been investigated by means of XPS, or X-ray photoelectron spectroscopy (Koudelka, Sanchez, and Augustynski 1982). The enhanced efficiency of polyphosphate inhibitor obtained by the conjoint use of the divalent Ca^{2+} or Zn^{2+} ions was related with the formation of a diffusion barrier layer on the cathodic areas. This barrier limits the access of oxygen to the iron surface. Analysis results suggested the reversion of polyphosphates to orthophosphates as being an essential step in the formation of the protective film. Reversion depends on factors such as pH, temperature, detention time, and poly-to-ortho ratio. In addition, reversion rate is quicker in diluted stock than in pure product (Yousef, Williams, and Rozza 1994).

Similar results were found in another analysis using Auger electron spectroscopy in which corrosion inhibiting films formed on iron in phosphate-containing solutions were depth profiled. In calcium-free solutions, orthophosphate is not incorporated into the film in significant amounts. When divalent calcium ion is added to the solution, the resulting film contains significant amounts of phosphate, and the film composition depends on the applied potential (Kamrath, Mrozek, and Wieckowski 1993).

The calcium phosphate deposition on iron in oxygen-containing neutral aqueous solutions has been studied with respect to the corrosion inhibition properties of the deposited layer. Results of this study show that calcium phosphate deposition on iron results from super saturation (pH increase). Calcium carbonate codeposition via heterogeneous nucleation accounts for up to 15% of inhibition. The apparent thickness of the calcium phosphate layer increases linearly with temperature and bulk solution pH (Jovancicevic and Bauer 1989).

The silica in sodium silicate solutions carries a negative charge and will, in theory, migrate to anodic areas where it can react with metallic ions. The PQ Corporation (2001) reported that monomeric silica represented by $(SiO_3)^{-2}$ is adsorbed onto pipe surfaces at anodic areas, forming a thin monomolecular film on the interior of the pipe, and prevents further corrosion at the anode. Microscopic and X-ray examination of the film formed at the metal surface showed it had two layers, with most of the silica in the surface layer adjacent to the water. During lab or pilot testing, a relatively high passivation dose of SiO_2 is required initially for at least 30 days in order to establish the protective film on metal surfaces and then subsequently reduce the dosage. After the film has been created, the protection is maintained as long as silicate treatment is continued; if stopped, the protection is gradually lost. They also reported that naturally occurring silica in the form of SiO_2 has probably already reacted with cations in bulk water and may not be effective in reacting with pipe surfaces.

Electrochemical kinetic parameters on copper plumbing surfaces were investigated under hydraulic and water quality conditions typical of distribution systems. Orthophosphates were found to significantly reduce corrosion rates on copper surfaces; however, the film formed because of phosphate addition can dissolve at a pH below 6, which can lead to higher corrosion rates comparable to new copper surfaces (Reiber 1989). Feng et al. (1996a, b) conducted very thorough experiments in order to understand the rate-limiting processes of copper corrosion. Using XPS, the study found that under their experimental conditions, the rate-limiting process was the diffusion of copper ions through the oxide film.

INHIBITION OF METAL RELEASE

The inhibition capabilities of orthophosphate and polyphosphates on corrosion by-product release have been studied under various conditions of water qualities and hydraulic flows. Release of copper and lead are of interest because of the Lead and Copper Rule (LCR), in addition to iron release from pipe distribution systems.

Inhibition of Iron

Red water, or rusty water, is generally attributed to ferric hydroxide. Ferric hydroxide forms from the reaction of insoluble (ferric) iron with water after the oxidation of soluble (ferrous) iron to the insoluble ferric in water. It has been shown in the past that polyphosphate is able to reduce red water when dosed at about 2–4 ppm per 1 ppm of iron. The effectiveness of polyphosphate in corrosion control of municipal water is greatly affected by flow velocity. Slow velocities (around 0.5 ft/sec = 0.15 m/sec) have shown increased corrosion rates. However, turbulent flow velocities (2–5 ft/sec = 0.6–1.5 m/sec) are not continually maintained in all parts of a typical distribution system (Larson 1957).

Klueh and Robinson (1988) investigated the effectiveness of sequestering iron in groundwater by polyphosphate addition. Polyphosphate dosing proved beneficial in sequestering iron, compared to no polyphosphate addition, but the presence of calcium adversely limited iron sequestration by polyphosphate. Higher turbidity was recorded at high polyphosphate doses, which was attributed to exceeding the solubility of calcium phosphate. Boffardi (1993) has noted that overdosing polyphosphates can cause old loose deposits, especially iron oxide deposits, to dislodge from pipe surface and disperse, increasing red water problems.

McNeill and Edwards (2000) investigated phosphate inhibitors in new iron pipes in up to 72 hours of stagnant conditions. Five different water qualities (variable pH and alkalinity) were used and dosed with orthophosphate, polyphosphate, or none. Examination of iron release measurements showed that both ortho- and polyphosphates either increased or had no effect on iron concentrations. The only tested conditions that showed reduced iron release was at high alkalinity (300 mg/L as $CaCO_3$) and pH of 7.2. This observation of phosphate inhibitors influence in stagnant pipes is opposite to common experiences in flowing water conditions. On average, other pH and alkalinity conditions showed an increase in iron release during stagnation when polyphosphate is used, compared to pipes receiving no inhibitor (McNeill and Edwards 2000).

Davis, Chen, and Edwards (2002) studied silicate sorption at 0–200 mg/L SiO_2 on ferric hydroxide at pH from 5 to 9.5. They concluded that at all pH values studied, sorption densities exceeding monolayer sorption were observed at silica levels typical of natural waters. The authors proposed a soluble dimeric silica $Si_2O_2(OH)_5$ sorption model which fitted the data from zero to 0.40 mol SiO_2/mol Fe (iron) (i.e., a weight ratio of about 0.2 for SiO_2/Fe). They also reported that the iron surfaces become more negatively charged with increasing SiO_2 dose and pH.

During a 4-month-long experiment, Rushing, McNeill, and Edwards (2003) reported that higher levels of silica caused more iron release to the water and decreased the size of suspended iron particles. More silica was released into the water from cast iron during corrosion and was removed from the water by incorporation into the scale layer. Silica also affected the type of scale that formed on the iron coupons. Scale at the lower silica concentrations was fairly uniform and easy to remove from the coupons, whereas the scale from the high-silica reactors was denser and was more difficult to remove. Iron corrosion occurring via hydrogen evolution was significant under all experimental conditions. The authors report a significant effect of initial silica concentration on iron release. Initially, more silica released more iron; however, iron release decreased with higher silica dose over time. In all cases, silica consistently caused more iron release and red water release and was attributed to the short duration of the study.

Inhibition of Lead and Copper

Dodrill and Edwards (1995) conducted a survey of about 360 utilities to examine their strategies in controlling lead and copper release, in response to the LCR of 1991. It was observed that without phosphate inhibitors, higher alkalinity resulted in lower lead release. At low alkalinity, the use of inhibitors reduced lead release compared to no inhibitors. The survey revealed that utilities with low alkalinity and pH below 7.4 benefited from using orthophosphate but not polyphosphate. Moreover, polyphosphate increased lead release at higher alkalinity. For copper, the survey showed that at high pH, copper release is reduced with and without inhibitors. At pH <7.8, copper release was high at high alkalinity, but inhibitor use mitigated that release well. However, at pH >7.8, inhibitors had variable and adverse effects on copper corrosion by-product release.

Edwards, McNeill, and Holm (2001) compared the benefits of orthophosphate versus polyphosphate in controlling copper corrosion by-product release in aged copper pipes at variable pH and alkalinity values. Polyphosphate seems to perform less favorably than orthophosphate at comparable concentrations of 1 mg/L as P (phosphorus). It was believed that orthophosphate reduced copper solubility by forming a cupric phosphate scale. While in the absence of any phosphate inhibitors, an insoluble malachite scale formed over a period of years. Polyphosphate, however, increased copper release in comparison to orthophosphate because it complexed copper, increasing soluble copper release.

Holm and Schock (1991) studied the effects of polyphosphate products on lead (Pb) solubility in plumbing systems. Three BOP products were tested in household plumbing at pH values of 8.0, 40 mg/L Ca (calcium), and approximately 1 mg/L total P. Results showed elevated soluble lead concentrations compared to no inhibitors use at a wide range of alkalinity values. It is believed that Pb complexation by polyphosphates would reduce free Pb in water, which promotes the dissolution of hydrocerussite [$Pb_3(OH)_2(CO_3)_2$], the most common Pb mineral in plumbing systems. This explains the elevated Pb concentrations found when polyphosphates are used.

Hozalski, Esbri-Amador, and Chen (2005) studied the benefit of inhibitors in controlling lead release from lead pipe. They used OP, polyphosphate, BOP, and stannous chloride ($SnCl_2$). All inhibitors performed better than the no-action alternative at reducing lead release with various degrees. The ranking of inhibitors based on total lead concentrations, from lowest to highest, in the tested pipe loops was as follows: OP < $SnCl_2$ < BOP < PP < control. The authors also observed that phosphate-based inhibitors resulted in increased bio-growth compared to the stannous chloride and control pipes.

Solubility models are used to predict maximum soluble concentration of a metal in water, given a controlling solid phase. The controlling solid phase for copper is believed to be [$Cu(OH)_2$] or [$Cu_3(PO_4)$] (cuprous phosphate) in the presence of phosphate inhibitors, whereas for lead it could be cerrusite [$Pb(CO_3)$], hydrocerrusite [$Pb_3(OH)_2(CO_3)_2$] and hydroxypyromorphite [$Pb_5(PO_4)_3OH$] (Edwards, Jacobs, and Dodrill 1999). However, the study indicated that solubility modeling did not accurately predict copper and lead release because of the lack of accuracy of constants in the model, misunderstanding of scale types, and the influence of other mechanisms such as colloid detachment from the surface.

Use of ZOP and sodium silicate has shown promising results in the control of copper corrosion (MacQuarrie, Mavinic, and Neden 1997). A study in the Greater Vancouver Water District (Canada) was conducted to evaluate the effectiveness of ZOP, sodium silicate, and a commercial blend of the two as corrosion inhibitors. The results of the weight loss determinations from the pipe inserts indicated that all of the inhibitors, particularly the ZOP, worked very favorably with copper, but they offered no additional benefit over that obtained from the pH and alkalinity adjustments alone in the case of the cast-iron coupons.

A 12-month pilot-plant study was conducted to evaluate several corrosion control treatments to reduce metal leaching from typical household plumbing materials in the Greater Vancouver Water District (Churchill et al. 2000). A pipe loop system was used to test six treatment options, with a control loop of the characteristically soft, acidic water. Four pH–alkalinity combinations and two concentrations of ZOP (0.37 and 1.0 mg/L as Zn) were studied. Results of this study showed the pH–alkalinity treatments exacerbated metal leaching in standing samples relative to the control loop. The ZOP-treated loops were effective at reducing lead and copper leaching to below control levels.

Indian Hill Water Works (IHWW) in Ohio supplies water to several suburban communities to the east of Cincinnati. In the late 1980s and early 1990s, various coupon and field tests were carried out using ZOP chemicals of different Zn/PO_4 (zinc/phosphate) ratios and dosages. The ZOP treatment was mostly successful for lead and inconsistently successful for copper throughout the early and mid-1990s, yet the 90th percentile was still at 1.54 mg/L for copper in 1997. This prompted withdrawal of the ZOP treatment from IHWW (Schock and Fox 2001).

pH Controls

When inhibitors are not considered, copper release can be controlled best by raising pH (Edwards, Hidmi, and Gladwell 2003). It was demonstrated that bicarbonates adversely affected copper release and that a slight pH increase (7.0–8.0) showed significant reduction in copper release. Carbon dioxide (CO_2) stripping was the recommended method for raising the pH without raising alkalinity.

Maddison, Gagnon, and Eisnor (2001) report comparison of polyphosphate addition and pH elevation by sodium hydroxide in a pilot-scale distribution system. Hundred-year-old cast-iron sections of pipe with inside diameters of 4 in. (0.1 m) were dug out and exposed to flowing water (0.2 ft/sec = 0.06 m/sec) and then later to water stagnant up to 12 hours. Polyphosphate was dosed at 0.75 mg/L and resulted in average increases of 0.3 ntu for turbidity and 2 cpu for apparent color, whereas the respective values for pH control were 1.2 ntu and 15 cpu. Full-scale distribution studies followed and resulted in an approximately linear increase in turbidity from 0.05 to about 0.17 ntu and color from 2 to 6 cpu for pH control for an HRT of 55 hours. Polyphosphate dosing resulted in a negligible increase in turbidity or color for the same time. Loss of free chlorine residual was similar for both treatments and was shown to be statistically independent of the other water quality parameters. In summary, full-scale sampling showed that water quality deteriorated with increasing HRT when pH (LSI = 0) adjustment was used.

Electrochemical Noise

Electrochemical noise (ECN) is simply a methodology for monitoring corrosion activity in water. ECN parameters are specific for a metal and provide rapid monitoring of corrosion activity. Because of the reduced time response, ECN monitoring represents a potential improvement for corrosion control in drinking water distribution systems.

ECN measurements are literally made at the speed of light. The transfer of electrons from the anodic areas to the cathodic areas in a metal is initiated by a potential difference and generally arises because of surface imperfections, crystal structure scales, and trace impurities. These factors lead to a difference in metallic activity that drives the electrochemical potential. Theoretically, the relationship between the noise resistance and corrosion current depends on the potential, pH, and flow rate. However, in general, the surface coverage of the anodic and cathodic sites is relatively independent of potential or bears a simple relationship with potential. Consequently, noise resistance can yield accurate corrosion rates (Eden 1998, NACE 1993, and Cottis 2000).

In potable water, there are at least three dominant electron acceptors: DO, chlorine, and hydrogen ion. These compete for the electrons accumulated on the cathode, depending on their relative electrode potentials and their activities. ECN estimates the charge transfer resistance from potential and current noise, which provides proactive monitoring for corrosion control.

Duranceau, Townley, and Bell (2004) reported ECN as a means of monitoring iron, copper, and lead corrosion in distribution systems. This study found that ECN was instantly sensitive to corrosion in iron, copper, and lead electrodes in contact with drinking water in a distribution system and accurately monitored the effects of inhibitors. Norton and LeChevallier (1997) reported corrosion rates from 0.5 to 6.1 mpy (0.01 to 0.15 mm/yr) and 1.2 to 11.4 mpy (0.03 to 0.29 mm/yr) in two distribution systems. The corrosion rate exhibited a close relationship with temperature and lagged by a month or two with maximum and minimum temperatures. The corrosion rate from plant effluent dropped from a range of 4.9–5.6 mpy (0.12–0.14 mm/yr) to 1.3–1.4 mpy (0.03–0.04 mm/yr) after increasing ZOP levels from 0.5 to 2.0 mg/L as P in the distribution system, which decreased the corrosion rate below 4.0 mpy (0.10 mm/yr) with no coliform detection. Norton and LeChevallier (1997) also reported that proactive monitoring of daily corrosion rates using online ECN monitoring equipment would reduce the operating expenses by optimizing corrosion inhibitor feed rates.

Duranceau, Townley, and Bell (2004) conducted a corrosion inhibitor screening evaluation using ECN techniques. Four poly-orthophosphate formulations were tested one at a time at various doses (0.4 to 1.2 mg/L as P) for a period of 28 to 31 days. The study examined the correlation coefficient between water quality and noise resistance and made recommendations on a qualitative basis. All inhibitors showed evidence of reduced localized corrosion for iron;

mitigating iron corrosion with increases in alkalinity and calcium; and decreases in sulfate, DO, chlorine residual, and chlorides. Copper corrosion was increased by higher alkalinity and mitigated by a decrease in sulfate, whereas lead corrosion was mitigated by decreases in sulfate and chlorides.

Passivity has been related to the formation of a duplex structure Cu_2O/CuO (cuprous oxide/tenorite), $Cu{\sim}OH_2$. It is believed that at pH higher than 7.0, copper forms a passivation duplex layer. A laboratory study by Tan, Bailey, and Kinsell (1996) provided evidence that the trend of ECN effectively followed inhibitor film formation and a destruction process. The authors state that ECN is a valid technique for continuous monitoring of inhibitor film performance and evaluation of inhibitor film persistency. They also confirmed the strong correlation between ECN and linear polarization resistance (LPR) or the sum of charge transfer resistance and inhibitor film resistance. This may prove advantageous for ECN over LPR since film resistance effects can be a serious limitation in cases of oxide/inhibitor films.

SURFACE ROUGHNESS

Pipe surface roughness affects hydraulic efficiency and system water quality of drinking water distribution systems. Surface roughness considers the magnitude and spacing of microscale surface features such as peaks, valleys, ridges, and grooves. Currently, there are several statistics describing different aspects of the surface profile (Lippold and Podlesny 1998). Most of these parameters describe the height, shape, or spacing of peaks and valleys in the profile.

The roughness average is the arithmetic mean of the surface deviations from the mean plane. As such, it is a statistic describing the height of peaks and valleys on a surface. Roughness average is a robust parameter, capable of providing an overall picture of the roughness of the surface while limiting the effect of any atypical peaks or valleys on the roughness value. It is not sensitive to differences in the spacing of roughness peaks and valleys; thus, a surface with many peaks packed closely together could have the same roughness average as a surface with its peaks spread apart.

Surface Roughness and Hydraulic Capacity

The hydraulic capacity of a pipe describes the pressure losses that occur when water flows under typical pressure gradients. Pipes with low hydraulic capacity have higher friction losses per unit of flow than pipes with high hydraulic capacity. Surface roughness has been linked to hydraulic capacity (Nikuradse 1950) and is measured indirectly in a distribution system by a flow test (Walski et al. 2003).

In the literature on hydraulic capacity in distribution systems, the most common formula used to relate hydraulic capacity and head loss is the empirical Hazen–Williams equation, which is presented in Equation 2.5.

$$V = 1.318 C_{HW} R_h^{0.63} S^{0.54} \tag{2.5}$$

where V = average velocity, ft/sec
C_{HW} = Hazen–Williams coefficient
R_h = hydraulic radius, ft
S = slope of the energy grade line, ft/ft

The hydraulic radius, R_h, is the ratio of the cross-sectional area to the cross-sectional wetted perimeter. The energy grade line slope, S, is the ratio of the hydraulic head loss to the length of pipe over which that loss occurred. The Hazen–Williams coefficient, C_{HW}, is the measure of pipe hydraulic capacity. A higher C_{HW} corresponds with a higher hydraulic capacity, such as that found in new pipe. A lower C_{HW} corresponds with a lower hydraulic capacity, such as that found in old, highly tuberculated, unlined iron pipe. Thus, the C_{HW} is inversely proportional to pipe surface roughness. The C_{HW} is also a function of fluid velocity, fluid kinematic viscosity, and pipe diameter (Sharp and Walski 1988).

Effect of Bulk Water Quality on Surface Roughness

Hudson (1966) examined the decline in carrying capacity of water distribution systems in seven cities. The decline in the Hazen–Williams coefficient varied in each city, suggesting that roughness growth rates, α, may vary with water quality.

Larson and Sollo (1967) examined the relationship between water quality and the corrosion rate of UCI. They recommended that the pH be adjusted to maintain a zero saturation index in order to avoid losses in pipe carrying capacity, as indicated by declines in the Hazen–Williams coefficient, C_{HW}.

Sharp and Walski (1988) developed a model for predicting the increase in roughness of unlined metal pipes and water quality. The model assumed that roughness increased linearly with time and the rate of roughness increase was dependent on the LSI, with more negative LSIs corresponding with greater rates of roughness increase. By incorporating the LSI into the roughness growth model, Sharp and Walski, demonstrated that calcium hardness and alkalinity, in addition to pH, are factors affecting the roughness growth rate. These studies suggest that roughness growth rate is largely influenced by water quality.

Effect of Surface Roughness on Water Quality Phenomena Occurring at the Pipe Wall

Historically, surface roughness in the distribution system has been considered for its effect on flow capacity. However, surface roughness may also be a factor influencing various water quality phenomena occurring at the pipe wall, such as biofilms and chlorine dissipation. Fletcher and Marshall (1982) found the rate of biofilm formation to be dependent on the physicochemical properties of a surface, surface roughness, and the physiological characteristics of the organism.

LePuil et al. (2005) measured biofilm density using the protein exoproteolytic activity (PEPA) assay by Laurent and Servais (1995) and found that PEPA was greatest for UCI pipe followed by LCI and GS pipe, which had similar PEPA, and the least biofilm density was observed for PVC pipe. Visual observations of the roughness of these materials led to the conclusion that differences in biofilm density by material can be related to differences in material surface roughness.

Surface roughness is a proposed factor affecting the wall rate constant of disinfectant residuals. The literature relates the rate of chlorine decay at the pipe wall with the pipe material, and it might be possible to describe how the wall rate constant varies as a function of pipe roughness.

Rossman, Clark, and Grayman (1994) developed a mass-transfer–based model for chlorine decay, assuming first-order kinetics in the bulk flow and at the pipe wall. The overall first-order decay constant, K_i, is given in Equation 2.6.

$$K_i = k_b + \frac{k_w k_f}{r_h(k_w + k_f)} \tag{2.6}$$

where k_b = decay rate constant in the bulk flow, $time^{-1}$
k_w = decay rate constant at the wall, $L \times time^{-1}$
k_f = mass transfer coefficient, $L \times time^{-1}$
r_h = hydraulic radius, L

Disinfection residual decay at the pipe wall must consider the reactions at the wall, the mass transfer limitations, and the available surface-area-to-volume geometry (Frateur et al. 1999; Vikesland, Ozekin, and Valentine 2001; Vikesland and Valentine 2002a, b; Hallam et al. 2002). Arévalo (2003) and Kiéné, Lu, and Lévi (1998) found that old cast-iron and steel pipes had a higher chlorine demand than pipes made of synthetic materials. The difference on decay rates between iron-based and synthetic materials could be due to chemical reactions with the iron and iron corrosion by-products or higher surface area in the corroded pipes.

Doshi, Grayman, and Guastella (2003) conducted a series of simultaneous chlorine loss and head loss tests within the water distribution system of Detroit, Mich. Most of the pipes were UCI, from 70 to 135 years old. Using a first-order mass-transfer model, Doshi, Grayman, and Guastella reported that the wall decay coefficient of chlorine, k_w ($L \times time^{-1}$), did not appear to be related to the Hazen–Williams coefficient (i.e., pipe roughness), but rather k_w increased directly with increasing flow.

SURFACE STRUCTURE OF IRON CORROSION SCALES

The surface structure of iron corrosion scales plays a significant role in iron release. Iron corrosion by-products tend to accumulate in large mounded tubercles that often merge to cover the entire pipe surface (Sarin et al. 2004). Scales in iron pipes are often characterized by a corroded floor, a porous core, a dense shell-like layer, and a surface layer loosely attached to the shell-like layer (Sarin et al. 2001). The formation of uneven surface deposits, such as tubercles, increases the pipe surface roughness (Schock 1999). Iron corrosion occurs beneath the tubercle at the corroded metal surface of the pipe where iron(0) is oxidized to iron(II).

Iron release occurs when dissolved Fe(II) diffuses into anoxic bulk water, where it is oxidized to Fe(III) particulates that may precipitate back on the surface or pass through the distribution system to the consumer. The porosity of the iron tubercle may affect the corrosion rate by limiting outward diffusion of Fe(II) and inward diffusion of anions for charge neutralization. The dense shell-like layer can also play a role in controlling iron corrosion, and consequently iron release by acting as a diffusion barrier.

Chlorides and sulfates may penetrate protective metal oxide films, accelerating corrosion and increasing the surface roughness of iron pipes. This increase in surface roughness might then increase the number of available reaction sites at the film–water interface, possibly further aggravating water quality problems. Gotpagar et al. (1999) showed that pretreating zero-valent iron with chlorides enhanced the rate of reductive dehalogenation of trichloroethane (TCE). The authors used atomic force microscopy, optical profiling, and scanning electron microscopy (SEM) to show that chloride-pretreated iron had greater surface roughness than the untreated iron. It was proposed that lattice defects and grooves helped to catalyze the transformation of

TCE. Orthophosphate inhibitors might control iron corrosion by plugging holes in the iron oxide film with an iron phosphate compound (Pryor and Cohen 1951).

SURFACE STRUCTURE OF GALVANIZED STEEL SCALES

GS pipe has an internal zinc coating that limits iron corrosion by acting as a sacrificial anode. GS pipes tend to have less severe incrustation and tuberculation than unlined iron pipe. Using jar tests and GS pipe coupons, Pisigan and Singley (1985) found DO to be the most significant factor in GS corrosion, with LSI and pH being less significant. The authors also said that hydrozincite ($Zn_5(CO_3)_2(OH)_6$) may be the dominant zinc scale on GS pipe, although they did not detect it as a corrosion by-product.

Corrosion products of GS water pipes may be similar to those observed on unlined iron pipes. Graham and Cohen (1976) used Mossbauer spectroscopy and found goethite, hematite, lepidocrocite, magnetite, and even a zinc iron oxide ($Zn_xFe_{3-x}O_4$).

RESIDUAL

The implementation of USEPA's Stage 2 Disinfectants/Disinfection By-Products Rule has forced utilities to implement alternative disinfection techniques to reduce the formation of disinfection by-products (DBPs). Chloramines are widely used in the water industry because they produce less regulated DBPs than free chlorine (Guay, Rodriguez, and Sérodes 2005).

The concentration of chloramines in water is reduced by their reactions with organic and inorganic matter. Monochloramines are considered more stable than free chlorine but have been shown to decay by auto decomposition (Valentine and Jafvert 1988, Duirk et al. 2002) as well as in the presence of humic materials by direct reactions with natural organic matter (NOM) (Vikesland, Ozekin, and Valentine 1998; Duirk et al. 2005). The auto decomposition of monochloramines starts with hydrolysis reactions that produce free chlorine and ammonia. The free chlorine molecules would then react with the monochloramines to form dichloramines, which rapidly dissociate into chlorides and ammonia (Duirk et al. 2002). In the presence of NOM, monochloramines react directly with the NOM particles by oxidation–reduction (redox) or substitution reactions (Vikesland, Ozekin, and Valentine 1998). Duirk et al. (2005) found that some of the free chlorine formed by monochloramine hydrolysis reacts with the NOM, and the reactive fraction of the NOM structure that is reactive with free chlorine correlates with specific ultraviolet absorbance at 280 nm. Water temperature and pH have also been found to have an effect on chloramine decay. Residual decay accelerates when the water temperature increases (Kiéné, Lu, and Lévi 1998; Powell et al. 2000; Vikesland, Ozekin, and Valentine 2001). Monochloramine stability increases with the increase of pH, and slower rates of decay have been observed at high pH (Duirk et al. 2002, 2005). Vikesland, Ozekin, and Valentine (2001) related the increase on monochloramine loss to the increase on dichloramine formation rate at lower pH.

In addition to organic matter, chloramine reacts with inorganic compounds such as iron, bromides, nitrites, and sulfates (Vikesland and Valentine 2002a; Margerum et al. 1994; Vikesland, Ozekin, and Valentine 2001). The presence of nitrites has been found to increase the decay of chloramines (Margerum et al. 1994; Vikesland, Ozekin, and Valentine 2001). Iron is usually present in solution in its reduced form: ferrous iron Fe(II), which reacts with chloramine, as shown in Equation 2.7 (Vikesland and Valentine 2002a).

$$2Fe^{2+} + NH_2Cl + H^+ \rightarrow 2Fe^{3+} + Cl^- + NH_3 \quad (2.7)$$

From a distribution system viewpoint, the reactions between chlorine/chloramines and iron are also important. Kiéné, Lu, and Lévi (1998) found that old cast-iron and steel pipes had a higher chlorine demand than pipes made of synthetic materials. Arévalo et al. (2004) had similar results for chloramines, where UCI and GS pipes had a much larger chloramine residual loss than PVC and lined cast-iron pipes.

The decay of chlorine and chloramine residuals in distribution systems have also been correlated to the flow velocity (Rossman, Clark, and Grayman 1994; Vasconcelos et al. 1997; Hallam et al. 2002; Doshi, Grayman, and Guastella 2003; Clark and Haught 2005; Mutoti et al. 2007a, b). Some authors use a mass-transfer limited approach that limits the rate of chlorine decay at the pipe wall to the transfer of chlorine from the bulk liquid to the pipe wall (Rossman, Clark, and Grayman 1994; Vasconcelos et al. 1997; Hallam et al. 2002), whereas recent studies show that the rate of decay is directly dependent on the Reynolds number and specifically the flow velocity for free chlorine (Doshi, Grayman, and Guastella 2003; Clark and Haught 2005) and chloramines (Mutoti et al. 2007a, b).

Limited literature is available relating chloramine decay and the use of corrosion inhibitors. Volk et al. (2000) used two annular reactors made of cast iron to compare control water (plant conditions) with the same water dosed with ZOP at 0.7 mg/L as PO_4. No significant difference was observed in the biofilm density, bulk HPC, or the effluent chloramine residual between the control and the test reactor. Batté et al. (2003) showed that the addition of phosphate inhibitors to the water doubled the phosphate concentration in the biofilm but did not affect the biofilm densities or its composition. Butterfield et al. conducted a study in 2002 and found that the use of PO_4 inhibitors reduced free chlorine demand by 10% in the presence of iron corrosion products but had no effect on glass beads coated with iron oxide.

Biological Characteristics

Typically, in biologically unstable water, bacterial populations proliferate as a biofilm becomes attached to the pipe wall. In the biofilm, immobilized cells frequently are embedded in an organic polymer matrix of microbial origin (Characklis and Marshall 1990). In general, the population of suspended cells in distribution systems is orders of magnitude less than the population of biofilm cells (Camper 1996; Van der Wende, Characklis, and Smith 1989; Characklis 1988). Because suspended cells are often considered to be introduced to the liquid phase from the biofilm, the concentration of suspended cells is best achieved by minimizing the number of biofilm cells (Ollos, Huck, and Slawson 2001). However, regression analysis suggests low or very low correlation between biofilm and suspended HPC numbers (Ollos, Slawson, and Huck 1998). The accumulation and proliferation of fixed bacteria under the form of biofilms is usually controlled by a large number of parameters, more or less well studied at bench scale and more rarely in the field at full scale (hydraulics, temperature, nature and concentration of nutrients, bacterial density and species introduced in the system, nature of pipe materials, predators). Figure 2.1 shows the mechanisms taking place in the accumulation of biofilms on a surface in contact with potable water.

It has not yet been demonstrated that the concentration of biodegradable organic matter (BOM) in distributed water directly controls the bacterial density in the biofilm. Ndiongue, Huck, and Slawson (2005) suggested that for the water with no BOM added, there appeared to be no

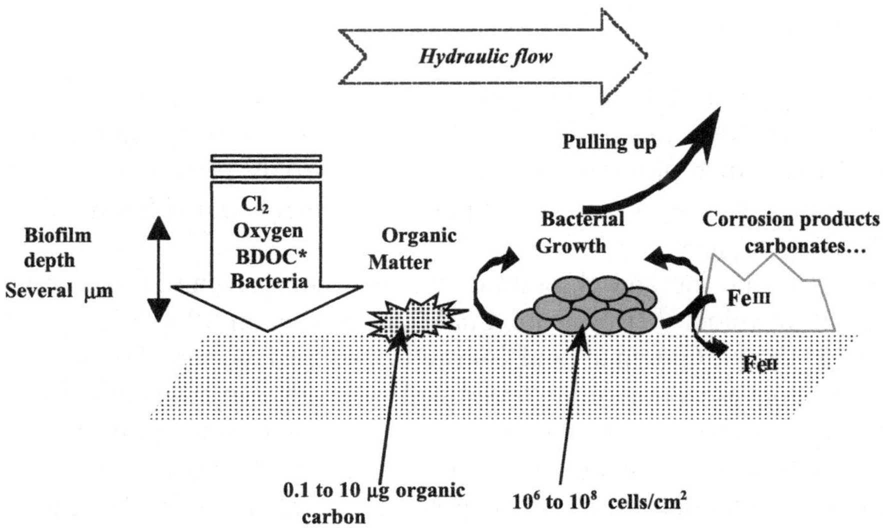

*BDOC = biodegradable dissolved organic carbon.

Figure 2.1 Scheme of several mechanisms taking place in the accumulation of biofilms on a surface in contact with potable water

clear relationship between the temperature and the free chlorine residual required to reduce the biofilm heterotrophic plate count (BFHPC) to the detection limit (3.2 log cfu/m^2). However, for water with BOM added, a strong linear correlation was found between the temperature and the free chlorine residual required to control the biofilm.

The biofilm appears to be a metastable system, in one hand "fed" by the entrance of cells from the bulk (the velocity of deposit/fixation of cells on the surface of materials is correlated to the cell density of moving waters) and by the multiplication of bacteria, and in the other hand stabilized by the constant pulling up and release of cells from the biofilm to the bulk.

Resistance of fixed biomass to oxidants (Morin et al. 1999) can be partially explained by the reducing ability of the biofilm, the fixed organic matter, and sometimes the supporting material, as well as poor diffusion of the oxidant (de Beer, Srinivasan, and Stewart 1994; Stewart and Raquepas 1995). Microbes in biofilm will cause taste and odor in water (Nagy and Olson 1985, Armon and Kott 1995) and also increase corrosion of piping materials (LeChevallier, Babcock, and Lee 1987; Beck 1995; Holden et al. 1995). The accumulation of microorganisms on the surfaces and the formation of biofilm depend on many factors prevailing in the distribution system (e.g., type of surface material, microbial occurrence in water, concentration and quality of nutrients and disinfectants, temperature) (Zacheus et al. 2000). BFHPC is the parameter being used to quantify biofilm density in this study, and the biofilms are being quantified using four materials: PVC, LCI, UCI, and GS.

CHAPTER 3
METHODS

PILOT DISTRIBUTION SYSTEM DESIGN

The PDS and two pilot treatment systems used in the implementation of this project were designed and built by UCF and TBW-MG personnel for the previous project (Taylor et al. 2005). The previous project is sometimes referred to as Tampa Bay Water 1 (TBW I), and this current project, which is the continuation of the previous TCP, is sometimes referred to as Tampa Bay Water 2 (TBW II). The PDS was designed to allow flexibility to study water quality changes resulting from blends of significantly different source waters in old distribution pipe systems. The feedwater blends represented the typical water chemistry likely to be experienced by TBW member governments. The pipes used in the PDS traditionally received conventionally treated groundwater. Both the physical systems and pipe geometries selected represent typical scenarios experienced in a real distribution system.

The PDSs are composed of 18 distribution lines. Lines 1 to 14 are hybrid lines that contain segments of four materials: PVC, UCI, LCI, and GS pipes. Lines 15 to 18 contain multiple segments of each of the single materials. The PDS was constructed of aged pipes that were obtained from existing utility distribution systems to represent the pipe materials used in the TBW-MG distribution systems. The PDS has been maintained with a feed of 100% GW since completion of field activity for TBW I in the summer of 2005. Introduction of a blend of GW, SW, and RO was initiated in December 2005. Prior to introduction of the blend, field monitoring was resumed in the fall of 2005 to verify uniformity between the PDS effluent for all PDS lines. As learned during the equilibration phase from TBW I, apparent color is a useful parameter for definition of the attainment of an equilibrated condition in the PDS and was used for this purpose. The project was divided into four phases; each of 3 months' duration. A different blend was used during each phase to evaluate the effect of water quality. Similar blends were used during phases I and III to evaluate the effect of seasonal conditions on the PDS.

PILOT DISTRIBUTION SYSTEM COMPONENTS

Relevant characteristics of the individual lines include pipe identification, material of construction, pipe length and diameter, and feed rates. The PDSs were identified sequentially (PDS01 to PDS18). The pilot distribution lines were operated to maintain a 2-day HRT. Standpipes were located at the beginning and end of each PDS. The standpipes were made from translucent plastic pipe that were 60 in. (1.5 m) long and had a diameter of 4 in. (0.1 m). The retention time in the PDS feed standpipe was 3.1 hours because of the low velocities associated with the 2-day HRTs. To avoid bacterial growth, the standpipes were wrapped in a nontransparent material to eliminate direct light exposure and regularly cleaned with a plastic brush and a 0.1% solution of sodium hypochlorite. The sodium hypochlorite solution was allowed a 4-hour contact time in the standpipes. All PDSs were constructed with a sampling port after each pipe segment to allow an assessment of water quality changes associated with each pipe material.

PDS01 to PDS14 were composed of four materials, laid out sequentially as

- Approximately 20 ft (6.1 m) of 6-in. (0.15-m)-diameter PVC pipe,
- Approximately 20 ft (6.1 m) of 6-in. (0.15-m)-diameter LCI pipe,
- Approximately 12 ft (3.7 m) of 6-in. (0.15-m)-diameter UCI pipe, and
- Approximately 40 ft (12.2 m) of 2-in. (0.05-m)-diameter GS pipe.

PDS15 to PDS18 were composed of a single material, each as follows:

- PDS15: Eight pipe reaches of approximately 12 ft (6.1 m) of 6-in. (0.15-m)-diameter cast iron each,
- PDS16: Four pipe reaches of approximately 20 ft (6.1 m) of 6-in. (0.15-m)-diameter LCI plus 10 ft (3.0 m) of 6-in. (0.15-m) lined cast iron,
- PDS17: Five pipe reaches of approximately 20 ft (6.1 m) of 6-in. (0.15-m)-diameter PVC each, and
- PDS18: Eight pipe reaches totaling 135 ft (41.1 m) of 2-in. (0.05-m)-diameter GS pipe.

The structures built for the field study are shown in Figures 3.1 through 3.12. As shown in Figure 3.1, a truck and stainless-steel food-grade trailer was used to haul surface water to the pilot site. The surface water was obtained from the TBW regional surface water treatment plant. The surface water was stored in two 7,000-gal (26.5 m^3) storage tanks, as shown in Figure 3.2, before being transferred to the finished water tank. The large process area used to prepare the finished waters is shown in Figure 3.3. The ground on which the equipment is placed was covered by a 4,400-ft^2 (409-m^2) by 6-in. (0.15-m)-thick cement pad and shielded from the sun by a hurricane-rated roof. The trailers shown in Figure 3.4 contained an RO pilot plant, an EN monitoring facility, a storage facility, and a field laboratory.

Figure 3.5 shows the influent standpipes that allowed direct input from the finished storage and inhibitor tanks. Peristaltic pumps were used to control the flow of water and inhibitors to the PDS as well as the flow to the cradles. The inhibitor tanks and the pumps used to feed the PDS and the cradles are shown in Figure 3.6. Figure 3.7 shows the PDS, which was followed by coupon cradles, shown in Figure 3.8. The cradles were 4-in. (0.1-m) PVC pipes that housed 6-in. (0.15-m) PVC pipes, which had been cut in half and supported pipe coupons for surface characterization and microbiological studies (Figure 3.9). Finally, the majority of the PDS effluent was directed to a corrosion shed, shown in Figure 3.10, which contained 18 loops of copper pipes and lead coupons, as shown in Figure 3.11, for the copper and lead corrosion study. The rest of the PDS effluent was sent to the ECN trailer to feed the Nadles (noise cradles), shown in Figure 3.12, which contained iron, copper, and lead electrodes and coupons.

PILOT-PLANT OPERATION

Finished Source Waters

Three source waters were blended and fed to the PDS: conventionally treated groundwater (GW), enhanced coagulation–sedimentation–filtration treated surface water (SW), and desalinated water by reverse osmosis (RO). The surface water was actual finished surface water from the TBW Regional Surface Water Treatment Plant that was transported to the project site, and the

Figure 3.1 Truck and stainless-steel trailer used to haul raw surface water

Figure 3.2 Raw surface water storage

Figure 3.3 Covered tanks for process treatment

Figure 3.4 Field trailers

Figure 3.5 Influent standpipes

Figure 3.6 Inhibitor tanks and feed pumps

Figure 3.7 Pilot distribution systems

Figure 3.8 Cradles for housing coupons

Figure 3.9 Mounted coupons

Figure 3.10 Corrosion shed

Figure 3.11 Copper lines with lead coupons

Figure 3.12 Electrochemical noise trailer

groundwater and desalinated water were obtained from the pilot water treatment systems at the project site. The description of the three finished source waters and the finished water treatment goals for the project are presented in Tables 3.1 and 3.2.

Groundwater Pilot Plant

The GW pilot unit was designed to simulate the finished water of member governments that utilize conventional treatment of GW. The GW unit used the Cypress Creek well field as a raw water source. The main unit processes were aeration, disinfection, and pH stabilization. Before disinfection, GW was aerated to reduce hydrogen sulfide, produce an aerobic water supply, and stabilize the water with respect to calcium carbonate. Aeration was achieved in GW by pumping the raw water to the top of the finished water tank through a nozzle that sprayed the water inside the tank.

Sodium hypochlorite was used as the source of free chlorine for primary disinfection and was dosed to provide a 5-mg/L residual after a 5-minute contact time. Afterwards, ammonium chloride was added to produce a 5-mg/L monochloramine residual. Ammonia was added in the form of ammonium chloride (NH_4Cl) at a 5:1 ratio. The NH_3/Cl_2 (ammonia/free chlorine) ratio was initially 4:1 to protect against DBP formation. This ratio was increased to 5:1 in phase III to reduce free ammonia. Five thousand gallons (19.0 m^3) of GW were produced every week.

Table 3.1
Finished source water descriptions

Source water	Source	System description
GW	Groundwater	Treatment by aeration; disinfection by free chlorine with a residual of 5 mg/L after a 5-minute contact time. 5.0 mg/L chloramine residual.
SW	Surface water	TBW treatment plant: treatment by ferric sulfate coagulation, flocculation, settling, filtration; disinfection by ozonation and chloramination. Project site: adjustment of chloramine residual to 5.0 chloramine residual.
RO	Groundwater	Treatment by membrane reverse osmosis, aeration; disinfection by free chlorine with a residual of 5 mg/L after a 5-minute contact time. 5.0 mg/L chloramine residual.

Table 3.2
Pilot plant finished water treatment goals

Parameter	Standard	Target
pH units	7.4 minimum	0.2 above pH_s
Alkalinity (mg/L as $CaCO_3$)	40 minimum	50 minimum
Calcium (mg/L as $CaCO_3$)	50 minimum, 250 maximum	60 mg/L
Total chlorine (mg/L Cl_2 [chlorine])		5 mg/L
Sodium (mg/L)	80 maximum	<80
Chloride (mg/L)	100 maximum	<100
Sum of sulfate and chloride	3.8 meq/L maximum	<3.8 meq/L
Total dissolved solids (mg/L)	500 maximum	<500
Iron (mg/L)	0.15 maximum	<0.15
Color (cpu)	15 maximum	<15
Total organic carbon (mg/L)	3.6 maximum average/6.5 maximum	<3.6
Ammonia (mg/L as N [nitrogen])	1 maximum	<0.5
Turbidity (ntu)	At filter 0.3 maximum/0.1 (95%), 0.25(100%)	<0.2
	Finished 1 maximum average	<0.3

Surface Water Pilot Plant

SW was treated at the TBW Regional Surface Water Treatment Plant by enhanced coagulation, ozonation, biologically activated carbon filtration, aeration, and chloramination. The SW was hauled weekly to the field facility for use and temporarily stored in two 7,000-gal (26.5-m^3) storage tanks, shown in Figure 3.2, before being transferred to the SW finished water tank. In the SW finished tank, the chloramine residual was adjusted to 5 mg/L as Cl_2.

Table 3.3
Blend ratios of GW, SW, and RO waters used during project

Phase	Time period	% GW	% SW	% RO
I	February–May 2006	62	27	11
II	May–August 2006	27	62	11
III	August–November 2006	62	27	11
IV	November 2006–February 2007	40	40	20

Reverse Osmosis Pilot Plant

The RO (desalination) treatment pilot system produced a finished water that was described as RO. The RO pilot plant was housed in the second trailer from the right, as shown in Figure 3.4, and utilized raw groundwater for the feed stream. The RO treatment pilot system required the addition of TDS, calcium, and alkalinity to the RO permeate to represent the finished water produced by the TBW regional desalination facility. RO pretreatment consisted of 2.7 mg/L anti-scalant addition (Hypersperse MDC700TM, Aqua-Chem, Milwaukee, Wis.), followed by 5-µm cartridge filtration.

The RO (high-rejection) membrane filtration unit was operated at 72%–73% recovery, producing 9.3 gpm (0.6 L/sec) permeate flow, which was aerated by a 10-in. (0.25-m)-diameter aeration tower filled with tripack plastic packing, as previously shown in Figure 3.3. After aeration, 50 mg/L of sea salt was added to the aerated permeate stream to simulate the TBW desalination process. Calcium chloride and sodium bicarbonate were also added to meet the calcium and alkalinity specifications. The finished water was stabilized with sodium hydroxide to 0.1–0.3 pH units above pH_s, and 3,000 to 4,000 gal (11.3 to 15.1 m^3) of RO were produced every week.

Blended Finished Water

Three blends of GW, SW, and RO were investigated during the project. All PDSs received the same blend composition for a 3-month period or phase. At the end of each 3-month period, the blend composition was changed, as shown in Table 3.3.

The blends had different water qualities because of the variation in the ratios of the main water sources used in the blends. But the blend ratios during phases I and III were identical. The reason for repeating the blend was to evaluate seasonal effects. The timing of the phases with the repeated blend was designed to capture the largest temperature variation during the year.

The flow diagram for the finished waters is presented in Figure 3.13. For all the finished waters, a new batch was prepared each week. The water quality analysis schedule required in the field for monitoring finished water storage and process operation is presented in Table 3.4. Each of the three finished source water storage tanks and the blend storage tank were measured for water stability parameters, disinfection residual, color, and UV_{254} using field methods. Additionally, each of the storage tanks was measured daily for disinfection residual maintenance. The feed, permeate, and concentrate from the membrane process were measured prior to posttreatment to evaluate membrane performance and productivity. Additional water quality monitoring was conducted to evaluate specific unit processes within a treatment system.

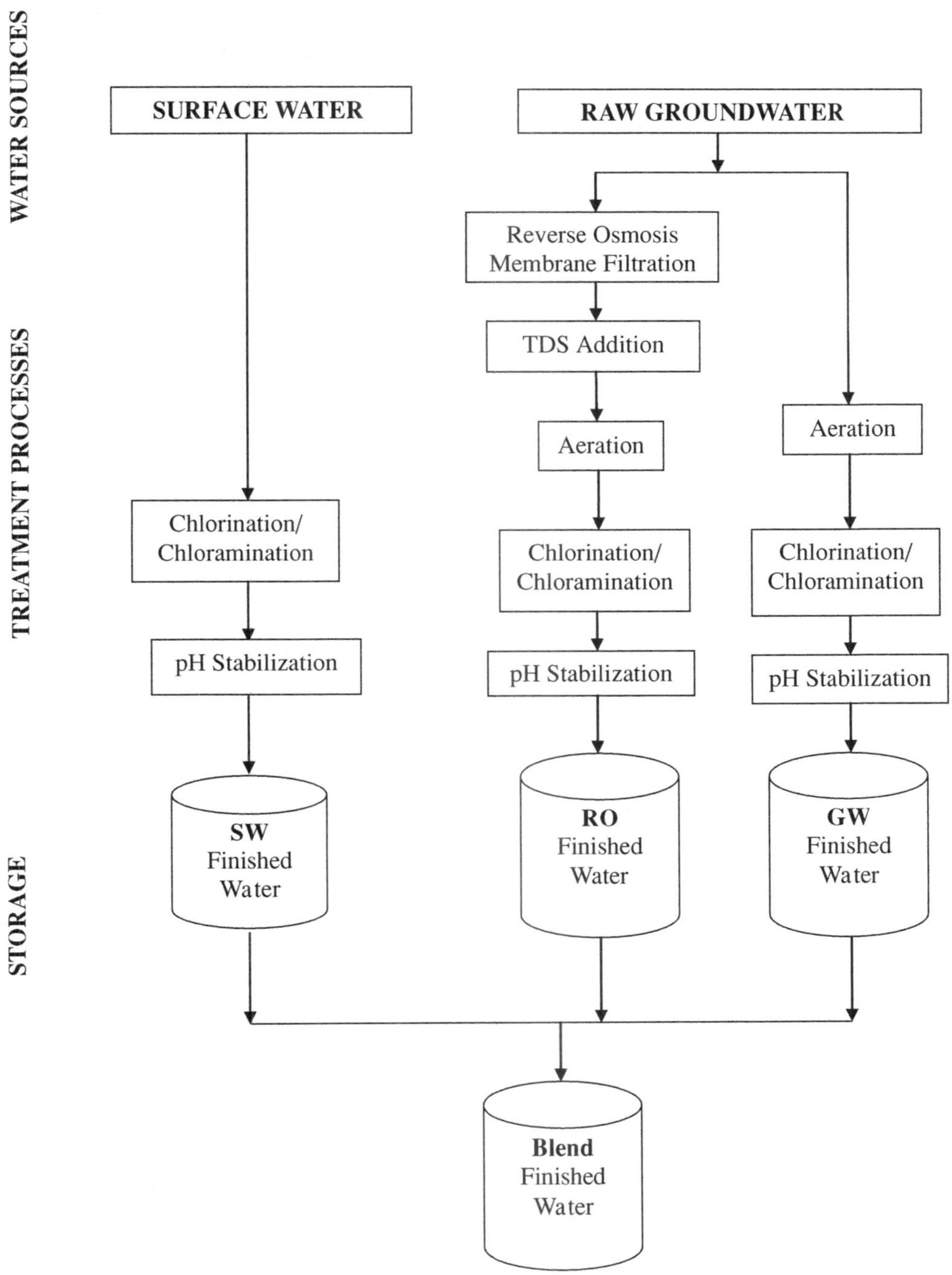

Figure 3.13 Process train layout

Table 3.4
Field analysis schedule for water treatment processes

	GW raw	GW finished	SW finished	RO finished	Blend
Free chlorine	N/A	D	D	D	D
Total chlorine	N/A	D	D	D	D
Ammonia	N/A	D	D	D	D
Temperature	B	D	D	D	D
pH	B	B	B	B	B
Alkalinity	B	B	B	B	B
Calcium hardness	B	B	B	B	B
Total hardness	B	B	B	B	B
Magnesium hardness	B	B	B	B	B
UV_{254}	B	B	B	B	B
Color	B	B	B	B	B
Turbidity	B	B	B	B	B
Dissolved oxygen	B	B	B	B	B
Conductivity	B	B	B	B	B
Total dissolved solids	B	B	B	B	B
Chlorides	B	B	B	B	B
Sulfates	B	B	B	B	B
Iron	B	B	B	B	B
Heterotrophic plate count	N/A	B	B	B	B

NOTES: N/A = not applicable, D = daily, B = batch (after each batch is finished). UV_{254} is filtered. Magnesium hardness is determined by the difference between calcium hardness and total hardness. pH_s is determined after all chemicals are added to adjust water quality to targeted goals. TDS was measured by conductivity probe.

Process Finished Water Quality

Water quality was monitored after each batch was produced to ensure that they met target water quality objectives designed to mimic the actual and expected TBW blended water quality shown in Table 3.5. Some variations in average water quality were observed for the surface water. The variations somehow follow the dry season (December–April) and wet season (May–November) that are characteristic of Florida. During the dry season (February–March), the average alkalinity and calcium were 89 and 250 mg/L as $CaCO_3$, respectively. During part of the rainy season (April–August), the alkalinity dropped to 76 mg/L as $CaCO_3$ and calcium dropped to 195 mg/L as $CaCO_3$. At the end of the rainy season (September–November), the alkalinity stayed at 78 mg/L as $CaCO_3$ while calcium returned to almost dry season levels of 234 mg/L as $CaCO_3$. The water quality of the surface water during the following dry season was different from the previous with lower alkalinity (73 mg/L as $CaCO_3$) and lower calcium (160 mg/L as $CaCO_3$). Sulfates followed a similar pattern, with averages of 208 mg/L, 166 mg/L, 195 mg/L, and 175 mg/L for the mentioned time periods.

Table 3.5
Water quality objectives for finished process waters

Parameters	Finished water type			
	GW	SW	RO	Blend
pH	pH_s+0.3	>pH_s	>pH_s	>pH_s
Total hardness (mg/L as $CaCO_3$)	240	60 minimum	60 minimum	60 minimum
Alkalinity (meq/L)	4	1 minimum	1 minimum	1 minimum
Total chlorine (mg/L as Cl_2)	4.0	4.0	4.0	4.0
Sodium (mg/L)	80 maximum	80 maximum	80 maximum	80 maximum
SO_4 and Cl^- (chloride) (meq/L)	3.8 maximum	3.8 maximum	3.8 maximum	3.8 maximum
Total dissolved solids (mg/L)	500 maximum	500 maximum	500 maximum	500 maximum
Color (cpu)	15 maximum	15 maximum	15 maximum	15 maximum
Total organic carbon (mg/L)	3.6 maximum	3.6 maximum	3.6 maximum	3.6 maximum

Table 3.6
Seasonal variations in water quality for finished surface water

Time period	Water quality parameters									
	pH	Alkalinity (mg/L)	Calcium (mg/L)	Color (cpu)	UV_{254} (cm^{-1})	Turbidity (ntu)	DO (mg/L)	TDS (mg/L)	SO_4 (mg/L)	Temp. (°C)
01/25/06–03/30/06	7.7	89	250	1	0.051	0.3	9.7	444	208	18.3
04/06/06–08/18/06	7.8	76	195	1	0.058	0.2	8.2	396	166	25.0
08/25/06–11/30/06	7.8	78	234	1	0.063	0.2	8.3	463	195	23.1
12/07/06–02/08/07	7.6	73	160	1	0.060	0.4	9.7	416	175	18.5

NOTES: Alkalinity and calcium are in milligrams per liter as $CaCO_3$. DO is in milligrams per liter as O_2.

Selected water quality parameters for the surface water are presented in Table 3.6. The average water quality for the different source waters are enumerated in Table 3.7.

The water quality of the raw groundwater had little variation during the project. The average alkalinity was 215 mg/L as $CaCO_3$ with a standard deviation of 3 mg/L as $CaCO_3$, calcium hardness was 216 mg/L as $CaCO_3$ with a standard deviation of 8 mg/L as $CaCO_3$, color was 8 cpu with a standard deviation of 3 cpu, and TDS was 318 mg/L with a standard deviation of 26 mg/L. The variation in water quality for the finished source waters was minimal, as shown by the standard deviation values presented in Table 3.7. The average alkalinity of GW, SW, and RO was 211, 79, and 70 mg/L as $CaCO_3$ with standard deviations of 7, 8, and 7 mg/L as $CaCO_3$, respectively. The average total hardness was 242 mg/L as $CaCO_3$ for GW and SW, while RO had an average of 63 mg/L as $CaCO_3$. Turbidity and TDS averaged 0.2 ntu and 357 mg/L for GW, 0.3 ntu and 428 mg/L for SW, and 0.1 ntu and 285 mg/L for RO, respectively. The chloramine residual averaged 5 mg/L as Cl_2 for GW, SW, and RO. Table 3.8 shows results of one-time samples analyzed for chlorides, sulfates, and NPDOC at the UCF laboratory on Jan. 6, 2006, as

Table 3.7
Water quality averages for the water pilot treatment processes

Process statistic	Free Cl_2 (mg/L Cl_2)	Total Cl_2 (mg/L Cl_2)	NH_3 (mg/L NH_3)	pH	Alkalinity (mg/L $CaCO_3$)	Calcium hardness (mg/L $CaCO_3$)	Total hardness (mg/L $CaCO_3$)	HPC (cfu/mL)	UV_{254} (cm^{-1})	Color (cpu)	Turbidity (ntu)	DO (mg/L)	Conductivity (µS/cm)	TDS (mg/L)	Temp. (°C)	Cl^- (mg/L)	SO_4 (mg/L)	Fe (mg/L)
GW																		
Average	0.07	4.9	0.15	7.7	211	214	242	218	0.073	4	0.2	7.4	540	357	23	36	29	0.06
Std. dev.	0.02	1.1	0.07	0.2	7	9	7	510	0.016	1.4	0.1	0.5	32	20	2.2	7.4	4.3	0.04
SW																		
Average	0.07	5.1	0.16	7.8	79	210	242	1016	0.058	1	0.3	8.8	647	428	22	51	184	0.03
Std. dev.	0.02	1.2	0.07	0.2	8	4	4	2041	0.014	1.1	0.1	0.9	57	38	3.8	9.5	22.9	0.01
RO																		
Average	0.07	5.0	0.10	7.9	70	63	69	309	0.029	0	0.1	8.3	431	285	23	92	2	0.01
Std. dev.	0.02	0.9	0.05	0.2	7	6	7	1207	0.009	0.2	0.1	0.6	34	22	2.8	9.4	2.1	0.01
pH_s+0.3 (Phase I)																		
Average	0.07	5.0	0.11	7.9	163	202	231	700	0.071	3	0.3	8.7	549	365	21	47	72	0.07
Std. dev.	0.02	0.7	0.05	0.1	4	8	8	1514	0.009	1.1	0.1	0.6	20	13	1.8	5.3	5.2	0.05
pH_s (Phase I)																		
Average	0.07	4.9	0.10	7.4	129	192	216	43	0.078	3	0.2	8.1	617	407	24	83	44	0.03
Std. dev.	0.01	0.6	0.04	0.0	15	7	3	N/A	0.002	0.7	0.1	0.2	43	29	0.1	20.0	34.7	0.01
pH_s+0.3 (Phase II)																		
Average	0.08	6.2	0.13	7.9	109	185	213	142	0.069	2	0.2	8.0	588	388	26	59	112	0.04
Std. dev.	0.02	0.7	0.05	0.1	4	8	9	78	0.012	0.7	0.0	0.2	24	16	1.36	6.29	6.76	0.01
pH_s (Phase II)																		
Average	0.08	6.1	0.14	7.5	101	185	211	66	0.069	2	0.2	8.1	613	405	26	72	111	0.04
Std. dev.	0.02	0.6	0.06	0.1	6	8	6	97	0.012	1.2	0.0	0.4	39	26	1.6	10.4	5.1	0.01

(continued)

Table 3.7 (Continued)

Process statistic	Free Cl_2 (mg/L Cl_2)	Total Cl_2 (mg/L Cl_2)	NH_3 (mg/L NH_3)	pH	Alkalinity (mg/L $CaCO_3$)	Calcium hardness (mg/L $CaCO_3$)	Total hardness (mg/L $CaCO_3$)	HPC (cfu/mL)	UV_{254} (cm^{-1})	Color (cpu)	Turbidity (ntu)	DO (mg/L)	Conductivity (μS/cm)	TDS (mg/L)	Temp. (°C)	Cl^- (mg/L)	SO_4 (mg/L)	Fe (mg/L)
pH_s+0.3 (Phase III)																		
Average	0.08	6.8	0.10	7.9	154	206	229	19	0.077	2	0.2	8.0	626	413	25	65	74	0.05
Std. dev.	0.02	0.4	0.06	0.3	3	9	7	16	0.012	0.8	0.1	0.3	29	19	1.8	8.3	7.6	0.01
pH_s (Phase III)																		
Average	0.08	6.7	0.12	7.5	151	203	226	22	0.074	2	0.2	8.1	674	445	24	83	74	0.05
Std. dev.	0.02	0.5	0.07	0.1	9	8	7	21	0.013	0.8	0.1	0.5	43	28	2.0	11.2	6.4	0.01
pH_s+0.3 (Phase IV)																		
Average	0.09	6.9	0.08	7.8	127	168	194	151	0.063	2	0.2	9.1	573	378	20	56	85	0.04
Std. dev.	0.04	0.9	0.04	0.1	6	14	12	144	0.012	0.8	0.1	0.3	19	13	1.7	6.3	2.9	0.01
pH_s (Phase IV)																		
Average	0.09	6.6	0.08	7.5	124	169	195	745	0.058	2	0.2	9.1	597	394	19	65	85	0.04
Std. dev.	0.04	0.8	0.04	0.1	6	11	11	824	0.011	0.8	0.1	0.4	31	20	2.1	9.6	2.5	0.00
Raw GW																		
Average				7.4	215	216	244		0.077	8	0.1	0.3	476	318	23	14	30	0.06
Std. dev.				0.1	3	8	6		0.016	2.6	0.3	0.2	14	26	3.4	2.2	4.8	0.02

Std. dev. = standard deviation.

Table 3.8
Average water quality HPCs on Jan. 6, 2006

Parameter	GW	SW	RO
Chlorides (mg/L)	22*	38*	81*
Sulfates (mg/L)	27*	242*	6*
NPDOC (mg/L as C)	2.2*	1.5*	0.1*
Silica (mg/L as SiO_2)	14.1†	1.2†	0.4†
HPC (cfu/mL)	218	1,016	309

*Measured once at UCF laboratory. Should not be compared to average values.
†Measured once on project site.

well as silica results obtained from the finished source waters on May 25, 2006. The HPC values shown in Table 3.8 are the average for the project.

The averages for the control waters are shown by phase because the blending percentages were changed, as was presented in Table 3.3. As can be seen in Table 3.7, the main difference between the pH_s and the pH_s+0.3 waters is the pH. The average pH for the pH_s water was 7.4 to 7.5 and was always 0.3 to 0.4 pH units below the average pH for the pH_s+0.3 water (7.8 to 7.9). The alkalinity and chlorides concentration were also slightly different because of the addition of HCl (hydrochloric acid) to the pH_s water to lower the pH. For the pH_s and pH_s+0.3 waters, the average chloramine residual was 5 mg/L as Cl_2 during phase I, but it was increased above 6 mg/L as Cl_2 for the remainder of the project to ensure residual maintenance.

PILOT DISTRIBUTION SYSTEM OPERATION

The PDSs were maintained with a feed of 100% GW since completion of field activity for TBW I in the summer of 2003. Introduction of a blend of GW, SW, and RO was initiated in December 2005; this blend was identical to the blend required for phase I, as shown in Table 3.3. The corrosion inhibitors feed was started on Jan. 19, 2006, and phase I was started on Jan. 30, 2006. The water quality was changed every 3 months for all PDSs. Fourteen hybrid PDSs were used for this investigation. These PDSs are identical and consist of increments of PVC, LCI, UCI, and GS pipes connected in series. PDS01–PDS12 contained three different doses of the four corrosion inhibitors, while PDS13 and PDS14 were adjusted for pH control. One set of water quality conditions was repeated in one phase to evaluate seasonal effects.

The selection of inhibitor for optimization of distribution system water quality investigated corrosion inhibitor, dose, change of blend, and season. It was a 2-year project involving 1 year of data collection and PDS operation. Four corrosion inhibitors were investigated: orthophosphates, blended ortho- and polyphosphates, zinc orthophosphate, and silicates. These inhibitors provided a wide spectrum of currently available corrosion control inhibitors for various water quality conditions. Each of these inhibitors has been shown to successfully control copper, lead, and iron corrosion, and corrosion products under specific environments. After the inhibitor and dose was established for a given PDS, the inhibitor and dose was held constant for 1 year of investigation.

The nominal HRT in the hybrid pipe system was set to 2 days. Results from the prior study documented difficulties in maintenance of a combined chlorine residual in the PDS during elevated temperature conditions if the HRT exceeded 2 days (Taylor et al. 2005). This configuration is intended to mimic dead zones that would be encountered in a prototype system. The PDS feed and inhibitor pumps were fitted with backpressure devices to ensure consistent flows and associated maintenance of steady HRTs in the PDS. Input flow measurements and adjustments were completed every 2 weeks to regulate the HRT at the desired value.

The PDS feed was produced as a blend of treated groundwater, reverse osmosis, and treated surface water from the TBW regional facility. The individual sources were chloraminated to produce a residual of approximately 4.5 to 5.0 mg/L. The sources were blended in a storage tank to obtain the required blend to feed the PDS. Regular chlorine additions were made to the blend to maintain the desired combined chlorine residual at a value slightly in excess of 5 mg/L.

The PDS lines were flushed at 2-week intervals. The flushing event achieved a velocity of 1 ft/sec (0.3 m/sec) for several minutes. This duration was selected to pass three pipe volumes of flush water through the hybrid lines. The nominal length of the hybrid lines is only 100 ft (30.5 m); thus the normal system velocities are very small. The objective of the flushing exercise is periodic removal of excessive films by hydraulic scour that may otherwise accumulate on the pipe interior during the nearly stagnant conditions associated with the 2-day HRT. The influent and effluent standpipes were cleaned at the same frequency to remove biological growths. Flushing and standpipe cleaning operations were concluded at the end of the week (Thursday or Friday).

The PDS sampling activity was undertaken at the beginning of the week. PDS influent samples were also obtained at the beginning of the week (typically a Monday), corresponding with a new batch of process water. PDS effluent samples were not collected until at least 60 hours had elapsed after a flushing event in order to provide the desired 2-day HRT prior to sample collection.

Corrosion Inhibitors

The four types of inhibitors selected for project use were orthophosphates (OP), blended ortho- and polyphosphates (BOP), zinc orthophosphate (ZOP), and silicates (SiO_2). The doses utilized were selected by combining information of recommended doses from the inhibitor manufacturers specific for the blends of GW, SW, and RO used in the project as well as a survey of utilities that use the same types of inhibitors. However, the methods used by the manufacturers and utilities appeared to be based on experience and were qualitative rather than quantitative. This variation of independent variables provided an evaluation of inhibitor, dose, and season in a varying water quality environment for control of color, turbidity and total iron release, residual maintenance, biofilm stability, and bulk water biostability. The variation of water quality and inhibitors by PDS is shown in Table 3.9. The target doses of the inhibitors were defined as 0.5, 1.0, and 2.0 mg/L as P for the phosphorus-based inhibitors and 10, 20, and 40 mg/L as SiO_2 above the background concentration for the silicon-based inhibitor. After 1 month of operation, the doses of the silicate inhibitor had to be changed to 3, 6, and 12 mg/L because the 20- and 40-mg/L doses were causing $CaCO_3$ precipitation within the PDS.

The stock solutions of the corrosion inhibitors were diluted in a 40-gal (0.15-m^3) tank prepared twice a week, and the feed rates of the diluted inhibitor solution were calibrated to deliver the low, medium, and high inhibitor target concentration. The water used during the flushes was also dosed with the required type and amount of inhibitor to match the feed of the respective PDS. The diluted inhibitor tanks were cleaned every week and maintained with a free chlorine residual enough to minimize bacteriological growth.

Table 3.9
Variations of water quality, inhibitor, and dose by project phase

Independent variable							**Water Quality 1**							
Phase I	PDS01	PDS02	PDS03	PDS04	PDS05	PDS06	PDS07	PDS08	PDS09	PDS10*	PDS11*	PDS12*	PDS13	PDS14
Inhibitor	BOP	BOP	BOP	OP	OP	OP	ZOP	ZOP	ZOP	Si	Si	Si	pH_s	pH_s+0.3
Dose (mg/L)	0.5 - P	1.0 - P	2.0 - P	0.5 - P	1.0 - P	2.0 - P	0.5 - P	1.0 - P	2.0 - P	3 - SiO_2	6 - SiO_2	12 - SiO_2	0	0
							Water Quality 2							
Phase II	PDS01	PDS02	PDS03	PDS04	PDS05	PDS06	PDS07	PDS08	PDS09	PDS10	PDS11	PDS12	PDS13	PDS14
Inhibitor	BOP	BOP	BOP	OP	OP	OP	ZOP	ZOP	ZOP	Si	Si	Si	pH_s	pH_s+0.3
Dose (mg/L)	0.5 - P	1.0 - P	2.0 - P	0.5 - P	1.0 - P	2.0 - P	0.5 - P	1.0 - P	2.0 - P	3 - SiO_2	6 - SiO_2	12 - SiO_2	0	0
							Water Quality 1							
Phase III	PDS01	PDS02	PDS03	PDS04	PDS05	PDS06	PDS07	PDS08	PDS09	PDS10	PDS11	PDS12	PDS13	PDS14
Inhibitor	BOP	BOP	BOP	OP	OP	OP	ZOP	ZOP	ZOP	Si	Si	Si	pH_s	pH_s+0.3
Dose (mg/L)	0.5 - P	1.0 - P	2.0 - P	0.5 - P	1.0 - P	2.0 - P	0.5 - P	1.0 - P	2.0 - P	3 - SiO_2	6 - SiO_2	12 - SiO_2	0	0
							Water Quality 3							
Phase IV	PDS01	PDS02	PDS03	PDS04	PDS05	PDS06	PDS07	PDS08	PDS09	PDS10	PDS11	PDS12	PDS13	PDS14
Inhibitor	BOP	BOP	BOP	OP	OP	OP	ZOP	ZOP	ZOP	Si	Si	Si	pH_s	pH_s+0.3
Dose (mg/L)	0.5 - P	1.0 - P	2.0 - P	0.5 - P	1.0 - P	2.0 - P	0.5 - P	1.0 - P	2.0 - P	3 - SiO_2	6 - SiO_2	12 - SiO_2	0	0

*Original dosages of 10, 20, and 40 mg/L were changed to reported dosages after 3 weeks.

Table 3.10
Inhibitor product properties

Parameter	BOP	OP	ZOP	SiO_2
Percent active product	36%	36%	44%	37.5%
Bulk density (lb/gal)*	11.5	11.25	10.8	11.6
Specific gravity (at 22°C [72°F])	1.3	1.35	1.45	1.39
pH 1% solution (at 22°C [72°F])	6.3–6.6	5.1–5.4	<1	11.3
Recommended dose (mg/L)	1–2 as P	1–4 as P	2–3 as P	4–12 as SiO_2
Recommended pH range	6.0–8.5	6.8–7.8	7–8	N/A†
Solubility in water (g/g H_2O)	60/100	N/A	N/A	Miscible
Shelf life (months)	6	None	6	None
Storage limitation	Indoors	None	None	None

*lb/gal = 0.12 g/mL.
†N/A = not available.

The relevant properties of each inhibitor used during the project are shown in Table 3.10. The selected BOP product is called Sodium Polyphosphate SK-7641 (Stiles-Kem/Met-Pro Corporation, Waukegan, Ill.). The manufacturer claimed that when produced, the BOP product contained approximately 40% OP and 60% polyphosphate. However, monitoring of the BOP dose administered during operation indicated it contained 60%–80% OP. Periodic determination of the actual ratio was conducted to administer the correct dose of the product. The OP inhibitor was Inhibit-All WSF-36 (SPER Chemical Corporation, Clearwater, Fla.). The ZOP inhibitor was CP630 (Sweetwater Technologies, Temecula, Calif.). ZOP is made by dissolving zinc sulfate into a phosphoric acid solution in a 1:5 zinc-to-PO_4 ratio. The silica inhibitor, product name N, was a sodium silicate solution (PQ Corporation, Valley Forge, Pa.).

Data Collection

Samples were collected and analyzed in the field and at the UCF laboratory. Monitoring and analyses of many physical, chemical, and biological water quality parameters were carried out on the influent and effluent to each of the 14 PDSs, cradles, and corrosion lines. Table 3.11 shows the different parameters that were monitored during the study. Data was collected for 1 year of operation. Data collection for 1 year is critical to understanding the seasonal variation of the blends, microbial activity, and water quality parameters.

Weekly analyses were completed in the field laboratory and samples were also brought to the UCF laboratory for analysis. Quality assurance and quality control of both the laboratory and field determinations of water quality parameters were established by duplicating analyses of at least 10% of the samples. Where appropriate standards were available, 10% of the samples were spiked with known concentrations of the parameter being analyzed and the recovery measured. Blind duplicates and spikes were also used to determine the accuracy of measurements. Dynamic control charts were used to determine whether the results were acceptable.

Table 3.11
Influent and effluent chemical monitoring for each of 14 PDSs and cradles

			Frequency in phase I		Frequency in phases II, III, & IV			
			PDS01–PDS14		PDS01–PDS12		PDS13 & PDS14	
Parameter	Location	Method	Influent	Effluent	Influent	Effluent	Influent	Effluent
Alkalinity	Field lab	Titration	W	W	BW	BW	W	W
Ammonia	Field lab	NH_3 probe	W	W	BW	BW	W	W
Chlorine, free	Field lab	Spectrophotometer	W	W	W	W	W	W
Chlorine, total	Field lab	Spectrophotometer	W	W	W	W	W	W
Color, apparent	Field lab	Spectrophotometer	W	W	BW	BW	W	W
Conductivity	Field lab	Conduct probe	W	W	BW	BW	W	W
Nitrite-N	Field lab	Spectrophotometer	No	W	No	W	No	W
Oxidation–reduction potential	Field lab	Redox probe	W	W	BW	BW	W	W
Oxygen, dissolved	Field lab	DO probe	W	W	W	W	W	W
pH	Field lab	pH probe	W	W	W	W	W	W
Phosphorus, ortho	Field lab	Spectrophotometer	3/W	3/W	3/W	3/W	3/W	3/W
Silica	Field lab	Spectrophotometer	3/W	3/W	3/W	3/W	3/W	3/W
Total dissolved solids	Field lab	TDS probe	W	W	BW	BW	W	W
Temperature	Field lab	Probe	W	W	W	W	W	W
Turbidity	Field lab	Turbidimeter	W	W	W	W	W	W
UV_{254}	Field lab	Spectrophotometer	W	W	W	W	W	W
Aluminum	UCF lab	ICP	W	W	BW	BW	W	W
Calcium	UCF lab	ICP	W	W	BW	BW	W	W
Chloride	UCF lab	Ion chromatograph	W	W	BW	BW	W	W
Copper, dissolved	UCF lab	ICP	No	W	No	BW	N/A	W
Copper, total	UCF lab	ICP	No	W	No	BW	N/A	W
Iron, dissolved	UCF lab	ICP	W	W	BW	BW	W	W
Iron, total	UCF lab	ICP	W	W	BW	BW	W	W
Lead, dissolved	UCF lab	ICP	No	W	No	BW	No	W
Lead, total	UCF lab	ICP	No	W	No	BW	No	W
Magnesium	UCF lab	ICP	W	W	BW	BW	W	W
NPDOC	UCF lab	Total organic carbon instrument	W	W	BW	BW	W	W
Phosphorus, total	UCF lab	ICP	W	W	W	W	W	W
Silica	UCF lab	ICP	W	W	W	W	W	W
Sodium	UCF lab	ICP	W	W	BW	BW	W	W
Sulfate	UCF lab	Ion chromatograph	W	W	BW	BW	W	W
Total Kjeldahl nitrogen	UCF lab	Digestion	Q	Q	Q	Q	Q	Q
Zinc, dissolved	UCF lab	ICP	W	W	BW	BW	W	W
Zinc, total	UCF lab	ICP	W	W	BW	BW	W	W
Assimilable organic carbon	UCF lab		Q	Q	Q	Q	Q	Q
HPC, bulk	UCF lab		W	W	W	W	W	W

NOTES: BW = every 2 weeks; Q = four times a year; 3/W = three times a week; W = once a week; ICP = inductively- (or ion-) coupled plasma spectroscopy; N/A = not applicable. All the methods of analysis are described later.

PDS Chemical and Physical Water Quality Characterization

The 14 PDSs and stored source waters were monitored for physical and chemical water quality, as shown in Table 3.11. The general water quality varied in the four project phases. The feedwater quality was similar for all PDSs except for pH, inhibitor type, and concentration, and sometimes alkalinity. Parameters such as free and total chlorine, free ammonia, ortho-phosphorus, and silica were monitored at least three times a week, and more often if necessary.

Corrosion Loops

A corrosion shed was built that housed 18 corrosion loops, shown in Figures 3.10 and 3.11, respectively. A portion of the flow from each PDS discharged to a designated corrosion loop to simulate a kitchen environment and tap monitoring for LCR compliance. The corrosion loop consisted of 30 ft (9.1 m) of copper tubing with a diameter of 1.6 cm (5/8 in.), which could hold approximately 1.8 L of water. One lead/tin (Pb/Sn) coupon was placed within the copper tubing between two standard copper tube fittings (brass) for each loop system. All other fittings and materials were PVC or other plastic polymers. The coupons had a surface area of 3.38 in.^{2} (21.8 cm^2). Assuming that a 1/8-in. (3.2-mm) bead on the inside of each 0.5-in. (1.27-cm)-diameter joint was the surface area exposed to water using solder, the coupon surface area was equivalent to 17 joint ends or 7 to 8 fittings and was reasonable for a kitchen sink.

To simulate the water that is delivered to customers, the corrosion loop units received water after passing through the associated pipe distribution system. The corrosion loops were flushed with 2 gal (7.6 L) of water every morning. The effluent of the copper loops was monitored regularly for copper and lead release. Samples were usually taken weekly, as shown in Table 3.11, with extra samples taken when needed for specific purposes. To sample the copper loops, a 1-L sample was collected after 6 hours of water stagnation, shaken to mix completely, and then 125 mL of it was used for metal measurements. Various water quality parameters such as pH, alkalinity, chloride, sulfate, silicate, calcium, sodium, DO, and so forth were also monitored.

PIPE COUPON STUDY

Biofilm and surface characterization investigations were conducted using coupons made from the same MG pipes that were used to construct the PDSs. The coupons were housed in cradles following every PDS, as shown in Figures 3.8 and 3.9, each receiving one of the 14 PDS waters in parallel to its corresponding PDS. Each cradle consisted of 4-in. (0.1 m) PVC pipe housing and sliding pipe coupon holders made of 6-in. (0.15-m)-diameter PVC pipe cut lengthwise (Figure 3.9). The length of the cradles was approximately 12 ft (3.7 m). The pipe coupon holders were easily removed from the cradles and replaced after each experimental phase for biofilm and chemical deposit analysis. The holders were transported in a large plastic bin with the coupons still in place. The humidity in the bin was elevated by placing a wet sponge inside. Upon arrival at the UCF labs, the coupons were harvested for surface or biofilm studies.

The coupons for biofilm studies were circular, whereas the surface characterization coupons were rectangular. The surface characterization coupon holders (containing much larger coupons) were placed downstream of the biological cradles to minimize transport of corrosion products onto the biofilm coupons. Cu (copper) and Pb/Sn coupons, were placed in the PVC electrochemical noise Nadles. These two materials were not originally part of the experimental plan for roughness but were added in phases III and IV when it was found that they could be measured

Figure 3.14 Physical Electronics 5400 ESCA XPS

even more quickly than other pipe materials. Examination of Cu and Pb/Sn surface roughness may provide insight into changes in the surfaces of Cu and Pb/Sn under exposure to varying source water blends and corrosion inhibitor treatments.

Surface Scale Characterization

Pipe surface characterizations were conducted at the Material Characterization Facility at the Advanced Materials Processing and Analysis Center of UCF. To protect the corrosion layers formed in the PDS, the coupon samples were stored in clean plastic boxes containing the same water in which the coupons had been incubated. These boxes were covered with plastic wrap (polyethylene film) to avoid reactions with the atmosphere after the coupon samples were harvested, and then the coupons were transported to UCF.

The techniques employed for surface characterization analysis were XPS, SEM, and energy dispersive X-ray spectroscopy (EDS). The XPS spectra patterns identify chemical components of the top surface of the corrosion layer. SEM visually magnifies the physical structure of the scale surface for identification of the corrosion products. EDS identifies the elemental composition of the corrosion layer.

Surface Characterization by XPS

Chemical compositions of corrosion scales on the surface of metal coupons (iron, copper, lead, and galvanized steel) exposed to the various inhibitors were characterized by XPS (Physical Electronics 5400 ESCA, Chanhassen, Minn.) following each phase of operations. In XPS, an Mg (magnesium) anode (1,254 eV kα) bombards the coupon surface with high-energy X-rays that interact with atoms in the top 5–6 nm of the surface (Seal and Barr 2001). Then, according to the photoelectric effect, these atoms emit high-energy electrons. An electron analyzer counts the number of electrons emitted over a range of energies, producing scan spectra that may be used for surface composition analysis. Figure 3.14 shows the 5400 ESCA (electron spectroscopy for chemical analysis) XPS used in this study.

Coupons were inserted into the PDSs at the beginning of each phase of pilot-plant operation and then retrieved at the end of the phase. The XPS scanning consists of a two-step process:

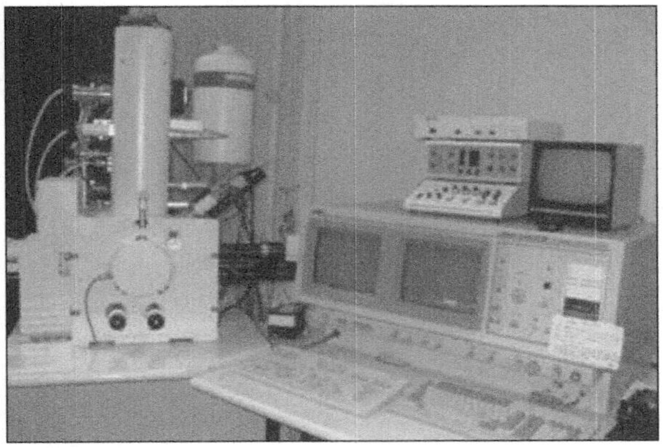

Figure 3.15 SEM JEOL 6400F

an initial "survey" scan followed by a "high-resolution" scan. The survey scan, conducted over a broad range of energy levels, is useful for confirming the presence or absence of elements on the coupon surface. In contrast, a high-resolution scan, conducted over a narrow range of energy levels, is useful for establishing the chemical states present for a given element.

The elements are indicated on the survey scan by pronounced peaks above the background of the survey, and these peaks are then selected individually for a second high-resolution scan for each element. Additionally, elements of interest, such as those associated with inhibitors, coupon material, or water quality, were analyzed with high-resolution scans. Possible compounds associated with a given element were determined through deconvolution of the high-resolution scan.

Surface Characterization by SEM

The scanning electron microscope is a type of microscope capable of producing high-magnification images of solid materials with a high depth of field. In SEM, the sample is scanned with an electron beam thermionically produced by a tungsten gun. When the electron beam reaches the sample surface, electrons and electromagnetic radiation are emitted and can be used to produce images. The secondary electrons emitted from the atoms because of the incident electron beam bombardment occupy the sample top surface and produce a readily interpretable surface image. The corrosion layer can be differentiated by analyzing the difference of physical structure. Under certain circumstances, the crystalline structure could be observed clearly, resulting in the identification of chemical components, thus indirectly verifying the results of other detection techniques as an auxiliary approach. The corrosion layer formed in water solution may not exhibit the legible crystalline structures but an amorphous state within short-term incubation. In this case, SEM can still provide information to determine whether the corrosion is homogeneous or pitting. The physical morphology can also clarify whether the corrosion layer formed is compact or porous and can help make the correct decision about protecting strategy. The SEM JEOL 6400 F (JEOL, Peabody, Mass.) that was used for SEM and EDS scans is shown in Figure 3.15.

In order to work out the elemental composition of corrosion layer, EDS was used. The principle behind EDS is that the interaction of the primary beam with atoms in the sample causes shell transitions that result in the emission of an X-ray along with other continuous spectrum. The emitted X-ray has an energy characteristic of the parent element. EDS can provide rapid

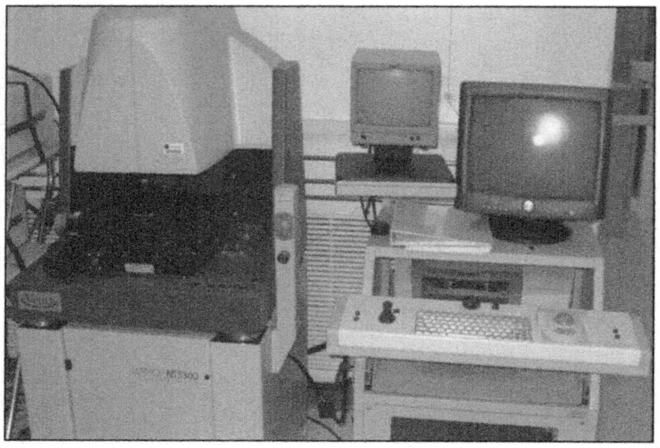

Figure 3.16 WYKO NT 3300 optical profiler

qualitative or quantitative analysis, with adequate standards, of elemental composition with a sampling depth of 1–2 μm. X-rays may also be used to form maps or line profiles showing the elemental distribution in a sample surface.

Surface Roughness

An optical profiler, WYKO NT 3300 (Veeco Instruments, Woodbury, N.Y.), was used to measure the surface roughness of six plumbing materials examined in this project. The WYKO NT 3300 is a noncontact optical profilometer capable of producing high-resolution and accurate three-dimensional surface measurements with vertical resolution to 0.1 nm. Surface roughness GS, LCI, PVC, and UCI was measured using coupons cut from pipes like those making up the PDSs. Surface roughness of Cu and a 50/50 Pb/Sn alloy was measured using flat metal corrosion coupons measuring ½ in. × 3 in. × 1/16 in. (1.27 cm × 7.6 cm × 0.16 cm). The WYKO NT 3300 optical profiler is shown in Figure 3.16.

The surface roughness of each coupon was measured both before and after incubation in its corresponding cradle. Table 3.12 shows the number of surface roughness coupons by material and phase. In phase I, three surface roughness measurements were made per coupon, both before and after incubation in the PDSs. In phase II, because of increased efficiency, five measurements per coupon were done for all materials. In phases III and IV, eight measurements were made per metal coupon and five measurements were made per nonmetal coupon.

Table 3.13 shows the number of measurements made per coupon by material and phase both before and after incubation. For example, in phase II, five measurements were made of each GS coupon before incubation and after incubation, totaling 10 measurements. Only for the coupons of phase I, a stripe of reflective paint was applied to the edge of each pipe coupon to help bring the coupon surface into focus, improving scan quality. Surface roughness was only measured on the unpainted portions of the coupons. This paint coat may have interfered with the surface chemistry and surface roughness of the pipe coupons. After phase I, no paint was applied to any coupon. Instead, a flashlight was used to help bring the surface into focus.

Table 3.12
Number of surface roughness coupons by material and phase

Material	Phase I	Phase II	Phase III	Phase IV	Total
Pipe coupons					
Galvanized steel	14	14	14	14	56
Lined cast iron	6	14	14	14	48
Polyvinyl chloride	6	14	14	14	48
Unlined cast iron	14	14	14	14	56
Metal coupons					
Copper	—	—	14	14	28
Lead/tin	—	—	14	14	28
Total by phase	40	56	98	98	292

Table 3.13
Number of measurements per coupon by material and phase

Material	Phase I*	Phase II	Phase III	Phase IV
Pipe coupons				
Galvanized steel	3	5	8	8
Lined cast iron	3	5	5	5
Polyvinyl chloride	3	5	5	5
Unlined cast iron	3	5	8	8
Metal coupons				
Copper	—	—	8	8
Lead/tin	—	—	8	8

* "3" indicates a total of six measurements, with three before incubation and three after incubation.

Biological Film Characterization

Bioassays were conducted on the PDSs and cradles using coupons from all pipe materials, as shown in Table 3.14. PDS pipe coupons for biological sampling were approximately 3 cm in diameter with a small PVC peg made from PVC welding rod attached to what was the outer surface of the pipe (Figure 3.9). The material was drilled from aged or pristine pipe and then deburred to give smooth edges. In each integrated PDS cradle, there were duplicate aged coupons of each of the four materials. The sequence of the coupons in these cradles from upstream to downstream was PVC, LCI, UCI, and finally GS coupons.

The coupons were placed in this order to avoid transport of corrosion materials downstream to contaminate the less easily corroded materials such as PVC and LCI. Thus, the two more easily corroded materials (i.e., UCI and GS) were placed in the downstream locations. The aged pipe coupons were obtained from used pipe segments from actual MG networks. Biofilms were generated on the coupons that were housed in the cradles and were harvested after 5–6 weeks of growth. The biofilms were assessed using BFHPC with units of HPCs per square centimeter and bulk biological stability was assessed using HPCs and AOC.

Table 3.14
Biological monitoring

Material	Location'	Weekly	Biweekly	Every 6 weeks	Quarterly
Cradle Coupons					
Cast iron					
BFHPC	UCF-Micro			X	
BF-TC/*Escherichia coli*	UCF-Micro			X	
Galvanized steel					
BFHPC	UCF-Micro			X	
BF-TC/*E. coli*	UCF-Micro			X	
Lined cast iron					
BFHPC	UCF-Micro			X	
BF-TC/*E. coli*	UCF-Micro			X	
Polyvinyl chloride					
BFHPC	UCF-Micro			X	
BF-TC/*E. coli*	UCF-Micro			X	
PDS Bulk					
AOC	UCF-Micro	All finished waters and four PDS I/O at startup for 15 total AOC			
HPC	UCF-Micro	X*			
TC/*E. coli*	UCF-Micro				X
Culture-AOB (ammonia-oxidizing bacteria)	UCF-Micro	AOB done when nitrite indicator warrants >0.2 mg/L as N; in/out on four PDSs, 1 month maximum			

*HPC sampled daily during some special periods.

Figure 3.17 The Nadles layout with the probes connected

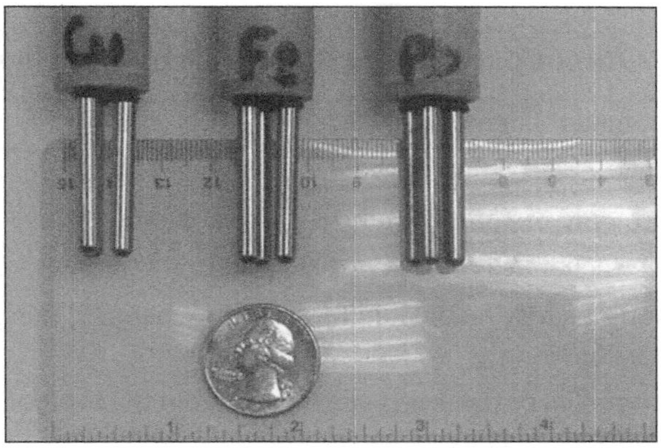

Figure 3.18 Detail showing new electrode probes

Table 3.15
Numbering of electrodes for ECN monitoring

Probe no.	Electrode identification	Element	Atomic weight	No. of electrons*	Density (g/cm^3)
01–14	A, B, C	Cu	63.54	2	8.89
15–28	A, B, C	Fe	55.46	2	7.84
29–42	A, B, C	Pb/Sn	162.95	2	7.86

*Number of electrons transferred in electrochemical reaction for corrosion.

ELECTROCHEMICAL NOISE

A portion of the effluent of each of the PDSs was sent using a ⅛-in. (0.3-cm)-diameter flexible PVC tubing to the noise trailer into the individual flush tanks for each of the 14 respective Nadles. This ensured that the water quality for the copper loops in the corrosion shed and the Nadles was the same. The Nadles housed the probes for EN monitoring as well as corrosion coupons for surface characterization. Figure 3.17 shows the housing where the EN probes were installed. The Nadles were flushed with 5 pipe volumes of water at an average velocity of 1 ft/sec (0.3 m/sec), and then corrosion data was collected continuously for at least a 6-hour standing period to simulate conditions applicable according to the LCR (USEPA 1991).

There were three probes in each Nadle, which allowed simultaneous monitoring of Cu, Fe, and Pb corrosion data. The probes were placed with the more noble metal upstream, that is, Cu > Fe > Pb, to minimize any galvanic corrosion due to possible particulate release and subsequent deposition on the downstream metal. Figure 3.18 shows a picture of the three electrodes (iron, copper, and lead) in new condition before exposure to any blends.

Each of the Fe, Cu, and Pb probes has three electrodes (A, B, and C). All electrodes had an area of 0.72 in.^{2} (4.67 cm^2). Table 3.15 shows the characteristics of each electrode as well as the numbering scheme used to identify each individual electrode. Because electrochemical reactions are spontaneous and have a random nature, it is not possible or meaningful to identify the anode or cathode before corrosion occurs without forcing a particular electrode to corrode preferentially.

Table 3.16
Output parameters from ECN monitoring

Symbol	Parameter name	Units
LPRCR	Linear polarization resistance corrosion rate	mpy
HMCR	Harmonic distortion analysis corrosion rate	mpy
ECNCR	Electrochemical noise corrosion rate	—
PF	Pitting factor—evaluated from electrochemical noise measurements	Volts
B	Stearn–Geary constant—evaluated from the harmonic distortion analysis	Volts/decade
Ba	Anodic Tafel slope	Volts/decade
Bc	Cathodic Tafel slope	—
PSKW	Skew of the potential	—
PKRT	Kurtosis of the potential	—
ISKW	Skew of the current	—
IKRT	Kurtosis of the current	Ohms
RSOL (Rs)	Solution resistance	mpy

ECN monitoring was continuous, with cycles of 430 seconds. For each of cycles, 300 seconds are for ECN, 100 seconds are for LPR and harmonic distortion analysis (which uses time-lagged AC [alternating current] impedance), and 30 seconds are for the solution resistance. A description of the typical parameters obtained from ECN monitoring output is displayed in Table 3.16 as well as the respective measurement units.

The corrosion monitoring system is called SmartCET (CET) and was connected to the probes using a 15-ft (4.6-m)-long probe cable. The CET transferred the data to a computer using an interface through a regular communications port. Computer interface software called Field CETTM (Honeywell International, Inc., Houston, Texas) was used to manipulate the "raw" data and provided calculated user data. The software allowed creation and storage of data from each probe into separate files that were converted to user-friendly file formats, such as Microsoft Excel (Microsoft Corporation, Redmond, Wash.), for data analysis.

Only one CET device was used during the project, which was switched manually between the Nadles every 6 hours to allow data collection from all Nadles. The probes remained in each EN cradle and were not disturbed. The tips were not changed during the year of investigation in order to document the effect of water quality change on EN parameters.

ADDITIONAL STUDIES

A series of experiments were completed to investigate the effect of flow velocity on chloramine residual decay for varying water quality, with and without inhibitors for different pipe materials. The single-material PDSs (PDS15 to PDS18) were used to conduct experiments in which the Reynolds number was varied from 50 to 200 using two types of water, GW and RO. A set of experiments included the addition of BOP inhibitor at a dose of 1.0 mg/L as P to evaluate the effect of the inhibitor on chloramine decay. Table 3.17 shows the principal characteristics of the experiments conducted using PDS15 (UCI), PDS16 (LCI), PDS17 (PVC), and PDS18 (GS).

Table 3.17
Principal characteristics of Reynolds number studies for chloramine decay

Experiment no.	Reynolds no.	Water type	Pipe material	Flow rate (gpm)	Velocity (ft/min)	HRT (hr)	Inhibitor?
1	50	RO	UCI	0.10	0.07	24	No
2	50	GW	UCI	0.10	0.07	24	No
3	50	GW	UCI	0.10	0.07	24	Yes
4	100	RO	UCI	0.19	0.13	12	No
5	100	GW	UCI	0.19	0.13	12	No
6	100	RO	UCI	0.19	0.13	12	Yes
7	150	RO	UCI	0.29	0.20	8	No
8	150	GW	UCI	0.29	0.20	8	Yes
9	200	GW	UCI	0.38	0.26	6	No
10	200	RO	UCI	0.38	0.26	6	Yes
11	50	RO	PVC	0.10	0.07	24	No
12	50	RO	PVC	0.10	0.07	24	Yes
13	100	RO	PVC	0.19	0.13	12	No
14	100	GW	PVC	0.19	0.13	12	Yes
15	150	RO	PVC	0.29	0.20	8	Yes
16	200	RO	PVC	0.38	0.26	6	No
17	200	GW	PVC	0.38	0.26	6	Yes
18	50	GW	LCI	0.10	0.07	24	No
19	50	RO	LCI	0.10	0.07	24	No
20	50	GW	LCI	0.10	0.07	24	Yes
21	100	GW	LCI	0.19	0.13	12	No
22	100	RO	LCI	0.19	0.13	12	Yes
23	150	RO	LCI	0.29	0.20	8	No
24	150	GW	LCI	0.29	0.20	8	Yes
25	200	GW	LCI	0.38	0.26	6	No
26	200	RO	LCI	0.38	0.26	6	Yes
27	50	RO	GS	0.06	0.44	12	No
28	50	RO	GS	0.06	0.44	12	Yes
29	100	GW	GS	0.13	0.87	6	No
30	100	GW	GS	0.13	0.87	6	No
31	100	RO	GS	0.13	0.87	6	No
32	100	GW	GS	0.13	0.87	6	Yes
33	150	RO	GS	0.19	1.31	4	No
34	150	RO	GS	0.19	1.31	4	Yes
35	200	RO	GS	0.26	1.74	3	No
36	200	GW	GS	0.26	1.74	3	Yes

NOTES: 1 gpm = 0.063 L/sec; 1 ft/min = 0.005 m/sec.

Several water quality parameters such as free and total chlorine, free ammonia, temperature, UV_{254}, DO, apparent color, turbidity, nitrites, and pH were monitored to determine the effect of turbulence on overall chloramine decay in the PDS. The effect of the Reynolds number on the wall decay rate constant (W) for PVC, LCI, UCI, and GS pipes was assessed for different water quality and temperatures.

CHAPTER 4
QUALITY ASSURANCE

This portion of the report is intended to provide a comprehensive overview of quality control procedures, analytical results, and supplementary discussion of associated PDS sampling events and process sampling events over the duration of this entire project.

Quality control as discussed in this document refers to the practices of the UCF civil and environmental engineering laboratories, both main and field facilities for sample bottle preparation, sample storage after receipt from the field, analysis, and data reporting. General laboratory practices include the use of suitable grades of reagents, gases, glassware, and standard materials. All reagents were of at least reagent grade for the inorganic and NPDOC analyses. Gases and standards were ultra-high-purity grade or equivalent.

Primary standard materials were purchased fresh at least every 6 months (or more often as needed) from an external vendor. Working standard solutions were replaced in accordance with the particular analytical methods. All volumetric glassware was of class A grade. Periodic checks on performance of the laboratory equipment were performed regularly as a part of the quality control program. Likewise, the performance of the analytical balances was monitored on at least an annual basis. They were calibrated annually by weighing a series of class S weights. If there was a deviation of more than 0.0005 g or another malfunction was observed, an external maintenance firm specializing in the calibration of analytical equipment was to be consulted. No such errors or malfunctions were observed in the use of any of our analytical balances throughout the course of this project.

The total number of samples, spiked samples, duplicates, and blind analyses varied by parameter, mainly because diverse parameters were analyzed at different locations of the experimental system (i.e., PDS ports, corrosion loops, process waters, stock tanks, etc.). The following section addresses analytical quality assurance/quality control (QA/QC) issues and other laboratory function and information issues pertinent to those samples that were analyzed in the field as well as those that were brought back to the main Environmental Systems Engineering Institute lab facility for analysis.

QUALITY CONTROL METHODS

The following subsections describe QA/QC measures that were always taken to ensure that all data were of known and acceptable quality throughout the entire study. The master QA/QC plan for all data analysis and handling has followed all guidelines for chemical sampling and analysis as presented in the 20th edition of *Standard Methods for the Examination of Water and Wastewater* (APHA, AWWA, and WEF 1998), hereafter referred to as *Standard Methods*.

Chemical Data—UCF Main Laboratory

The procedures used in the main UCF laboratory, along with their analytical specifications for the duration of this project, are summarized in Tables 4.1 and 4.2.

Table 4.1
Methods and reporting limits for inorganic analysis in the main laboratory during the entire project

Parameter	Method*	Reporting limit
Aluminum	SM 3120 B, pages 3-38 to 3-43, ICP† method	0.001 mg/L
Bicarbonate	SM 2320 B, pages 2-26 to 2-29, titration method	5 mg/L
Calcium	SM 3120 B, pages 3-38 to 3-43, ICP method	0.1 mg/L
Chloride	SM 4110, pages 4-2 to 4-8, ion chromatography with chemical suppression of eluent conductivity	0.1 mg/L
Copper	SM 3120 B, pages 3-38 to 3-43, ICP method	0.001 mg/L
Iron	SM 3120 B, pages 3-38 to 3-43, ICP method	0.001 mg/L
Lead	SM 3120 B, pages 3-38 to 3-43, ICP method	0.001 mg/L
Magnesium	SM 3120 B, pages 3-38 to 3-43, ICP method	0.1 mg/L
Nitrogen (NH_3, total Kjeldahl nitrogen)	SM 4500-Norg, pages 4-123 to 4-125, macro-Kjeldahl method	0.1 mg/L
pH	SM 4500-H+ B, pages 4-86 to 4-91, electrometric method	±0.01 pH units
Phosphorus	SM 3120 B, pages 3-38 to 3-43, ICP method	0.001 mg/L
Silica	SM 3120 B, pages 3-38 to 3-43, ICP method	0.001 mg/L
Sodium	SM 3120 B, pages 3-38 to 3-43, ICP method	0.1 mg/L
Sulfate	SM 4110, pages 4-2 to 4-8, ion chromatography with chemical suppression of eluent conductivity	0.1 mg/L
Zinc	SM 3120 B, pages 3-38 to 3-43, ICP method	0.001 mg/L

*All techniques designated with "SM," referenced in *Standard Methods*, are from the 20th edition (APHA, AWWA, and WEF 1998).
†ICP = inductively- (or ion-) coupled plasma spectroscopy.

Table 4.2
Methods and reporting limits for aggregate properties and NPDOC measured in the main laboratory during the entire project

Parameter	Method*	Reporting limit
Color	SM 2120 A, pages 2-1 to 2-3 or Hach 8025† cobalt–platinate method (with spectrophotometer)	1 cpu
Conductivity	SM 2510 B, pages 2-44 to 2-47, laboratory method	1 μmho/cm
NPDOC	SM 5310 C, pages 5-22 to 5-24	0.1 mg C/L
Solids (TDS)	Estimation of TDS by major ion sum	1 mg/L
Turbidity	SM 2130 B, pages 2-8 to 2-10, nephelometric method	0.01 ntu
UV_{254}	SM 5910 pages 5-65 to 5-68, UV absorption at 254 nm	0.0001 cm^{-1}

*All techniques designated with "SM," referenced in *Standard Methods*, are from the 20th edition (APHA, AWWA, and WEF 1998).
†Hach Company, Loveland, Colo.

Table 4.3
Methods and reporting limits for inorganic analysis in the field laboratory during the entire project

Parameter	Method	Method reference*	Approximate range
Alkalinity	Titration	SM 2320 B	5–500 ppm
Ammonia-N	Membrane probe method	SM 4500-NH_3 C	0.1–3 ppm
Chloride	Argentometric titration	SM 4500-Cl^- B	1–100 mg/L
Chlorine, free	DPD† colorimetric	SM 4500-Cl G or Hach 8021‡	0.1–2 ppm
Chlorine, total	DPD colorimetric	SM 4500-Cl G or Hach 8167‡	0.1–2 ppm
Color, apparent	Visual comparison (by spectrometer)	SM 2120 B	1–50 cpu
Conductivity	Conductivity bridge	SM 2510 B	Variable
Hardness (total, calcium)	EDTA§ titration	SM 2340 C	5–500 mg/L
Nitrate	Cadmium reduction	Hach 8192‡	0.1–0.5 mg/L
Nitrite	Diazotization	Hach 8507‡	0.1–0.3 mg/L
Oxygen, dissolved	Membrane probe	SM 4500-O G	0.1–20 mg/L
pH	Electrometric	SM 4500-H^+ B	2–13
Phosphate-P (reactive)	Ascorbic acid method	SM 4500-P E or Hach 8048‡	0.1–2.5
Silica, SiO_2 (reactive)	Molybdosilicate method	SM 4500-SiO_2 or Hach 8185‡	0.1–100 mg/L as SiO_2
Temperature	Direct reading		0–100°C
Turbidity	Nephelometric	SM 2130 B	0.02–200 ntu
UV_{254}	UV spectrometry	SM 5910 A	0–0.5200 cm^{-1}

*All techniques designated with "SM," referenced in *Standard Methods*, are from the 20th edition (APHA, AWWA, and WEF 1998).
†DPD = N,N-diethyl-p-phenylenediamine.
‡Hach Company, Loveland, Colo.
§EDTA = Ethylenediaminetetraacetic acid.

Chemical Data—Field Laboratory

Maintenance of good analytical practice in the field was carried out according to the same relevant procedural and statistical principles that were used in the main laboratory. While facilities available in a field laboratory are somewhat less sophisticated than those in a full-scale laboratory, the same guidelines were applied to obtain the maximum quality possible from the data gathered in the field. The methodologies used in the field are summarized in Table 4.3. Many of these techniques were the same as those used in the laboratory (such as the alkalinity determination). Others were modifications of known standard laboratory methods that were commercially available for application to field usage (such as the Hach Company application for chlorine, reactive phosphorus, and reactive silica, among others).

In all cases involving water quality analysis during the course of this project, the commercial modifications cited and used here were widely accepted and available applications of existing standard protocols described in *Standard Methods*. They were not novel, obscure, or untested methods; they are broadly accepted and used by utilities throughout the United States.

Statistics (accuracy, precision, etc.) were monitored in the same fashion as in the main laboratory, where applicable. Some parameters (UV_{254}, pH, etc.) cannot be monitored for accuracy under field conditions (i.e., cannot be accurately spiked with a known increment of a primary standard). In such cases, either a primary standard does not exist (UV_{254}), is not applicable (turbidity, color), or is not sufficiently stable, particularly under field conditions (DO, sulfide). Precision was monitored for these parameters. For reasons such as these, surrogate or secondary standards were also used for alkalinity, turbidity, and pH to further assist with assessment. External primary standards were purchased for all relevant parameters, including those measured onsite. These were used for instrument calibration as well as for check standards (when two separate vendors/sources were available).

Biological Data

In order to ensure quality of the biological data for this study, several measures were taken to minimize and quantify the variability and analytical errors in the bioassays. These measures can be divided into three categories: sampling protocol, preparation of reagents and microorganisms, and analysis of the samples. In each category, the fate of both bulk liquid and biofilm samples is further discussed.

For all the bulk liquid assays (i.e., AOC and HPC) samples were taken using sterile gloves and sterile containers (either prepurchased sterile, disposable, 100-mL coliform test sampling containers or 1-L Pyrex [Corning Incorporated, Lowell, Mass.] bottles that had been muffled at 525°C for 6 hours). After collection, sample bottles were capped with persulfate-washed caps. The caps were water-bathed in 20% persulfate solution at 70°C for 1 hour, then washed with distilled water four times, dried at room temperature, and then autoclaved for 20 minutes. Bulk liquid samples were then placed in a cooler with ice for transport and brought to the laboratory where they were stored in a 4°C walk-in cooler until preservation and/or analysis. Storage of bulk liquid samples did not exceed 24 hours prior to analysis unless proper preservation measures were taken. For biofilm assays (i.e., BFHPC and BF-TC/*Escherichia coli*), coupons affixed to PVC holders were handled with sterile gloves and transported in airtight PVC containers. The moisture level in the containers was maintained with sponges soaked with the same PDS water the coupons had been immersed in prior to sampling. The soaked sponges prevented desiccation of the biofilms during transport. For the preparation of reagents in contact with bulk liquid samples and/or microorganisms, a strict aseptic technique was used. Sterilization of agar and reagents (such as distilled water, dilution buffer, and control reagents) was performed by autoclaving at 121°C and 15 psi (103.4 kPa) for 20 minutes. Preservation of AOC samples was carried out by pasteurization in a water bath at 70°C for 1 hour or with the alternate pasteurization technique described by Escobar and Randall (2000) for extended storage. Aliquots of AOC samples were added to 40-mL USEPA vials, previously muffled at 525°C for 6 hours, and capped with persulfate-washed caps. To prevent contamination, handling and manipulating of samples and reagents requiring aseptic conditions was carried out under a laminar flow hood, previously wiped with bactericidal detergent (IDEXX Laboratories, Inc., Westbrook, Maine) and irradiated with an ultraviolet (UV) lamp for 5 minutes. For biofilm samples, a sterilized (with 70% ethanol

Table 4.4
Summary of incubation temperature and time

Sample type	Incubation temperature (°C)	Incubation time
HPC	26	3–5 days
BFHPC	26	3 days
BF-TC/*E. coli*	35	24 hours
AOC	26	3 days

solution) cell scraper and tissue homogenizer were used for cell detachment and homogenization of suspended materials, respectively.

Analysis of samples was mainly carried out under a laminar flow hood (kept sterile as described previously). In order to check aseptic conditions and proper sterility of the reagents, control measures included the following three steps:

1. The laminar flow hood air sterility was checked using triplicate open-agar plates during spreading.
2. The dilution buffer sterility was checked by spreading triplicate plates with dilution buffer alone as blanks during the same time as the spreading of the actual samples.
3. The sterility of the agar was verified by inclusion of triplicate-covered noninoculated plates in each run.

These nine plates (triplicate for each control) were incubated along with the samples. A glass spreader (i.e., bent glass rod) was sterilized with 70% ethanol solution and flamed with a Bunsen burner. Sterile disposable pipette tips, sterilized powder-free gloves, and sterilized dilution vials were used to prevent contamination during experiments. For AOC, sterile needles and syringes were used to inoculate samples with *Pseudomonas fluorescens* P17 and *Spirillum* NOX (LeChevallier et al. 1993). Time series enumeration data were used to quantify the time to stationary phase, and purity of the inocula was checked by inspecting the uniformity of observed colonies. P17 and NOX spread-plates were made to monitor growth of P17 and NOX working cultures. P17 and NOX pure cultures were previously frozen in a cryogenic freezer (–80°C) in 1–2-mL aliquots mixed with 2% peptone and 20% glycerol solution. Incubation of samples was performed in a temperature-controlled incubator. The incubation times and temperatures are listed in Table 4.4. For bioassays involving plate-spreading techniques (HPC, BFHPC, and AOC), serial dilutions were always performed in duplicate and plate counts in the range of 30–300 cfu were selected whenever possible. However, in the case of low (<30 cfu at 10^{-1} dilution) or high (>300 cfu at highest dilution) density samples, plate counts out of this range have been used for interpretation of samples, but only counts >30 cfu were used for quality control (QC) purposes. The variability of counts between 30 and 300 and >300 were not statistically different, but variability for counts <30 cfu was significantly greater. Because the sample analysis and assessment required 7 to 10 days to complete, it became impracticable to do a second analysis of the sample if the first analysis run failed to produce an acceptable number. This was due to the excessive storage time that would have resulted.

QUALITY ASSESSMENT

The discussion and respective tables that follow provide a summary of the accuracy and precision of laboratory analysis during the entire project, in the main UCF laboratory facilities, as well as at the field site laboratory.

Chemical Data—UCF Main Laboratory

The mean recoveries for the inorganic parameters ranged from about 90% to 101%, as shown in Table 4.5. Metal recoveries ranged from 90% to 101% recovery. Recoveries of the other inorganics fell over a range of 95% to 101%. NPDOC showed about 93% recovery. Spike recoveries were tracked using Shewart control charts in accordance with *Standard Methods*. The total number of PDS samples taken and measured for most of these parameters throughout the entire project was approximately 1,500 analyses of each parameter. The actual number of spikes taken for each relevant parameter, as shown in the "N" column, exceeds the required 10% (150 spikes in the majority of cases) called for by the quality assurance plan in all cases. Those parameters, which measured less than 1,500 times (such as total Kjeldahl nitrogen [TKN]), also complied with this policy and had at least 10% of their samples spiked and another 10% duplicated. Table 4.5 also summarizes the warning limit (WL) and control limit (CL) information gathered in the accuracy control charts for all applicable parameters analyzed in the main laboratory during the entire project.

The most typical source of error with accuracy in both lab and field was analyst error, usually from use of a faulty spiking solution, or some other error due to faulty standard curve or spike preparation. For example, many metal concentrations encountered in this project were in the low-ppb range. This requires several dilutions (for standards, spiking solutions, etc.) to prepare. It is expected that an analyst would occasionally make dilution errors or easily cross-contaminate a sample or standard of low concentration. When any accuracy error was encountered, the first step taken was to prepare an extra spiked sample with a freshly prepared spiking solution. This new spiked sample would be included at the end of a batch. If this new value was within acceptable limits, the data set was accepted and the next batch was analyzed. If this new value was faulty, results were rejected and an investigation into instrument function or maintenance would be carried out. The vast majority of accuracy errors were corrected with the use of a new spiking solution.

Errors not corrected with the new spike preparation were investigated further. A new standard curve (i.e., new set of diluted standards) was prepared and analyzed. If the new curve was acceptably linear ($R > 0.99$), spiked samples were run to verify correction of the problem. If the results of the spiked samples were within control, the batch was rerun. Errors beyond this would result in the purchase of a new external standard, but this need was never encountered.

Errors that revealed instrument malfunction issues were immediately dealt with before any further samples were run. This occurred once during the project with the inductively- (or ion-) coupled plasma spectroscopy (ICP), which needed a new nebulizer. The part was replaced and the unit was returned to functional status within 2 weeks. Accuracy of analysis was restored to previous levels by the instrument repair.

Average RPD (relative percent difference) values for NPDOC as well as all inorganic parameters analyzed in the laboratory during the entire project are shown in Table 4.6. The precision control chart data for these parameters are also provided in Table 4.6.

Table 4.5

Accuracy summary for inorganic parameters and NPDOC analysis in the UCF main laboratory during the entire project*

Parameter	Method	Units	N	Average % Recorded	% RSD	Maximum % Recorded	Minimum % Recorded	No. >WL	No. >CL	% <WL	% <CL
Metals											
PDS											
Aluminum	ICP	mg/L	167	101.4	13.0	133.1	76.4	8	0	95.2	100
Calcium	ICP	mg/L	171	100.1	14.5	138.8	78.2	8	0	95.3	100
Iron, dissolved	ICP	mg/L	56	92.2	6.2	109.3	81.0	3	0	94.6	100
Iron, total	ICP	mg/L	169	97.9	11.4	131.7	76.6	6	0	96.4	100
Magnesium	ICP	mg/L	168	96.6	11.4	127.5	79.9	8	0	95.2	100
Sodium	ICP	mg/L	161	95.0	12.3	132.4	73.0	7	0	95.7	100
Zinc, total	ICP	mg/L	160	97.8	10.9	139.7	73.0	4	1	97.5	99.4
Corrosion Loops											
Copper, dissolved	ICP	mg/L	100	99.0	8.6	124.7	78.7	8	0	92.0	100
Copper, total	ICP	mg/L	100	96.9	8.0	121.5	78.7	9	0	91.0	100
Lead, dissolved	ICP	mg/L	100	90.4	8.5	118.6	75.2	4	0	96.0	100
Lead, total	ICP	mg/L	100	92.5	8.2	117.5	78.0	1	0	99.0	100
Other Inorganics											
Chloride	IC	mg/L	177	95.1	6.5	117.6	80.7	9	0	94.9	100
PO_4-P (total)	ICP	mg/L	188	96.4	13.7	144.0	72.5	6	0	96.8	100
Silica	ICP	mg/L	208	89.7	7.3	111.4	72.0	7	3	96.6	98.6
Sulfate	IC	mg/L	167	96.5	6.2	112.0	83.4	10	0	94.0	100
TKN†	Kjeldahl	mg/L	17	100.6	12.9	120.1	73.5	N/A	N/A	N/A	N/A
Organics											
NPDOC	UV/oxidation	mg/L	177	92.9	8.2	111.1	74.6	10	0	94.4	100

*IC = ion chromatograph; ICP = inductively- (or ion-) coupled plasma spectroscopy; N/A = not applicable; RSD = relative standard deviation.

†Insufficient data to construct control chart for this parameter.

Table 4.6
Precision summary for inorganic parameters and NPDOC analysis in the UCF main laboratory during the entire project*

Parameter	Method	Units	N	Mean Range (mg/L)	Range RSD (mg/L)	Maximum (mg/L)	Minimum (mg/L)	Total >WL	Total >CL	% <WL	% <CL
Metals											
PDS											
Aluminum	ICP	mg/L	170	0.001	0.001	0.007	<0.001	8	3	95.3	98.3
Calcium	ICP	mg/L	170	1.4	1.2	5.5	<0.1	13	0	92.3	100
Iron, dissolved	ICP	mg/L	56	0.001	0.001	0.005	<0.001	4	3†	92.9	94.6
Iron, total	ICP	mg/L	172	0.006	0.009	0.449	<0.001	18	10‡	89.7	94.3
Magnesium	ICP	mg/L	170	0.1	0.1	0.8	<0.1	9	1	94.8	99.4
Sodium	ICP	mg/L	165	0.7	1.0	5.5	<0.1	11	10‡	93.3	93.9
Zinc, total	ICP	mg/L	163	0.002	0.004	0.029	<0.001	9	5	94.5	96.9
Corrosion Loops											
Copper, dissolved	ICP	mg/L	101	0.012	0.012	0.065	<0.001	10	1	90.1	99.0
Copper, total	ICP	mg/L	101	0.013	0.016	0.088	<0.001	5	2	95.0	98.0
Lead, dissolved	ICP	mg/L	101	0.000	0.000	0.000	0.001	10	9†	90.1	91.1
Lead, total	ICP	mg/L	101	0.000	0.001	0.005	<0.001	4	1	96.0	99.0
Other Inorganics											
Chloride	Titration	mg/L	177	0.5	0.5	2.6	<0.1	11	1	93.8	99.0
PO_4-P(total)§	ICP	mg/L	195	0.03	0.03	0.19	<0.01	13	2	93.3	99.0
Silica	ICP	mg/L	193	0.08	0.08	0.44	<0.1	10	4	94.8	97.9
Sulfate	IC	mg/L	168	1.3	0.5	3.1	<0.1	9	3	94.6	98.2
TKN§	Kjeldahl	mg/L	17	0.1	0.1	0.5	<0.1	N/A	N/A	N/A	N/A
Organics											
NPDOC	UV/oxidation	mg/L	177	0.12	0.11	0.57	<0.1	9	2	94.9	98.9

* IC = ion chromatograph; N/A = not applicable; RSD = relative standard deviation.
†Average range extremely narrow (usually zero); violations physically insignificant (<0.001 mg/L).
‡Drastic concentration change (due to phase change) resulted in the majority of control violations.
§Insufficient data to construct control chart for this parameter.

Most of the warning and control violation incidents that were encountered were explainable by one of a few common sources. One of the key error sources for precision involved values that were measured at concentrations near or at the detection limit for a given technique and could not be accurately resolved further. For example, PDS influent zinc measurements usually ranged between 0 and 0.005 mg/L in magnitude. Duplicate results such as these inevitably gave rise to a high RPD between analyses.

Another source of positive error typically arose around the time of a phase change. The magnitude of some parameters would change 5–10 (or more) times over what they were in the previous phase. This would send the limits of the precision control charts on a sharp upward or downward trend, resulting in the analyst exceeding the control limit, even though this was most likely not a true violation. Up to 6 weeks were necessary for the limits to stabilize for the new ambient matrix concentrations of these parameters. A good example of this is provided by sodium during the change from phase I to phase II (around May 9, 2006). At this time, over the entire PDS, sodium increased from about 8–9 mg/L to about 53–55 mg/L, a six-fold change. The control chart showed seven precision violations during this change. Two months (another 20–22 duplicate sets) were required for restabilization of the warning and control limits to values more representative of the data set. Even so, from phase II onward (where this expansion occurred), an average duplicate range of 0.7 mg/L was obtained. The maximum range value was 5.5 mg/L for a sample containing about 30 mg/L sodium. More than 75% (126/165) of the known sodium duplicates had a range of 1.0 mg/L or below. The vast majority (163/165) had %RPD values below 10%. All of these values indicate that sodium precision was most likely in control (i.e., acceptable) during this time. The precision control chart, however, continued to experience the mathematical effects of the large range change (on May 9) until that value was no longer averaged in to obtain the mean range. This became an unavoidable artifact of the water quality change at the phase change and did not originate from a problem with quality control.

Similar trends during the same date/event were clearly visible in several other parameters as well, including total iron and calcium. This information was consistent with good QA/QC for sodium in spite of the range-changing issue and further suggests that most of the violations were the results of the drastic limit change.

As an added measure of precision, blind duplicates were collected onsite with every sampling event. With each batch of samples, usually two to four blind duplicates were also collected. These samples were submitted as routine samples, along with their respective sample batch in the same work order, and analyzed accordingly. This practice ensured that blind duplicates were analyzed using the same method, handling, and treatment over the same time period as the corresponding known samples. After the results for each parameter were submitted to the lab manager for the entire work order, the blind duplicate results were assessed. A summary of the projectwide assessment of blind duplicates precision for all laboratory parameters is shown in Table 4.7.

Blind duplicates served many useful purposes. First, instead of merely demonstrating repeatability of a measurement on replicate aliquots from the same bottle (i.e., an analytical duplicate), they also helped monitor the consistency of the sample custody and treatment process as a whole. A second purpose they served was to demonstrate the ability to deal with the aggregate nature of some of the sample parameters. For example, the values of some of these parameters will differ from one collected aliquot to the next because they are aggregate in nature. Total metals (copper, iron, lead, zinc) provide clear examples of this phenomenon, because by virtue of the term *total*, we understand that some of the metal in question may be in solution (dissolved) and

Table 4.7
Blind duplicates precision summary for all parameters for UCF main laboratory during the entire project*

Parameter	Method	Units	N	Mean Range (mg/L)	Range RSD (mg/L)	Maximum (mg/L)	Minimum (mg/L)	No. >WL	No. >CL	% <WL	% <CL
Metals											
PDS											
Aluminum	ICP	mg/L	170	0.003	0.004	0.013	<0.001	8	4	95.3	97.8
Calcium	ICP	mg/L	170	1.8	2.0	17.3	<0.1	11	7	93.5	95.9
Iron, dissolved	ICP	mg/L	53	0.002	0.004	0.020	<0.001	2	1	96.2	98.1
Iron, total	ICP	mg/L	170	0.016	0.038	0.319	<0.001	11	9	93.5	94.7
Magnesium	ICP	mg/L	170	0.2	0.2	0.9	<0.1	10	7	94.1	95.9
Sodium	ICP	mg/L	170	1.0	1.5	9.8	<0.1	10	7	94.1	95.9
Zinc, total	ICP	mg/L	170	0.008	0.023	0.159	<0.001	9	7	94.7	95.9
Corrosion Loops											
Copper, dissolved	ICP	mg/L	98	0.030	0.059	0.359	<0.001	4	4	95.9	95.9
Copper, total	ICP	mg/L	102	0.042	0.082	0.474	<0.001	7	4	93.1	96.1
Lead, dissolved	ICP	mg/L	101	<0.001	0.001	0.006	<0.001	4	3	96.0	97.0
Lead, total	ICP	mg/L		<0.001	0.001	0.006	<0.001	4	3	96.0	97.0
Other Inorganics											
Chloride	IC	mg/L	228	1.0	1.0	6.7	<1.0	15	7	93.4	96.9
PO_4-P, total	ICP	mg/L	170	0.02	0.03	0.22	<0.01	12	3	92.9	98.2
SiO_2, reactive	ICP	mg/L	170	0.17	0.29	2.61	<0.01	8	4	95.3	97.6
Sulfate	IC	mg/L	204	1.3	1.2	7.6	<1.0	11	6	94.6	97.1
TKN†	Kjeldahl	mg/L	15	0.2	0.1	0.4	<0.1	N/A	N/A	N/A	N/A
Organics											
NPDOC	UV/oxidation	mg/L	205	0.30	0.46	3.57	<0.1	9	3	95.6	98.5

*IC = ion chromatograph; N/A = not applicable; RSD = relative standard deviation.
†Insufficient data to construct control chart for this parameter.

some may be present in particulate form. When a measured species can be present (even partially) in particulate form, it is then, by nature, not homogeneous throughout the bulk of the sample. This is where its aggregate nature arises from. So, during an iron release event, if several fragments of oxidized iron particulate dislodge from a pipe wall, they create heterogeneity in the bulk sample. One bottle may catch some of this particulate matter; another in succession may not. Furthermore, one bottle may capture a different number of iron particles that happen to be traveling through the pipe at a given time, while its blind duplicate might capture a different number of particles. These particles do not have to be the same size or shape. When each blind duplicate is digested for total metals, this will inevitably introduce error. Blind duplicates help with assessing repeatability in the face of this phenomenon, in the same sense that an experimental duplicate does, with the additional advantage of not introducing bias on behalf of the analyst (i.e., foreknowledge of what the "correct" concentration might be).

Chemical Data—Field Laboratory

Table 4.8 provides a summary of the accuracy (when applicable) of field analysis during the entire project. Table 4.9 provides a similar summary of precision in the field. The information provided here is analogous to that provided for the data analyzed in the main UCF laboratory. The only prevailing difference is that data summarized here were gathered directly at the field site. Error sources were similar and were responded to in similar fashions.

Biological Data

For plate-counting techniques (HPC, BFHPC, and AOC), all samples were quantified using triplicate plates for each dilution on each day of spread plating. For each individual plate within each sample, the counts were assessed and then the triplicate plate counts were averaged. An additional 10% of the samples were run in replicate (two sets of triplicate plates from two identical aliquots coming from the same sample container).

In the USEPA guidelines for biological assays such as microbial enumeration, no quantitative assessment of precision is required or cited, and no QC charts have been established in the literature for bioassays. This aspect of the project was developed as an original contribution by the project team. A summary of biological quality control completed during the project is presented in Table 4.10, which shows that 599 samples were analyzed for quality control of HPCs, BFHPCs, TC/*E. coli*, and AOC. Fifty-six QC samples, or 9.3% of the total, were in violation based on the control and warning limits defined using the same method developed for nonbiological assays. Biological analyses have sources of variability that nonbiological assays do not, and 9.3% violation was considered acceptable given the nature of these analyses. These violations are discussed in the following subsections.

Both blind duplicates from the field (sample identity unknown to the lab analyst) and lab replicates selected by the analyst were used to quantify precision. The values of raw plate counts usable for QC purposes were the same for HPCs, AOC, and BFHPC, that is, ≥30 cfu per plate. Dilutions were selected to increase the likelihood of counts in the optimal range of 30 to 300 cfu. To ensure that this occurred, from two to four dilutions (i.e., 10^{-1} to 10^{-4} for HPC, 10^{-2} to 10^{-4} for BFHPC, 3×10^{-3} and 10^{-4} for AOC) were used for all duplicates and lab replicates. Counts greater than 300 were accepted for QC purposes. Analysis early in the study showed that high

Table 4.8
Accuracy summary for field parameters during the entire project

Parameter	Method	Units	N	Average % Recorded	RPD	Maximum	Minimum	Total >WL	Total >CL	% <WL	% <CL
Alkalinity	Titration	mg/L $CaCO_3$	182	100.0	6.7	173.0	88.2	14	5	92.3	97.3
Ammonia	Probe	mg/L N	251	101.7	16.7	148.4	43.2	12	1	95.2	99.6
Chloride	Titration	mg/L	53	100.7	3.5	120.5	95.8	3	1	94.3	98.1
Cl_2-free*	Spectrometry	mg/L	357	99.6	2.0	105.6	94.4	37	21	89.6	94.1
Cl_2-total*	Spectrometry	mg/L	468	99.7	2.1	105.6	88.9	40	11	91.5	97.6
Color, apparent†	Spectrometry	cpu	184	99.6	4.2	113.6	80.0	25	14	86.4	92.4
Conductivity*	Probe	µmho/cm	462	98.8	1.8	104.1	94.5	29	5	93.7	98.9
Hard, Ca	Titration	mg/L $CaCO_3$	60	99.5	3.1	108.3	91.7	5	0	91.7	100
Hard, total	Titration	mg/L $CaCO_3$	61	99.9	9.4	160.0	77.0	6	4	90.2	93.4
Iron, dissolved	Spectrometry	mg/L	62	96.9	8.9	110.6	80.4	4	0	93.5	100
Nitrite-N	Spectrometry	mg/L N	129	95.9	6.8	112.1	78.4	12	2	90.7	98.4
Oxidation–reduction potential*	Probe	mV	163	97.3	4.8	117.5	83.0	13	5	92.0	96.9
Oxygen, dissolved	Probe	mg/L	N/A†	N/A	N/A	N/A	N/A	N/A	N/A	N/A	N/A
pH*	Probe	(unitless)	536	96.4	5.1	117.5	81.3	28	5	94.8	99.1
Phosphate, *o*-	Spectrometry	mg/L P	275	101.0	9.2	166.3	39.5	19	4	93.1	98.5
Silica	Spectrometry	mg/L SiO_2	126	100.1	15.8	129.4	80.6	5	2	96.0	98.4
Sulfate	Spectrometry	mg/L	60	104.5	5.2	114.4	84.6	5	1	91.7	98.3
Turbidity	Turbidity	ntu	542	98.0	11.6	153.3	59.0	67	33	87.6	93.9
UV_{254}	Spectrometry	cm^{-1}	N/A	N/A	N/A	N/A	N/A	N/A	N/A	N/A	N/A

*Accuracy by matrix spike not valid for this parameter; assessed by use of check standards/"distilled water spikes."
†N/A = not applicable.

Table 4.9
Precision summary for field parameters during the entire project

Parameter	Method	Units	N	Average range	RPD	Maximum	Minimum	Total >WL	Total >CL	% <WL	% <CL
Alkalinity	Titration	mg/L $CaCO_3$	235	2	3	26	<1	14	6	94.0	97.4
Ammonia	Probe	mg/L N	301	0.02	0.03	0.52	<0.01	29	14	90.4	92.0
Chloride	Titration	mg/L	53	0.7	0.5	2.0	<0.1	2	0	96.2	100
Cl_2-free	Spectrometry	mg/L	372	0.02	0.09	1.65	<0.01	47	31	87.4	91.7
Cl_2-total	Spectrometry	mg/L	447	0.04	0.08	0.96	<0.01	41	25	90.8	94.4
Color, apparent	Spectrometry	cpu	242	0.29	0.48	2.00	<0.01	15	3	93.8	98.8
Conductivity	Probe	μmho/cm	252	2	2	12	<1	10	3	96.0	98.8
Hard, Ca	Titration	mg/L $CaCO_3$	57	1	1	4	<1	2	0	96.5	100
Hard, total	Titration	mg/L $CaCO_3$	60	1	1	4	<1	3	0	95.0	100
Iron, dissolved	Spectrometry	mg/L	69	0.003	0.002	0.009	<0.001	1	0	98.6	100
Nitrite-N	Spectrometry	mg/L N	155	0.001	0.001	0.007	<0.001	10	2	96.7	98.7
Oxidation–reduction potential	Probe	mV	187	4.2	6.0	50.2	0.1	24	17	87.2	90.9
Oxygen, dissolved	Probe	mg/L	222	0.03	0.02	0.13	<0.01	13	4	94.1	98.2
pH	Probe	(unitless)	338	0.03	0.11	1.22	<0.01	22	7	93.5	97.9
Phosphate, *o*-	Spectrometry	mg/L P	379	0.013	0.034	0.408	<0.001	39	27	89.7	92.9
Silica	Spectrometry	mg/L SiO_2	135	0.277	0.246	1.800	<0.001	4	2	97.0	98.5
Sulfate	Spectrometry	mg/L	59	0.65	0.99	7.20	<0.01	2	1	96.6	98.3
Turbidity	Turbidity	ntu	262	0.012	0.022	0.275	<0.001	15	9	94.3	96.6
UV_{254}	Spectrometry	cm^{-1}	240	0.0008	0.0013	0.0120	<0.0001	14	7	94.2	97.1

Table 4.10
Summary of biological quality control sampling

Parameter	Type	Total QC samples	Violations
HPC	Lab replicate	183	31
	Blind duplicate	190	21
BFHPC	Lab replicate	45	2
	Duplicate coupon	36	1
TC/*E. coli*	Lab control	66	0
	Lab replicate	36	0
AOC	Lab replicate	32	0
	Blind duplicate	11	1
Total		599	56

Table 4.11
Mann–Whitney rank sum test for HPC duplicates and replicates

Group	N	Missing	Median	25%	75%
Duplicate	168	0	0.183	0.0859	0.352
Replicate	165	0	0.100	0.0462	0.194

counts had similar variability to counts between 30 and 300, whereas low counts had a significantly higher variability.

Heterotrophic Plate Count

Duplicate and replicate QC samples were analyzed in separate control charts since it was shown early in the study that there was a significant difference in the range values and variability of these two types of QC samples. The analysis of the total project data confirmed this observed difference. The mean of the blind duplicate data range was 0.17 log higher than the replicate data range, which was 0.16 log. Because the normality test failed ($p < 0.05$), the Mann–Whitney rank sum test was used to compare the difference between these two sets of data. The results are shown in Table 4.11. The results show that the difference in the median values between the two groups is greater than would be expected by chance with a p-value of 0.001. As a result, it was concluded that the duplicate data had higher variability than the replicate data, which was logical given that duplicates have additional sources of variability (e.g., different sample containers). Pooled analysis of the duplicate and replicate data was not appropriate, and separate control charts for blind duplicates and lab replicates were used to monitor method precision (Figures 4.1 and 4.2). In total, 21 duplicates were found to be control violations, as shown in Figure 4.1, and 31 replicates were also found to be control violations (Figure 4.2).

The QC data for HPC showed increased variability during the last 2 to 3 months of the project, which was shown in the QC charts. Some of the samples were stored longer than 48 hours. Excessive storage times were eliminated. The precision was decreased by roughly 20%

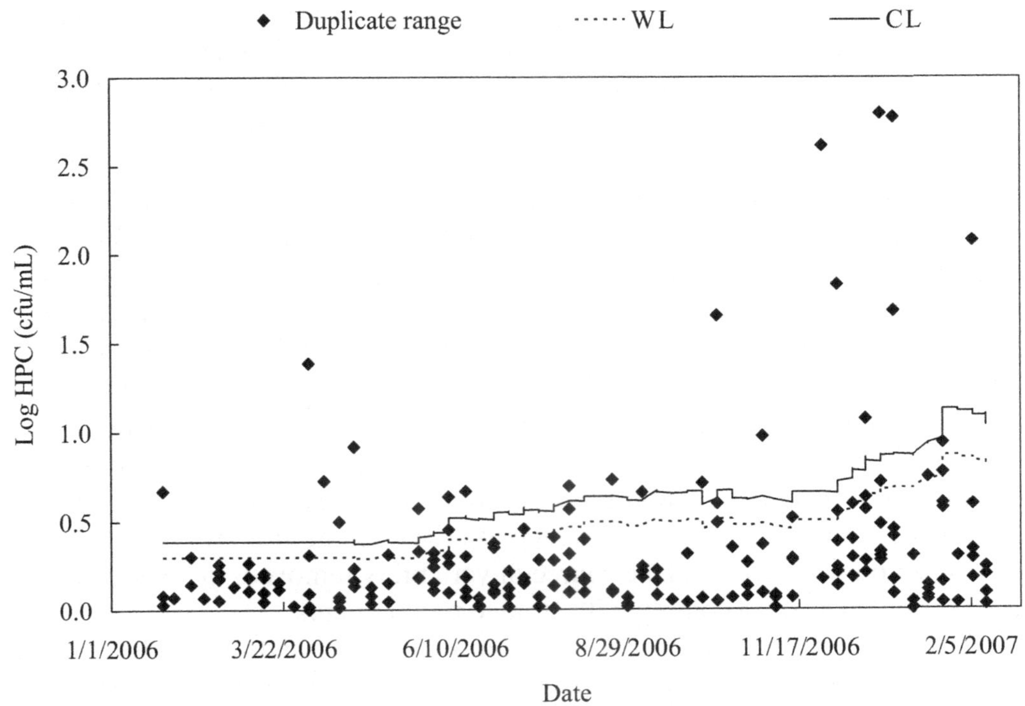

Figure 4.1 HPC precision duplicate data control chart

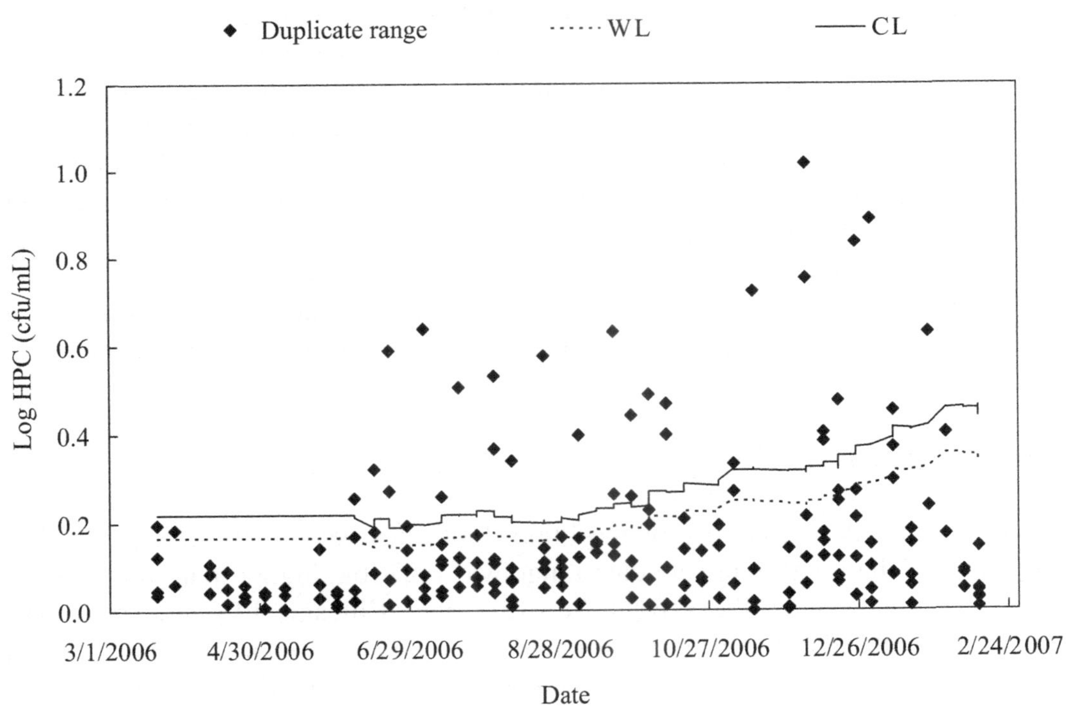

Figure 4.2 HPC precision replicate data control chart

during this final period, but fortunately the increased variability did not significantly affect data analysis and interpretation. In this analysis, data arising from samples with storage times of 48 hours or less were separated from data from samples whose storage exceeded 48 hours (*Standard Methods* specified a storage time of 24 hours; whereas other published literature indicated that storage times up to 48 hours will not typically affect enumerations). These two sets of data were cross-compared (i.e., data with normal storage times were modeled and applied to predict the suspect data, and vice versa). The results from the comparison between predicted data and actual data indicated no statistical difference between the two data sets. Moreover, several additional tests between suspect data and the whole data set as well as subsequent sensitivity analysis all verified the fact that all data could be grouped in spite of the unintended excess storage time.

The distribution of the entire QC data was not normally distributed. As a result, the control and warning limits cannot be assigned a probability based on a normal distribution (which is typically done for nonbiological assays). However, the control chart was useful for monitoring, which allowed identification of source variability twice. In the first case, aliquots were not immediately spread after pipetting onto agar, and in the second case, storage time was exceeded. In the first case, corrective action was taken and the source of variability removed. In the second case, corrective action was implemented to comply with storage requirements.

The distribution of the QC data as shown in Figures 4.1 and 4.2 implies that there was a periodic source of variability that causes between 11% and 17% of the replicate pairs to violate the upper control limit (UCL). In contrast, warning limit violations were only about half as frequent. In many instances, the range of the UCL violations was twice that of the corresponding range of samples that were not control violations. It is probable that this type of behavior was caused by nonhomogeneity of dispersions of microbial cells. With dissolved constituents, the assumption of homogeneity is sound, and, thus, this is not a source of variability for replicate or duplicate analysis. In this case, however, there are colloidal dispersions, and cells per particle vary significantly because of interparticle bridging resulting from extracellular polymers. In addition, some bacteria are mobile and there are a host of other phenomena that would not normally affect other colloidal dispersions. Analysis indicated that the distribution of the data was not normal. At the end of this section, the distribution of the data from bulk HPC, BFHPC, and AOC is evaluated to determine what distribution fits each data set and what the National Institute of Science and Technology (NIST) upper warning limit (UWL) and UCL calculations mean in terms of probability (95% and 99.7% for a normal distribution but the data was not normal). In addition, a modified approach is given for bioassays.

Biofilm Heterotrophic Plate Count

Differences in the variability of the blind duplicates versus lab replicates were investigated. The mean of the blind duplicate data range was 0.27 log, higher than the replicate data at 0.21. Statistical tests by SigmaPlot (Systat Software, Inc., Chicago, Ill.) confirmed that the QC data were not normally distributed (i.e., normality test failed, $p < 0.050$). A Mann–Whitney rank sum test (Table 4.12) was used to compare the data. It showed that the difference in the median values between the two groups was greater than would be expected by chance (i.e., there was a statistically significant difference ($p = 0.001$) between the replicate and duplicate data). The variability was much lower for replicates than duplicates because of the additional sources of error applicable to duplicates (e.g., variability between duplicates due to sampling, different sample containers, different coupons, etc.).

Table 4.12
Mann–Whitney rank sum test for BFHPC duplicates and replicates

Group	N	Missing	Median	25%	75%
Duplicate	32	0	0.263	0.13	0.387
Replicate	45	0	0.0952	0.0544	0.233

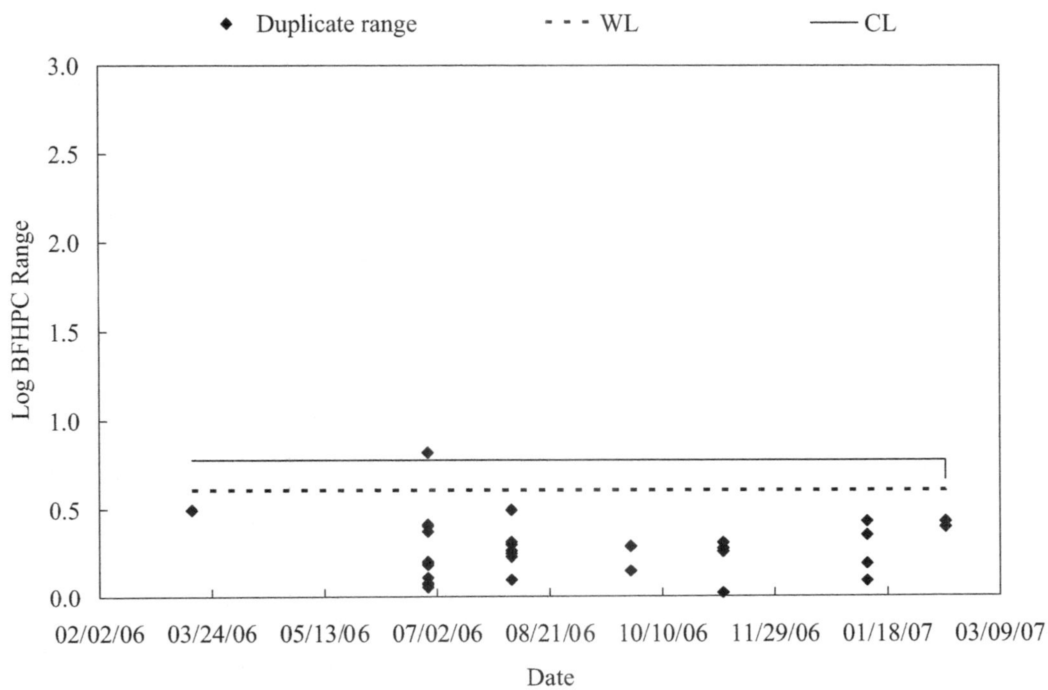

Figure 4.3 BFHPC precision duplicate data control chart

Separate control charts were developed to monitor method precision (Figures 4.3 and 4.4). One violation was observed in the duplicate data (Figure 4.3) and two for the replicate data (Figure 4.4). The number of violations is so small in both data sets that there is probably no significance to the fact that the replicate controls had more violations than the duplicate controls. As listed in Table 4.13, the percentages of the violations out of observations for replicates and duplicates are 4.44% and 3.13%, respectively.

There is no significant difference between the percentages of violations listed in Table 4.13. The most notable difference between the two data sets is that the warning and control limits (and thus the variability) of the replicates was about 0.3 log lower than the duplicate QC data.

Biofilm Total Coliform and E. coli

No biofilm samples tested positive for coliforms during the entire study until phase IV. The first coupon sampling event for phase IV tested positive for coliforms in the PDS11 UCI coupon biofilm and the PDS14 galvanized coupon. Both were negative for *E. coli*. The coliform densities of the coupons were 154 and 1.1 MPN/cm^2 which was an extremely small fraction of the

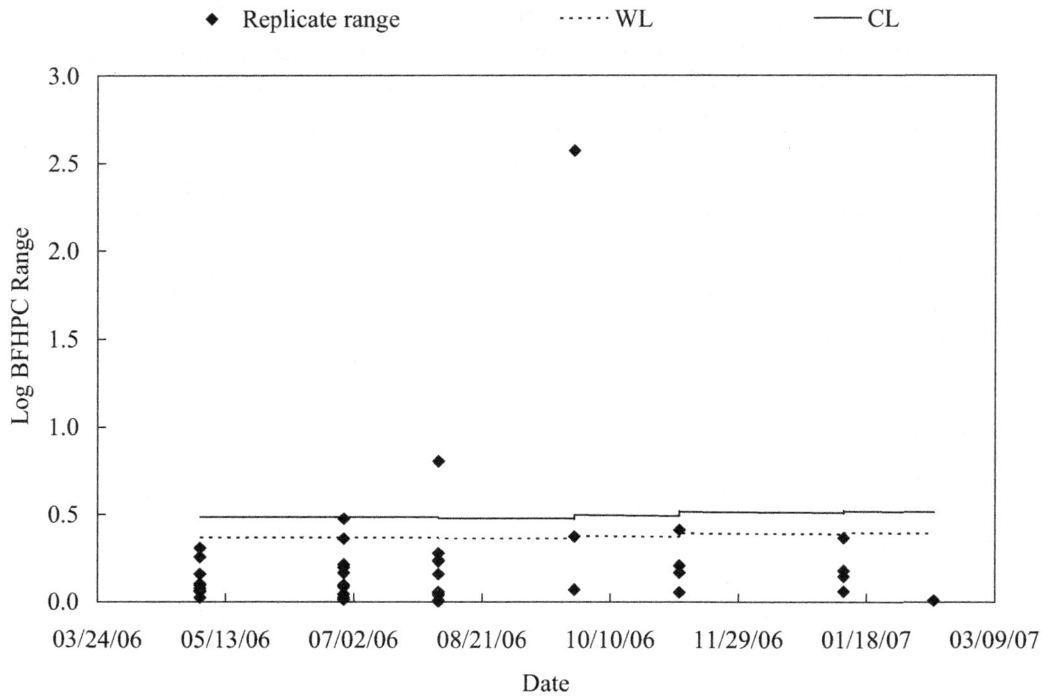

Figure 4.4 BFHPC precision replicate data control chart

Table 4.13
Violations for BFHPC duplicates and replicates

Group	Violations	Total observations	Percentage
Duplicate	1	32	3.13
Replicate	2	45	4.44

total biofilm biomass (several orders of magnitude less). Table 4.14 shows that control data for the run behaved perfectly.

Assimilable Organic Carbon

The AOC precision QC chart is shown in Figure 4.5. Both blind duplicates from the field (sample identity unknown to the lab analyst) and lab replicates selected by the analyst were used to quantify precision. Unlike the other biological parameters, the AOC QC replicates and duplicates were analyzed together (i.e., pooled QC data). The statistics test by SigmaPlot confirmed that the QC data were normally distributed (normality test passed, $p = 0.110$). An F-test showed that the two sets of data had equal variances (Table 4.15). A t-test was used to compare the means of the two data sets, and it showed there was not enough evidence to conclude that the duplicate and replicate means were different. Because there were only 24 duplicate and 17 replicate data points, it was less likely that any difference between the replicates and duplicates would be statistically significant (Table 4.16). It is unknown whether they would have been different had there been more AOC QC data. The values of raw plate counts (both P17 and NOX) usable for QC

Table 4.14
Biofilm total coliform/*E. coli* control data from phase IV sampling event 2 (02/12/2007)

		Without UV (coliforms)			With UV (*E. coli*)		
Control	Degrees of freedom		Diluted MPN* (MPN/100 mL)	MPN of sample (MPN/100 mL)		Diluted MPN (MPN/100 mL)	MPN of sample (MPN/100 mL)
E. coli	1.00E+02	>	2419.2	2.42E+05	>	2419.2	2.42E+05
(+/+)	1.00E+04	>	2419.2	2.42E+07	>	2419.2	2.42E+07
	1.00E+06	=	35	3.50E+07	=	35	3.50E+07
Pseudomonas aeruginosa;	1.00E+02	<	1	1.00E+02	<	1	1.00E+02
(–/–)	1.00E+04	<	1	1.00E+04	<	1	1.00E+04
	1.00E+06	<	1	1.00E+06	<	11	1.00E+06
Klebsiella pneumoniae	1.00E+02	>	2419.2	2.42E+05	<	1	1.00E+02
(+/–)	1.00E+04	>	2419.2	2.42E+07	<	1	1.00E+04
	1.00E+06	=	54.5	5.45E+07	<	1	1.00E+06

* MPN = most probable number.

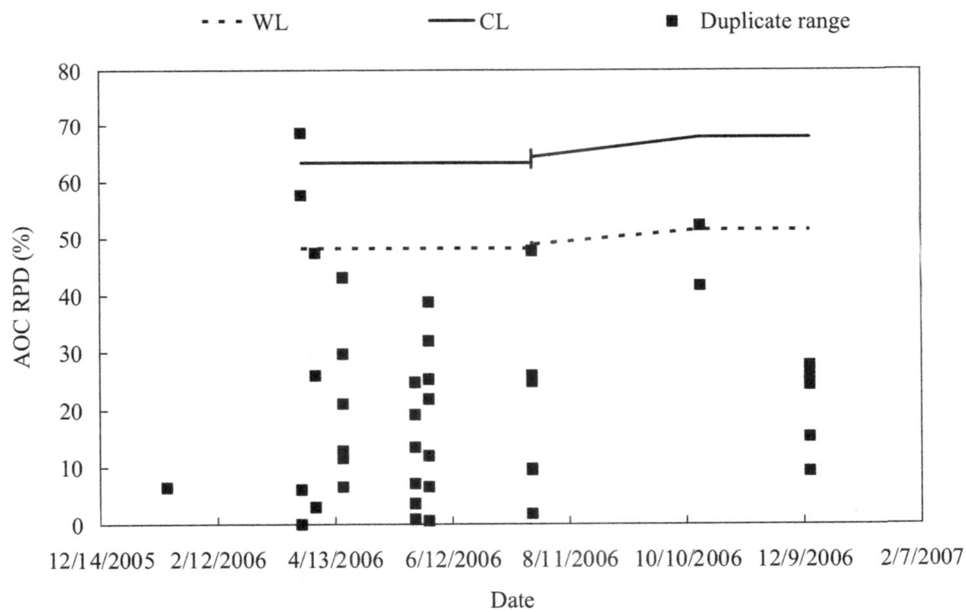

Figure 4.5 AOC precision pooled data control chart

purposes were the same as those adopted for HPC (i.e., ≥30 cfu per plate). Dilutions were selected to increase the likelihood of counts in the optimal range of 30 to 300 cfu. To ensure that this occurred, dilutions of 3×10^{-3} and 10^{-4} were used for all lab replicates.

There was only one violation in the AOC control chart (Figure 4.5). The average RPD for AOC QC data was 21%. The RPD was adopted as the statistic for AOC rather than log range because it allows comparison with prior research where the source water AOC can vary from 10 to

Table 4.15
F-Test two-sample test for unequal variances

	Duplicate	Replicate
Mean	22.01	18.96
Variance	395.44	181.10
Observations	24	17
df	23	16
F	2.18	
p(F ≤ f) one-tail	0.06	
F Critical one-tail	2.24	

Table 4.16
t-Test comparing means of duplicates and replicates: two-sample assuming equal variance

	Duplicate	Replicate
Mean	22.01	18.96
Variance	395.44	181.10
Observations	24	17
Pooled variance	307.51	
Hypothesized mean difference	0	
df	39	
t Stat	0.55	
p(T ≤ t) one-tail	0.29	
t Critical one-tail	1.68	
p(T ≤ t) two-tail	0.59	
t Critical two-tail	2.02	

250 µg/L, while a log transformation was not appropriate for this parameter because it does not vary by orders of magnitude as do HPC and BFHPC.

HPC Duplicates and Replicates Distribution Fit Tests

After project completion, the type of distribution for bulk HPC, BFHPC, and AOC QC data was determined. The NIST-calculated UWL and UCL did not correspond to 95% and 99.7% of the data (e.g., only a 5% probability that a valid observation would exceed the UWL) as they would for a normal distribution. This made it desirable to develop a method to select a UWL and UCL for these nonnormal distributions where the probability was the same as the NIST UWL and UCL for a normal distribution.

Statistical tests during the study showed that the log range (our QC statistic for all but AOC) of duplicates was more variable than those of replicates. As a result, the QC analysis was conducted separately for each class of replicate. Separate precision control charts are provided for

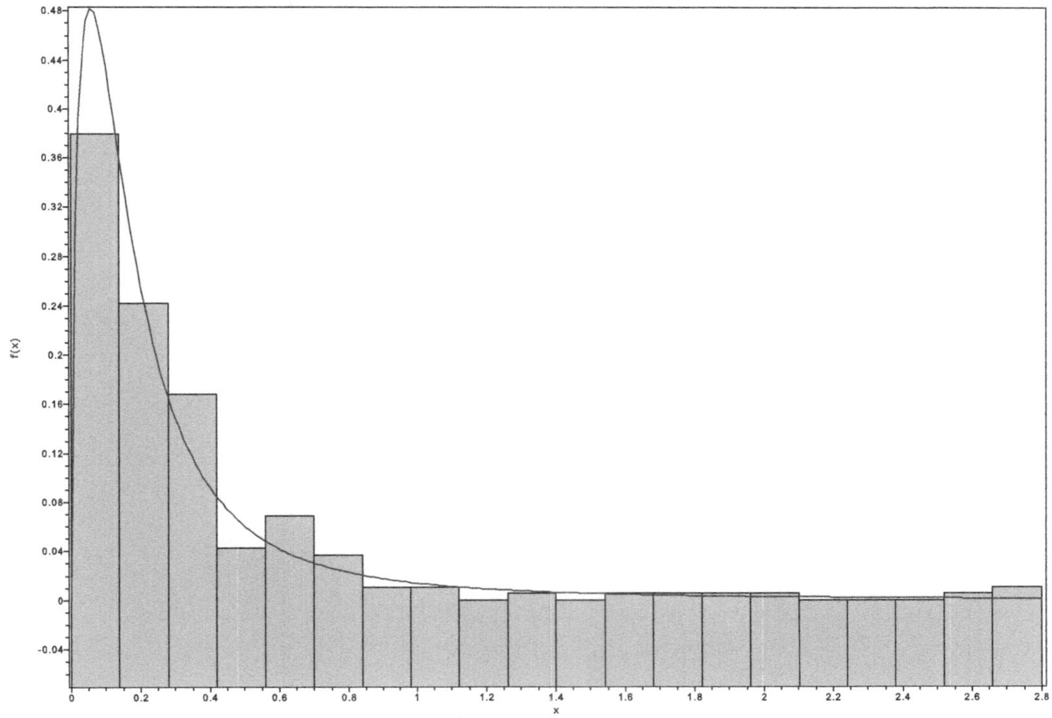

Figure 4.6 HPC duplicate QC data probability density distribution fits log-logistic

blind duplicates and lab replicates in Figures 4.1 and 4.2, respectively. As indicated in these figures, the warning and control limits were established at constant values until 30 determinations were obtained. Thereafter, the limits were based on the most recent set of 30 observations that did not violate the control limit.

EasyFit software (MathWave Technologies, San Diego, Calif.) was employed to test the distribution of HPC QC data. Duplicate and replicate samples gave log HPC ranges that were used in the distribution test. Through Kolmogorov–Smirnov, Anderson–Darling, and chi-squared distribution fit tests, both duplicate and replicate log HPC ranges fitted a log-logistic distribution, as shown in Figures 4.6 and Figure 4.7. In these two figures, the empirical probability density function (PDF) was displayed as a histogram consisting of equal-width vertical bars, each representing the number of sample data values (falling into the corresponding interval) divided by the total number of data points. The theoretical PDF is displayed as a continuous curve properly scaled, depending on the number of intervals. The scaling means multiplying the PDF values by the interval width. Duplicate and replicate data were divided into 20 intervals in the following graphs.

NIST precision guidelines (ASTM International 1977) specified warning limits and control limits at 2 and 3 standard deviations above the average range (range equals the absolute value of the difference between two duplicate or replicate measurements). Thus, the formulas specified for the limits are as follows:

WL = average range + 2× standard deviation of range

CL = average range + 3× standard deviation of range

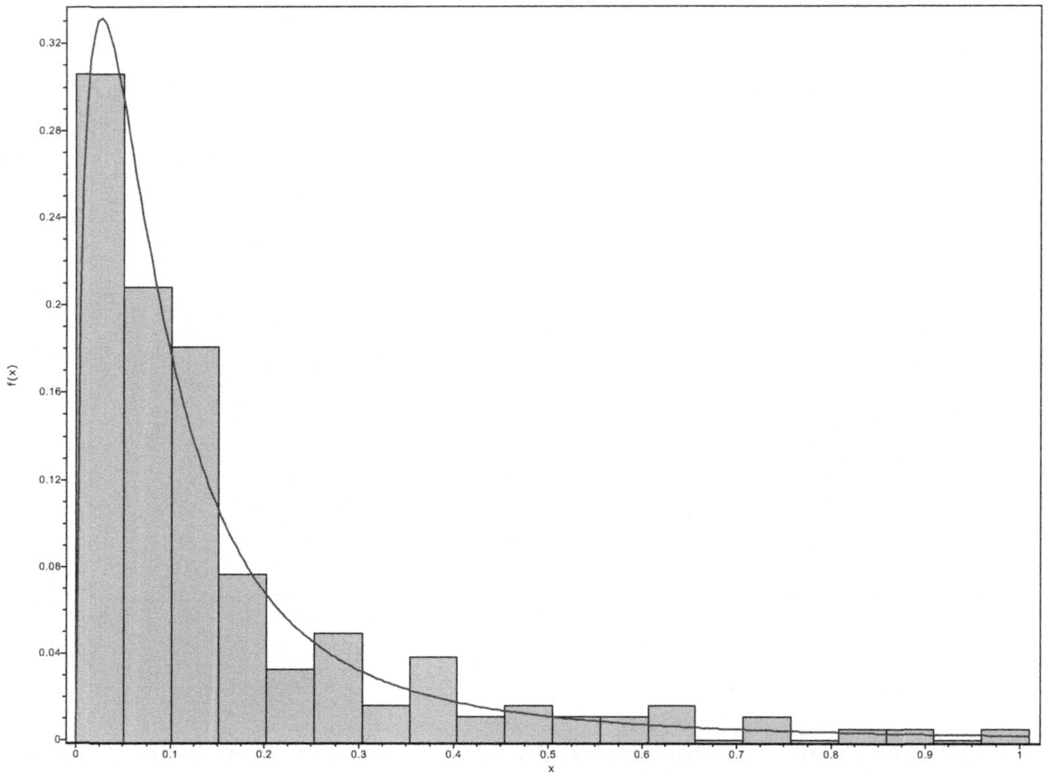

Figure 4.7 HPC replicate QC data probability density distribution fits log-logistic

For application of these warning and control limits, it has been assumed that 5% of valid observations will exceed the warning limit and 0.3% of the valid observations will exceed the control limit. Control violations occur when a single event exceeds the control limit or when two consecutive events exceed the warning limit. For a normal distribution, it is assumed that 5% of valid observations will exceed the warning limit and 0.3% of the valid observations will exceed the control limit.

Using the full complement of duplicate and replicate QC analysis for the project according to the NIST standard, the associated control limits and warning limits for duplicate ranges were 1.037 and 0.787 (log transform). The control limits and warning limits for replicate ranges were 0.440 and 0.334 (log transform). The cumulated distribution function plots are shown in Figures 4.8 and 4.9. From the figures, 6.25% of valid observations exceeded the warning limit and 4.38% of valid observations exceeded the control limit with duplicate HPC ranges. With replicates QC samples, the percentage over the warning limit and control limit was 14.44% and 9.38%, respectively.

The NIST-specified limits may be adopted independently of any assumed data distribution. If there is no knowledge of the data distribution, the limits serve as an indicator of relative change in the variability of a measurement system. If the distribution is known, then it is possible to determine the actual probability that corresponds to the limits as defined by NIST, or to determine the UWL and UCL based on probability rather than the NIST calculation.

Table 4.17 describes the total number of duplicate and replicate samples and the numbers under the warning limit, under the control limit, and between these two limits. The UWL and UCL were determined based on probabilities of 99.7% and 95% from the cumulative distribution

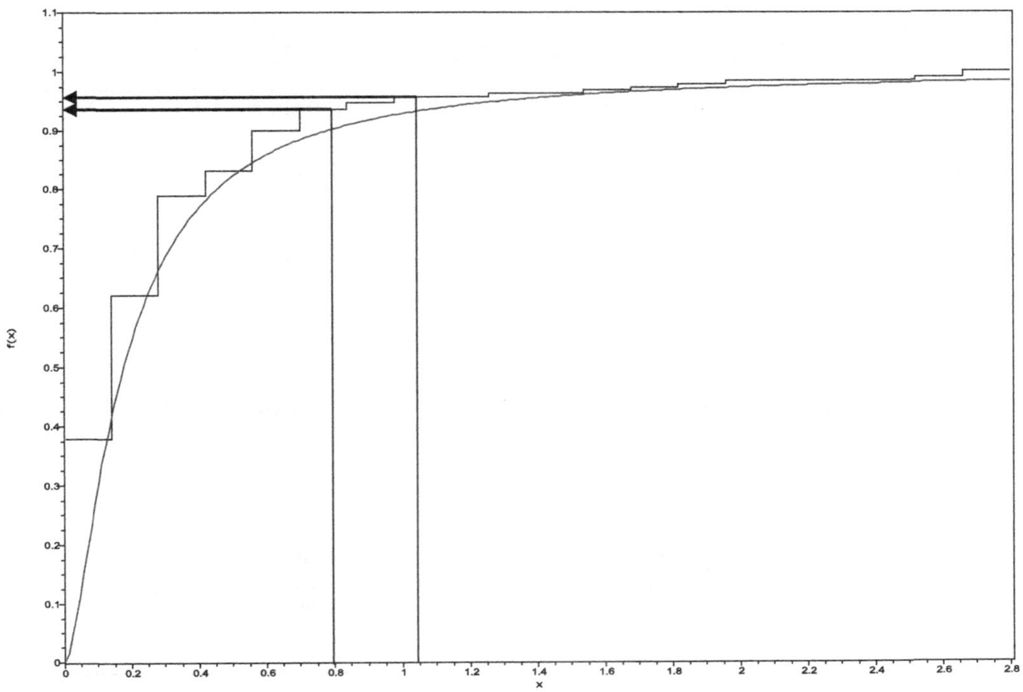

Figure 4.8 HPC duplicate QC data cumulative distribution function plot

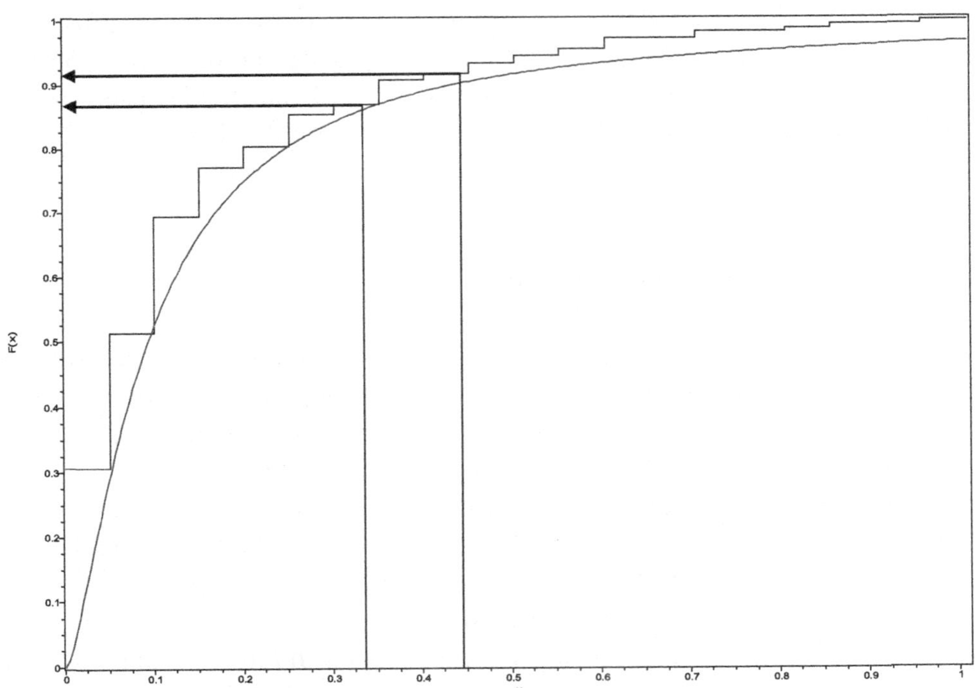

Figure 4.9 HPC replicate QC data cumulative distribution function plot

Table 4.17
HPC duplicate and replicate UWL and UCL data summary

	Total number of observations	Number > UWL		UWL < number < UCL		Number > UCL	
		(no.)	(%)	(no.)	(%)	(no.)	(%)
HPC duplicates	190	38	20.0	17	9.0	21	11.0
HPC replicates	183	45	24.6	14	7.7	31	17.0

Table 4.18
Comparison of normal distribution UWL and UCL with real distribution UWL and UCL

	UWL from chart (±2 std dev)	UCL from chart (±3 std dev)	UWL from CPDF	UCL from CPDF	Percent change of UWL vs. NIST UWL	Percent change of UCL vs. NIST UCL
HPC duplicates	0.787	1.037	0.988	2.660	20.30	61.02
HPC replicates	0.334	0.440	0.569	0.970	41.27	54.64

function. The warning limit and control limit from the NIST calculations used during the project were compared with those obtained from the cumulative PDF (CPDF) and are shown in Table 4.18. The CPDF suggested that if 5% of the observations were assumed to exceed the warning limit, the warning limit should be 0.988 for duplicate samples and 0.569 for replicate samples, which is significantly different from the NIST UCL and UWL used during the study (0.787 and 0.334, respectively). The CPDF warning limits were 20.30% and 41.27% greater than the NIST warning limits for duplicates and replicates. The CPDF control limits were 61.02% and 54.64% more than the NIST limits.

BFHPC Pooled QC Data Distribution Fit Tests

Although duplicate and replicate warning limits and control limits were statistically different for BFHPC, the data sets were not large enough to separate them and still be able to test the distribution. The pooled samples (duplicates plus replicates) were tested, and they fit the Wakeby probability density distribution best, as shown in Figure 4.10.

The actual warning limit and control limit according to NIST formulas were 0.469 and 0.609 (log transform). However, from Figure 4.11, 2% of the observations would exceed both the actual warning limit and control limit, and thus they had very different probabilities than a normal distribution (0.3% and 5% from the NIST standard).

BFHPC QC data was analyzed using the pooled total number of duplicate and replicate samples. The range values were categorized with respect to the NIST UWL and UCL used in the study (Table 4.19). The UWL and UCL from the cumulative distribution function were 0.733 and 2.148, respectively, which were 36.02% and 71.65% greater than the NIST limits.

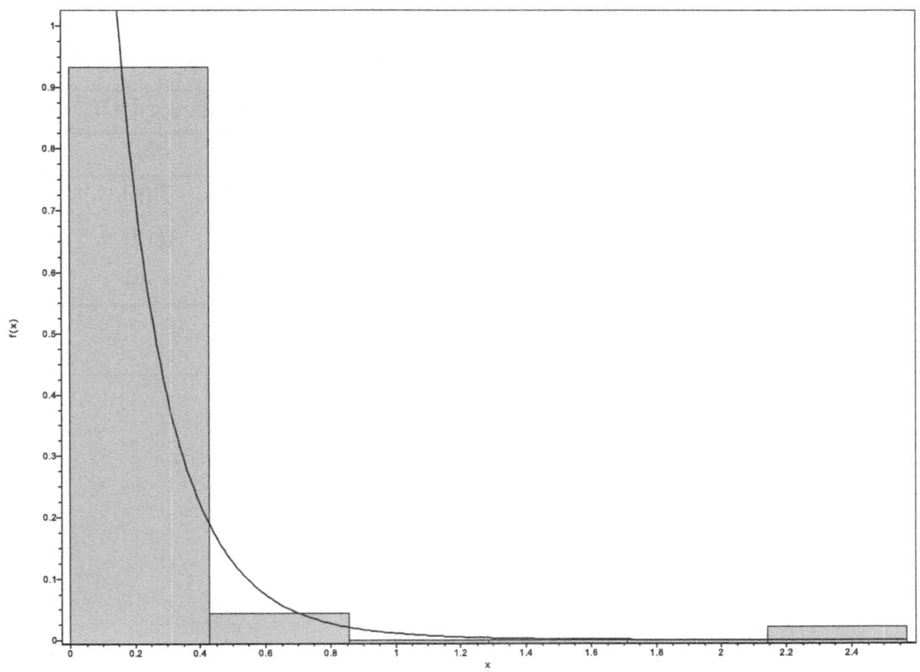

Figure 4.10 BFHPC pooled QC data probability density distribution fit Wakeby distribution

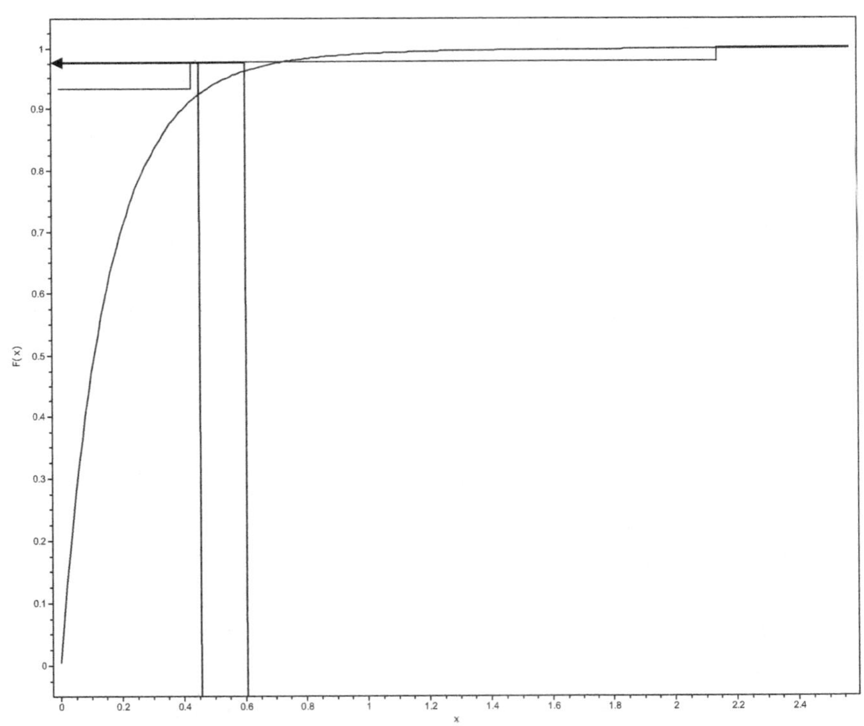

Figure 4.11 BFHPC pooled QC data cumulative distribution function plot

Table 4.19
BFHPC pooled, duplicate, and replicate UWL and UCL data summary

	Total number of observations	Number > UWL		UWL < number < UCL		Number > UCL	
		(no.)	(%)	(no.)	(%)	(no.)	(%)
BFHPC pooled	77	5	6.49	2	2.60	3	3.90
BFHPC duplicates	32	1	3.13	0	0.00	1	3.13
BFHPC replicates	45	4	8.89	2	4.44	2	4.44

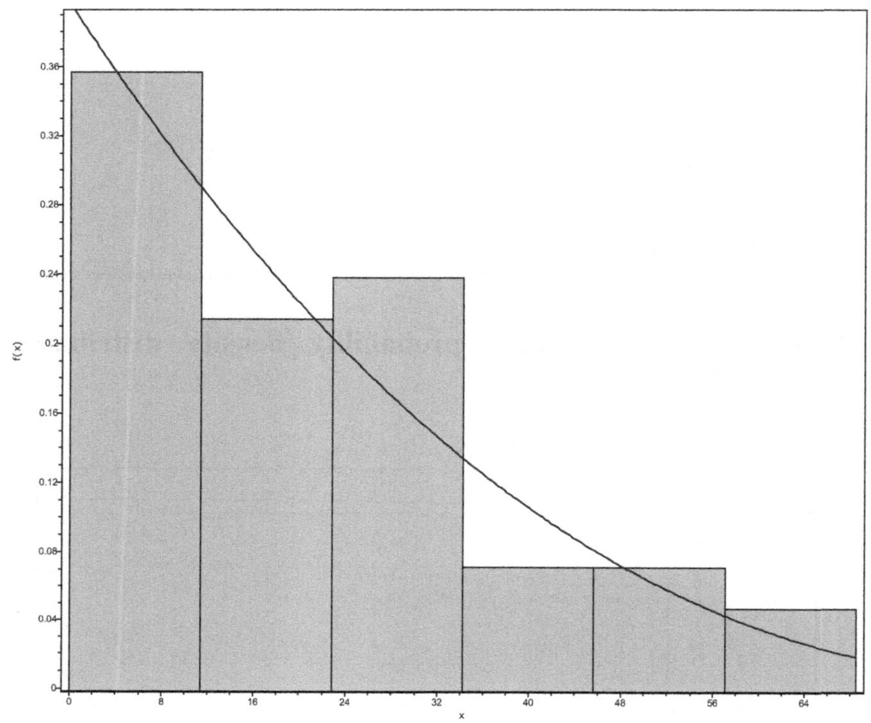

Figure 4.12 AOC pooled QC data probability density distribution fit Wakeby distribution

AOC Pooled QC Data Distribution Fit Tests

Because there were only 24 duplicate and 17 replicate data points, it was very unlikely that any difference between the replicates and duplicates would be statistically significant. Unlike the other biological parameters, the AOC QC replicates and duplicates were analyzed together (i.e., pooled QC data) throughout the study. In addition, the RPD was used to build the QC chart instead of the log-transformed range. The pooled AOC QC data were tested, and they fit the Wakeby probability density distribution best, as shown in Figure 4.12.

The actual warning limit and control limit according to NIST formulas were 52% and 68%. However, from Figure 4.11, 4.9% (instead of 5% from NIST) of the observations would exceed the warning limit and there was no observation that exceeded the control limit, as shown in Figure 4.13.

There were 42 observations in the whole AOC QC chart. Four observations, or 9.52% of all observations, exceeded the NIST warning limit. One observation (2.38%) exceeded the NIST control limit.

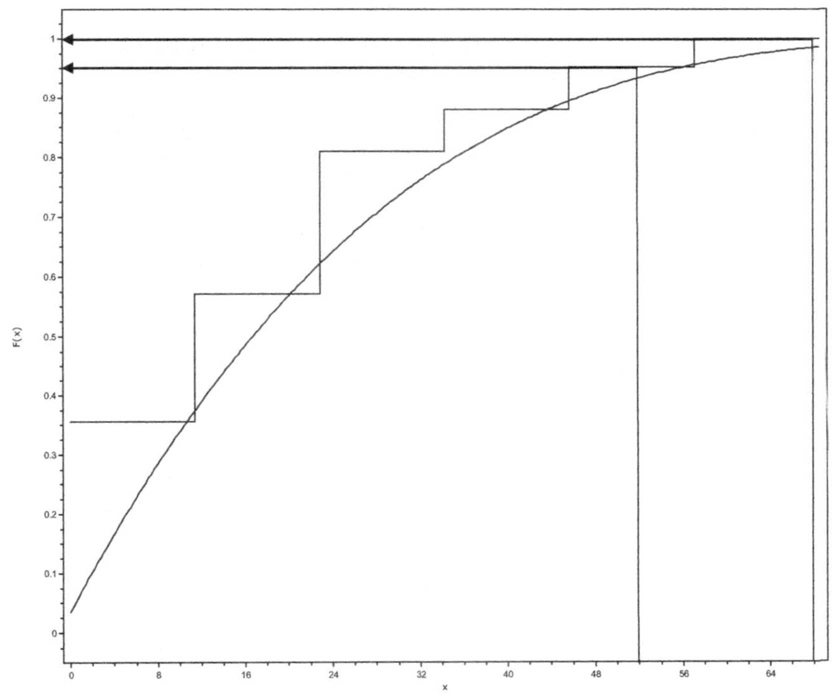

Figure 4.13 AOC pooled QC data cumulative distribution function plot

The UWL and UCL from the cumulative distribution function were 50% and 67%, which were 13.0% and 19.3% lower than the actual limits. This result suggested that when there was a QC data set with small number of observations, the NIST formulas were less stringent rather than more stringent as they were with the bulk HPC data, for example. This result may also be affected by the distribution of the data, which is different for AOC data and bulk HPC data.

Suggested HPC QC Chart Setup

Instead of using the NIST formulas (average ±2 or 3 times the standard deviation), warning and control limits for an extensive data set like bulk HPC can be based on the cumulative distribution function plot to find log HPC range values at 99.7% and 95% based on the most recent set of 30 observations that do not violate the control limit. The cumulative distribution function plot calculation was listed in Table 4.20 by using 30 randomly chosen observations from the duplicate HPC QC data. Thirty observations were separated into seven groups, and the bins were 0.14.

The warning limit and control limit can be obtained from the cumulative distribution function plot, as shown in Figure 4.14. The warning limit was 0.41 (log transformed) and the control limit was 0.83 (log transformed). If NIST formulas were used to calculate the UWL and UCL, the UWL would be 0.50 and the UCL would be 0.65, which would not exclude 5% and 0.3% violations of the valid observations. Thus, if sufficient historical data exists, this approach can be used for QC purposes. However, in many cases it will be simpler and easier to compare with other labs if the NIST formulas are used. Nevertheless, with microbial enumeration and other nonnormal distributed data, it is erroneous to associate the NIST UWL and UCL with a known probability level.

Table 4.20
Simple example for HPC duplicate QC chart setup (cumulative distribution function calculation)

Raw data		Data characteristics	
0.16	0.03	Maximum	0.41
0.02	0.08	Minimum	0.02
0.30	0.22	Number of observations	30
0.11	0.15		
0.11	0.07		
0.13	0.21		
0.29	0.07		
0.05	0.07		
0.18	0.13		
0.04	0.28		
0.05	0.08		
0.41	0.72		
0.33	0.34		
0.14	0.35		
0.38	0.18		

Cumulative Distribution Function Estimation

Bins	Count	Total
0.00	0	0
0.14	15	0.5
0.28	7	0.733333
0.42	7	0.966667
0.56	0	0.966667
0.70	0	0.966667
0.84	1	1

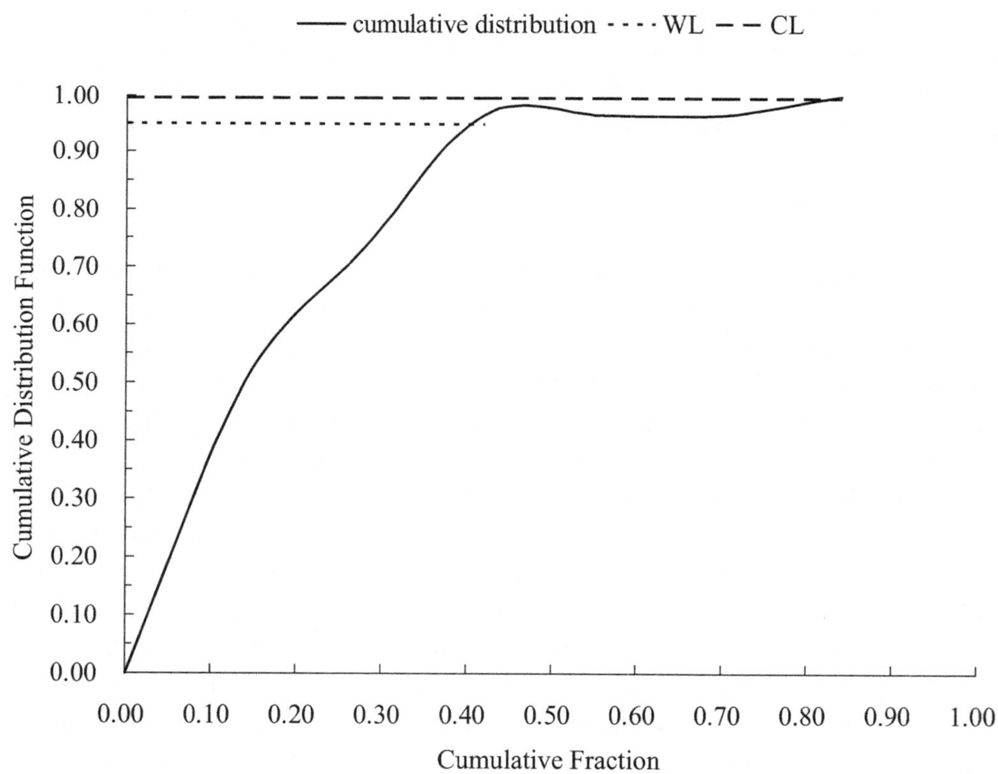

Figure 4.14 HPC duplicate data setup, new QC chart by cumulative distribution function plot

CHAPTER 5
RESULTS AND DISCUSSION

METAL RELEASE

Overview

This discussion of iron, copper, and lead control in the experimental system by application of corrosion control strategies (either inhibitor addition or pH elevation) is organized as follows:

- Review of empirical models from historical TBW I study.
- Identification of conflicting criteria for control of iron and copper.
- Empirical modeling procedures for the current study.
- Performance of corrosion control strategies for mitigation of iron release.
- Performance of corrosion control strategies for mitigation of copper release.
- Performance of corrosion control strategies for mitigation of lead release.
- Sensitivity analysis and inhibitor performance using the resultant empirical models for varying blends of groundwater, surface water, and desalination finished water sources.

The corrosion control strategies are evaluated based on the empirical model predictions for variant inhibitor dose. Binary comparisons of the performance of each inhibitor/dose PDS with the pH_s control PDS are also provided. The binary comparisons are supported with a statistical verification of the significance of any observed reduction in total metal concentration associated with the particular corrosion control measure.

Empirical Modeling Results From Historical TBW I Study

The TBW I study included development of models for predicting total iron, total copper, and total lead release for the hybrid PDSs. Dissolved metals were not measured, so no modeling efforts were completed for dissolved metal species. All measured water quality parameters were considered for inclusion in the empirical models. Analysis of variance (ANOVA) methods were employed to identify those water quality variables that were significant in the empirical models at a confidence level of 95% ($\alpha = 0.05$). Independent variables that did not meet this level of significance were omitted from the models.

A brief discussion of limitations of this type of empirical modeling effort is provided to guide application of the resulting empirical models. The procedure for selection and/or removal of variables from the final model was an ANOVA. Two forms of a model were compared: a version that included a potential variable in addition to the shared variables, and a version that included all variables (the shared variables) except the potential variable. If the expanded model, including the potential variable, achieved a statistically significant improvement in ability to describe the data relative to the model with shared variables only, then the variable was judged to be significant and would remain in the model. Thus, inclusion (or exclusion) of variables from the final model would be based on criteria other than a single goodness-of-fit statistic such as the correlation coefficient. This approach is fundamentally different from a mechanistic model that would presume the significance of variables in advance of parameter estimation.

Table 5.1
TBW I variable range for model development

Parameter	Minimum	Maximum
Alkalinity (mg/L as $CaCO_3$)	51	210
Calcium (mg/L)	23	88
Chloride (mg/L)	19	100
Sulfate (mg/L)	4	232
Sodium (mg/L)	12	65
Silica (mg/L as SiO_2)	1	16
Dissolved oxygen (mg/L)	6.9	12.2
UV_{254} (cm^{-1})	0.0233	0.0688
Conductivity (μS/cm)	331	707
pH	7.8	8.4
Temperature (°C)	16.8	27.0
Hydraulic residence time (days)	2	5

The empirical approach may be useful to suggest mechanistic relationships, but the presence of confounding effects between study variables may obscure proper identification of such mechanistic relationships. The study mandated examination of three specific water sources (GW, SW, and RO), with the result that the data set contained an inherent correlation between certain water quality parameters. For example, all blends that are enriched with groundwater would contain an elevated alkalinity and a depleted sulfate. All blends that are enriched with surface water would contain a depleted alkalinity and an elevated sulfate. Similarly, all blends that are enriched with the desalinated source would exhibit a depleted alkalinity. Although these factors may hinder a mechanistic interpretation of the data, the same type of confounding issues will be present in the prototype distribution system when varying blend ratios of the three principal sources are present. The approach would be expected to provide a satisfactory description of events that would be experienced in MG distribution systems when subjected to variations in water quality associated with differing blends of GW, SW, and RO.

Application of any empirical model should recognize the range of values of the study variables in the data set that was used to develop the empirical model. The ranges of water quality parameters used in model development for the TBW I study are reported in Table 5.1.

The variability in water quality represented by this data set is more extreme than would likely be experienced in the actual distribution system, as individual PDSs were operated with single-source inputs (e.g., 100% of either GW, SW, or RO) as well as blends of the principal sources. Consequently, the database for development of empirical models in TBW I was robust. Application of these empirical models to actual experienced blends would not require extrapolation to conditions outside of the database used for model development.

These factors provide confidence in the suitability of the models for description of metal release in the hybrid PDSs and the corrosion loops. Nevertheless, use of the empirical models to describe water quality not represented by the data would be inappropriate. Examples of such an inappropriate use of the model would include description of anaerobic systems, unstable waters (with respect to calcium carbonate solubility), and waters with a low pH relative to the data range defined in Table 5.1. During the equilibration period in TBW I, the change in apparent color was found to correlate well with total iron release in the system. Consequently, change in apparent color in the PDSs was used to show the effects of water quality on iron release. The model incorporates HRT but otherwise was developed specific to the geometry of the hybrid PDSs. The resulting model is shown in Equation 5.1. Increasing alkalinity has a beneficial effect on iron release. Sulfate, chloride, sodium, DO, temperature, and HRT were all found to have a positive correlation with iron release. The pH did not have a significant effect on iron release because of the water being stabilized with respect to $CaCO_3$ solubility.

$$\Delta \text{ apparent color (cpu)} = 0.04775(\text{alkalinity})^{-0.912}(\text{sulfate})^{0.118}(\text{chloride})^{0.485} \times (\text{sodium})^{0.561}(\text{DO})^{0.967}(\text{temperature})^{0.813}(\text{HRT})^{0.836} \quad (5.1)$$

where Δ apparent color = cpu
alkalinity = mg/L as $CaCO_3$
sulfate = mg/L
chloride = mg/L
sodium = mg/L
DO = dissolved oxygen, mg/L
temperature = °C
HRT = hydraulic residence time, days

The total copper release model was similarly developed for the corrosion loop data. This model is shown in Equation 5.2. Alkalinity, along with sulfate and temperature, was found positively correlated with copper release. This is notable because an increase in alkalinity for control of iron release could lead to increased copper concentrations. Silica and pH had negative correlations with copper release, and chloride was not a significant parameter. HRT is not a relevant variable because sampling of the corrosion loops was done after a uniform 6-hour stagnation period.

$$\text{Cu} = \text{temperature}^{0.72} \times \text{alkalinity}^{0.73} \times \text{pH}^{-2.86} \times \text{sulfate}^{0.10} \times \text{silica}^{-0.22} \quad (5.2)$$

where Cu = total copper concentration, mg/L
temperature = °C
alkalinity = mg/L as $CaCO_3$
pH = $-\log [H^+]$
sulfate = mg/L
silica = mg/L as SiO_2

Total lead release in the corrosion loops was also modeled in the TBW I study. The developed model is shown in Equation 5.3. Alkalinity, chloride, and temperature were all found to be positively correlated with lead release, while pH and sulfate were negatively correlated to lead release. HRT is not included as a variable because lead is also sampled in the corrosion loops after a uniform 6-hour stagnation period.

$$\text{Pb} = 1.027^{(T-25)} \times \text{alkalinity}^{0.677} \times \text{pH}^{-2.726} \times \text{chloride}^{1.462} \times \text{sulfate}^{-0.228} \quad (5.3)$$

where Pb = total lead concentration, mg/L
T = temperature, °C
alkalinity = mg/L as $CaCO_3$
pH = $-\log [H^+]$
chloride = mg/L
sulfate = mg/L

Conflicting Objectives for Iron, Copper, and Lead Control

The historical results from TBW I have identified a serious complication with simultaneous maintenance of target concentrations for iron, copper, and lead. In order to avoid red water issues associated with elevated iron concentrations, maintenance of an elevated alkalinity would be desired. This beneficial effect of alkalinity on iron (or color) is evident from the negative exponent on the alkalinity term in the empirical model (Equation 5.1). Supplementation of alkalinity in the low-alkalinity sources (SW and RO) to achieve a minimum target alkalinity of 80 to 100 mg/L as $CaCO_3$ may be appropriate to mitigate possible red water episodes associated with blends that are deficient in groundwater sources.

Based on the empirical models developed in TBW I, efforts to control iron release by maintenance of elevated alkalinity would be counterproductive for compliance with copper and lead action levels based on the empirical models in Equations 5.2 and 5.3, respectively. In both of these models, the exponent on the alkalinity term is positive, indicating that an increase in both copper and lead concentrations would be associated with increases in alkalinity. From a practical operations perspective, the results have identified a fundamental dilemma. Operation at blends that are deficient in groundwater (thus low alkalinity) may result in compliance difficulties for iron (or color). Operation at blends that are enriched in groundwater (thus high alkalinity) may produce compliance issues with copper and, to a lesser degree, lead.

An analysis of feasible blends of groundwater, surface water, and desalination sources was completed by Imran et al. (2006). This analysis revealed potential violations of the copper action level for blends that are enriched in groundwater, potential violations of the lead action level for blends that are enriched in RO, and potential violations of color standards for blends that are deficient in groundwater. Management of blend composition to simultaneously address all constraints would be difficult in the presence of constraints on the available output of finished water from the three principal sources operated by TBW. The current study objectives include an evaluation of the potential benefits of inhibitors to expand the feasible domain of blends that would maintain suitable distribution system water quality. This objective may be restated as a determination of whether addition of inhibitors would permit operation at elevated alkalinities associated with groundwater-enriched blends without violation of copper and lead action levels.

Empirical Metal Release Models for Current Study

Characterization of Database

Model development for this study was undertaken for prediction of total and dissolved iron, copper, and lead concentration in the hybrid PDSs and corrosion loops. A power form of the model was used, similar to those developed in the TBW I study. The water quality parameters monitored in the PDSs were evaluated using ANOVA procedures to identify statistically

Table 5.2
TBW II variable range for model development

Parameter	Minimum	Maximum
Total phosphorus (mg/L as P)	0.01	3.36
Silica (mg/L)	4	65
pH	7.4	9.1
Alkalinity (mg/L as $CaCO_3$)	84	175
Calcium (mg/L)	43	88
Zinc (mg/L)	0.001	0.793
Chloride (mg/L)	35	123
Sulfate (mg/L)	52	119
Sodium (mg/L)	5	92
Influent iron (mg/L)	0.003	0.358
Dissolved oxygen (mg/L)	6.6	10.9
UV_{254} (cm^{-1})	0.007	0.105
Temperature (°C)	10.4	29.7

significant parameters. Nonlinear least-squares regression techniques were used to estimate parameters in the empirical models. Those independent variables not shown to be significant at a 95% confidence level were eliminated from the models. The ranges of variables used in model development for the current study are shown in Table 5.2. Compared to the range of values available in the TBW I study, this data set indicates a smaller range of values.

The regression for model development was done on the data for the 14 hybrid PDSs, including all four phases. Metal concentrations below the detection limit were set at predetermined values so zero values would not cause complications in convergence of the models. The metals that had observations below detection include dissolved iron, dissolved and total lead, and dissolved and total zinc. These observations were set at 1 μg/L for the data analysis.

Model Development Using All Measured Water Quality Parameters

A series of attempts were made to develop empirical models for prediction of total and dissolved metal concentration as a function of water quality, corrosion control strategy, and inhibitor dose. The model development attempted to preserve the relationships that were previously identified in TBW I, where appropriate. Model specification was complicated, in some cases, by the existence of confounding relationships in the data between potential independent variables. The range of water quality parameters in the current study is not as robust as the data set for TBW I, with a result that determination of independent variable significance with the current data set is less likely than with a data set that exhibits a larger range in values of the independent variables.

The initial modeling effort incorporated all the monitored parameters from this study. These were done with power models similar to TBW I, but with additional dummy variables to segregate data by the individual corrosion control measures: BOP, OP, ZOP, Si, or pH elevation. The general form of the model, including dummy variables, is depicted in Equation 5.4.

$$\text{Metal} = (a \times BOP + b \times OP + c \times ZOP + d \times Si + e \times pH_s) \times pH^f \times alk^g \times Cl^h \ldots \quad (5.4)$$

where
BOP = blended ortho-polyphosphate inhibitor dummy variable (0, 1)
OP = orthophosphate inhibitor dummy variable (0, 1)
ZOP = zinc orthophosphate inhibitor dummy variable (0, 1)
Si = silica inhibitor dummy variable (0, 1)
pH_s = pH control dummy variable (0, 1)
pH = $-\log [H^+]$
alk = alkalinity, mg/L as $CaCO_3$
Cl = chloride, mg/L
a = BOP dummy variable coefficient
b = OP dummy variable coefficient
c = ZOP dummy variable coefficient
d = silica dummy variable coefficient
e = pH control dummy variable coefficient
f = pH exponent
g = alkalinity exponent
h = chloride exponent

The dummy variables are ones and zeros used to indicate the presence or absence of a corrosion inhibitor and which type. This allows a model coefficient to be adjusted to achieve sensitivity to specific inhibitor for a constant set of water quality parameters. Nonlinear least-squares regressions were completed on the entire data set to estimate all the parameter exponents and the dummy variable coefficients.

The database includes several water quality parameters that are confounded. Confounding was identified by examination of a linear correlation matrix for the entire set of water quality parameters. The existence of correlations between water quality parameters is unavoidable for blends of the three principal source waters. For example, the surface water contains high sulfate concentrations and low alkalinity. In contrast, the GW source has elevated alkalinity and reduced sulfate. Thus, blends that are enriched in groundwater will exhibit high alkalinity and low sulfate. Conversely, blends that are enriched in surface water will exhibit low alkalinity and high sulfate. For blends that are produced from groundwater and surface water, alkalinity and sulfate will be negatively correlated. Water quality effects that are observed for GW-enriched blends may properly result from high alkalinity, low sulfate, or a combination of both factors.

Resolution of a link to a specific water quality parameter may be difficult because of any confounding effect. Furthermore, inclusion of confounded variables in an empirical model may generate convergence difficulties for nonlinear least-squares parameter estimation. The linear correlation matrix for the database was examined to identify potential difficulties associated with confounded variables. Sodium was confounded with chloride. Calcium, sodium, and chloride were confounded with TDS. Color and turbidity were confounded. Temperature and dissolved oxygen were confounded. Elimination of selected independent variables from model development may be necessary to overcome convergence problems experienced during parameter estimation.

Model Development Using Only Significant Water Quality Parameters From TBW I

The second modeling effort utilized the form of models that were generated in TBW I. The water quality variables that were identified as significant in the TBW I models were alkalinity, chloride, DO, pH, sodium, sulfate, and temperature. In addition, total phosphorus, silica, and zinc were included to account for the addition of inhibitors to the PDSs in the current study. These models also incorporated dummy variables to segregate data by inhibitor type. The exponents on the parameters from the TBW I models were held constant at the value obtained in the previous study, and the regression was performed to allow estimation of the exponents on the added parameters and the coefficients on the dummy variables. These models did not offer the significance that was expected on the inhibitor variables. These models were abandoned.

Model Development Using Water Quality Parameters Selected to Reduce Confounding Effects

Specific independent variables were then selected for inclusion in a model so that confounding effects could be avoided while maintaining similarity to the TBW I modeling results. For example, the data set suffered from confounding effects between sodium and chloride. Sodium was eliminated from model development in favor of chloride. Similarly, DO was eliminated in favor of temperature. The remaining candidate independent variables considered were pH, alkalinity, chloride, sulfate, and temperature. The influent iron concentration was also considered, but only for the iron models. ANOVA procedures were employed to identify significance of variables for retention in the empirical models. Parameter estimates were obtained by nonlinear least-squares regression. The resulting models did not display sensitivity to the effect of inhibitor dose on metal release, so further modeling efforts were pursued.

Model Development Including the Effect of Phosphorus Speciation

The next modification of the models incorporated an additional term to account for the difference in phosphorus speciation in BOP inhibitor. A term was added as the ratio of ortho-phosphorus to total phosphorus. Two terms would be present to account for phosphorus speciation in this model specification: total phosphorus and the ratio of ortho-phosphorus to total phosphorus. The model modification was not productive, however, as the ratio (ortho-phosphorus/total phosphorus) was found to be statistically insignificant in all cases except for description of the total iron concentration. In this case (total iron), the total phosphorus term was not shown to be significant. Thus for all resultant models, only one term involving phosphorus was retained, suggesting that the phosphorus speciation is not of importance in relation to metal release. Retention of the dummy variables for the inhibitors provides some degree of sensitivity to phosphorus speciation by separation of data for the BOP inhibitor from the data for other phosphate-based inhibitors that rely solely on OP species.

Model Selection Using Unique Parameter Estimates for Inhibitor Concentration

The final modeling effort included the water quality terms used previously that were common to all hybrid PDSs. These include pH, alkalinity, chloride, sulfate, and temperature. The influent iron concentration was also incorporated for the iron models. The dummy variables and associated coefficients were included to account for differences associated with inhibitor type. All terms associated with the dose of individual inhibitor were collected with the dummy variable for that inhibitor. These collected terms include total phosphorus, silica, and zinc concentrations. This approach permitted individual parameter estimation of exponents on the inhibitor dose (e.g., total phosphorus concentration) that are unique for each inhibitor. ANOVA methods were used to identify statistically significant variables for inclusion in the model. Parameter estimates were obtained with nonlinear least-squares methods. This model specification significantly improved the accuracy of the empirical models, particularly with respect to the effect of inhibitor dose. This general form of model was specified for subsequent data analysis, as represented in Equation 5.5:

$$\begin{aligned}\text{Metal} = (&a \times BOP \times TP^f + b \times OP \times TP^g + c \times ZOP \times TP^h \times Zn^i + d \times Si \times SiO_2^{\,j} \\ &+ e \times pH_s) \times pH^k \times alk^l \times Cl^m \times SO_4^{\,n} \times temp^o \times Fe_{inf}^{\,p}\end{aligned} \tag{5.5}$$

where
- BOP = blended ortho-polyphosphate inhibitor dummy variable (0, 1)
- OP = orthophosphate inhibitor dummy variable (0, 1)
- ZOP = zinc orthophosphate inhibitor dummy variable (0, 1)
- Si = silica inhibitor dummy variable (0, 1)
- pH_s = pH control dummy variable (0, 1)
- TP = total phosphorus, mg/L
- Zn = zinc, mg/L
- SiO_2 = silica, mg/L as SiO_2
- pH = $-\log [H^+]$
- alk = alkalinity, mg/L as $CaCO_3$
- Cl = chloride, mg/ SO_4 = sulfate, mg/L
- $temp$ = temperature, °C
- Fe_{inf} = influent iron, mg/L
- a = BOP dummy variable coefficient
- b = OP dummy variable coefficient
- c = ZOP dummy variable coefficient
- d = silica dummy variable coefficient
- e = pH control dummy variable coefficient
- f = TP exponent associated with BOP dummy variable
- g = TP exponent associated with OP dummy variable
- h = TP exponent associated with ZOP dummy variable
- i = Zn exponent associated with ZOP dummy variable
- j = SiO_2 exponent associated with silica dummy variable
- k = pH exponent
- l = alkalinity exponent
- m = chloride exponent
- n = sulfate exponent
- o = temperature exponent
- p = influent iron exponent

The following discussion is separated into sections that address iron, copper, and lead individually. The final versions of the empirical models are presented in the appropriate section for each specific metal.

Release of Iron From the Hybrid Pilot Distribution System

Empirical Model for Total Iron Concentration

The empirical model developed for total iron concentration in the hybrid PDSs is presented in Equation 5.6. This model includes dummy variables for each of the inhibitor types, as previously described. The model suggests that alkalinity will decrease iron release with increasing concentration, as shown by the negative exponent on the term, whereas chloride and influent iron will contribute to iron release in the PDSs. Each corrosion control strategy (inhibitor or pH elevation) has different contributing terms, multiplied by the dummy variables, which adjust the predicted iron concentration as a function of the corrosion control strategy.

$$\begin{aligned}\text{total Fe} = {} & (0.495\text{BOP} \times \text{TP}^{-0.104} + 0.593\text{OP} \times \text{TP}^{0.047} \\ & + 0.470\text{ZOP} \times \text{TP}^{-0.073} \times \text{Zn}^{-0.180} + 0.122\text{Si} \times \text{SiO}_2^{0.528} + 0.661\text{pH}_s) \\ & \times \text{alk}^{-0.457} \times \text{Cl}^{0.345} \times \text{Fe}_{inf}^{0.136} \qquad (5.6)\end{aligned}$$

where total Fe = total iron, mg/L
BOP = blended ortho-polyphosphate inhibitor (0, 1)
OP = orthophosphate inhibitor (0, 1)
ZOP = zinc orthophosphate inhibitor (0, 1)
Si = silica inhibitor (0, 1)
pH_s = pH control (0, 1)
TP = total phosphorus, mg/L
Zn = zinc, mg/L
SiO_2 = silica, mg/L as SiO_2
alk = alkalinity, mg/L as $CaCO_3$
Cl = chloride, mg/L
Fe_{inf} = influent iron, mg/L

As was seen in the TBW I study, pH did not have a significant effect on iron release because of the water being stabilized with respect to $CaCO_3$ solubility. Sulfate was not significant in this model but was in the previous study for iron release. HRT also was not used in the iron models in this study because a constant 2-day HRT was maintained in the hybrid PDSs for all phases. It is believed that the iron concentration would exhibit a linear dependence on the HRT for values other than 2 days. The *p*-values for the independent variables that were retained in the model are TP (BOP) – 0.2573, TP (OP) – 0.5332, TP (ZOP) – 0.3468, Zn – <0.0001, SiO_2 – <0.0001, alk – 0.0001, Cl – 0.0002, and Fe_{inf} – 0.0011. The terms associated with total phosphorus were retained in the model in spite of the large *p*-value (>0.05) in order to develop a model with some sensitivity to inhibitor dose. This relaxation of criteria for *p*-values was adopted in other instances to preserve sensitivity to the inhibitor dose.

Figure 5.1 shows the fit of the total iron model predictions with the actual observations for all phases. A line is indicated on the plot that would represent perfect agreement between the model and the observed data. The coefficient of determination (R^2) is not particularly impressive for the total iron model ($R^2 = 0.29$). The low value is indicative of the absence of a monotonic response of the iron data to increases in inhibitor dose, as discussed in detail in subsequent

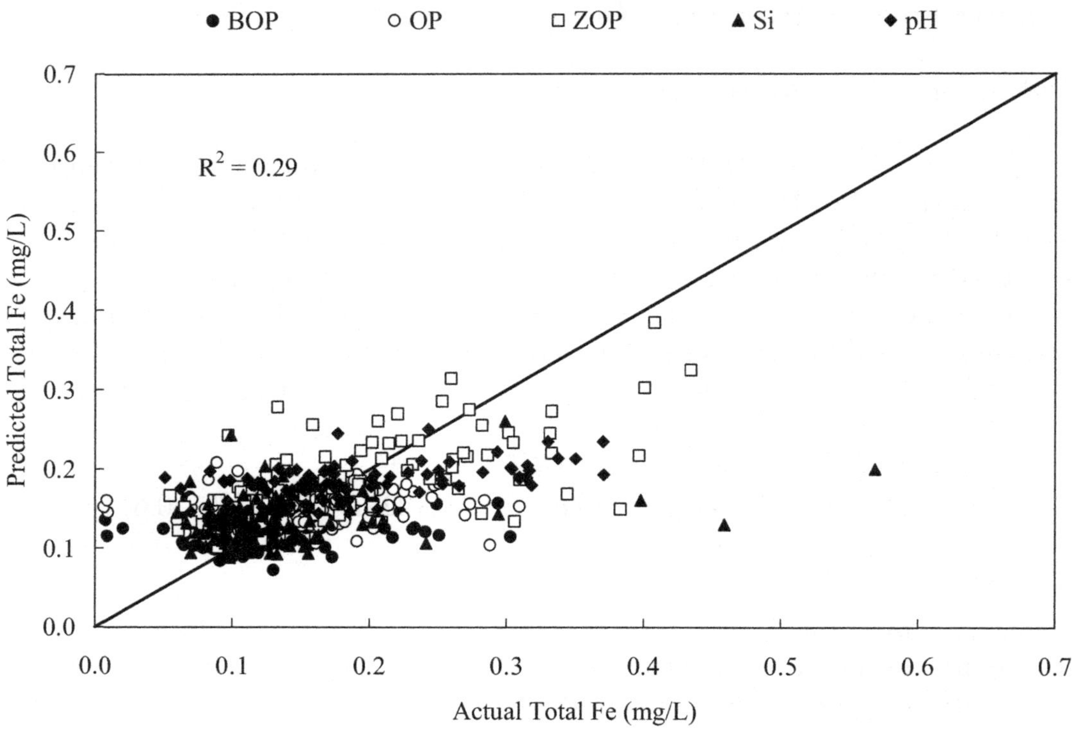

Figure 5.1 Model predicted versus actual total iron concentration

sections of this report. The model tends to underpredict the higher concentrations of iron. Several observations that exceed the secondary iron standard of 0.3 mg/L are not always predicted to violate this standard.

Empirical Model for Dissolved Iron Concentration

In this study, dissolved metal models were also developed using the same form as was used in development of the total metal models. The dissolved iron model is shown in Equation 5.7. Alkalinity and influent iron contribute to dissolved iron concentrations, as shown by the positive exponent on those terms in the model. The model also suggests that increases in chloride concentration and temperature will mitigate dissolved iron concentrations as they increase.

$$\text{diss Fe} = (3.82\text{BOP} \times \text{TP}^{-0.027} + 2.91\text{OP} \times \text{TP}^{0.178} + 7.94\text{ZOP} \times \text{TP}^{-0.760} \times \text{Zn}^{0.313} + 2.38\text{Si} \times \text{SiO}_2^{0.012} + 3.62\text{pH}_s) \times \text{alk}^{2.10} \times \text{Cl}^{-2.76} \times \text{Fe}_{\text{inf}}^{0.245} \times \text{temp}^{-1.36} \quad (5.7)$$

where diss Fe = dissolved iron, mg/L
BOP = blended orthopolyphosphate inhibitor (0, 1)
OP = orthophosphate inhibitor (0, 1)
ZOP = zinc orthophosphate inhibitor (0, 1)
Si = silica inhibitor (0, 1)
pH_s = pH control (0, 1)
TP = total phosphorus, mg/L

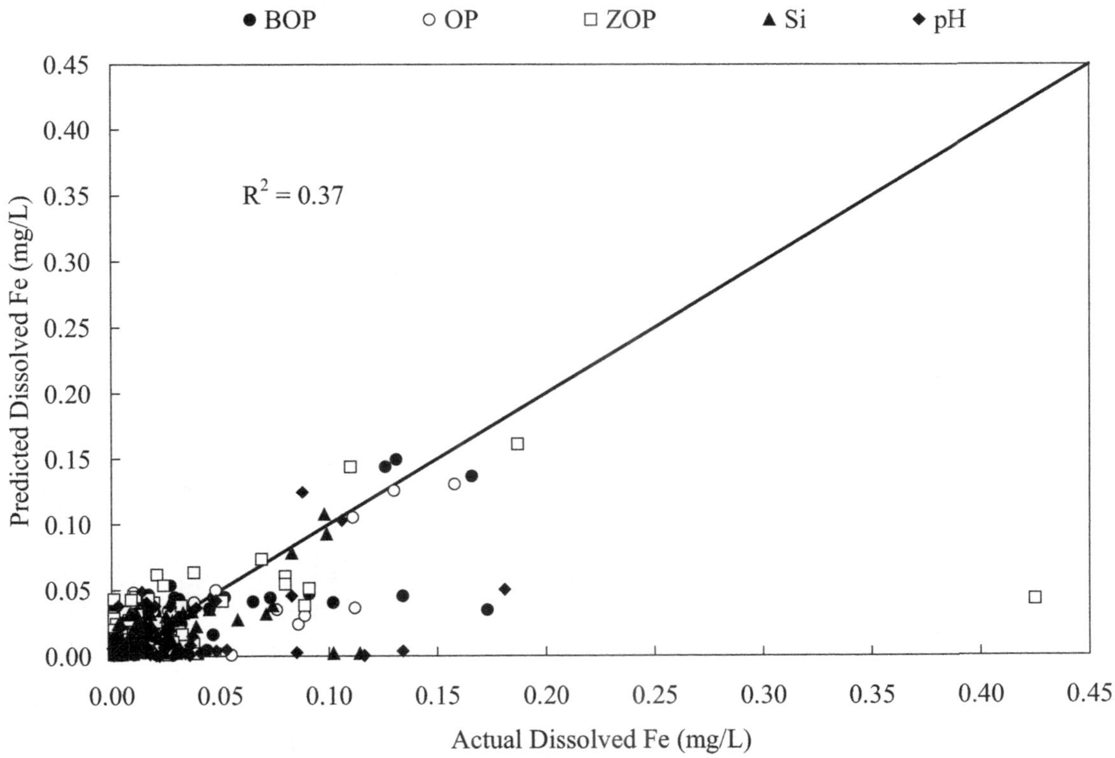

Figure 5.2 Model predicted versus actual dissolved iron concentration

Zn = zinc, mg/L
SiO_2 = silica, mg/L as SiO_2
alk = alkalinity, mg/L as $CaCO_3$
Cl = chloride, mg/L
Fe_{inf} = influent iron, mg/L
temp = temperature, °C

As in the total iron model, pH and sulfate did not significantly contribute to the dissolved iron concentration. The trend for alkalinity and chloride are opposite of what was seen in the total iron model, and temperature is significant in this model. The exponents on the inhibitor-specific terms are comparable to what was seen with the total iron model, but the zinc term is positive in the dissolved iron model. The ZOP coefficient also has a TP term that is larger and negative to account for a higher zinc concentration in those PDSs. The *p*-values for the independent variables that were retained in the model are TP (BOP) – 0.8542, TP (OP) – 0.2809, TP (ZOP) – 0.0253, Zn – 0.0786, SiO_2 – 0.9600, alk – 0.0477, Cl – 0.0003, Fe_{inf} – 0.0062, and temp – <0.0001.

Figure 5.2 shows the fit of the predictions of the dissolved iron model to the actual observations. The straight line represents perfect agreement between the model and the study observations. The dissolved iron model exhibits a slight improvement in fit to the data ($R^2 = 0.37$) when compared to the total iron model, although there are many low concentrations in the dissolved iron, because of the high percentage of the total iron being in particulate form.

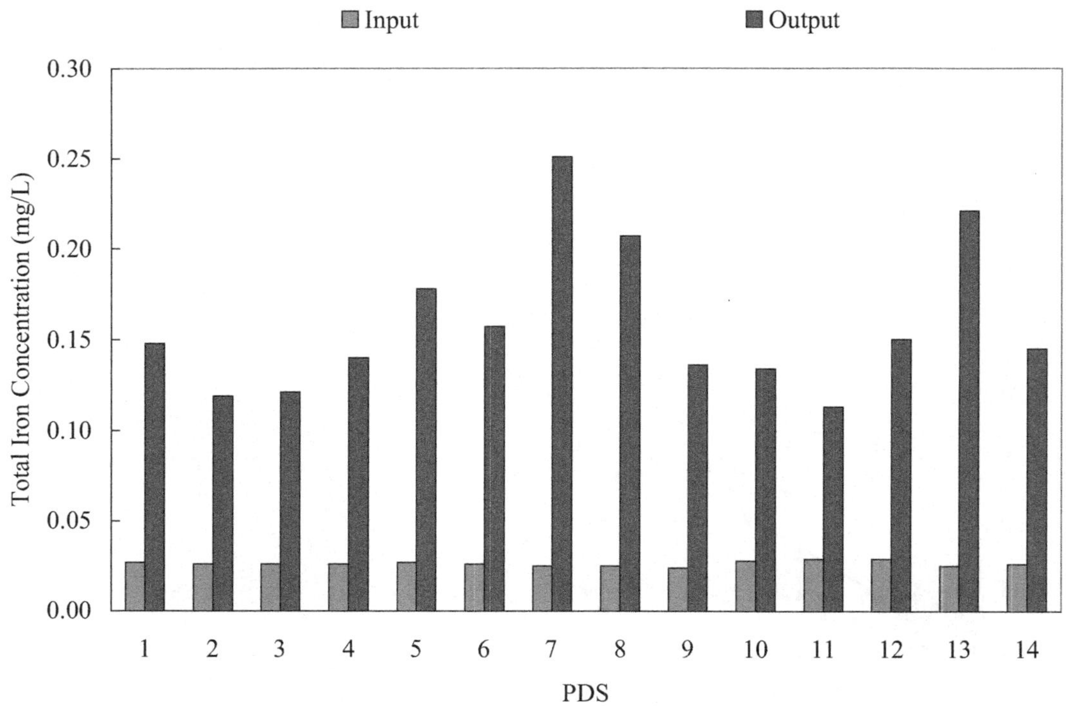

Figure 5.3 Comparison of PDS input and output iron concentration

Comparison of PDS Input and Output Iron Concentration

The empirical model for iron release that was developed in TBW I (Equation 5.1) indicated that the increase in iron concentration was virtually proportional to the HRT in the hybrid PDS. A proportional relationship was also developed by Mutoti et al. (2007a, b) based on a material balance approach that assumed a constant iron mass flux from the pipe interior wall surface for laminar flow conditions. This relationship between iron concentration and HRT indicates that red water problems would be experienced, regardless of water quality, if the HRT in UCI or GS pipe materials were sufficient. The occurrence of adverse water quality (red water) has been associated with stagnant regions in distribution systems. The data collection for TBW I did not employ any type of corrosion inhibitors.

Data from the current project for the PDS influent and corrosion loop iron concentrations were compared to evaluate the ability of the corrosion control strategies (inhibitor or pH elevation) to mitigate release of iron corrosion products from the surface of UCI or GS pipes in the hybrid PDS. The HRT was maintained at 2 days for the duration of the current study.

The input and output concentration data are displayed in Figure 5.3. These iron concentrations represent the average of all determinations for the entire project duration (phases I to IV). The data clearly indicate an increase in iron concentration as a result of the 2-day exposure to the pipe materials for all 14 hybrid PDSs. The annual average influent iron concentrations were slightly less than 0.03 mg/L, while the annual average output concentration varied from 0.11 to 0.25 mg/L, depending on the corrosion control strategy. A statistical comparison of the input and output concentrations was completed using a one-tailed, paired-data t-test. The statistical analysis is summarized in Table 5.3. The *p*-values are provided for the hypothesis tests for equality of

Table 5.3
Statistical comparison of PDS input and output iron concentration

	PDS01	PDS02	PDS03	PDS04	PDS05	PDS06	PDS07	PDS08	PDS09	PDS10	PDS11	PDS12	PDS13	PDS14
Treatment	BOP	BOP	BOP	OP	OP	OP	ZOP	ZOP	ZOP	Silicate	Silicate	Silicate	pH	pH
Dose	Low	Medium	High	Low	Medium	High	Low	Medium	High	Low	Medium	High	pH_s	pH_s+0.3
Phase I	PDS In	PDS In	PDS In	PDS In	PDS In	PDS In	PDS In	PDS In	PDS In	PDS In	PDS In	PDS In	PDS In	PDS In
Average	0.028	0.028	0.028	0.027	0.028	0.028	0.028	0.028	0.025	0.030	0.035	0.033	0.027	0.029
Std. dev.	0.022	0.028	0.031	0.021	0.021	0.022	0.022	0.026	0.020	0.020	0.029	0.027	0.018	0.021
N	13	14	14	14	14	14	14	14	13	14	14	14	14	14
	CorrLoop	CorrLoop	CorrLoop	CorrLoop	CorrLoop	CorrLoop	CorrLoop	CorrLoop	CorrLoop	CorrLoop	CorrLoop	CorrLoop	CorrLoop	CorrLoop
Average	0.141	0.097	0.124	0.123	0.138	0.114	0.179	0.142	0.163	0.140	0.122	0.171	0.151	0.113
Std. dev.	0.067	0.022	0.048	0.042	0.026	0.059	0.034	0.049	0.118	0.098	0.040	0.136	0.031	0.039
N	13	14	14	14	14	14	14	14	13	14	14	14	14	14
Significant	Yes	Yes	Yes	Yes	Yes	Yes	Yes	Yes	Yes	Yes	Yes	Yes	Yes	Yes
p-value	<0.001	<0.001	<0.001	<0.001	<0.001	<0.001	<0.001	<0.001	<0.001	<0.001	<0.001	<0.001	<0.001	<0.001
Phase II	PDS In	PDS In	PDS In	PDS In	PDS In	PDS In	PDS In	PDS In	PDS In	PDS In	PDS In	PDS In	PDS In	PDS In
Average	0.025	0.026	0.026	0.026	0.027	0.026	0.020	0.021	0.019	0.025	0.021	0.021	0.018	0.021
Std. dev.	0.008	0.010	0.008	0.009	0.010	0.011	0.005	0.004	0.006	0.007	0.003	0.002	0.004	0.004
N	7	7	7	7	7	7	7	7	7	7	7	7	10	10
	CorrLoop	CorrLoop	CorrLoop	CorrLoop	CorrLoop	CorrLoop	CorrLoop	CorrLoop	CorrLoop	CorrLoop	CorrLoop	CorrLoop	CorrLoop	CorrLoop
Average	0.207	0.149	0.147	0.159	0.210	0.174	0.340	0.230	0.164	0.166	0.134	0.191	0.277	0.188
Std. dev.	0.117	0.050	0.053	0.037	0.057	0.040	0.128	0.056	0.044	0.042	0.024	0.068	0.065	0.068
N	7	7	7	7	7	7	7	7	7	7	7	7	10	10
Significant	Yes	Yes	Yes	Yes	Yes	Yes	Yes	Yes	Yes	Yes	Yes	Yes	Yes	Yes
p-value	<0.001	<0.001	<0.001	<0.001	<0.001	<0.001	<0.001	<0.001	<0.001	<0.001	<0.001	0.001	<0.001	<0.001

(continued)

Table 5.3 (Continued)

	PDS01	PDS02	PDS03	PDS04	PDS05	PDS06	PDS07	PDS08	PDS09	PDS10	PDS11	PDS12	PDS13	PDS14
Treatment	BOP	BOP	BOP	OP	OP	OP	ZOP	ZOP	ZOP	Silicate	Silicate	Silicate	pH	pH
Dose	Low	Medium	High	Low	Medium	High	Low	Medium	High	Low	Medium	High	pH_s	pH_s+0.3
Phase III	PDS In	PDS In	PDS In	PDS In	PDS In	PDS In	PDS In	PDS In	PDS In	PDS In	PDS In	PDS In	PDS In	PDS In
Average	0.029	0.028	0.027	0.029	0.030	0.027	0.030	0.029	0.028	0.028	0.029	0.030	0.029	0.030
Std. dev.	0.005	0.007	0.006	0.006	0.007	0.006	0.004	0.005	0.003	0.003	0.004	0.004	0.006	0.006
N	7	7	7	7	7	7	6	6	6	6	6	6	13	13
	CorrLoop	CorrLoop	CorrLoop	CorrLoop	CorrLoop	CorrLoop	CorrLoop	CorrLoop	CorrLoop	CorrLoop	CorrLoop	CorrLoop	CorrLoop	CorrLoop
Average	0.125	0.105	0.123	0.144	0.230	0.144	0.215	0.222	0.107	0.115	0.097	0.114	0.268	0.140
Std. dev.	0.033	0.019	0.040	0.032	0.043	0.026	0.057	0.048	0.015	0.018	0.011	0.014	0.058	0.026
N	7	7	7	7	7	7	6	6	6	6	6	6	13	13
Significant	yes	yes	yes	yes	yes	yes	yes	yes	yes	yes	yes	yes	yes	yes
p-value	<0.001	<0.001	<0.001	<0.001	<0.001	<0.001	0.001	<0.001	<0.001	<0.001	<0.001	<0.001	<0.001	<0.001
Phase IV	PDS In	PDS In	PDS In	PDS In	PDS In	PDS In	PDS In	PDS In	PDS In	PDS In	PDS In	PDS In	PDS In	PDS In
Average	0.021	0.020	0.020	0.020	0.020	0.020	0.022	0.022	0.021	0.023	0.024	0.024	0.022	0.022
Std. dev.	0.008	0.006	0.005	0.006	0.006	0.006	0.009	0.008	0.008	0.007	0.007	0.008	0.007	0.007
N	6	6	6	6	6	6	6	6	6	6	6	6	12	12
	CorrLoop	CorrLoop	CorrLoop	CorrLoop	CorrLoop	CorrLoop	CorrLoop	CorrLoop	CorrLoop	CorrLoop	CorrLoop	CorrLoop	CorrLoop	CorrLoop
Average	0.117	0.126	0.089	0.135	0.134	0.202	0.278	0.239	0.107	0.114	0.095	0.118	0.192	0.141
Std. dev.	0.053	0.054	0.039	0.057	0.057	0.126	0.079	0.126	0.049	0.030	0.025	0.043	0.087	0.047
N	6	6	6	6	6	6	6	6	6	6	6	6	12	12
Significant	yes	yes	yes	yes	yes	yes	yes	yes	yes	yes	yes	yes	yes	yes
p-value	0.009	0.005	0.007	0.006	0.005	0.008	<0.001	0.001	0.002	<0.001	<0.001	0.002	<0.001	<0.001

(continued)

Table 5.3 (Continued)

	PDS01	PDS02	PDS03	PDS04	PDS05	PDS06	PDS07	PDS08	PDS09	PDS10	PDS11	PDS12	PDS13	PDS14
Treatment	BOP	BOP	BOP	OP	OP	OP	ZOP	ZOP	ZOP	Silicate	Silicate	Silicate	pH	pH
Dose	Low	Medium	High	Low	Medium	High	Low	Medium	High	Low	Medium	High	pH_s	pH_s+0.3
Composite	PDS In	PDS In	PDS In	PDS In	PDS In	PDS In	PDS In	PDS In	PDS In	PDS In	PDS In	PDS In	PDS In	PDS In
Average	0.027	0.026	0.026	0.026	0.027	0.026	0.025	0.025	0.024	0.028	0.029	0.029	0.025	0.026
Std. dev.	0.015	0.019	0.020	0.015	0.015	0.015	0.015	0.017	0.014	0.014	0.020	0.018	0.011	0.012
N	33	34	34	34	34	34	33	33	32	33	33	33	49	49
	CorrLoop	CorrLoop	CorrLoop	CorrLoop	CorrLoop	CorrLoop	CorrLoop	CorrLoop	CorrLoop	CorrLoop	CorrLoop	CorrLoop	CorrLoop	CorrLoop
Average	0.148	0.119	0.121	0.140	0.178	0.157	0.251	0.207	0.136	0.134	0.113	0.150	0.221	0.145
Std. dev.	0.081	0.043	0.049	0.044	0.063	0.077	0.101	0.083	0.073	0.060	0.031	0.086	0.081	0.054
N	33	34	34	34	34	34	33	33	32	33	33	33	49	49
Significant	Yes	Yes	Yes	Yes	Yes	Yes	Yes	Yes	Yes	Yes	Yes	Yes	Yes	Yes
p-value	<0.001	<0.001	<0.001	<0.001	<0.001	<0.001	<0.001	<0.001	<0.001	<0.001	<0.001	<0.001	<0.001	<0.001

input and output mean iron concentrations. A reported p-value of 0.05 or less indicates there is sufficient evidence to conclude that the output iron concentration is greater than the input iron concentration (significance level of 95%). The analysis confirmed that the increase in iron concentration was statistically significant for all corrosion control strategies for all four study phases. These results confirm the historical conclusions from TBW I which suggest that red water episodes would be expected in stagnant regions of the distribution system regardless of water quality or corrosion control strategy. Management of the HRT in the distribution system by periodic flushing may be required to avoid excessive release of iron corrosion products and associated increase in color from UCI or GS pipe materials.

Control of Iron Release With BOP Inhibitor Addition

Polyphosphate Reversion Rate Characterization. Addition of the BOP inhibitor introduces a mixture of poly-phosphorus (PP) and ortho-phosphorus species. Reversion reactions that transform the PP species into ortho-phosphorus species are expected during residence in the distribution system. The empirical model development did not demonstrate a significant dependence of this speciation on the inhibitor performance; consequently, all of the final models rely on the TP concentration for prediction of BOP performance. Nevertheless, data were obtained to permit a limited characterization of the reversion rate. For those applications where there is interest in the persistence of PP forms in a distribution system, this analysis may have some relevance.

Phosphorus Speciation Data. Data were collected for total phosphorus (TP) and ortho-phosphorus concentration for the BOP hybrid PDSs (PDS01 to PDS03) and for the single-material PDSs to support a characterization of the reversion rate of PP to ortho-phosphorus. Samples were collected from intermediate ports in the hybrid PDSs to permit a rate calculation that was specific to the pipe material. Samples were also obtained from intermediate ports in the single-material PDSs to support estimation of reaction order and rate constants. The relevant TP and ortho-phosphorus data are presented in Tables 5.4, 5.5, and 5.6 for PDS01, PDS02, and PDS03, respectively. Data for the single-material PDSs are provided in Table 5.7.

Reaction-Order Models. Various rate expressions were evaluated for the kinetics of the reversion of PP to ortho-phophorus, including zero-order, irreversible first-order, reversible first-order, second-order, and first-order with wall reaction. A first-order kinetic model produced reasonable agreement with the data for the single-material PDSs, but evaluation of the data from the hybrid PDSs indicated apparent differences in the first-order rate constant that were associated with different pipe materials. Therefore, the first-order model with wall reaction was selected for characterization of the reversion kinetics. The appropriate kinetics are summarized in Equation 5.8.

$$\frac{dPP}{dt} = -kPP = -k(TP - OP) \tag{5.8}$$

where PP = poly-phosphorus concentration, mg/L P
TP = total phosphorus concentration, mg/L P
OP = orthophosphate concentration, mg/L P
k = first-order reversion rate constant, $time^{-1}$

Table 5.4
Phosphorus speciation data for PDS01

	TP data (mg/L as P)					OP data (mg/L as P)					
Date	Influent	Port 4	Port 5	Port 6	Effluent	Influent	Port 4	Port 5	Port 6	Effluent	Temp. (°C)
04/09/2006	0.798	0.600	0.577	0.593	0.620	0.400	0.448	0.476	0.469	0.492	25.1
04/16/2006	0.633	0.762	0.792	0.728	0.611	0.386	0.398	0.495	0.508	0.529	25.0
04/23/2006	0.494	0.460	0.511	0.513	0.899	0.341	0.374	0.412	0.423	0.446	25.6
05/02/2006	0.813	0.625	0.505	0.475	0.852	0.272	0.358	0.325	0.327	0.361	24.8
05/23/2006	0.028	0.162	0.282	0.395	0.388	0.024	0.108	0.204	0.318	0.379	25.1
05/30/2006	0.666	0.246	0.515	0.518	0.433	0.398	0.203	0.426	0.428	0.391	24.9
06/06/2006	0.750	0.830	0.818	0.836	0.616	0.443	0.561	0.578	0.552	0.519	25.4
06/15/2006	0.661	0.801	0.786	0.776	0.737	0.461	0.850	0.865	0.863	0.632	23.9
06/20/2006	0.739	0.695	0.687	0.704	0.708	0.482	0.512	0.494	0.481	0.531	27.8
06/27/2006	0.650	0.698	0.720	0.781	0.720	0.473	0.483	0.493	0.495	0.516	28.5
07/03/2006	0.627	0.628	0.647	0.631	0.616	0.442	0.467	0.480	0.469	0.506	27.2
07/11/2006	0.480	0.542	0.569	0.555	0.621	0.322	0.399	0.437	0.532	0.465	28.5
07/25/2006	0.678	0.554	0.571	0.560	0.540	0.405	0.423	0.435	0.407	0.432	28.2
08/01/2006	0.360	0.460	0.397	0.382	0.403	0.279	0.357	0.384	0.408	0.377	28.5
08/08/2006	0.588	0.323	0.300	0.274	0.315	0.370	0.249	0.230	0.206	0.262	25.4
08/21/2006	0.450	0.477	0.453	0.452	0.467	0.323	0.394	0.393	0.393	0.424	27.8
08/28/2006	0.374	0.427	0.394	0.414	0.464	0.278	0.339	0.349	0.377	0.437	27.6
09/04/2006	0.518	0.442	0.455	0.476	0.418	0.390	0.364	0.383	0.389	0.402	24.9
09/11/2006	0.510	0.354	0.373	0.376	0.386	0.266	0.341	0.320	0.294	0.359	28.0
09/18/2006	0.506	0.378	0.430	0.457	0.413	0.395	0.312	0.352	0.428	0.365	28.3
09/25/2006	0.488	0.406	0.418	0.406	0.389	0.372	0.380	0.358	0.344	0.405	28.0
10/02/2006	0.548	0.566	0.543	0.506	0.465	0.379	0.480	0.466	0.463	0.420	24.8
10/09/2006	0.425	0.453	0.443	0.416	0.420	0.349	0.412	0.415	0.401	0.394	25.0
10/16/2006	0.411	0.466	0.442	0.427	0.425	0.297	0.352	0.348	0.359	0.376	26.1
10/23/2006	0.529	0.434	0.434	0.439	0.410	0.442	0.392	0.382	0.396	0.381	24.2
10/29/2006	0.488	0.498	0.477	0.451	0.482	0.386	0.392	0.404	0.411	0.421	24.5
11/05/2006	0.646	0.494	0.451	0.445	0.498	0.480	0.444	0.439	0.444	0.457	22.9
11/13/2006	0.522	0.583	0.534	0.507	0.519	0.406	0.443	0.461	0.484	0.456	20.6
11/27/2006	0.534	0.526	0.484	0.467	0.468	0.408	0.418	0.438	0.435	0.438	21.0
12/04/2006	0.565	0.651	0.655	0.615	0.515	0.289	0.351	0.385	0.425	0.431	24.2
12/11/2006	0.751	0.752	0.720	0.697	0.674	0.393	0.434	0.488	0.500	0.527	20.9
12/17/2006	0.847	0.768	0.751	0.703	0.731	0.400	0.418	0.501	0.507	0.534	23.1
12/23/2006	0.745	0.814	0.830	0.726	0.704	0.348	0.478	0.535	0.537	0.565	23.2
12/30/2006	0.761	0.695	0.725	0.696	0.643	0.383	0.366	0.407	0.464	0.453	23.8
01/08/2007	0.597	0.650	0.672	0.644	0.594	0.334	0.387	0.446	0.467	0.523	23.8
01/14/2007	0.912	0.639	0.662	0.636	0.768	0.361	0.374	0.460	0.472	0.500	22.4
01/22/2007	0.723	0.917	0.861	0.767	0.704	0.279	0.351	0.416	0.446	0.537	22.1
01/28/2007	0.631	0.684	0.717	0.703	0.644	0.289	0.360	0.403	0.431	0.482	16.4
02/05/2007	0.609	0.512	0.508	0.508	0.473	0.269	0.267	0.319	0.340	0.413	17.1
02/12/2007	0.452	0.492	0.528	0.519	0.466	0.205	0.280	0.320	0.350	0.362	17.5

Table 5.5
Phosphorus speciation data for PDS02

	TP data (mg/L as P)					OP data (mg/L as P)					
Date	Influent	Port 4	Port 5	Port 6	Effluent	Influent	Port 4	Port 5	Port 6	Effluent	Temp. (°C)
04/09/2006	1.386	1.245	1.237	1.270	1.180	0.716	0.857	0.911	0.900	1.037	25.2
04/16/2006	1.163	1.406	1.466	1.390	1.141	0.601	0.790	0.957	0.969	1.040	24.8
04/23/2006	0.878	0.846	0.873	0.876	1.371	0.652	2.273	0.707	0.705	0.706	25.6
05/02/2006	1.760	1.241	1.012	0.963	1.689	0.612	0.668	0.712	0.711	0.757	24.7
05/23/2006	1.070	1.240	1.213	1.105	0.941	0.656	0.741	0.865	0.827	0.888	25.1
05/30/2006	0.980	0.894	0.923	0.899	0.799	0.606	0.659	0.782	0.799	0.775	25.2
06/06/2006	0.941	1.074	1.044	1.007	0.861	0.641	0.750	0.767	0.788	0.842	25.4
06/15/2006	1.075	0.948	0.936	0.952	0.951	0.599	0.850	0.865	0.863	0.901	23.8
06/20/2006	0.922	0.906	1.008	1.054	0.999	0.603	0.766	0.738	0.689	0.764	27.8
06/27/2006	0.760	0.884	0.871	0.940	0.892	0.508	1.826	0.733	0.746	0.661	28.6
07/03/2006	0.950	0.952	0.916	0.935	0.932	0.648	0.717	0.640	0.664	0.734	27.2
07/11/2006	0.797	0.888	0.925	0.936	1.010	0.553	0.678	0.715	0.719	0.771	28.9
07/25/2006	1.197	1.017	1.019	0.992	1.030	0.801	0.796	0.761	0.741	0.865	28.7
08/01/2006	0.838	0.674	0.819	0.781	0.850	0.560	0.668	0.775	0.782	0.736	28.5
08/08/2006	1.111	0.973	0.887	0.858	0.903	0.713	0.742	0.669	0.625	0.627	25.6
08/21/2006	0.791	0.941	0.935	0.933	0.959	0.682	0.771	0.826	0.785	0.9024	27.5
08/28/2006	0.897	0.871	0.875	0.906	0.876	0.603	0.714	0.745	0.741	0.856	27.6
09/04/2006	0.899	0.924	0.953	1.017	0.916	0.643	0.757	0.764	0.766	0.865	26.1
09/11/2006	0.960	0.779	0.775	0.828	0.781	0.504	0.682	0.693	0.677	0.696	28.0
09/18/2006	0.850	0.744	0.811	0.808	0.827	0.645	0.671	0.722	0.754	0.798	28.2
09/25/2006	0.720	0.684	0.695	0.712	0.701	0.565	0.633	0.673	0.680	0.688	28.0
10/02/2006	0.720	0.777	0.807	0.795	0.751	0.595	0.676	0.732	0.732	0.814	25.0
10/09/2006	0.859	0.738	0.745	0.722	0.737	0.684	0.677	0.700	0.713	0.731	25.0
10/16/2006	0.835	0.899	0.846	0.822	0.826	0.651	0.736	0.709	0.726	0.771	26.1
10/23/2006	0.864	0.812	0.827	0.830	0.795	0.775	0.373	0.363	0.359	0.714	24.2
10/29/2006	0.972	0.967	0.926	0.899	0.931	0.781	0.780	0.812	0.806	0.856	24.7
11/05/2006	1.215	1.008	0.900	0.891	0.984	0.886	0.908	0.880	0.888	0.916	22.9
11/13/2006	1.072	1.172	1.060	1.041	1.032	0.802	0.980	0.957	1.041	0.947	20.8
11/27/2006	1.073	1.047	0.996	0.935	0.938	0.832	0.878	0.924	0.901	0.923	21.0
12/04/2006	1.244	1.219	1.171	1.123	0.974	0.633	0.853	0.916	0.912	0.878	24.2
12/11/2006	1.303	1.280	1.237	1.144	1.105	0.670	0.708	0.817	0.871	0.939	20.9
12/17/2006	1.471	1.363	1.281	1.224	1.227	0.671	0.737	0.862	0.889	0.982	23.1
12/23/2006	1.316	1.370	1.380	1.258	1.188	0.704	0.709	0.887	0.917	0.981	23.2
12/30/2006	1.550	1.570	1.424	1.418	1.151	0.796	0.802	0.893	0.893	0.933	23.8
01/08/2007	1.448	1.479	1.489	1.415	1.251	0.910	0.841	0.968	1.022	1.139	23.7
01/14/2007	1.274	1.257	1.356	1.256	1.375	0.731	0.687	0.894	0.964	1.007	22.4
01/22/2007	1.611	1.482	1.402	1.307	1.135	0.565	0.773	0.865	0.933	0.950	22.2
01/28/2007	1.097	1.150	1.251	1.238	1.139	0.529	0.600	0.780	0.740	0.925	16.4
02/05/2007	1.316	0.999	1.053	1.053	1.047	0.593	0.541	0.702	0.725	1.488	16.8
02/12/2007	1.029	1.077	1.130	1.081	1.048	0.467	0.560	0.700	0.800	0.952	17.5

Table 5.6
Phosphorus speciation data for PDS03

Date	TP data (mg/L as P)					OP data (mg/L as P)					
	Influent	Port 4	Port 5	Port 6	Effluent	Influent	Port 4	Port 5	Port 6	Effluent	Temp. (°C)
04/09/2006	2.189	2.244	2.246	2.304	1.897	1.213	1.280	1.409	1.519	1.629	25.4
04/16/2006	1.871	2.498	2.633	2.868	2.119	1.206	1.172	1.369	1.669	1.901	25.3
04/23/2006	1.626	1.681	1.778	1.803	2.510	1.207	1.263	1.328	1.334	1.335	25.0
05/02/2006	2.899	2.323	1.987	1.621	2.776	1.066	1.222	1.184	1.177	1.358	25.2
05/23/2006	1.940	2.133	2.031	1.958	0.446	1.179	1.242	1.311	1.409	0.501	25.1
05/30/2006	1.882	1.135	1.335	1.522	1.358	1.043	0.853	1.031	1.341	1.268	25.2
06/06/2006	1.888	2.079	2.137	2.052	1.682	1.338	1.368	1.379	1.325	2.299	25.4
06/15/2006	2.316	1.950	1.984	1.956	1.826	1.553	1.748	1.729	1.569	1.656	23.9
06/20/2006	2.010	3.364	2.402	1.993	2.040	1.316	2.168	1.660	1.401	1.506	27.8
06/27/2006	1.790	1.742	1.821	1.986	1.820	1.178	1.182	1.290	1.339	1.389	28.5
07/03/2006	1.671	1.807	1.840	1.768	1.754	1.065	1.217	1.252	1.344	1.393	27.2
07/11/2006	1.582	1.711	1.768	1.707	1.762	1.124	1.149	1.358	1.243	1.344	29.1
07/25/2006	2.183	1.828	1.838	1.717	1.942	1.228	1.299	1.255	1.279	1.457	28.6
08/01/2006	1.879	1.626	1.595	1.557	1.667	1.084	1.227	1.428	1.317	1.392	28.5
08/08/2006	2.122	1.864	1.752	1.722	1.786	1.181	1.251	1.277	1.353	1.367	23.5
08/21/2006	1.475	1.690	1.696	1.682	1.760	1.018	1.421	1.400	1.455	1.5544	27.3
08/28/2006	1.869	1.583	1.566	1.625	1.648	1.180	1.266	1.302	1.378	1.338	27.6
09/04/2006	1.638	1.654	1.775	1.837	1.581	1.200	1.295	1.290	1.358	1.413	25.7
09/11/2006	1.678	1.569	1.579	1.601	1.414	1.192	1.301	1.263	1.207	1.320	28.0
09/18/2006	1.523	1.506	1.581	1.650	1.622	1.112	1.361	1.391	1.432	1.572	29.0
09/25/2006	1.579	1.479	1.529	1.463	1.501	1.288	1.247	1.219	1.256	1.321	28.0
10/02/2006	1.394	1.505	1.517	1.541	1.508	1.145	1.198	1.306	1.377	1.584	25.5
10/09/2006	1.435	1.405	1.406	1.366	1.370	1.255	1.230	1.230	1.253	1.248	25.0
10/16/2006	1.631	1.670	1.639	1.603	1.601	1.287	1.236	1.236	1.269	1.389	26.1
10/23/2006	1.773	1.586	1.600	1.600	1.497	1.557	1.377	1.511	1.386	1.342	24.2
10/29/2006	2.032	2.046	1.989	1.931	1.931	1.586	1.584	1.608	1.595	1.681	24.7
11/05/2006	2.318	1.928	1.900	1.882	2.025	1.703	1.717	1.721	1.743	1.801	22.9
11/13/2006	2.136	2.362	2.206	2.188	2.036	1.627	1.925	1.951	2.058	1.800	20.7
11/27/2006	2.029	2.050	2.004	1.969	1.948	1.554	1.671	1.868	1.806	1.889	21.2
12/04/2006	2.378	2.515	2.582	2.466	2.174	1.249	1.498	1.728	1.728	1.931	24.2
12/11/2006	2.788	2.662	2.510	2.477	2.485	1.405	1.466	1.501	1.701	1.903	21.0
12/17/2006	3.120	2.759	2.757	2.626	2.661	1.254	1.393	1.639	1.770	1.845	23.1
12/23/2006	3.363	2.745	2.877	2.724	2.659	1.550	1.437	1.694	1.844	1.993	23.0
12/30/2006	2.775	2.956	3.149	3.071	2.753	1.453	1.347	1.260	1.799	1.791	23.8
01/08/2007	2.894	2.942	2.926	2.809	2.558	1.569	1.567	1.818	1.835	2.135	23.5
01/14/2007	2.742	2.543	2.668	2.559	2.579	1.378	1.360	1.532	1.780	1.970	22.4
01/22/2007	3.087	3.120	2.979	2.791	2.623	1.235	1.444	1.733	1.738	1.969	22.2
01/28/2007	1.990	1.997	2.256	2.193	2.032	0.907	0.960	1.320	1.320	1.560	16.4
02/05/2007	2.591	2.232	2.019	2.019	1.993	1.123	1.077	1.072	1.242	1.454	16.8
02/12/2007	2.188	1.801	1.969	2.020	2.037	1.038	0.960	1.080	1.320	1.433	17.5

Table 5.7
Phosphorus speciation data for single-material PDSs

Date	Pipe material	Sample port	TP (mg/L as P)	OP (mg/L as P)	Temp. (°C)
01/17/2007	UCI	Feed	1.31	0.612	19.6
		Port 5			19.9
		Port 6	1.27	0.644	19.8
		Port 8		0.659	19.7
		Effluent	1.19	0.646	20.0
	GS	Feed	1.05	0.534	19.8
		Port 5	1.04	0.611	20.7
		Port 6			20.4
		Port 8	1.04	0.702	20.8
		Port A		0.713	20.8
		Effluent	1.00	0.730	20.5
01/19/2007	PVC	Feed	1.40	0.66	18.6
		Port 5		0.67	18.0
		Effluent	1.36	0.67	17.3
01/24/2007	LCI	Feed	1.51	0.715	17.9
		Port 5	1.50	0.721	16.7
		Port 7	1.48	0.746	16.3
		Effluent	1.42	0.751	16.5
01/26/2007	GS	Feed	1.44	0.690	12.9
		Port 4	1.41	0.720	12.9
		Port 5	1.41	0.740	13.0
		Port 6	1.39	0.750	12.9
		Port 8	1.39	0.780	12.9
		Port A	1.35	0.790	12.9
		Effluent	1.34	0.800	12.2
02/01/2007	UCI	Feed	1.18	0.538	17.5
		Port 4	1.14	0.548	17.5
		Port 5	1.09	0.570	17.2
		Port 6	1.11	0.581	17.1
		Port 8	1.04	0.569	16.9
		Effluent	1.12	0.574	17.0
02/02/2007	LCI	Feed	1.16	0.576	19.8
		Port 4	1.12	0.626	20.2
		Port 6	1.15	0.588	20.9
		Effluent	1.15	0.626	20.9
02/07/2007	GS	Feed	1.23	0.579	14.0
		Port 4	1.25	0.595	14.4
		Port 5	1.21	0.620	15.2
		Port 6	1.20	0.640	16.7
		Port 8	1.12	0.688	17.6
		Port A	1.11	0.725	18.1
		Effluent	1.12	0.746	18.6

(continued)

Table 5.7 (Continued)

Date	Pipe material	Sample port	TP (mg/L as P)	OP (mg/L as P)	Temp. (°C)
02/09/2007	UCI	Feed	1.34	0.591	15.8
		Port 4	1.29	0.618	16.5
		Port 5	1.27	0.622	16.8
		Port 7	1.22	0.622	16.8
		Port 9	1.22	0.623	16.7
		Effluent	1.19	0.628	16.5
02/14/2007	PVC	Feed	1.28	0.611	19.2
		Port 5	1.29	0.610	19.9
		Effluent	1.27	0.613	19.5
02/16/2007	LCI	Feed	1.15	0.520	13.0
		Port 5	1.14	0.530	12.1
		Effluent	1.13	0.558	11.6
	PVC	Feed	1.04	0.521	12.7
		Effluent	1.07	0.521	12.2

The first-order overall rate constant (k) may be resolved into a bulk reaction and a wall reaction component as described in Equation 5.9:

$$k = k_b + \frac{4}{D}k_x \qquad (5.9)$$

where k_b = bulk polyphosphate reversion rate constant, time^{-1}
D = pipe diameter, L
k_x = wall reaction polyphosphate reversion rate constant, L/time

Construction of the PDS utilized the same pipe diameter for each pipe material; thus, data were not available for multiple diameters of an individual pipe material. It was not possible to estimate individual values for the bulk and wall rate constants in Equation 5.9. Estimates of the overall rate constant (k) were obtained, and it is recognized that these values would depend on the pipe material in order to account for differing wall reaction rate constants. The reported values for k would be specific for the pipe material, diameter, and laminar flow conditions employed in this study.

Integration of Equation 5.8 yields a time-domain equation that can be readily used for determining the rate constant. Equation 5.10 was used for nonlinear least-squares regression for data from the single-material lines to estimate the first-order rate constant (k). Equation 5.10 was also used for calculation of individual values of the first-order rate constant (k) between successive sample ports in the hybrid lines. The latter approach provides estimates of the rate constant that are associated with individual pipe materials.

$$OP = OP_0 + (TP - OP_0)(1 - \exp(-kt)) \qquad (5.10)$$

where OP_0 = initial orthophosphate concentration, mg/L P

Table 5.8
Average first-order rate constant determinations for single-material PDSs (reversion of poly- to orthophosphate)

Pipe material	Rate constant k (hr^{-1})	Normalized rate constant k_{20} (hr^{-1})
Polyvinyl chloride	0.00065	0.00086
Lined cast iron	0.00807	0.01659
Unlined cast iron	0.00877	0.01339
Galvanized steel	0.07453	0.13306

Table 5.9
Average first-order rate constant determinations for hybrid PDS01 (reversion of poly- to orthophosphate)

Pipe material	Rate constant k (hr^{-1})	Normalized rate constant k_{20} (hr^{-1})
Polyvinyl chloride	0.01797	0.00762
Lined cast iron	0.00543	0.00391
Unlined cast iron	0.01631	0.00683
Galvanized steel	0.04834	0.01998

Pipe Material Effects. Average values for the first-order rate constants (and values normalized to 20°C using Equation 5.12) from the single-material PDSs are reported in Table 5.8. The parameter estimates were developed by nonlinear least-squares methods using Equation 5.10.

The rate constants appear to have substantially different values, depending on the pipe material. This discrepancy may be explained by a difference in the wall reaction term in Equation 5.9. These results suggest that the persistence of the PP species may differ substantially depending on the materials of construction in the distribution system. The galvanized pipe exhibits the greatest value of the reversion rate constant, whereas the PVC pipe appears to be less reactive. The galvanized pipe employed in this study had a smaller diameter than the other pipe materials. Some portion of this greater observed rate constant may be properly explained by the difference in diameter per Equation 5.9.

First-order rate constant values for the individual pipe materials in the hybrid lines were calculated from an algebraic manipulation of Equation 5.10 using data from successive sample ports and the associated residence time in that reach of pipe. The average first-order rate constants (and values normalized to 20°C) for PDS01, PDS02, and PDS03 are reported in Tables 5.9, 5.10, and 5.11, respectively.

The results from the hybrid PDSs substantially agree with the results obtained from the single-material lines, with the exception of the PVC material. The data from the hybrid lines confirms that the reversion rate constant is greatest for exposure to galvanized pipe materials, but the hybrid data indicate that the rate constant associated with PVC is comparable to the constants determined for either LCI or UCI.

Table 5.10
Average first-order rate constant determinations for hybrid PDS02 (reversion of poly- to orthophosphate)

Pipe material	Rate constant k (hr^{-1})	Normalized rate constant k_{20} (hr^{-1})
Polyvinyl chloride	0.01750	0.00737
Lined cast iron	0.01491	0.00823
Unlined cast iron	0.00817	0.00603
Galvanized steel	0.08596	0.03916

Table 5.11
Average first-order rate constant determinations for hybrid PDS03 (reversion of poly- to orthophosphate)

Pipe material	Rate constant k (hr^{-1})	Normalized rate constant k_{20} (hr^{-1})
Polyvinyl chloride	0.00995	0.00537
Lined cast iron	0.01061	0.00658
Unlined cast iron	0.00755	0.00882
Galvanized steel	0.03517	0.01168

Water Quality Effects on Polyphosphate Reversion. The data were examined to identify any potential dependence of the first-order rate constant on water quality parameters. A power law relation was selected per Equation 5.11:

$$k = \alpha(WQ)^{\beta} \tag{5.11}$$

where α, β = parameter estimates
WQ = water quality parameter (e.g., pH)

Data were segregated for each of the hybrid PDSs for analysis. ANOVA methods were applied to identify water quality variables that exerted a significant effect on the rate constant. The results failed to identify any water quality parameters that were significant. This lack of significance may result from the relatively small variation in water quality for the blends used in the current study. It is possible that with a more robust data set, it may have been possible to identify cases where the water quality significantly influenced the reversion rate constant.

Dependence of Reversion Rate Constant on Temperature. Based on a venerable understanding of chemical reaction kinetics, it would be expected that an increase in the reversion rate constant would accompany an increase in temperature. The hybrid data were examined for a relation between the first-order reversion rate constant and temperature, assuming a dependence as specified in Equation 5.12. There are two parameters in Equation 5.12 that must be estimated from the data: k_{20} and θ.

Table 5.12
Estimation of temperature correction factor (θ) for hybrid PDS data (reversion of poly- to orthophosphate)

	PDS01	PDS02	PDS03
k_{20} (hr^{-1})	0.00523	0.00362	0.00305
θ	1.12	1.30	1.08

$$k_T = k_{20} \times \theta^{(T-20)} \tag{5.12}$$

where k_T = reversion rate constant at a temperature = T, hr^{-1}
k_{20} = rate constant at a reference temperature of 20°C, hr^{-1}
T = temperature, °C
θ = temperature correction factor, dimensionless

ANOVA methods were used to determine the significance of variables in the model. It was determined that there is a significant effect of temperature on the reversion rate constant ($p < 0.001$). The parameter estimates for the constants in Equation 5.12 are summarized in Table 5.12. The average value of the temperature correction factor estimate is 1.17. This value of θ suggests that the rate of polyphosphate reversion is very sensitive to changes in temperature. An increase in temperature from 20°–25°C would result in more than a doubling of the first-order reaction rate constant.

BOP Data Summary (PDS01 to PDS03)—Iron. Summary data for the corrosion loop total and dissolved iron concentrations for PDS01–PDS03 that receive the BOP inhibitor are provided in Table 5.13. This data is segregated by phase and by PDS. The table presents the average, maximum, and minimum values observed for iron concentration for every PDS for every phase. Data shows that, on average, the majority of the iron is present in the particulate form. In addition, the results suggest that control of total iron release improves with BOP doses higher than 0.5 mg/L TP. Water quality data for the independent variables in the empirical models for metal release are summarized in Table 5.14.

Analysis of BOP Inhibitor Performance—Iron. When applying the developed total iron model (Equation 5.6) to the BOP PDSs only, the dummy variables have zero values for all observations except the BOP PDSs. The resulting model, applicable only to the BOP PDSs, is shown in Equation 5.13. Total iron release is influenced by total phosphate through BOP dosing, alkalinity, chloride, and influent iron in the raw water.

$$\text{total Fe} = 0.495 \times TP^{-0.104} \times alk^{-0.457} \times Cl^{0.345} \times Fe_{inf}^{0.136} \tag{5.13}$$

Exponents on terms in the model provide an indication of the degree of influence each water quality parameter has on total iron release. The addition of BOP inhibitor controls the extent of phosphorus concentrations in PDSs, which is represented by the TP term. The model suggests that BOP addition decreases iron release because of the negative exponent on the TP term. However, the magnitude of the exponent is the smallest in the model, indicating BOP has only a modest impact on total iron release. This finding is supported by average iron release data

Table 5.13
Iron release summary for BOP PDSs

			Dissolved Fe (mg/L)			Total Fe (mg/L)		
Phase	PDS	Dose	Average	Maximum	Minimum	Average	Maximum	Minimum
I	01	0.57	0.05	0.17	0.00	0.14	0.30	0.07
	02	1.08	0.03	0.13	0.00	0.10	0.15	0.06
	03	1.82	0.03	0.17	0.00	0.12	0.25	0.07
II	01	0.57	0.01	0.05	0.00	0.21	0.56	0.07
	02	1.00	0.01	0.02	0.00	0.15	0.25	0.07
	03	1.95	0.01	0.02	0.00	0.15	0.25	0.07
III	01	0.49	0.01	0.03	0.00	0.13	0.22	0.08
	02	0.90	0.01	0.03	0.00	0.11	0.13	0.07
	03	1.73	0.01	0.03	0.00	0.12	0.23	0.08
IV	01	0.68	0.00	0.01	0.00	0.12	0.18	0.01
	02	1.31	0.00	0.02	0.00	0.13	0.21	0.02
	03	2.66	0.01	0.04	0.00	0.09	0.13	0.01
All	01	0.58	0.02	0.17	0.00	0.15	0.56	0.01
	02	1.07	0.01	0.13	0.00	0.12	0.25	0.02
	03	2.03	0.02	0.17	0.00	0.12	0.25	0.01

Table 5.14
Average influent water quality parameters for BOP PDSs

PDS	Phase	pH	Alkalinity (mg/L as $CaCO_3$)	Chloride (mg/L)	Sulfate (mg/L)	Influent iron (mg/L)	Temperature (°C)
01	I	8.1	162	43.2	62.2	0.029	21.3
	II	7.9	106	67.5	104.4	0.025	26.1
	III	8.0	151	63.3	65.5	0.029	25.3
	IV	7.9	123	57.3	76.5	0.021	21.4
	Average	8.0	142	54.8	74.1	0.027	23.1
02	I	8.0	163	43.7	62.6	0.028	21.2
	II	7.9	106	67.7	104.9	0.026	26.3
	III	8.0	151	64.2	66.5	0.028	25.5
	IV	7.9	125	57.5	76.8	0.020	21.3
	Average	7.9	142	55.3	74.6	0.026	23.2
03	I	8.0	163	44.0	62.8	0.028	21.1
	II	7.8	106	68.6	105.8	0.026	26.0
	III	8.0	151	65.7	68.0	0.027	25.6
	IV	7.8	125	58.8	78.3	0.020	21.3
	Average	7.9	142	56.2	75.4	0.026	23.0

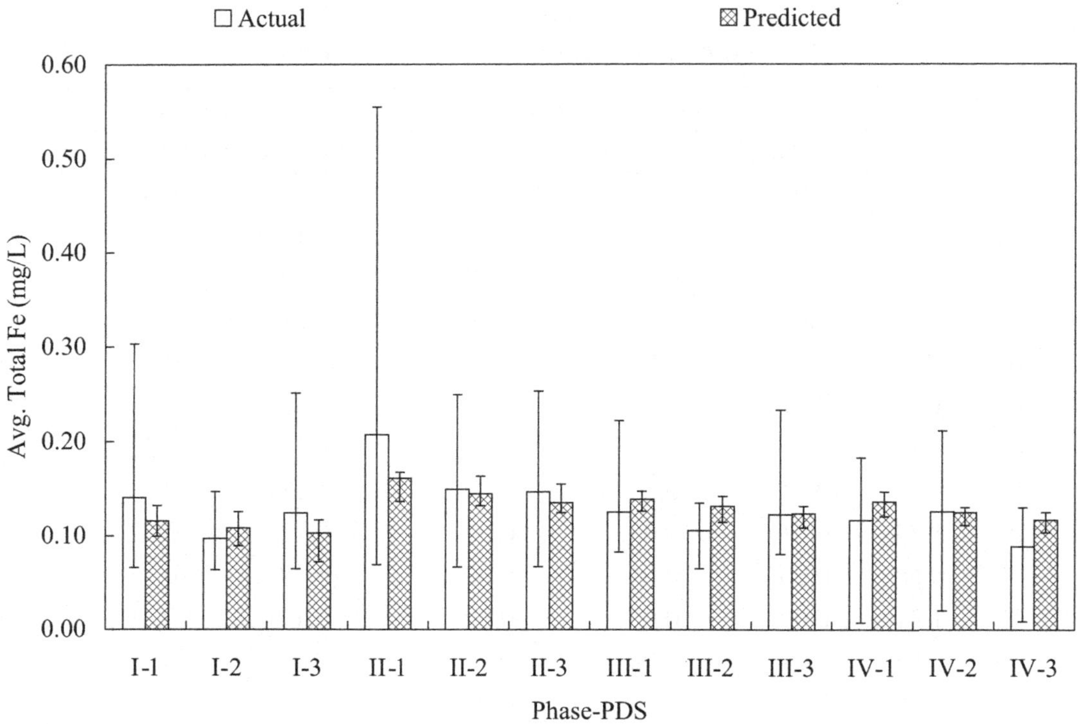

Figure 5.4 Total iron release model summary for BOP PDSs by phase

that shows minimal mitigation of iron release with increasing dose, especially above 1.0 mg/L as TP (Table 5.13).

A summary of the data for all study phases is presented in Figure 5.4. Separate bars are provided for the average predicted (using Equation 5.13) and average actual concentrations of total iron in each of the BOP PDSs for each phase. Whiskers are also shown to define the maximum and minimum measured and predicted concentrations. The model does reasonably well at predicting iron release for the BOP PDSs. Figure 5.4 shows that the medium BOP dose (1.0 mg/L P) had the best control of total iron release in phases I and III. The medium and high doses (1.0 and 2.0 mg/L P, respectively) were better than the low dose (0.5 mg/L P) in phase II, while the high dose in phase IV produced the best performance. Figure 5.4 suggests that decreased total iron release is not always related to increased BOP dose.

The remaining terms in the model (alkalinity, chloride, and influent iron) are related to the water quality of the source water blend. Phase II utilized a blend with a greater surface water composition (62%) than was used in either phase I or III (27%). The average total iron concentrations are 0.12 mg/L, 0.17 mg/L, and 0.12 mg/L for phases I, II, and III, respectively. The elevated surface water component resulted in a reduced alkalinity for phase II. The increase in iron concentration at reduced alkalinity is consistent with the inclusion of alkalinity in the empirical model with a negative exponent on that variable. Actual total iron release shown in Figure 5.4 supports this conclusion, where concentrations from PDSs in phase II were higher than those from any PDS in phases I or III.

The model is also sensitive to chloride, where it has the largest positive exponent in the model. This suggests that an increase in chloride results in an increase in total iron release. Phase I has the lowest average chloride concentration among phases, which would result in lower iron release for that phase. This is supported by the low levels of actual iron release in phase I, as shown in Figure 5.4.

Influent iron is also a significant term in the developed total iron model and contributes to the iron release because of its positive exponent in the empirical model. However, the influence of chloride may be more pronounced because of its larger exponent, as shown in Equation 5.13.

Because BOP is a blend of orthophosphate and polyphosphate, a term accounting for the difference in phosphorus speciation in BOP was introduced during model development. However, this term was not found to be statistically significant. Therefore, accounting for phosphate blend in BOP was integrated in the total iron model through the dummy variable coefficient and the associated TP term. The dummy variable coefficients for BOP and OP are different (0.495 and 0.593, respectively) in Equation 5.6, and the exponents on the TP terms associated with BOP and OP dummy variables are distinguishable in sign and magnitude (–0.104 and 0.047, respectively). This difference is indicative of the different effect BOP and OP have on iron release, even though actual measurements of the delivered dose of both inhibitors are similar in terms of TP measurements.

With respect to the secondary standard for iron concentration of 0.3 mg/L, the performance data in Figure 5.4 indicate that, on average, BOP was beneficial in controlling iron release to levels below the secondary standard. The empirical total iron model predicted average total iron release reasonably well, as shown in Figure 5.4, with the greatest predicted average iron concentration in PDS01 in phase II at 0.16 mg Fe/L. There were individual observations for PDS01, phase II, that exceeded the secondary standard.

Control of Iron Release With OP Inhibitor Addition

OP Data Summary (PDS04 to PDS06)—Iron. A summary of the total and dissolved iron release data for each of the PDSs receiving OP inhibitor is shown in Table 5.15 by phase. The majority of the iron in the PDSs is in the particulate form. There is not an obvious improvement in mitigation of iron release with the increase in OP dose. The lowest overall iron release appears in phase I, corresponding with a blend that was enriched in groundwater that exhibited the highest alkalinity concentration. Water quality data for the independent variables in the empirical models for metal release are summarized in Table 5.16.

Analysis of OP Inhibitor Performance—Iron. The total iron model developed for this study was presented previously in Equation 5.6. When the OP inhibitor is applied, the model is of the form shown in Equation 5.14, where all terms have been previously defined. The OP inhibitor addition contributes a significant concentration of phosphorus to the PDSs, which the model suggests will contribute to iron release by the positive exponent on the TP term. However, the low values of the OP dose used and the near-zero value of the exponent (0.047) on the TP term yields predictions of total iron concentration that are not very sensitive to the inhibitor dose for the OP PDSs. The data and the empirical model are in general agreement, as it is seen in Table 5.15 that increasing the dose of OP inhibitor does not offer improved iron control.

$$TFe = 0.593 \times TP^{0.047} \times alk^{-0.457} \times Cl^{0.345} \times Fe_{inf}^{0.136} \tag{5.14}$$

Table 5.15
Iron release summary for OP PDSs

Phase	PDS	Dose	Dissolved Fe (mg/L)			Total Fe (mg/L)		
			Average	Maximum	Minimum	Average	Maximum	Minimum
I	04	0.49	0.03	0.11	0.00	0.12	0.20	0.06
	05	1.04	0.03	0.16	0.00	0.14	0.19	0.10
	06	1.84	0.03	0.13	0.00	0.11	0.29	0.06
II	04	0.55	0.00	0.01	0.00	0.16	0.22	0.08
	05	0.90	0.01	0.06	0.00	0.21	0.33	0.10
	06	1.91	0.01	0.03	0.00	0.17	0.24	0.09
III	04	0.54	0.01	0.03	0.00	0.14	0.21	0.10
	05	0.99	0.01	0.03	0.00	0.23	0.31	0.15
	06	1.87	0.01	0.03	0.00	0.14	0.18	0.08
IV	04	0.47	0.00	0.00	0.00	0.13	0.20	0.01
	05	0.82	0.00	0.01	0.00	0.13	0.20	0.01
	06	1.69	0.00	0.01	0.00	0.20	0.51	0.01
All	04	0.51	0.01	0.11	0.00	0.14	0.22	0.01
	05	0.94	0.01	0.16	0.00	0.18	0.33	0.01
	06	1.83	0.01	0.13	0.00	0.16	0.51	0.01

Table 5.16
Average influent water quality parameters for OP PDSs

PDS	Phase	pH	Alkalinity (mg/L as $CaCO_3$)	Chloride (mg/L)	Sulfate (mg/L)	Influent iron (mg/L)	Temperature (°C)
04	I	8.0	164	43.2	61.8	0.027	21.0
	II	7.9	106	67.2	104.2	0.026	25.6
	III	8.0	152	63.6	65.1	0.029	25.6
	IV	7.9	125	56.6	76.3	0.020	21.2
	Average	7.9	143	54.7	73.8	0.026	22.9
05	I	8.0	164	43.1	62.0	0.028	21.0
	II	7.9	107	67.4	104.3	0.027	26.0
	III	8.0	152	62.8	65.1	0.030	25.6
	IV	7.8	125	56.4	76.0	0.020	21.0
	Average	7.9	143	54.5	73.8	0.027	23.0
06	I	8.0	163	43.2	61.5	0.027	20.9
	II	7.8	107	67.8	104.8	0.026	26.0
	III	8.0	151	65.0	65.7	0.027	25.8
	IV	7.8	125	57.9	77.7	0.019	21.2
	Average	7.9	142	55.4	74.1	0.026	23.0

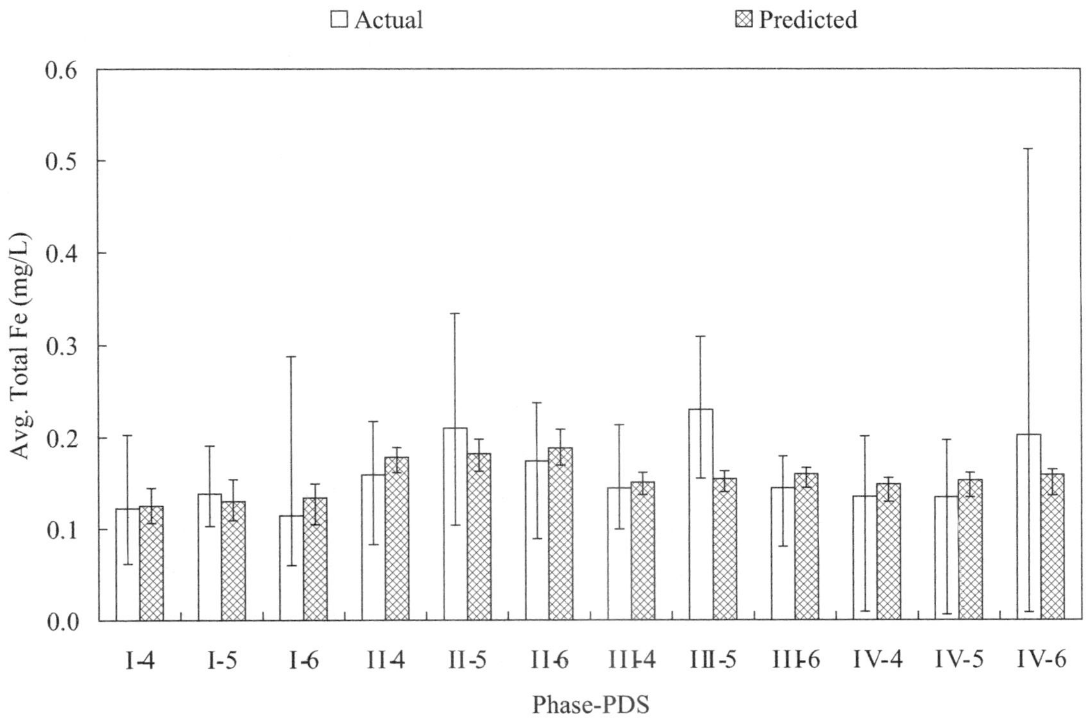

Figure 5.5 Total iron release model summary for OP PDSs by phase

Figure 5.5 shows the average iron release in each of the OP PDSs for each phase compared to the predicted iron release by the model. The maximum and minimum error bars are also shown. The model does reasonably well at predicting iron release for the OP PDSs. The performance data summarized in Figure 5.5 clearly confirms the general absence of a benefit of an increased OP inhibitor dose for control of iron release.

All other terms in the model are related to the water quality of the blend. Phases I and III were mostly groundwater, which is characterized by high alkalinity. Phase II had the lowest percentage of groundwater and therefore the lowest alkalinity. The negative exponent on the alkalinity term suggests that reduction of iron release can be achieved by increasing the alkalinity. This is confirmed with Figure 5.5, especially for the phase I data, which has the lowest iron release, and phase II, which has the highest iron release for both actual data and model predictions. The model is sensitive to the alkalinity of the blend because of the high concentration of alkalinity and the larger exponent.

The model is also sensitive to chloride, but increases contribute to iron release, as suggested by the positive exponent on the chloride term in the model. Phase I has the lowest average chloride concentration. The reduced chloride concentration would be another explanation for the low levels of iron observed in phase I.

Influent iron also is a significant term in the developed total iron model. Influent iron was relatively constant regardless of the study phase. This term also has a positive exponent in the model, suggesting that it contributes to the iron in the effluent of the PDSs. Influent iron concentrations were consistently low, around 0.03 mg/L, so the model is not as sensitive to this term.

Table 5.17
Iron release summary for ZOP PDSs

			Dissolved Fe (mg/L)			Total Fe (mg/L)		
Phase	PDS	Dose	Average	Maximum	Minimum	Average	Maximum	Minimum
I	07	0.64	0.04	0.19	0.00	0.18	0.27	0.13
	08	0.99	0.05	0.43	0.00	0.14	0.28	0.09
	09	1.78	0.02	0.09	0.00	0.16	0.41	0.06
II	07	0.54	0.01	0.10	0.00	0.34	0.64	0.12
	08	0.89	0.01	0.04	0.00	0.23	0.29	0.11
	09	1.77	0.01	0.04	0.00	0.16	0.24	0.07
III	07	0.49	0.01	0.03	0.00	0.22	0.30	0.10
	08	0.88	0.01	0.03	0.00	0.22	0.30	0.12
	09	1.82	0.01	0.03	0.00	0.11	0.13	0.08
IV	07	0.46	0.00	0.01	0.00	0.28	0.40	0.11
	08	0.88	0.00	0.00	0.00	0.24	0.37	0.01
	09	1.58	0.00	0.00	0.00	0.11	0.19	0.00
All	07	0.54	0.02	0.19	0.00	0.25	0.64	0.10
	08	0.91	0.02	0.43	0.00	0.21	0.37	0.01
	09	1.74	0.01	0.09	0.00	0.14	0.41	0.00

Control of Iron Release With ZOP Inhibitor Addition

ZOP Data Summary (PDS07 to PDS09)—Iron. A summary of total and dissolved iron release data for each of the PDSs receiving ZOP inhibitor is shown in Table 5.17. This data is segregated by phase and by PDS.

Water quality data for the independent variables in the empirical models for metal release are summarized in Table 5.18. PDS07, PDS08, and PDS09 were treated with a low, medium, and high dose of ZOP inhibitor, respectively. As shown in Table 5.17, the majority of the iron in ZOP PDSs is present in a particulate form. The data show that, in most cases, the total iron release decreased as the inhibitor dose increased. Lower iron release was observed in both phase I and phase III, which can be explained by the higher alkalinity in these two phases.

Analysis of ZOP Inhibitor Performance—Iron. The total iron model developed for this study was presented previously in Equation 5.6. When the ZOP inhibitor is applied, the model is of the form shown in Equation 5.15, where all terms have been previously defined.

$$TFe = 0.47 \times TP^{-0.073} \times Zn^{-0.18} \times alk^{-0.457} \times Cl^{0.345} \times Fe_{inf}^{0.136} \quad (5.15)$$

Significant amounts of phosphorus and zinc were added to PDS07–PDS09 in conjunction with addition of the ZOP inhibitor. The negative exponents on the TP term and zinc term in the empirical model indicate that both total phosphorus and zinc were beneficial for reducing iron release. Although the low values of the ZOP dose used and the near-zero value of the exponent on the phosphorus term would not cause the model to display pronounced sensitivity to the inhibitor

Table 5.18
Average influent water quality parameters for ZOP PDSs

PDS	Phase	pH	Alkalinity (mg/L as $CaCO_3$)	Chloride (mg/L)	Sulfate (mg/L)	Influent iron (mg/L)	Temperature (°C)
07	I	8.0	163	43.5	63.2	0.028	21.0
	II	7.8	104	64.0	104.8	0.020	26.2
	III	8.0	149	67.2	68.1	0.030	25.9
	IV	7.8	121	56.5	76.6	0.022	21.3
	Average	7.9	140	54.7	76.2	0.025	23.1
08	I	7.9	162	44.0	64.2	0.028	21.1
	II	7.7	104	64.6	105.4	0.021	26.3
	III	7.9	149	66.6	68.7	0.029	26.0
	IV	7.8	119	59.5	77.3	0.022	21.2
	Average	7.8	139	55.6	77.0	0.025	23.1
09	I	7.8	159	44.3	65.3	0.025	21.2
	II	7.6	101	65.0	107.5	0.019	26.3
	III	7.8	146	68.9	71.5	0.028	25.9
	IV	7.6	119	60.4	78.9	0.021	21.3
	Average	7.7	137	56.4	78.7	0.024	23.1

dose for ZOP, the impact of zinc in the model enhances the effect of the inhibitor dose significantly. The end result of the dual impact of TP and Zn suggests that increasing the dose of ZOP should offer improved iron control.

Figure 5.6 shows the average total iron release in each of the ZOP PDSs for each phase compared to the predicted iron release by model. The maximum and minimum values are also shown. PDS07, PDS08, and PDS09 were treated with a low, medium, and high dose of ZOP inhibitor, respectively. As shown in the graph, the actual total iron release decreased as inhibitor dose increased in most cases, indicating improved performance with increasing inhibitor dose. The model suggests the same trend in Figure 5.6 and does reasonably well at predicting iron release for the ZOP PDSs.

All other terms in the model are related to the water quality of the blend. The water quality of phases I and III was characterized by high alkalinity due to the reduced percentage of surface water. Phase II had the lowest percentage of groundwater and therefore the lowest alkalinity. The negative exponent on the alkalinity term would explain why phases I and III had lower iron release than phase II. The model also suggests that a reduction of iron release can be achieved by increasing the alkalinity.

The positive exponent on the chloride term in the model suggests chloride will contribute to iron release. The average chloride concentration for phases I–IV is 43.9 mg/L, 64.5 mg/L, 67.6 mg/L, and 58.8 mg/L, respectively. The lowest average chloride concentration in phase I would also contribute to a reduction in the iron concentration (observed or predicted) in phase I, as shown in Figure 5.6.

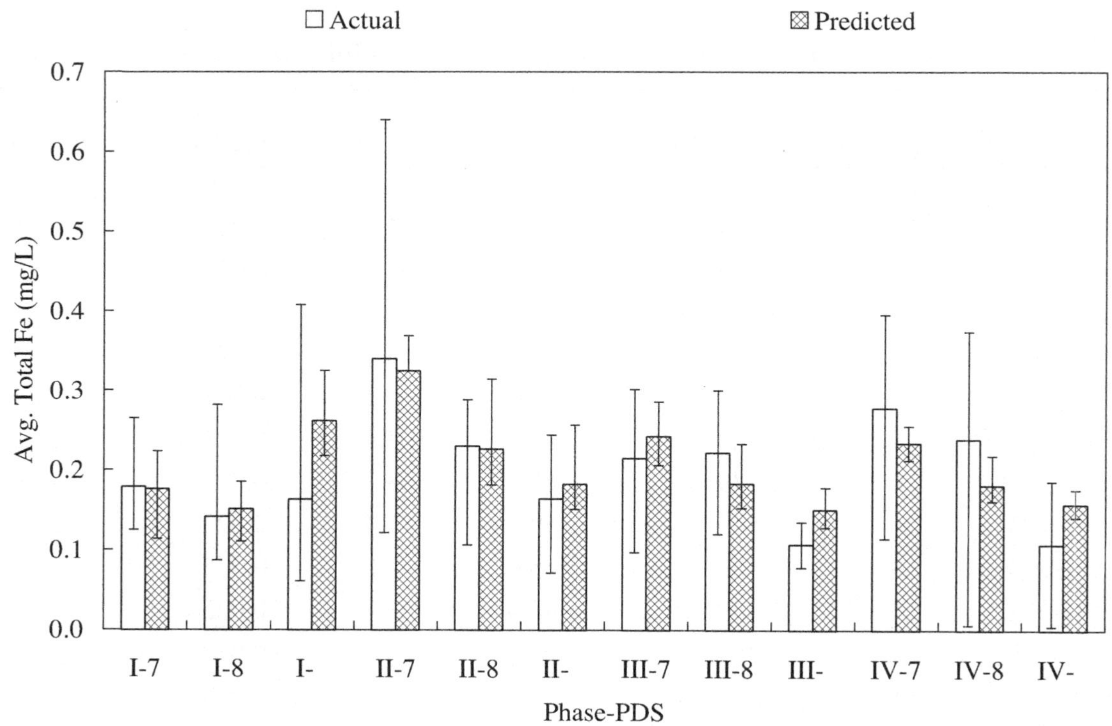

Figure 5.6 Total iron release model summary for ZOP PDSs by phase

Control of Iron Release With Silicon Inhibitor Addition

Si Data Summary (PDS10 to PDS12)—Iron. This section provides an evaluation of iron release observed within the PDS following treatment with silica as a corrosion inhibitor. Table 5.19 reports a summary of the iron release data collected throughout four phases of operation. These results clearly demonstrate that iron was released within the system predominantly in the form of particulate iron. Variations of inhibitor dosage are presented within Table 5.19 as well. The reported silica concentrations correspond to silica associated with both the addition of the inhibitor and silica found in the blended source water.

Excluding early sampling events during phase I in which the silica dose was 10, 20, and 40 mg/L SiO_2, the target concentrations for the inhibitor addition were 3, 6, and 12 mg/L SiO_2 for PDS10, PDS11, and PDS12, respectively. Water quality data for the independent variables in the empirical models for metal release are summarized in Table 5.20.

During operation and data collection, phase changes denoted a distinct change in water quality allowing for the analysis of main effects associated with both dose and water quality. Identification of the variation in iron concentration as a result of variation in inhibitor dose and water quality was explored through the development of empirical models.

Main effects associated with silica dose are somewhat ambiguous when analyzed within a common phase. For instance, a monotonic increase or decrease in total iron concentration with inhibitor dose was not observed. Furthermore, any pattern to relate iron concentration and inhibitor dose that existed for a given phase lacked replication when analyzed during other phases.

Table 5.19
Iron release summary for Si PDSs

Phase	PDS	Dose	Dissolved Fe (mg/L)			Total Fe (mg/L)		
			Average	Maximum	Minimum	Average	Maximum	Minimum
I	10	15.2	0.03	0.08	0.00	0.14	0.46	0.07
	11	21.5	0.02	0.10	0.00	0.12	0.20	0.07
	12	31.5	0.03	0.10	0.01	0.17	0.57	0.06
II	10	7.8	0.02	0.11	0.00	0.17	0.24	0.10
	11	11.7	0.02	0.10	0.00	0.13	0.20	0.10
	12	17.2	0.01	0.03	0.00	0.19	0.40	0.13
III	10	13.3	0.01	0.02	0.00	0.11	0.14	0.09
	11	17.0	0.01	0.03	0.00	0.10	0.12	0.08
	12	22.6	0.01	0.03	0.00	0.11	0.14	0.10
IV	10	9.2	0.00	0.00	0.00	0.11	0.16	0.04
	11	13.1	0.00	0.00	0.00	0.10	0.13	0.04
	12	17.7	0.00	0.01	0.00	0.12	0.21	0.04
All	10	11.5	0.01	0.11	0.00	0.13	0.46	0.04
	11	16.0	0.01	0.10	0.00	0.11	0.20	0.04
	12	22.5	0.01	0.10	0.00	0.15	0.57	0.04

Table 5.20
Average influent water quality parameters for Si PDSs

PDS	Phase	pH	Alkalinity (mg/L as $CaCO_3$)	Chloride (mg/L)	Sulfate (mg/L)	Influent iron (mg/L)	Temperature (°C)
10	I	8.2	162	42.3	60.9	0.030	21.1
	II	8.0	105	62.9	101.0	0.025	26.3
	III	8.2	149	63.5	64.7	0.028	25.9
	IV	8.0	122	56.9	73.3	0.023	21.1
	Average	8.1	140	53.5	73.2	0.028	23.1
11	I	8.3	161	42.8	60.0	0.035	20.9
	II	8.2	106	62.5	99.8	0.021	26.6
	III	8.3	150	63.8	64.5	0.029	26.1
	IV	8.2	123	56.5	73.2	0.080	21.0
	Average	8.2	140	53.6	72.5	0.039	23.1
12	I	8.4	156	40.9	58.0	0.033	21.1
	II	8.4	106	61.5	98.3	0.021	26.1
	III	8.4	147	62.7	63.7	0.030	26.1
	IV	8.4	123	55.3	73.1	0.024	21.1
	Average	8.4	138	52.1	71.1	0.028	23.1

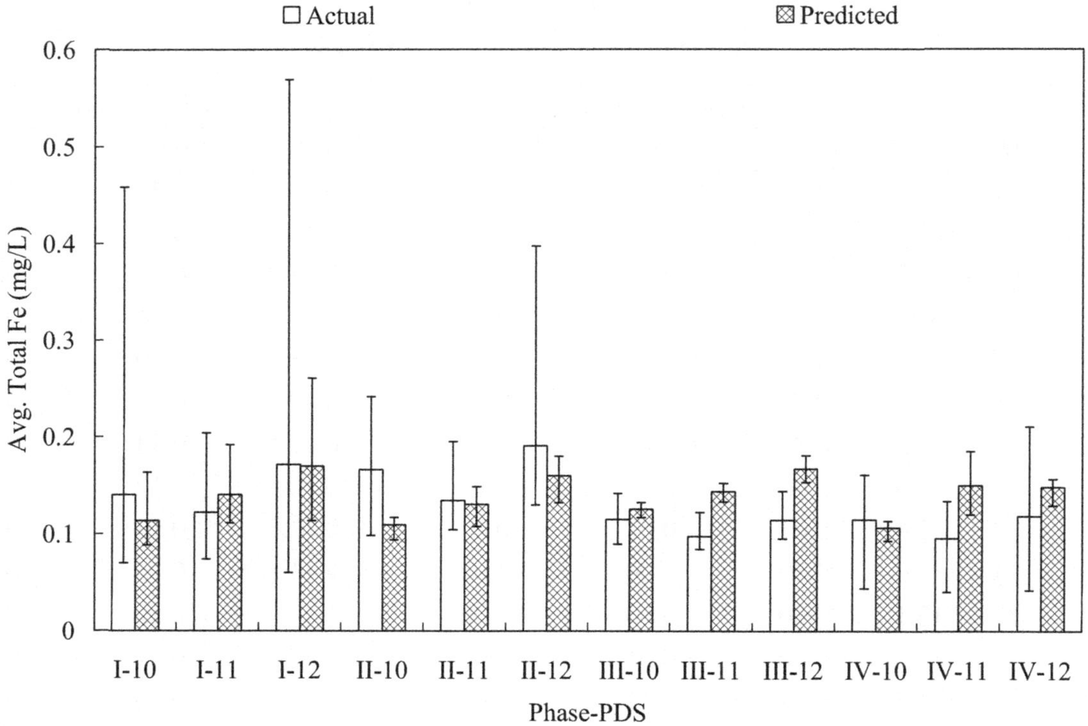

Figure 5.7 Total iron release model summary for Si PDSs by phase

Analysis of Silicon Inhibitor Performance—Iron. A model (Equation 5.6) describing expected total iron release was described earlier within the "Metal Release" section as a comprehensive model for all corrosion control strategies. This model was able to segregate effects of multiple inhibitors with the use of dummy variables. Equation 5.16 represents a simplified form of the comprehensive model, appropriate only for the effects of total iron release with the application of the Si inhibitor. The model suggests that total iron release does respond adversely to increases in the Si dose. Alkalinity, chlorides, and influent total iron were also significant to total iron release within the hybrid PDS.

$$TFe = 0.122 \times SiO_2^{0.528} \times alk^{-0.457} \times Cl^{0.345} \times Fe_{inf}^{0.136} \qquad (5.16)$$

Evaluation of each term within Equation 5.16 allows an assessment of the expected system response to the addition of the Si inhibitor. Also, the inclusion of terms descriptive of water quality renders the model suitable for various source water blends. The observed total iron release did not suggest any consistent response with variation of inhibitor dose. This relationship can be seen graphically within Figure 5.7. The actual average iron concentrations for the variant inhibitor doses are nearly indistinguishable. The empirical model suggests, however, that an increase in the inhibitor dose would release more iron. This relationship is demonstrated by the positive exponent associated with the silica term. The positive trend in total iron release that is suggested from the empirical model is not strongly corroborated by the actual data.

Variations in iron concentration between phases appear more pronounced than could be explained by changes in silica dose alone, indicating a significant dependence on water quality. Actual

Table 5.21
Iron release summary for pH control PDSs

		Dissolved Fe (mg/L)			Total Fe (mg/L)		
Phase	PDS	Average	Maximum	Minimum	Average	Maximum	Minimum
I	13	0.03	0.11	0.00	0.15	0.20	0.10
	14	0.04	0.18	0.00	0.11	0.21	0.07
II	13	0.02	0.12	0.00	0.28	0.37	0.17
	14	0.03	0.13	0.00	0.19	0.33	0.12
III	13	0.01	0.04	0.00	0.27	0.37	0.16
	14	0.01	0.04	0.00	0.14	0.18	0.09
IV	13	0.00	0.00	0.00	0.19	0.32	0.05
	14	0.00	0.00	0.00	0.14	0.24	0.06
All	13	0.02	0.12	0.00	0.22	0.37	0.05
	14	0.02	0.18	0.00	0.14	0.33	0.06

iron release diminishes during phases III and IV. Based on terms present in the empirical model, differences would have occurred following a change in alkalinity, chlorides, and/or influent total iron.

The negative exponent associated with alkalinity suggests that alkalinity decreases the corrosivity of the blended water source. Phases II and III illustrate the benefit of alkalinity with iron release. Both chlorides and influent iron were similar between these two phases; however, the average alkalinity during phase III was approximately 50 mg/L as $CaCO_3$ greater than during phase II. From Figure 5.7, the average iron concentration was reduced during phase III. The observed iron release during phase III also appears more stable, as indicated by the narrowing variation among the average for all silica doses.

From Figure 5.7, the predictions from the model are relatively uniform for all phases. Though the model may appear to render poor predictions, deviations from the model are likely attributed to random events of high iron release seen within phases I and II. As the study progressed into phases III and IV, the variation in observed iron release decreased notably. However, even during phases III and IV, examination of the data fails to define a distinct monotonic relationship between iron release and Si inhibitor dose.

Control of Iron Release by pH Elevation (pH Controls)

pH Controls Data Summary (PDS13 to PDS14)—Iron. A summary of the total and dissolved iron release data for the pH control PDSs is shown in Table 5.21 by phase. The majority of the iron released in the PDSs is in the particulate form. The pH_s control (PDS13) consistently experiences higher iron release than the PDS with an elevated pH (PDS14). The lowest iron release appears in phase I, which had the highest alkalinity, and the highest iron release appears in phase II, which had the lowest alkalinity. Water quality data for the independent variables in the empirical models for metal release are summarized in Table 5.22.

Analysis of pH Control Performance—Iron. The general total iron model for this study was presented previously in Equation 5.6. Modification of the empirical model for application to the pH control lines is shown in Equation 5.17, where all terms have been previously defined. The

Table 5.22
Average influent water quality parameters for pH control PDSs

PDS	Phase	pH	Alkalinity (mg/L as $CaCO_3$)	Chloride (mg/L)	Sulfate (mg/L)	Influent iron (mg/L)	Temperature (°C)
13	I	7.8	146	64.2	61.4	0.027	21.3
	II	7.9	92	82.8	100.3	0.018	26.4
	III	7.7	149	88.4	65.9	0.029	25.3
	IV	7.6	119	68.1	75.8	0.022	21.0
	Average	7.7	130	75.2	73.5	0.025	23.2
14	I	8.0	164	43.2	62.2	0.029	21.2
	II	7.9	106	65.1	102.2	0.020	26.5
	III	8.1	151	64.8	66.5	0.030	25.6
	IV	7.9	125	57.9	76.0	0.022	21.3
	Average	8.0	139	57.2	75.5	0.026	23.5

effect of increasing the pH is reflected in the model by an increase in alkalinity. Increasing pH, and therefore alkalinity, helps in reducing iron release, as shown in the total iron model by the negative exponent on the alkalinity term.

$$TFe = 0.661 \times alk^{-0.457} \times Cl^{0.345} \times Fe_{inf}^{0.136} \tag{5.17}$$

Figure 5.8 shows the average iron release in the pH control PDSs for each phase compared to the predicted iron release by the empirical model. Maximum and minimum observed and predicted concentrations are also shown. The model does reasonably well at predicting iron release for the pH control PDSs. It can be seen in Figure 5.8 that increasing the pH consistently achieves a reduction in iron release for both the observed data and the model predictions.

The alkalinity varies with the product water blend as well as with the increase in the pH in PDS14. Phases I and III were mostly groundwater, which is characterized by high alkalinity. Phase II had the lowest percentage of groundwater and therefore the lowest alkalinity. As discussed before, an increase in alkalinity would be expected to achieve a decrease in iron release. This, in part, explains why phase I has the lowest iron release and phase II has the highest, as seen in Figure 5.8 for both the observed iron release and that predicted by the model.

Chloride is also a significant water quality parameter in the total iron release model. The exponent on the chloride term in the model is positive, suggesting that an increase in chloride concentration would be associated with an increase in iron concentration. In addition to the highest alkalinity, the phase I blend has the lowest average chloride concentration, offering further explanation of the lower iron concentrations in phase I. When compared to the elevated pH PDS, chloride is also much higher in the pH_s control PDSs due to the maintenance of pH_s by addition of HCl.

Influent iron is another significant term in the total iron model. Influent iron was relatively constant in the pH control PDSs, regardless of study phase. This term also has a positive exponent in the model, suggesting that it contributes to the iron in the effluent of the PDSs. Influent iron concentrations were consistently low, around 0.03 mg/L, so the model is not as sensitive to this term.

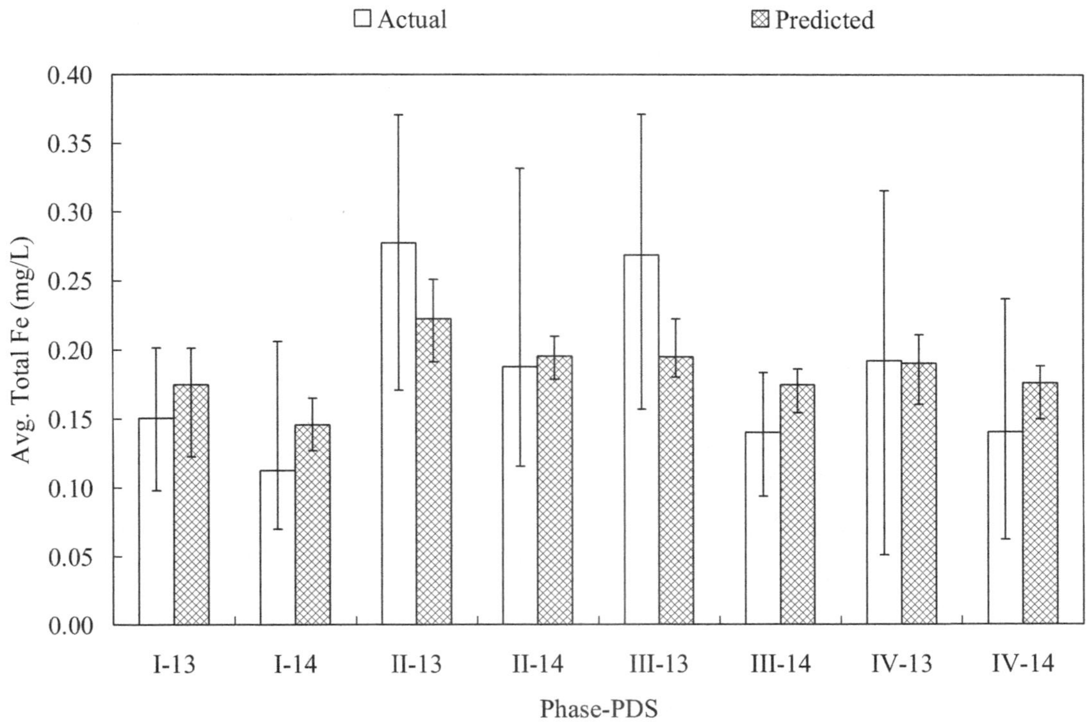

Figure 5.8 Total iron release model summary for pH control PDSs by phase

Addition of bicarbonate and hydroxide salts for increasing alkalinity increases the TDS, usually with sodium, in the system. The empirical model developed in the previous study (Equation 5.1) indicated that an undesired increase in iron release may be expected with an increase in sodium concentration. This model suggests that addition of sodium and/or chloride salts for pH regulation and/or alkalinity supplementation may produce an adverse response for iron concentration. Caution should be used when increasing alkalinity with chemicals that contain these ions. Alternate strategies that employ lime and carbon dioxide for alkalinity supplementation may be preferable.

Performance of PDS13 and PDS14 Relative to TBW I Models. From the onset of data collection, PDS13 experienced much greater water quality deterioration than PDS14. From the average values listed in Table 5.21, iron release in PDS13 was almost twice that of PDS14. Aggressive conditions were experienced during the study, including temperatures greater than 25°C and chloride and sulfate concentrations as large as 70 and 100 mg/L, respectively. Even under these conditions, PDS14 never violated the secondary iron standard of 0.3 mg/L during the entire study, whereas PDS13 violated the standard several times. Using this data, it is concluded that well-stabilized waters with sufficient alkalinity (>100 mg/L $CaCO_3$) were effective measures to control iron release in the PDS. These results were expected as they follow commonly accepted theories involving the LSI and the associated beneficial properties of alkalinity for iron corrosion control.

The data collected from PDS13 and PDS14 were used to assess the historical color release model that was developed in TBW I (Equation 5.1). The data obtained from the current study represent an independent set of data that could be used for verification of the previous empirical model. The model predictions and observed data are illustrated in Figures 5.9 and 5.10 for PDS13 and PDS14, respectively.

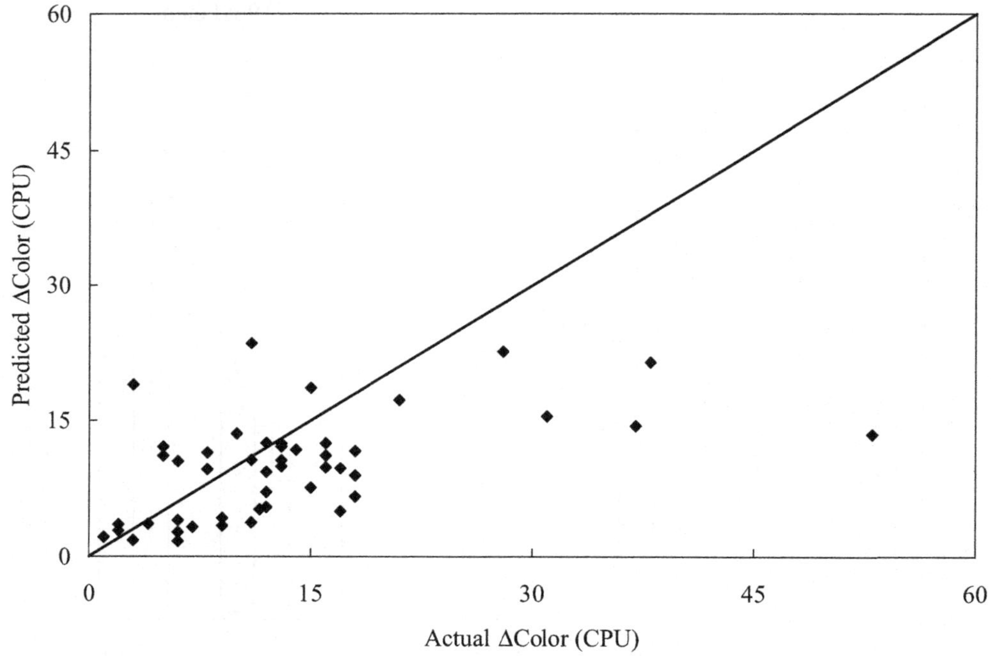

Figure 5.9 PDS13 predicted versus actual color using TBW I model

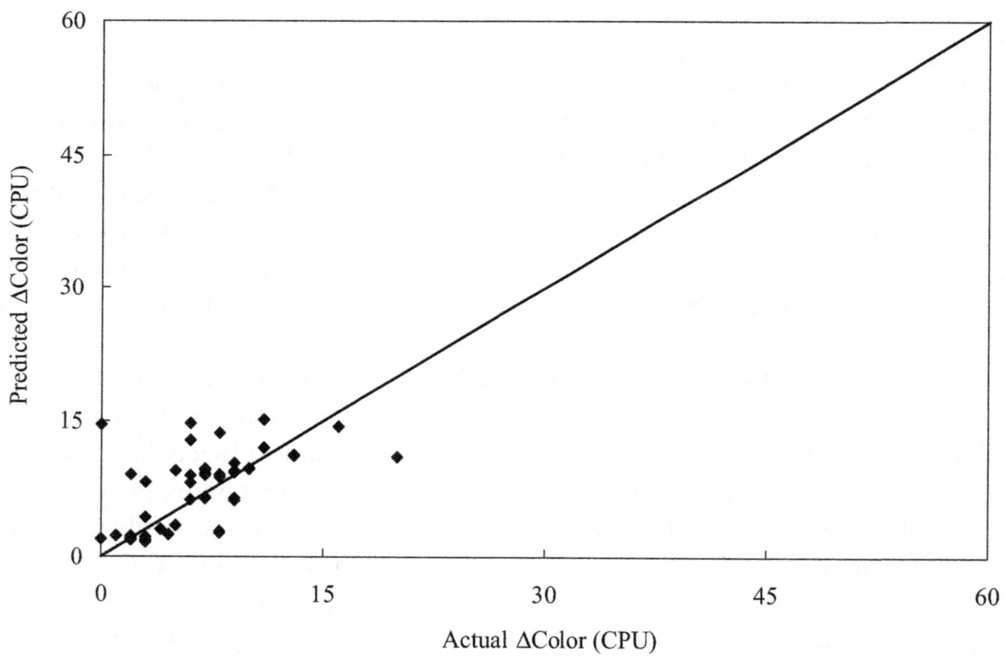

Figure 5.10 PDS14 predicted versus actual color using TBW I model

The plots suggest that the historical empirical model predicted the color in PDS14 with reasonable accuracy. The average difference in the predicted color values and the actual color data was less than 1 cpu, with no consistent trend of overprediction or underprediction as a function of the color (or iron concentration).

The response in PDS13 was quite different, as the model underpredicted the actual increase in color (or iron) for more than two-thirds of the data points. The historical model predicted only slightly higher color release for PDS13 than PDS14, but actual data for color release was substantially higher. The model averaged an underprediction of nearly 4 cpu, with greater underprediction of the larger color values. The historical color release model from TBW I does not appear to be applicable for the blends that are maintained at pH_S (PDS13). The model does appear to be appropriate for description of well-stabilized blends with a positive LSI (PDS14). This finding is not a surprise in light of the database that was used to develop Equation 5.1. All blends that were employed in the earlier study were well stabilized (LSI > 0.4), matching the conditions that were present for PDS14 in the current study. The empirical models developed in the current study (Equation 5.6 and 5.7) included data from PDS13. For a description of iron release for waters that exhibit a pH near pH_S, the current models would be expected to perform better than the historical model from TBW I.

Iron Release Summary

Review of Empirical Models. New empirical models were developed to predict total and dissolved iron concentration (Equations 5.6 and 5.7). The new models generally confirmed the relations that were defined in TBW I (Equation 5.1). Independent data from the current study from PDS14 were reasonably well described by the empirical model from TBW I. The blends from the historical study and the blends used in the current study for PDS14 were all well stabilized with respect to calcium carbonate solubility. In contrast, the current study data for operation at pH_S (PDS13) were not well described by the historical model. The historical model consistently underpredicted actual impairment of water quality. These observations confirm the benefits of elevated alkalinity for control of iron release that were documented in TBW I.

The dominant form of iron in the PDS effluent is particulate. Furthermore, consideration of adverse water quality impacts that are addressed by secondary drinking water standards (color and iron) relate to the total iron concentration. For these reasons, discussion of prediction and control of total iron concentrations would be more relevant than consideration of dissolved forms. The data for dissolved iron would be appropriate, however, for evaluation of thermodynamic relationships, including solubility of iron species.

The new empirical model for total iron provides support for increases in the release of iron as blends become enriched with surface water or RO at the expense of groundwater. The beneficial effect of alkalinity supplementation for mitigation of iron release from TBW I has been confirmed with the current study data. Based on the empirical model, a modest benefit may also be expected from elevation of pH due to the associated increase in alkalinity. Increases in the chloride concentration or the influent iron concentration are projected to result in increases in the output iron concentration.

The empirical model also indicates an expected benefit associated with addition of any of the inhibitors. Complete definition of the effect of inhibitor addition may require simultaneous adjustment of multiple terms in the empirical model. As an example, addition of the ZOP inhibitor would produce an increase in both the TP and the zinc (Zn) concentration terms in Equation 5.6.

Selection of a particular inhibitor would also introduce the corresponding dummy variable coefficient into the model. A direct interpretation of the benefit of inhibitor addition is complicated by these factors. Optimization simulations using the empirical models are presented later in this chapter to evaluate inhibitor performance for satisfaction of water quality objectives in the distribution system. This optimization exercise permits a proper assessment of the multivariable effects expressed by the empirical model.

Application of any of the inhibitors would not universally be expected to achieve mitigation of iron release relative to operation at an elevated pH (or alkalinity). Examination of the dummy variable coefficients indicates a lower coefficient for all of the phosphate-based inhibitors (BOP – 0.495, OP – 0.593, ZOP – 0.470) than for the pH control (0.661 mg Fe/L). When compared to the elevated pH treatment (PDS14), the adverse effect of insufficient TP and/or zinc concentrations associated with inhibitor addition may offset the effect of the smaller dummy variable coefficient associated with the inhibitor PDSs. Evaluation of the silicate inhibitor effects is similarly complicated by the competing influence of the dummy variable coefficient and increasing silicate concentrations. The dummy variable coefficient for Si (0.122) is somewhat smaller than the other coefficients. The exponent on the silicate concentration is positive, however, suggesting that silicate addition would elevate iron concentrations. It may be more instructive to rely on the optimization simulations presented later in this chapter to provide a definitive evaluation of the respective benefits of each control approach to mitigate iron release.

The impact of HRT on iron concentration must also be acknowledged. The current study of empirical models for iron concentration were developed from a data set with a single value of HRT (2 days). Consequently, it was not possible to identify an effect of HRT on iron release for this data. Prior empirical (Imran et al. 2005a) and theoretical (Mutoti et al. 2007a, b) analyses have supported a linear increase in iron concentration with HRT for laminar flow conditions. Any meaningful analysis of iron release must also consider HRT as well as water quality, inhibitor selection, and inhibitor dose.

Data Summary. A presentation of average output total iron concentrations for the 12-month study duration is shown in Figure 5.11. These data represent an average of the weekly determinations of corrosion loop iron concentrations. A brief explanation of the bar chart organization is provided to assist with data interpretation.

The intended comparison to be reviewed is between the pH_s control (PDS13) and the corrosion control strategies—either pH elevation (PDS14), BOP addition (PDS01 to PDS03), OP addition (PDS04 to PDS06), ZOP addition (PDS07 to PDS09), or Si addition (PDS10 to PDS12). The results for the pH_s control are shown at the left edge of the chart. The elevated pH data are reported adjacent to the pH_s data. The inhibitor results are then provided in the following order: BOP, OP, ZOP, and Si. There are three PDSs for each inhibitor, representing a low, medium, and high inhibitor dose. All data for a single inhibitor are clustered on the bar chart, and results are reported sequentially (left to right) for an increasing inhibitor dose. The bar chart format allows a comparison between corrosion control strategies (e.g., pH elevation vs. BOP addition). The presentation also permits an evaluation within a single corrosion control strategy to assess benefits associated with variable inhibitor dose.

The graphical summary of the data generally supports the previous synopsis that was based on the empirical model. There appears to be a clear benefit associated with operation at an elevated pH. The average project total iron concentrations for PDS13 and PDS14 were 0.22 and 0.14 mg/L, respectively. Inhibitor addition does not appear to offer any clear advantage in comparison to pH elevation.

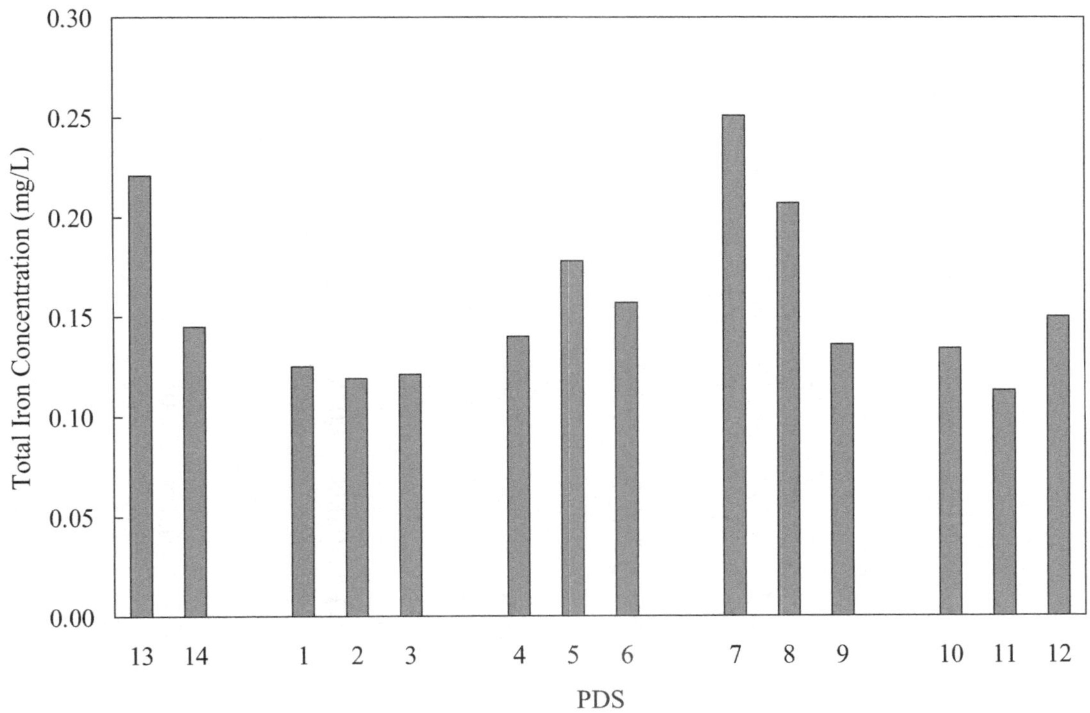

Figure 5.11 Data summary for total iron concentration

The average iron concentrations for all of the inhibitor lines displayed some improvement in comparison to the pH_s control. This benefit is not consistent for all of the ZOP inhibitor doses. A reduction in iron concentration of approximately 40% was achieved, with the exception of the ZOP applications in PDS07 and PDS08. There does not appear to be any consistent improvement in performance that could be expected with an increase in dose of any of the inhibitors except ZOP, within the range of values employed in this study.

A statistical comparison (one-tailed, paired-data t-test for equality of means) was completed to determine the statistical difference, if any, between the control (PDS13) and the corrosion control treatments (PDS14, PDS01 to PDS12). The results are summarized by phase and for the composite data set in Table 5.23. Several of the data entries are noted in boldface. These entries indicate the occurrence of an average concentration in the pH_s control that was exceeded by the average in the PDS that received a corrosion control measure. These events were confined to PDS07 (phase I, phase II, phase III, and composite), PDS06 (phase IV), and PDS08 (phase IV). The statistical analysis confirms that a significant improvement in iron release was achieved for all control strategies except the low and medium ZOP dose. This benefit was distributed throughout all study phases. Therefore, the benefit of the corrosion control measures is not restricted to a particular blend or water quality.

A similar statistical analysis was also conducted for comparison of the inhibitor-based treatments with the elevated pH control (PDS14), as summarized in Table 5.24. The statistical analysis identified only three inhibitor treatments (PDS02, PDS03, and PDS11) where the composite average iron concentration was significantly reduced by inhibitor addition, relative to a control strategy of pH elevation. Furthermore, in one half of the experimental units (PDS01, PDS05, PDS06, PDS07, PDS08, and PDS12), the composite average iron concentration in the

Table 5.23
Comparison of inhibitors to pH_s control for total iron

	PDS13	PDS14	PDS01	PDS02	PDS03	PDS04	PDS05	PDS06	PDS07	PDS08	PDS09	PDS10	PDS11	PDS12
Treatment	pH	pH	BOP	BOP	BOP	OP	OP	OP	ZOP	ZOP	ZOP	Silicate	Silicate	Silicate
Dose	pH_s	pH_s+0.3	Low	Medium	High	Low	Medium	High	Low	Medium	High	Low	Medium	High
Phase I														
Average	0.151	0.113	0.141	0.097	0.124	0.123	0.138	0.114	**0.179**	0.142	0.163	0.140	0.122	0.171
Std. dev.	0.031	0.039	0.067	0.022	0.048	0.042	0.026	0.059	**0.034**	0.049	0.118	0.098	0.040	0.136
N	14	14	14	14	14	14	14	14	**14**	14	14	14	14	14
Significant		Yes	No	Yes	Yes	Yes	No	Yes	**Yes**	No	No	No	Yes	No
p-value		0.008	0.315	<0.001	0.039	0.039	0.158	0.034	**0.005**	0.266	0.362	0.367	0.047	0.309
Phase II														
Average	0.277	0.188	0.207	0.149	0.147	0.159	0.210	0.174	**0.340**	0.230	0.164	0.166	0.134	0.191
Std. dev.	0.065	0.068	0.117	0.050	0.053	0.037	0.057	0.040	**0.128**	0.056	0.044	0.042	0.024	0.068
N	13	13	13	13	13	13	13	13	**13**	13	13	13	13	13
Significant		Yes	No	Yes	Yes	Yes	Yes	Yes	**Yes**	Yes	Yes	Yes	Yes	Yes
p-value		0.003	0.060	<0.001	<0.001	<0.001	<0.001	<0.001	**0.021**	0.008	<0.001	<0.001	<0.001	0.001
Phase III														
Average	0.268	0.140	0.125	0.105	0.123	0.144	0.230	0.144	0.215	0.222	0.107	0.115	0.097	0.114
Std. dev.	0.058	0.026	0.033	0.019	0.040	0.032	0.043	0.026	0.057	0.048	0.015	0.018	0.011	0.014
N	13	13	13	13	13	13	13	13	13	13	13	13	13	13
Significant		Yes	Yes	Yes	Yes	Yes	Yes	Yes	Yes	Yes	Yes	Yes	Yes	Yes
p-value		<0.001	<0.001	<0.001	<0.001	<0.001	0.018	<0.001	0.016	0.022	<0.001	<0.001	<0.001	<0.001

(continued)

Table 5.23 (Continued)

	PDS13	PDS14	PDS01	PDS02	PDS03	PDS04	PDS05	PDS06	PDS07	PDS08	PDS09	PDS10	PDS11	PDS12
Treatment	pH	pH	BOP	BOP	BOP	OP	OP	OP	ZOP	ZOP	ZOP	Silicate	Silicate	Silicate
Dose	pH_s	pH_s+0.3	Low	Medium	High	Low	Medium	High	Low	Medium	High	Low	Medium	High
Phase IV														
Average	0.192	0.141	0.117	0.126	0.089	0.135	0.134	**0.202**	**0.278**	**0.239**	0.107	0.114	0.095	0.118
Std. dev.	0.087	0.047	0.053	0.054	0.039	0.057	0.057	**0.126**	**0.079**	**0.126**	0.049	0.030	0.025	0.043
N	12	12	12	12	12	12	12	**12**	**12**	**12**	12	12	12	12
Significant		Yes	Yes	Yes	Yes	Yes	Yes	**No**	**Yes**	**Yes**	Yes	Yes	Yes	Yes
p-value		0.006	<0.001	0.002	<0.001	<0.001	0.003	**0.404**	**0.011**	**0.043**	<0.001	<0.001	<0.001	0.003
Composite														
Average	0.221	0.145	0.148	0.119	0.121	0.140	0.178	0.157	**0.251**	0.207	0.136	0.134	0.113	0.150
Std. dev.	0.081	0.054	0.081	0.043	0.049	0.044	0.063	0.077	**0.101**	0.083	0.073	0.060	0.031	0.086
N	52	52	52	52	52	52	52	52	**52**	52	52	52	52	52
Significant		Yes	Yes	Yes	Yes	Yes	Yes	Yes	**Yes**	No	Yes	Yes	Yes	Yes
p-value		<0.001	<0.001	<0.001	<0.001	<0.001	<0.001	<0.001	**0.016**	0.087	<0.001	<0.001	<0.001	<0.001

NOTE: Boldface entries indicate the occurrence of an average concentration in the pH_s control that was exceeded by the average in the PDS that received a corrosion control measure.

Table 5.24
Comparison of inhibitors to elevated pH control for total iron

	PDS14	PDS01	PDS02	PDS03	PDS04	PDS05	PDS06	PDS07	PDS08	PDS09	PDS10	PDS11	PDS12	PDS13
Treatment	pH	BOP	BOP	BOP	OP	OP	OP	ZOP	ZOP	ZOP	Silicate	Silicate	Silicate	pH
Dose	pH_s+0.3	Low	Medium	High	Low	Medium	High	Low	Medium	High	Low	Medium	High	pH_s
Phase I														
Average	0.113	**0.141**	0.097	**0.124**	**0.123**	**0.138**	**0.114**	**0.179**	**0.142**	**0.163**	**0.140**	**0.122**	**0.171**	**0.151**
Std. dev.	0.039	**0.067**	0.022	**0.048**	**0.042**	**0.026**	**0.059**	**0.034**	**0.049**	**0.118**	**0.098**	**0.040**	**0.136**	**0.031**
N	14	**14**	14	**14**	**14**	**14**	**14**	**14**	**14**	**14**	**14**	**14**	**14**	**14**
Significant		**No**	No	**No**	**No**	**Yes**	**No**	**Yes**	**Yes**	**Yes**	**No**	**No**	**No**	**Yes**
p-value		**0.103**	0.076	**0.211**	**0.205**	**0.014**	**0.460**	**<0.001**	**0.005**	**0.046**	**0.180**	**0.252**	**0.054**	**0.008**
Phase II														
Average	0.188	**0.207**	0.149	0.147	0.159	**0.210**	0.174	**0.340**	**0.230**	0.164	0.166	0.134	**0.191**	**0.277**
Std. dev.	0.068	**0.117**	0.050	0.053	0.037	**0.057**	0.040	**0.128**	**0.056**	0.044	0.042	0.024	**0.068**	**0.065**
N	13	**13**	13	13	13	**13**	13	**13**	**13**	13	13	13	**13**	**13**
Significant		**No**	No	Yes	No	**No**	No	**Yes**	**No**	No	No	Yes	**No**	**Yes**
p-value		**0.281**	0.051	0.039	0.086	**0.205**	0.268	**0.002**	**0.058**	0.082	0.186	0.003	**0.418**	**0.003**
Phase III														
Average	0.140	0.125	0.105	0.123	**0.144**	**0.230**	**0.144**	**0.215**	**0.222**	0.107	0.115	0.097	0.114	**0.268**
Std. dev.	0.026	0.033	0.019	0.040	**0.032**	**0.043**	**0.026**	**0.057**	**0.048**	0.015	0.018	0.011	0.014	**0.058**
N	13	13	13	13	**13**	**13**	**13**	**13**	**13**	13	13	13	13	**13**
Significant		No	Yes	No	**No**	**Yes**	**No**	**Yes**	**Yes**	Yes	Yes	Yes	Yes	**Yes**
p-value		0.140	<0.001	0.066	**0.366**	**<0.001**	**0.227**	**<0.001**	**<0.001**	<0.001	<0.001	<0.001	<0.001	**<0.001**

(continued)

Table 5.24 (Continued)

	PDS14	PDS01	PDS02	PDS03	PDS04	PDS05	PDS06	PDS07	PDS08	PDS09	PDS10	PDS11	PDS12	PDS13
Treatment	pH	BOP	BOP	BOP	OP	OP	OP	ZOP	ZOP	ZOP	Silicate	Silicate	Silicate	pH
Dose	pH_s+0.3	Low	Medium	High	Low	Medium	High	Low	Medium	High	Low	Medium	High	pH_s
Phase IV														
Average	0.141	0.117	0.126	0.089	0.135	0.134	**0.202**	**0.278**	**0.239**	0.107	0.114	0.095	0.118	**0.192**
Std. dev.	0.047	0.053	0.054	0.039	0.057	0.057	**0.126**	**0.079**	**0.126**	0.049	0.030	0.025	0.043	**0.087**
N	12	12	12	12	12	12	**12**	**12**	**12**	12	12	12	12	**12**
Significant		Yes	No	Yes	No	No	**No**	**Yes**	**Yes**	Yes	Yes	Yes	Yes	**Yes**
p-value		0.014	0.096	<0.001	0.294	0.250	**0.060**	**<0.001**	**0.003**	0.001	0.003	<0.001	0.041	**0.006**
Composite														
Average	0.145	**0.148**	0.119	0.121	0.140	**0.178**	**0.157**	**0.251**	**0.207**	0.136	0.134	0.113	**0.150**	**0.221**
Std. dev.	0.054	**0.081**	0.043	0.049	0.044	**0.063**	**0.077**	**0.101**	**0.083**	0.073	0.060	0.031	**0.086**	**0.081**
N	52	**52**	52	52	52	**52**	**52**	**52**	**52**	52	52	52	**52**	**52**
Significant		**No**	Yes	Yes	No	**Yes**	**No**	**Yes**	**Yes**	No	No	Yes	**No**	**Yes**
p-value		**0.390**	<0.001	0.002	0.248	**<0.001**	**0.148**	**<0.001**	**<0.001**	0.205	0.158	<0.001	**0.333**	**<0.001**

elevated pH control was less than the experienced concentration with addition of an inhibitor. It does not appear that the inhibitors achieved a consistent improvement in performance when compared to pH elevation. It must be acknowledged that elevation of pH may not be attractive when conditions are present that would produce excessive deposition of a calcium carbonate scale.

Release of Copper From the Corrosion Loop Systems

Empirical Model for Total Copper Concentration

The empirical model developed for total copper release in this study is presented in Equation 5.18. This model was developed using the same form as the iron models discussed previously. The model suggests that increasing pH will mitigate copper release, as shown by the negative exponent on the pH term. However, increases in alkalinity and chloride would be expected to increase copper release because of the positive exponents. Sulfate was not shown to be significant in this model but was found to be significant in the TBW I study. The additions of the different inhibitor types contribute total phosphorus, zinc, or silica to the system. These terms associated with the inhibitor dose all have negative exponents, indicating that addition of all inhibitors would have merit for reduction of the copper concentration.

$$\begin{aligned} TCu = (&6.13BOP \times TP^{-0.174} + 5.76OP \times TP^{-0.328} \\ &+ 5.24ZOP \times TP^{-0.205} \times Zn^{-0.069} + 20.9Si \times SiO_2^{-0.214} + 16.7pH_s) \\ &\times pH^{-3.48} \times alk^{0.569} \times Cl^{0.394} \end{aligned} \qquad (5.18)$$

where
- TCu = total copper, mg/L
- BOP = blended ortho-polyphosphate inhibitor (0, 1)
- OP = orthophosphate inhibitor (0, 1)
- ZOP = zinc orthophosphate inhibitor (0, 1)
- Si = silica inhibitor (0, 1)
- pH_s = pH control (0, 1)
- TP = total phosphorus, mg/L
- Zn = zinc, mg/L
- SiO_2 = silica, mg/L as SiO_2
- $pH = -\log \{H^+\}$
- alk = alkalinity, mg/L as $CaCO_3$
- Cl = chloride, mg/L

Figure 5.12 shows actual observed total copper concentrations versus the predictions of the model. The model fits the data well, with an R^2 of 0.76. The *p*-values for the independent variables that were retained in the model are TP (BOP) – 0.0337, TP (OP) – <0.001, TP (ZOP) – 0.0639, Zn – 0.2366, SiO_2 – 0.0053, pH – <0.0001, alk – <0.0001, and Cl – <0.0001. The terms for TP and Zn were retained in the model for the ZOP inhibitor in order to maintain sensitivity to the inhibitor dose.

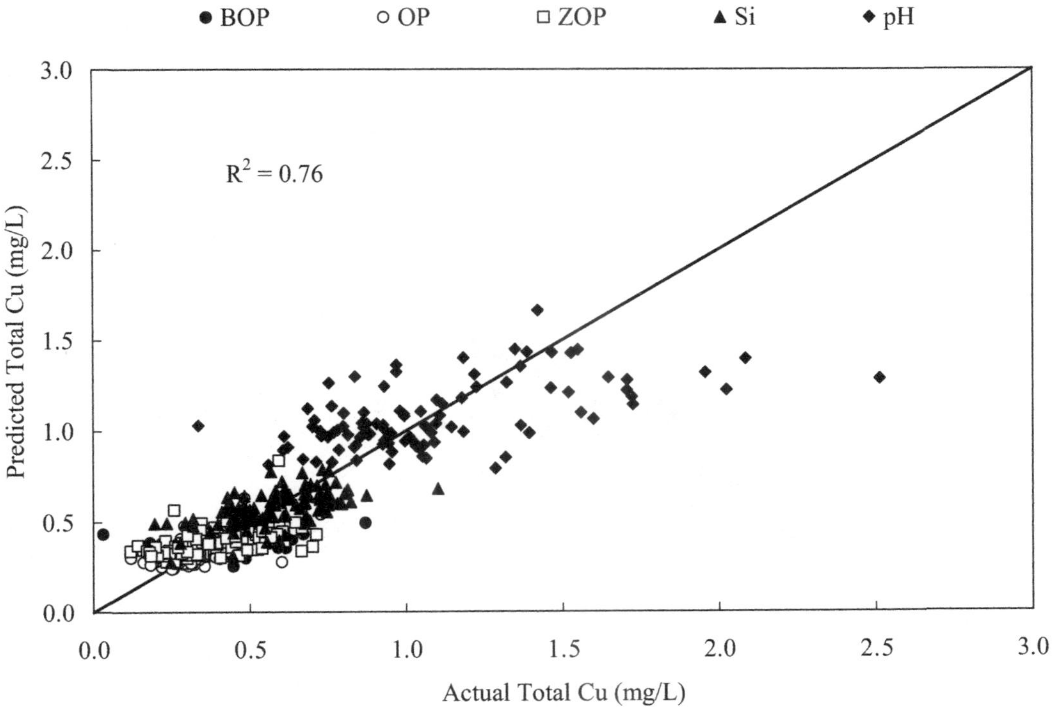

Figure 5.12 Model predicted versus actual total copper concentration

Empirical Model for Dissolved Copper Concentration

For this study, dissolved metal models were also developed, similar in form to the total models. The dissolved copper model is presented in Equation 5.19. Because the majority of the copper is present in a dissolved form, the model for dissolved copper is very similar to the total copper model. As was reported for the total copper model, adjustment (elevation) of pH would reduce dissolved copper concentrations. In contrast, elevation of alkalinity and chloride would be expected to increase dissolved copper concentrations. Sulfate was not demonstrated to be significant for prediction of dissolved copper. Also, all the model variables that quantify inhibitor dose (TP, Zn, SiO_2) have negative exponents, suggesting that addition of any of the inhibitors would reduce dissolved copper concentrations.

$$\begin{aligned} DCu = {} & (0.302BOP \times TP^{-0.140} + 0.2181OP \times TP^{-0.338} \\ & + 0.276ZOP \times TP^{-0.217} \times Zn^{-0.038} + 1.20Si \times SiO_2^{-0.285} + 0.831pH_s) \\ & \times pH^{-2.48} \times alk^{0.713} \times Cl^{0.421} \end{aligned} \tag{5.19}$$

where DCu = dissolved copper, mg/L
BOP = blended ortho-polyphosphate inhibitor (0, 1)
OP = orthophosphate inhibitor (0, 1)
ZOP = zinc orthophosphate inhibitor (0, 1)
Si = silica inhibitor (0, 1)
pH_s = pH control (0, 1)

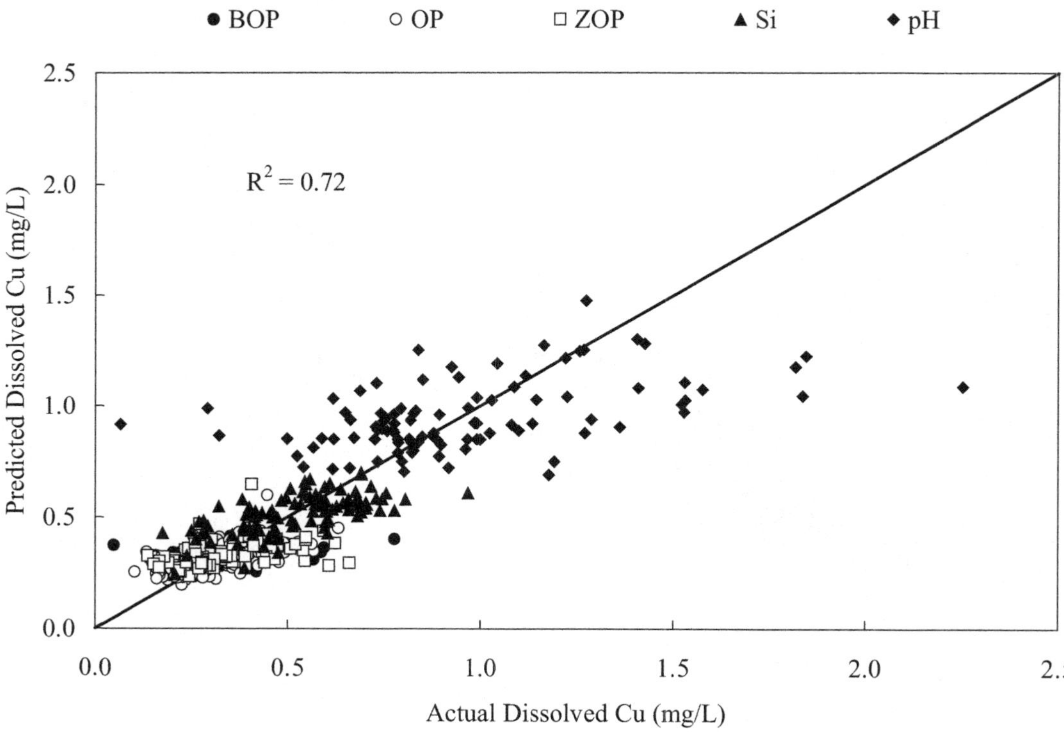

Figure 5.13 Model predicted versus actual dissolved copper concentration

TP = total phosphorus, mg/L
Zn = zinc, mg/L
SiO_2 = silica, mg/L as SiO_2
pH = –log $\{H^+\}$
alk = alkalinity, mg/L as $CaCO_3$
Cl = chloride, mg/L

A comparison between the predicted and observed dissolved copper concentrations is provided in Figure 5.13. The straight line represents a perfect agreement between the actual data and the model. The dissolved copper model also fits the data reasonably well, with an R^2 of 0.72. The *p*-values for the independent variables that were retained in the model are TP (BOP) – 0.1150, TP (OP) – <0.001, TP (ZOP) – 0.0842, Zn – 0.5770, SiO_2 – 0.0007, pH – <0.0001, alk – <0.0001, and Cl – <0.0001.

Evaluation of Copper Concentration Probability Distribution

An analysis was conducted to define the probability distributions for the copper concentration. As previously discussed, empirical models were developed to define the expected (or average) value of metal concentrations within the experimental system. However, consideration of the expected value would not be sufficient for determining regulatory compliance. Federal regulations mandate compliance with a copper action level of 1.3 mg/L that must not be exceeded in more than 10% of samples (corresponding to a 90th percentile). Thus, the identification of a suitable probability distribution would be necessary to relate the model output (defined in terms of the mean total copper concentration) to the regulatory standard (defined in terms of a 90th percentile event).

Multiple probability distributions were evaluated using the statistical software program EasyFit 3.2, developed by MathWave Technologies. All probability distributions provided with the software package were initially considered. Evaluation criteria were established to promote a thorough, yet practical, discrimination between the candidate probability distributions included in the software library. It was desired to select a distribution that was a good match for the data and that could be easily adapted for use. Any density function with more than three estimable parameters was rejected in order to promote a simpler approach. The selected distribution must be continuous; however, monotonic density functions were considered misrepresentative of metal release in a distribution system. In addition, a candidate distribution must satisfy a goodness-of-fit test for all data sets that were analyzed.

The goodness-of-fit test employed during the selection of a probability distribution was the Kolmogorov–Smirnov test. The test examines a sample from an unknown distribution and tests the null hypothesis that the unknown distribution is a specified function. The random sample is then compared with the specified distribution to determine whether the random sample is significantly different from the specified distribution. The test may be preferred for small sample sizes because of the ability of the Kolmogorov–Smirnov test to account for the sample size. The null hypothesis, stating that the sample follows the specified probability distribution, was evaluated using a level of significance of 0.2 (80% confidence level).

The normal probability distribution satisfied the established selection criteria. Data were partitioned by PDS and phase with the intent of allocating the source of variation in copper release to random fluctuations within the PDS. Of the 56 sample populations that were analyzed (14 PDSs operating during four phases), the normal distribution passed the Kolmogorov–Smirnov test for goodness-of-fit for all data sets.

The normal distribution was accepted as an adequate description of the random variation in copper release for the PDS. Although other probability distributions were just as successful during the analysis, the familiarity and simplicity of the normal distribution provided the basis for selection. Parameter estimates (mean and standard deviation) for the probability distribution are neither limited by software accessibility nor an advanced knowledge of statistics. Ratios defining the 90th percentile to the mean were calculated for all 56 populations. These ratios were then averaged to define a ratio of the 90th percentile concentration to the mean concentration of 1.3 for total copper release. This ratio (1.3) implies that for a 90th percentile copper action level of 1.3 mg/L, comparable average concentrations from the empirical model must not exceed 1.0 mg/L in order to remain in compliance.

Control of Copper Release With BOP Inhibitor Addition

BOP Data Summary (PDS01 to PDS03)—Copper. Table 5.25 shows summary data of the corrosion loop total and dissolved copper concentrations for PDS01 to PDS03 that receive the BOP. This data is segregated by phase and by PDS and lists the average, maximum, and minimum values observed for copper concentration. This data indicates that the majority (almost 90%) of the copper is found in the dissolved form. These results also suggest that control of copper release improves with increasing BOP dose. Water quality data for the independent variables in the empirical models for metal release were presented previously for the BOP PDSs in Table 5.14.

Table 5.25
Copper release summary for BOP PDSs

			Dissolved Cu			Total Cu		
Phase	PDS	Dose	Average	Maximum	Minimum	Average	Maximum	Minimum
I	01	0.57	0.36	0.46	0.05	0.40	0.52	0.03
	02	1.08	0.34	0.46	0.29	0.36	0.41	0.30
	03	1.82	0.33	0.43	0.22	0.37	0.62	0.23
II	01	0.57	0.35	0.57	0.16	0.43	0.59	0.18
	02	1.00	0.29	0.40	0.25	0.33	0.43	0.29
	03	1.95	0.24	0.32	0.17	0.27	0.33	0.21
III	01	0.49	0.38	0.51	0.26	0.43	0.57	0.30
	02	0.90	0.30	0.37	0.20	0.32	0.40	0.22
	03	1.73	0.26	0.33	0.22	0.29	0.35	0.23
IV	01	0.68	0.35	0.78	0.03	0.45	0.87	0.27
	02	1.31	0.31	0.59	0.20	0.37	0.67	0.25
	03	2.66	0.29	0.59	0.17	0.32	0.63	0.20
All	01	0.58	0.36	0.78	0.03	0.43	0.87	0.03
	02	1.07	0.31	0.59	0.20	0.34	0.67	0.22
	03	2.03	0.28	0.59	0.17	0.31	0.63	0.20

Analysis of BOP Inhibitor Performance—Copper. Application of the developed total copper model (Equation 5.18) to the BOP PDSs produces the reduced model shown in Equation 5.20. This simplified equation eliminates all dummy variables and associated data for corrosion control strategies other than addition of the BOP inhibitor. The total copper concentration is influenced by total phosphate dose, pH, alkalinity, and chloride.

The exponents on terms in the model define the influence of each variable on total copper release. The addition of BOP inhibitor determines the phosphorus concentrations in PDSs, represented by the TP term in Equation 5.20. The model suggests that BOP addition decreases copper release due to the negative exponent on the TP term. However, the magnitude of the exponent is the smallest in the model.

$$\mathrm{TCu} = 6.13 \times \mathrm{TP}^{-0.174} \times \mathrm{pH}^{-3.48} \times \mathrm{alk}^{0.569} \times \mathrm{Cl}^{0.394} \qquad (5.20)$$

Figure 5.14 shows separate bars for the average predicted and average actual concentrations of total copper concentration in each of the BOP PDSs for each phase. Whiskers define the maximum and minimum measured and predicted concentrations. The model does reasonably well at predicting copper release for the BOP PDSs, emulating the decreasing trend of actual copper release with increasing dose. Review of Figure 5.14 indicates that the model slightly overpredicts total copper release for the medium and high BOP doses (PDS02 and PDS03) in phases II, III, and IV, while the empirical model slightly underpredicts actual release for phase I. For the low BOP dose in PDS01, there is less variability between predicted and actual copper release, with the only noticeable underprediction occurring during phases I and II. Irrespective of phase changes, total

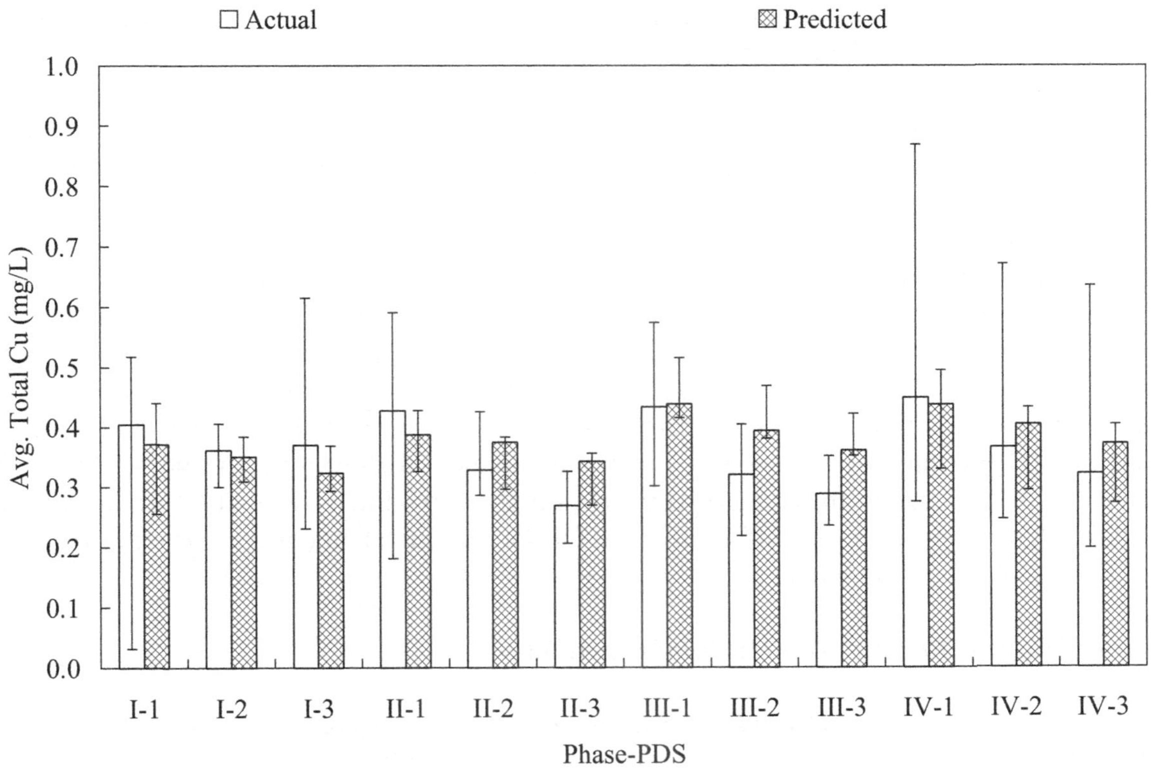

Figure 5.14 Total copper release model summary for BOP PDSs by phase

copper release appears to decrease with increasing dose, as evident in Figure 5.14 and Table 5.25. The results demonstrate a strong beneficial impact of increases in BOP dose on total copper release.

The remaining terms in the model (pH, alkalinity, and chloride) are related to the water quality of the source water blend. Phase II utilized a blend with a greater surface water composition (62%) than was used in either phase I or III (27%) or in phase IV (40%). The average total copper concentrations for all BOP PDSs in phases I through IV are 0.38 mg/L, 0.34 mg/L, 0.35 mg/L, and 0.38 mg/L, respectively. The elevated surface water component resulted in a reduced alkalinity (103 mg/L as $CaCO_3$) for phase II. Other phases had higher alkalinity (160, 150, and 123 mg/L as $CaCO_3$ for phases I, III, and IV, respectively) due to higher components of groundwater than surface water. Those phases with an elevated alkalinity coincide with the greater average copper concentration. The increase in copper concentration at elevated alkalinity is consistent with the inclusion of alkalinity in the empirical model with a positive exponent on that variable. Actual total copper release shown in Figure 5.14 supports this conclusion, where concentrations from PDSs in phase II were slightly lower than those in phases I, III, or IV.

The model in Equation 5.20 is very sensitive to pH given that its exponent has the largest magnitude of all exponents in the model (–3.48). This suggests that a slight change in pH would have a noticeable impact on total copper release. However, the average variation in pH between phases for all three BOP PDSs was limited to a range of 7.88 in phase IV to 8.04 in phase I (a pH range of 0.16). Even though the magnitude of the pH exponent is larger than the other exponents in Equation 5.20, the relative variations in actual pH values were minimal compared to the relative variation in alkalinity; therefore, the experienced impact of pH on actual copper release was less than that of alkalinity for the four phases.

The model is also sensitive to chloride, where it has a positive exponent in the model but a smaller magnitude than exponents of both pH and alkalinity. This suggests that an increase in chloride results in an increase in total copper release, but not to the extent caused by an increase in alkalinity.

Because BOP is a blend of orthophosphate and polyphosphate, a term accounting for the difference in phosphorus speciation in BOP was introduced during model development. However, this term was not found to be statistically significant. Therefore, phosphorus speciation associated with use of the BOP inhibitor was accounted for in the total copper model by the dummy variable coefficient and the associated TP term. The dummy variable coefficients for BOP and OP (Equation 5.18) are different (6.13 and 5.76, respectively), and the exponent on the TP term associated with BOP is larger (less negative) than the exponent for OP (–0.174 and –0.328, respectively). The differences in the numerical values for these parameter estimates are an indication that BOP affects copper release differently than OP for an equivalent delivered dose of both inhibitors in terms of TP measurements.

Control of Copper Release With OP Inhibitor Addition

OP Data Summary (PDS04 to PDS06)—Copper. A summary of the total and dissolved copper release data from the corrosion loops receiving OP inhibitor is presented in Table 5.26 by phase and PDS. Consistently, about 80% to 90% of the total copper is in the dissolved form. The copper concentration is reported to decrease with increasing OP dose, indicating an expected improvement in copper control with addition of the OP inhibitor. Also, the highest overall copper release was reported during phase I, which had the highest alkalinity concentration. Water quality data for the independent variables in the empirical models for metal release were presented previously in Table 5.16 for the OP PDSs.

Analysis of OP Inhibitor Performance—Copper. The total copper model developed for this study was presented previously in Equation 5.18. When the general model is restricted to the description of performance with the OP inhibitor, the model reduces to Equation 5.21, where all terms have been previously defined. The addition of OP inhibitor to the system adds a proportional quantity of phosphate, which is predicted by the model to achieve a reduction in the total copper concentration. The negative exponent on the TP term shows that an increase in OP inhibitor dose will cause a decrease in total copper release. This is consistent with the data summary presented in Table 5.26.

$$TCu = 5.76 \times TP^{-0.328} \times pH^{-3.48} \times alk^{0.569} \times Cl^{0.394} \qquad (5.21)$$

Figure 5.15 shows the total copper release in each of the OP corrosion loops. The averages are taken by phase and by PDS and compared to the total copper predicted by the empirical model. The maximum and minimum observed and predicted concentrations are also shown. The trend of decreasing copper with increasing OP dose is confirmed in Figure 5.15 with both the actual data and the model predictions.

The copper release for each phase is shown to be relatively consistent, with maximum concentrations never exceeding 0.73 mg/L. This concentration is well below the action level of

Table 5.26
Copper release summary for OP PDSs

			Dissolved Cu			Total Cu		
Phase	PDS	Dose	Average	Maximum	Minimum	Average	Maximum	Minimum
I	04	0.49	0.46	0.57	0.41	0.50	0.59	0.44
	05	1.04	0.38	0.48	0.26	0.41	0.48	0.27
	06	1.84	0.26	0.38	0.16	0.30	0.60	0.16
II	04	0.55	0.33	0.49	0.13	0.39	0.52	0.28
	05	0.90	0.28	0.38	0.15	0.35	0.43	0.18
	06	1.91	0.21	0.28	0.10	0.24	0.31	0.12
III	04	0.54	0.36	0.44	0.29	0.40	0.50	0.25
	05	0.99	0.29	0.37	0.24	0.35	0.43	0.28
	06	1.87	0.19	0.26	0.14	0.21	0.28	0.16
IV	04	0.47	0.34	0.63	0.20	0.40	0.73	0.26
	05	0.82	0.32	0.56	0.19	0.36	0.64	0.24
	06	1.69	0.21	0.34	0.12	0.23	0.37	0.15
All	04	0.51	0.38	0.63	0.13	0.42	0.73	0.25
	05	0.94	0.32	0.56	0.15	0.37	0.64	0.18
	06	1.83	0.21	0.38	0.10	0.25	0.60	0.12

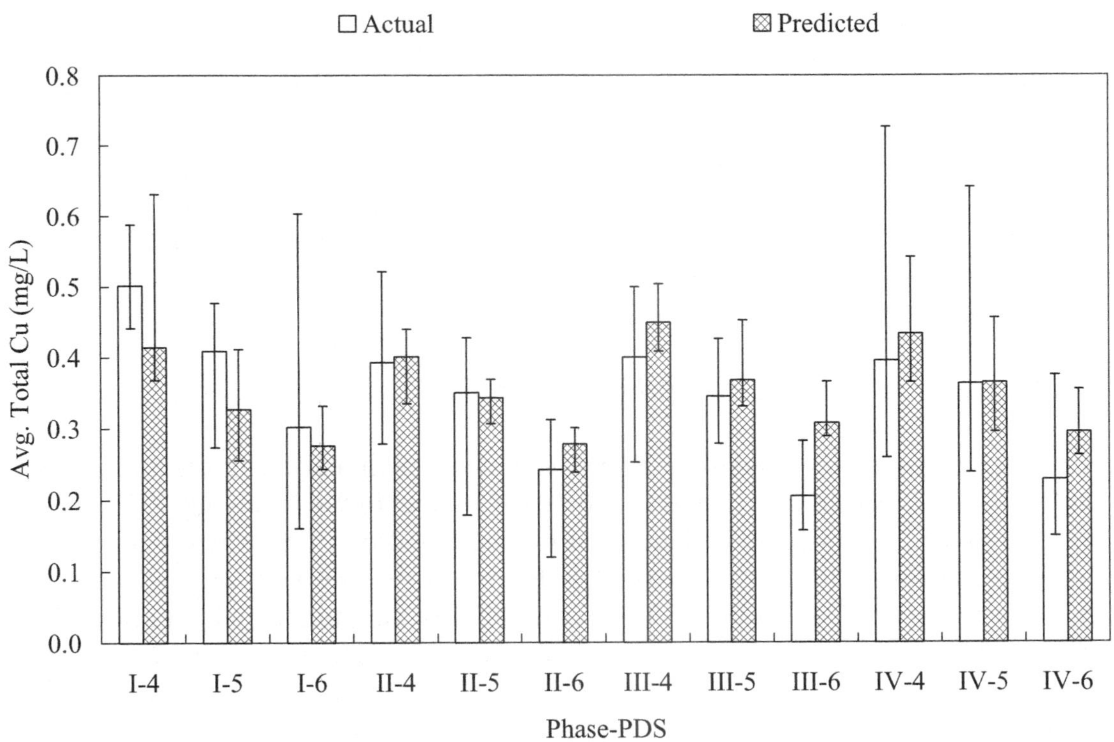

Figure 5.15 Copper release model summary for OP PDSs by phase

Table 5.27
Copper release summary for ZOP PDSs

			Dissolved Cu			Total Cu		
Phase	PDS	Dose	Average	Maximum	Minimum	Average	Maximum	Minimum
I	07	0.64	0.50	0.62	0.43	0.56	0.71	0.44
	08	0.99	0.34	0.44	0.23	0.36	0.44	0.23
	09	1.78	0.27	0.41	0.20	0.32	0.59	0.21
II	07	0.54	0.39	0.53	0.26	0.51	0.62	0.31
	08	0.89	0.32	0.52	0.19	0.39	0.58	0.25
	09	1.77	0.26	0.66	0.09	0.30	0.70	0.13
III	07	0.49	0.46	0.59	0.39	0.53	0.65	0.39
	08	0.88	0.30	0.41	0.24	0.34	0.45	0.28
	09	1.82	0.17	0.21	0.14	0.18	0.23	0.12
IV	07	0.46	0.40	0.73	0.27	0.49	0.84	0.34
	08	0.88	0.32	0.48	0.21	0.38	0.54	0.26
	09	1.58	0.29	0.61	0.15	0.32	0.67	0.18
All	07	0.54	0.44	0.73	0.26	0.52	0.84	0.31
	08	0.91	0.32	0.52	0.19	0.37	0.58	0.23
	09	1.74	0.25	0.66	0.09	0.28	0.70	0.12

1.3 mg/L. These results suggest that even the lowest dose of OP inhibitor is beneficial for reducing copper concentrations below the action level.

Phase I is shown to have the highest copper release for each dose in the actual data. This is likely due to the high alkalinity from the groundwater-enriched blend of phase I. The model supports an adverse effect of alkalinity, with a positive exponent on the alkalinity term. The impact of alkalinity on copper release is not beneficial, in marked contrast with the beneficial effect of alkalinity on the release of iron. The chloride term of the model also has a positive exponent, suggesting that it contributes to the copper release. The pH term is shown with a negative exponent, offering reduction in copper release with increasing values.

Control of Copper Release With ZOP Inhibitor Addition

ZOP Data Summary (PDS07 to PDS09)—Copper. Summary data for the corrosion loop total and dissolved copper concentrations for each of the PDSs receiving ZOP inhibitor are provided in Table 5.27. This data is segregated by phase and by PDS. Water quality data for the independent variables in the empirical models for metal release were presented previously in Table 5.18 for the ZOP PDSs. PDS07, PDS08, and PDS09 were treated with a low, medium, and high dose of ZOP inhibitor, respectively. As shown in Table 5.27, on average, the majority of the copper in ZOP PDSs is present in a dissolved form. The results confirm that the experienced copper concentration generally decreased as the ZOP inhibitor dose was increased.

Analysis of ZOP Inhibitor Performance—Copper. The total copper model developed for this study was presented previously in Equation 5.18. When applied to use of the ZOP inhibitor, the model is simplified to yield Equation 5.22, where all terms have been previously defined.

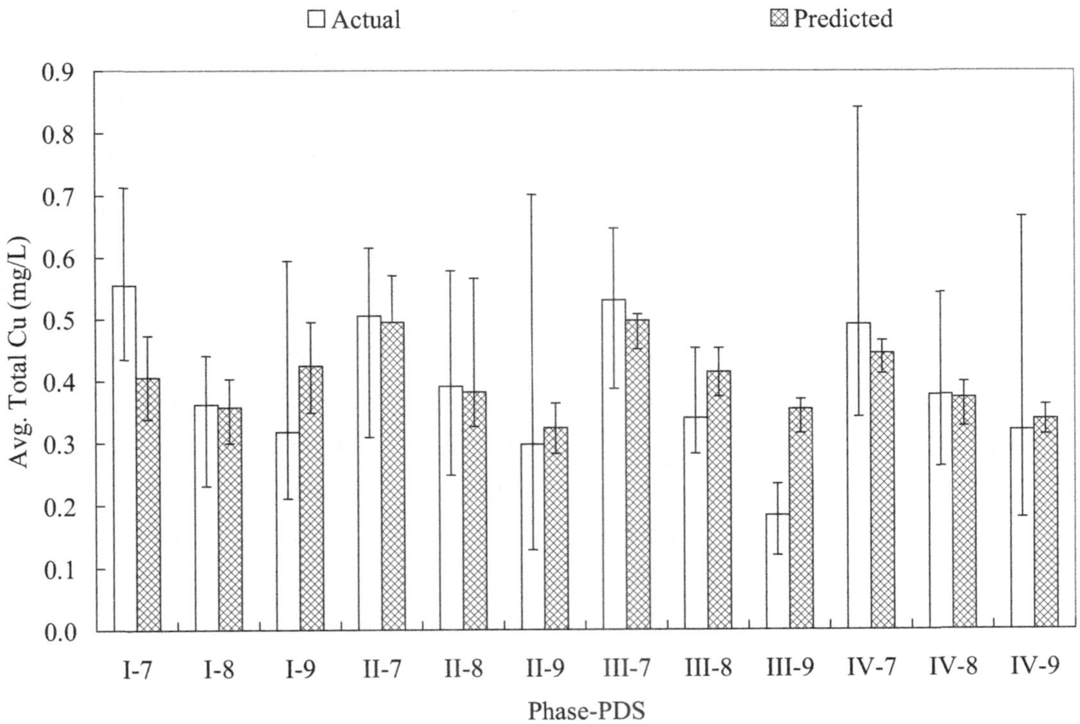

Figure 5.16 Total copper release model summary for ZOP PDSs by phase

$$TCu = 5.24 \times TP^{-0.205} \times Zn^{-0.069} \times pH^{-3.48} \times alk^{0.569} \times Cl^{0.394} \quad (5.22)$$

Significant amounts of phosphorus and zinc were added to PDS07–PDS09 in conjunction with the addition of ZOP inhibitor. The negative exponents on the TP and Zn terms in the model indicate that both total phosphorus and zinc were beneficial for reducing copper release. The small exponent on the zinc term (–0.069) generates a model that is not very sensitive to zinc for the ZOP PDSs. Nevertheless, the combined impact of total phosphorus and zinc in the empirical model indicate that addition of the ZOP inhibitor will result in improved copper release control.

Figure 5.16 shows the average total copper concentration in each of the ZOP PDSs for each phase compared to the predicted copper concentration for the empirical model. The maximum and minimum measured and predicted concentrations are also noted. As shown in the graph, the actual total copper release decreased as inhibitor dose increased in all cases, confirming a better inhibitor performance at higher inhibitor doses. As discussed before, the model suggests the same trend as in Figure 5.16, although the variation of predicted values is not as great as actual values.

All other terms in the model are related to the water quality of the blend. The blends in phases I and III were characterized by high alkalinity. The positive exponent on the alkalinity term indicates that increased alkalinity will result in higher copper concentrations. This pattern was confirmed by the data in Figure 5.16.

The model is also sensitive to chloride. The positive exponent on the chloride term in the model suggests that elevated chloride concentrations will aggravate copper release. The pH term in the model is shown with a negative exponent, which indicates that better control of copper would occur at higher pH values.

Table 5.28
Copper release summary for Si PDSs

			Dissolved Cu			Total Cu		
Phase	PDS	Dose	Average	Maximum	Minimum	Average	Maximum	Minimum
I	10	15.2	0.60	0.76	0.28	0.68	0.81	0.41
	11	21.5	0.61	0.81	0.30	0.68	0.87	0.37
	12	31.5	0.41	0.60	0.18	0.47	0.75	0.20
II	10	7.8	0.59	0.86	0.38	0.68	0.99	0.45
	11	11.7	0.51	0.67	0.37	0.60	0.78	0.40
	12	17.2	0.35	0.60	0.04	0.43	0.68	0.05
III	10	13.3	0.61	0.77	0.33	0.69	0.83	0.60
	11	17.0	0.58	0.71	0.35	0.65	0.78	0.57
	12	22.6	0.45	0.54	0.25	0.50	0.59	0.43
IV	10	9.2	0.62	1.08	0.38	0.70	1.26	0.45
	11	13.1	0.51	0.87	0.29	0.58	1.00	0.32
	12	17.7	0.37	0.55	0.25	0.42	0.63	0.29
All	10	11.5	0.60	1.08	0.28	0.69	1.26	0.41
	11	16.0	0.55	0.87	0.29	0.63	1.00	0.32
	12	22.5	0.39	0.60	0.04	0.46	0.75	0.05

Control of Copper Release With Silicon Inhibitor Addition

Si Data Summary (PDS10 to PDS12)—Copper. The copper release data using a silica inhibitor treatment are summarized in Table 5.28. Total copper control was consistently enhanced with any increase in the Si inhibitor dose. This observation is more pronounced between treatments associated with PDS11 and PDS12, perhaps suggesting a linear effect with silica dose. A similar response to silica was obtained for the total and dissolved forms of copper. Dissolved copper consistently accounted for nearly 90% of copper release within the system. Thus, total and dissolved copper would be expected to exhibit similar responses for a given source of variation.

Additional sources of variation were considered as water quality was altered between phases. These sources of variation are less conspicuous from simple assessment of Table 5.28. However, the development of empirical models describing copper release facilitated the identification of influential water quality parameters. Water quality data for the independent variables in the empirical models for metal release were presented previously in Table 5.20 for the Si PDSs.

Analysis of Silicon Inhibitor Performance—Copper. Total copper release may be described by Equation 5.23 for the Si PDSs. Equation 5.23 represents an alternative form of the comprehensive empirical model (Equation 5.18) for total copper release that applies specifically to treatment with silica as a corrosion inhibitor. As was suggested in the previous section, silica did affect the release of total copper. The model suggests that the water quality parameters (pH, alkalinity, and chlorides) affect copper release as well.

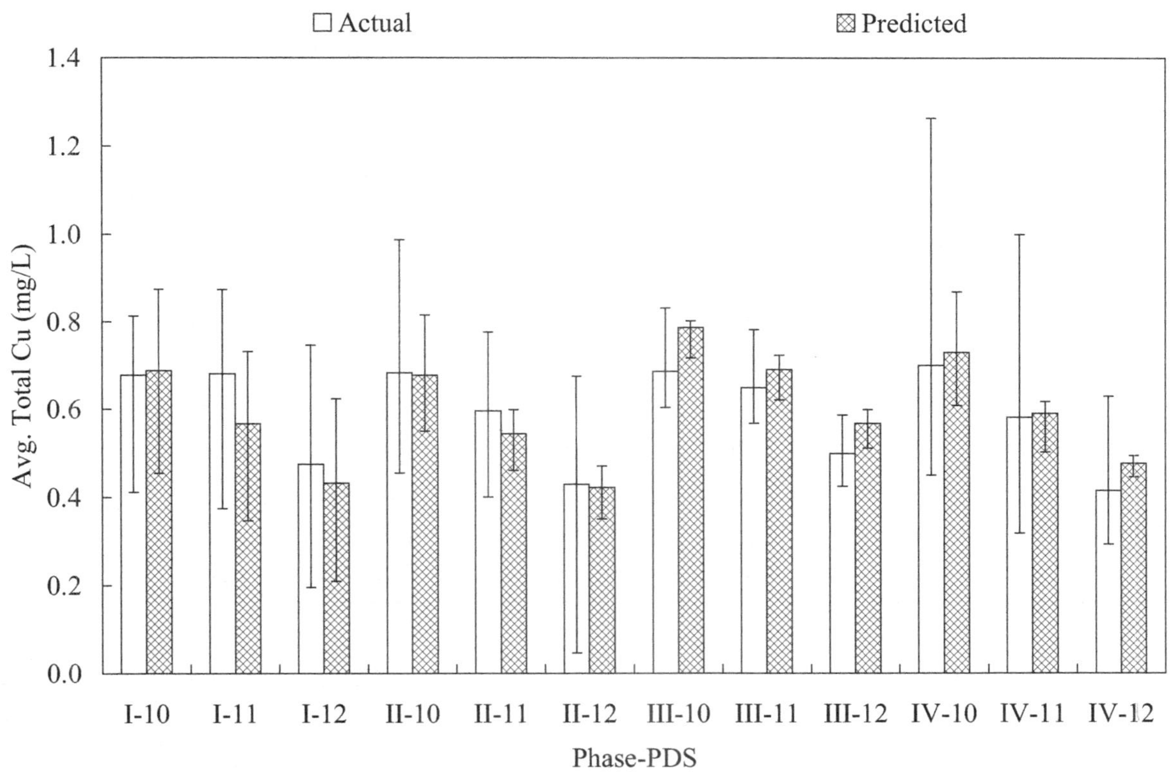

Figure 5.17 Total copper release summary for Si PDSs by phase

$$TCu = 20.9 \times SiO_2^{-0.214} \times pH^{-3.48} \times alk^{0.569} \times Cl^{0.394} \tag{5.23}$$

Simulation of the collected data using Equation 5.23 is shown in Figure 5.17, along with the corresponding actual total copper concentration. It is clear from the copper release observed in the system that increasing silica dose was beneficial to controlling copper release. The model demonstrates the same relationship due to the negative exponent on the SiO_2 variable.

Less obvious than the contribution of silica are the effects associated with the water quality parameters. However, it appears that the control of copper was more effective during phases II and IV. This observation becomes clear when comparing the high-dose treatments across all phases. Phases II and IV utilized blends in which groundwater was not the predominant source water, producing a blend with a reduced alkalinity. The effect of alkalinity is consistent with the associated exponent of Equation 5.23. The positive exponent suggests that increasing alkalinity would encourage copper release.

Equation 5.23 also identifies pH and chlorides as significant water quality parameters. The effect of chlorides may be demonstrated from phases II and IV. During phase II alkalinity was near 100 mg/L as $CaCO_3$, whereas during phase IV alkalinities were approximately 125 mg/L as $CaCO_3$. From Equation 5.23, the model would predict copper release to increase from phase II to phase IV. However, it can be seen from Figure 5.17 that both phases exhibit similar copper release. The beneficial effect associated with a reduction in chloride concentration counteracts the expected response while considering only alkalinity. Phase II experienced higher concentrations

Table 5.29
Copper release summary for pH control PDSs

		Dissolved Cu			Total Cu		
Phase	PDS	Average	Maximum	Minimum	Average	Maximum	Minimum
I	13	0.95	1.41	0.69	1.04	1.52	0.75
	14	0.89	1.29	0.29	1.00	1.37	0.76
II	13	1.04	1.53	0.73	1.16	1.71	0.67
	14	0.73	1.24	0.13	0.82	1.34	0.21
III	13	1.32	1.84	0.84	1.48	2.08	0.97
	14	0.79	1.13	0.62	0.85	1.14	0.69
IV	13	1.14	2.25	0.06	1.41	2.51	0.34
	14	0.83	1.53	0.50	0.93	1.72	0.56
All	13	1.11	2.25	0.06	1.27	2.51	0.34
	14	0.81	1.53	0.13	0.90	1.72	0.21

of chlorides than phase IV. Hence, the observations agree well with Equation 5.23, implying that chlorides promote copper release.

The empirical model also suggests that increasing pH will deter copper release. The effect of pH is complicated, however, by an interaction between SiO_2 concentration and pH. The sodium silicate inhibitor utilized during the study was highly alkaline, resulting in increases in pH with increasing inhibitor dose. The maximum pH increase between the silica-treated PDSs approached 0.2 pH units. The resulting data set suffered from a confounding relation between SiO_2 and pH, as both variables increased in value with greater inhibitor doses. A similar effect would be present for SiO_2 and alkalinity due to alkalinity elevation with greater inhibitor doses.

Addition of the Si inhibitor demonstrated a distinct benefit for control of copper within the PDS. For the inhibitor doses employed in this study, compliance with the 1.3 mg/L regulatory standard for total copper appears to be feasible. Although this standard represents the 90th percentile of the observed total copper release, the bars indicating the maximum event from Figure 5.17 suggests that most of the sampling events are in compliance.

Control of Copper Release by pH Elevation (pH Controls)

pH Controls Data Summary (PDS13 to PDS14)—Copper. A summary of the total and dissolved copper release data for the corrosion loops from the pH control PDSs is shown in Table 5.29 by phase and PDS. It can be seen that the majority of the copper is in the dissolved form, consistently about 80% to 90% of the total copper in all phases for each PDS. Also, the copper concentrations in the pH_s PDS (PDS13) are consistently and significantly greater than the concentrations for the elevated pH treatment (PDS14). This disparity suggests that increasing the pH in the system has beneficial effects on copper release. Water quality data for the independent variables in the empirical models for metal release were presented previously in Table 5.22 for the pH Control PDSs.

Analysis of pH Control Performance—Copper. The total copper model was presented previously in Equation 5.18. When the results for operation with an inhibitor (PDS01 to PDS12)

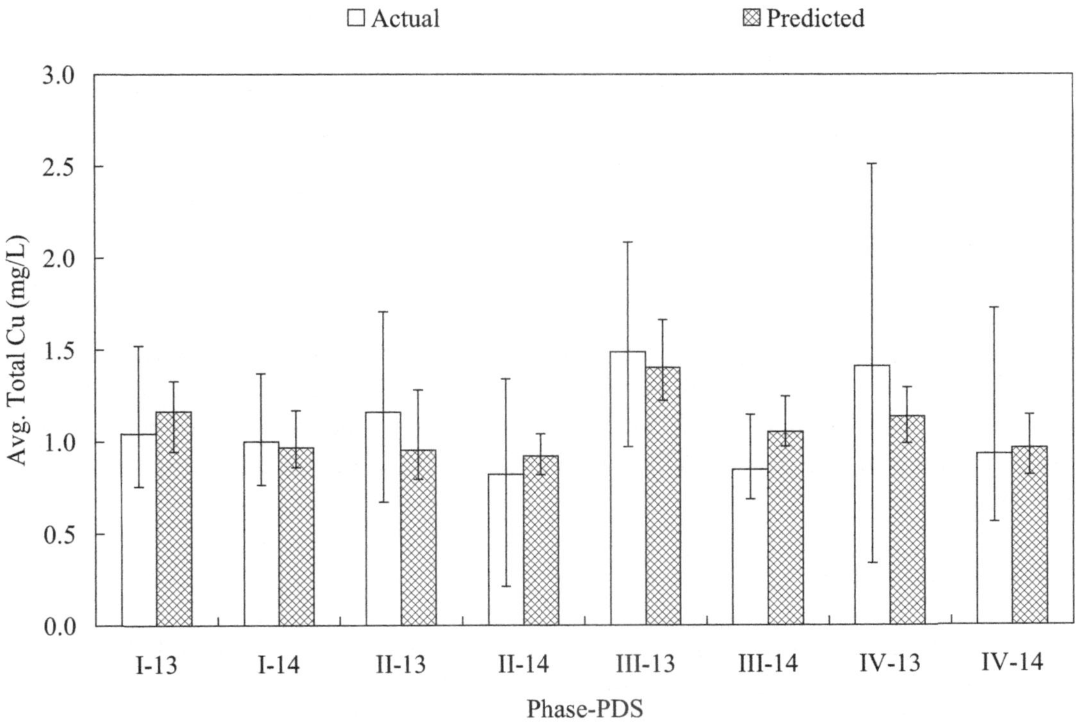

Figure 5.18 Total copper release model summary for pH control PDSs by phase

are removed, the empirical model reverts to Equation 5.24, where all terms have been previously defined. Any increase in pH between PDS13 and PDS14 would be accompanied by an increase in alkalinity. The increase in pH is beneficial in controlling copper release, as seen by the negative exponent on the pH term in the model. However, the increase in alkalinity contributes to copper release, shown by the positive exponent on the alkalinity term in the model. The actual data clearly demonstrate that the copper release is more sensitive to the increase in pH, as seen consistently by the lower copper concentration in PDS14 than PDS13.

$$TCu = 16.7 \times pH^{-3.48} \times alk^{0.569} \times Cl^{0.394} \tag{5.24}$$

Figure 5.18 shows the average copper concentration for each PDS by phase for the pH control PDSs compared to the model predictions. Observed and predicted maximum and minimum concentrations are also noted on the chart. Figure 5.18 further confirms that copper release is mitigated by the elevated pH treatment in PDS14. The actual data as well as the model predictions consistently support this trend.

The influence of variable alkalinity on copper concentration can be evaluated by comparison of data (observed or predicted) for an individual PDS for different phases. For PDS14, the greatest alkalinity was experienced in phase I. The greatest average total concentration (1.00 mg/L) was also experienced during phase I. The coincident occurrence of maximum observed values of alkalinity and copper concentration would also be predicted by the empirical model. A similar examination of the PDS14 data reveals the occurrence of the minimum observed average value of copper concentration during phase II (0.82 mg/L). The phase II blend represented the smallest GW component and exhibited the smallest alkalinity of all blends. For the

PDS13 data, the maximum observed average concentration of copper was experienced in phase III, corresponding with a high-alkalinity blend.

The effect of chloride concentration may also be examined by comparison of data for an individual PDS for different phases. For PDS13, chloride concentrations were the lowest in phase I. The minimum observed average copper concentration was also experienced during phase I. Because HCl was added to PDS13 to maintain pH_s, the chloride concentrations and the associated adverse effect on copper concentration are elevated in PDS13.

The effect of pH is more easily understood by comparison of results between PDS13 and PDS14. The pH remained relatively stable for each PDS across all phases of the study. The targeted pH for PDS13 was pH_s, and the set point pH for PDS14 was 0.3 units above pH_s. The beneficial effect of the elevated pH is readily apparent from lower average copper concentrations in all phases for PDS14 than in PDS13. PDS14 never exceeded the copper action level of 1.3 mg/L on average in any phase. PDS13 exceeded the action level on average in both phase III and phase IV.

Copper Release Summary

Review of Empirical Models. The empirical models that were developed for total and dissolved copper concentration (Equations 5.18 and 5.19) generally confirmed the relations that were defined in TBW I. For both empirical modeling efforts, elevation of pH was determined to have merit for mitigation of copper release. For both cases, elevation of alkalinity associated with GW-enriched blends was detrimental to compliance with copper standards. For the robust TBW I data set, temperature and sulfate were also identified to significantly accelerate copper release. For the current data set, temperature and sulfate were not demonstrated to have a significant impact on copper concentration. In contrast with the model results from TBW I, the current data support an adverse impact of chloride on the release of copper.

Data Summary. A presentation of average output total copper concentrations for the 12-month study duration is shown in Figure 5.19. These data represent an average of the weekly determinations of corrosion loop copper concentrations. The bar chart presentation follows the same organization as described previously for iron in Figure 5.11.

The graphical summary of the data provides a clear picture of the impact of the corrosion control measures for copper. Elevation of pH is shown to substantially reduce the copper concentration relative to operation at pH_s. The annual average corrosion loop concentrations for PDS13 and PDS14 were 1.27 and 0.90 mg/L, respectively. In the absence of a control strategy (either pH elevation or inhibitor addition), compliance with the action level (90th percentile concentration) of 1.30 mg/L would be problematic. The average (approximately 50th percentile) total copper concentration for the pH_s control exceeded the action level for phases III and IV. Elevation of the pH by 0.3 units was demonstrated to reduce the copper concentration by approximately 30%. The reported mitigation of copper release was even more dramatic when any of the inhibitors was used. Relative to operation at pH_s, the annual average corrosion loop concentrations were reduced by 45% to 80% with inhibitors. All of the inhibitors were effective at all doses that were investigated. The data in Figure 5.19 also clearly suggest a benefit associated with increasing the inhibitor dose.

A statistical comparison (one-tailed, paired-data t-test for equality of means) between the control (PDS13) and the corrosion control treatments (PDS14, PDS01 to PDS12) is reviewed in Table 5.30. The analysis demonstrated a statistically significant benefit for elevation of pH in four

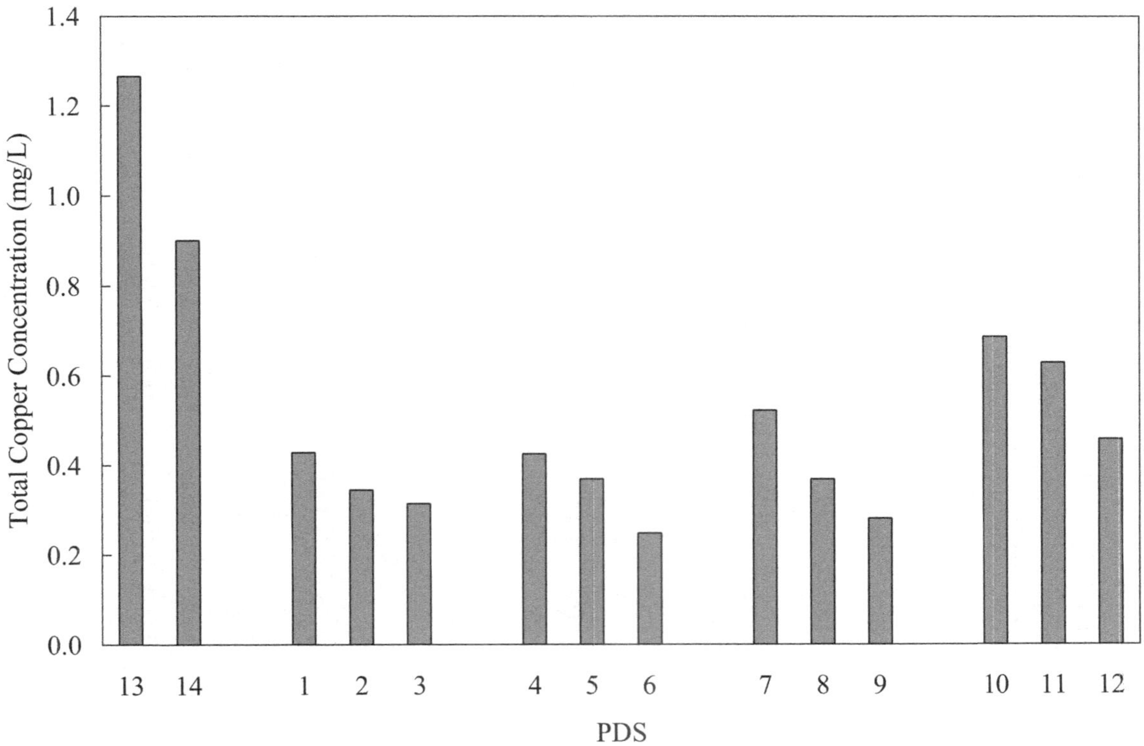

Figure 5.19 Data summary for total copper concentration

of five comparisons (phase II, phase III, phase IV, and composite). Inhibitor addition was shown to provide a statistically significant benefit for all inhibitors at all doses for all study phases (60 comparisons). The inhibitors proved to be effective for all blends and water qualities that were examined, including blends that were enriched in groundwater with an associated elevated alkalinity (phases I and III). These results provide important guidance for operation at alkalinities that would be desired to address red water episodes but would present copper compliance difficulties in the absence of inhibitor addition or pH elevation.

Release of Lead From the Corrosion Loop Systems

Empirical Model for Total Lead Concentration

The empirical model developed for total lead release in this study is presented in Equation 5.25. This model was developed in similar form to the iron and copper models discussed previously. As the model suggests, elevation of pH and sulfate concentration both reduce total lead concentration. This effect is expressed by the negative exponent on these water quality variables. Increases in chloride and temperature are associated with increases in total lead concentration due to the positive exponent on these variables. This model is similar to the model developed in the TBW I study in that the same water quality terms are significant and the effect of each variable is similar (same sign on exponents). However, in contrast with the results from TBW I, alkalinity was not demonstrated to have a significant effect on lead concentration in the current study.

Table 5.30
Comparison of inhibitors to pH_s control for total copper

	PDS13	PDS14	PDS01	PDS02	PDS03	PDS04	PDS05	PDS06	PDS07	PDS08	PDS09	PDS10	PDS11	PDS12
Treatment	pH	pH	BOP	BOP	BOP	OP	OP	OP	ZOP	ZOP	ZOP	Silicate	Silicate	Silicate
Dose	pH_s	pH_s+0.3	Low	Medium	High	Low	Medium	High	Low	Medium	High	Low	Medium	High
Phase I														
Average	1.042	0.999	0.405	0.362	0.370	0.501	0.409	0.302	0.555	0.362	0.318	0.678	0.680	0.475
Std. dev.	0.200	0.161	0.125	0.029	0.085	0.046	0.049	0.101	0.078	0.054	0.107	0.119	0.142	0.178
N	14	14	14	14	14	14	14	14	14	14	14	14	14	14
Significant		No	Yes	Yes	Yes	Yes	Yes	Yes	Yes	Yes	Yes	Yes	Yes	Yes
p-value		0.231	<0.001	<0.001	<0.001	<0.001	<0.001	<0.001	<0.001	<0.001	<0.001	<0.001	<0.001	<0.001
Phase II														
Average	1.158	0.820	0.427	0.328	0.269	0.392	0.350	0.243	0.505	0.391	0.298	0.683	0.596	0.429
Std. dev.	0.297	0.254	0.098	0.041	0.031	0.054	0.066	0.045	0.109	0.087	0.138	0.127	0.111	0.140
N	13	13	13	13	13	13	13	13	13	13	13	13	13	13
Significant		yes	yes	yes	yes	yes	yes	yes	yes	yes	yes	yes	yes	yes
p-value		<0.001	<0.001	<0.001	<0.001	<0.001	<0.001	<0.001	<0.001	<0.001	<0.001	<0.001	<0.001	<0.001
Phase III														
Average	1.484	0.846	0.434	0.320	0.288	0.400	0.345	0.206	0.525	0.340	0.184	0.686	0.648	0.499
Std. dev.	0.298	0.135	0.073	0.050	0.038	0.062	0.044	0.035	0.090	0.046	0.031	0.066	0.064	0.051
N	13	13	13	13	13	13	13	13	13	13	13	13	13	13
Significant		Yes	Yes	Yes	Yes	Yes	Yes	Yes	Yes	Yes	Yes	Yes	Yes	Yes
p-value		<0.001	<0.001	<0.001	<0.001	<0.001	<0.001	<0.001	<0.001	<0.001	<0.001	<0.001	<0.001	<0.001

(continued)

Table 5.30 (Continued)

	PDS13	PDS14	PDS01	PDS02	PDS03	PDS04	PDS05	PDS06	PDS07	PDS08	PDS09	PDS10	PDS11	PDS12
Treatment	pH	pH	BOP	BOP	BOP	OP	OP	OP	ZOP	ZOP	ZOP	Silicate	Silicate	Silicate
Dose	pH_s	pH_s+0.3	Low	Medium	High	Low	Medium	High	Low	Medium	High	Low	Medium	High
Phase IV														
Average	1.408	0.928	0.449	0.366	0.322	0.395	0.363	0.229	0.492	0.377	0.321	0.700	0.582	0.416
Std. dev.	0.574	0.342	0.166	0.123	0.121	0.136	0.123	0.065	0.133	0.101	0.145	0.254	0.193	0.118
N	12	12	12	12	12	12	12	11	12	12	12	12	12	11
Significant		Yes	Yes	Yes	Yes	Yes	Yes	Yes	Yes	Yes	Yes	Yes	Yes	Yes
p-value		<0.001	<0.001	<0.001	<0.001	<0.001	<0.001	<0.001	<0.001	<0.001	<0.001	<0.001	<0.001	<0.001
Composite														
Average	1.266	0.900	0.428	0.344	0.313	0.424	0.368	0.247	0.521	0.367	0.280	0.686	0.628	0.457
Std. dev.	0.396	0.238	0.117	0.070	0.084	0.092	0.078	0.075	0.103	0.075	0.124	0.150	0.137	0.132
N	52	52	52	52	52	52	52	51	52	52	52	52	52	51
Significant		Yes	Yes	Yes	Yes	Yes	Yes	Yes	Yes	Yes	Yes	Yes	Yes	Yes
p-value		<0.001	<0.001	<0.001	<0.001	<0.001	<0.001	<0.001	<0.001	<0.001	<0.001	<0.001	<0.001	<0.001

$$TPb = (6.64 \times 10^{-6}\ BOP \times TP^{-0.283} + 6.94 \times 10^{-6}\ OP \times TP^{-1.47} + 2.08 \times 10^{-7}\ ZOP \times TP^{-0.892} \times Zn^{-0.824} + 3.80 \times 10^{-4}\ Si \times SiO_2^{-1.16} + 4.14 \times 10^{-5}\ pH_s) \times pH^{-2.75} \times Cl^{1.98} \times SO_4^{-1.54} \times temp^{2.94} \quad (5.25)$$

where
TPb = total lead, mg/L
BOP = blended ortho-polyphosphate inhibitor (0, 1)
OP = orthophosphate inhibitor (0, 1)
ZOP = zinc orthophosphate inhibitor (0, 1)
Si = silica inhibitor (0, 1)
pH_s = pH control (0, 1)
TP = total phosphorus, mg/L
Zn = zinc, mg/L
SiO_2 = silica, mg/L as SiO_2
pH = $-\log [H^+]$
Cl = chloride, mg/L
SO_4 = sulfate, mg/L
temp = temperature, °C

Inhibitor addition would result in an increase in the concentration of total phosphorus, zinc, and and/or silica. The exponents associated with these water quality variables are all negative, supporting a beneficial effect of all study inhibitors. The *p*-values for the independent variables that were retained in the model are TP (BOP) – 0.6963, TP (OP) – <0.0001, TP (ZOP) – 0.4603, Zn – 0.1575, SiO_2 – 0.0049, pH – 0.0587, Cl – <0.0001, SO_4 – <0.0001, and temp – <0.0001. Even though the *p*-values for certain variables exceeded the normal criteria of 0.05, terms associated with inhibitor dose (TP, Zn, and pH) were retained to maintain sensitivity to inhibitor dose.

Figure 5.20 shows the empirical model predictions versus the actual observations in the entire data set. The straight line represents perfect agreement between the data set and the model predictions. Agreement between the model and the data is reasonable, evidenced by an R^2 of 0.55.

Empirical Model for Dissolved Lead Concentration

Dissolved metal models were similarly developed for this study. The dissolved lead model is shown in Equation 5.26. This model is very similar to the total lead model with the exception that alkalinity is a significant term. Elevation of pH and sulfate both are shown in the model to offer a reduction in dissolved lead concentration. This effect is evidenced by a negative exponent on those variables. In contrast, increases in alkalinity, chloride, and temperature all contribute to aggravation of the dissolved lead concentration.

$$DPb = (1.69 \times 10^{-6}\ BOP \times TP^{0.252} + 1.63 \times 10^{-6}\ OP \times TP^{-0.545} + 1.37 \times 10^{-6}\ ZOP \times TP^{0.083} \times Zn^{-0.108} + 8.45 \times 10^{-5}\ Si \times SiO_2^{-1.11} + 7.76 \times 10^{-6}\ pH_s) \times pH^{-4.35} alk^{0.882} \times Cl^{2.00} \times SO_4^{-1.52} \times temp^{2.84} \quad (5.26)$$

where
DPb = dissolved lead, mg/L
BOP = blended ortho-polyphosphate inhibitor (0, 1)
OP = orthophosphate inhibitor (0, 1)

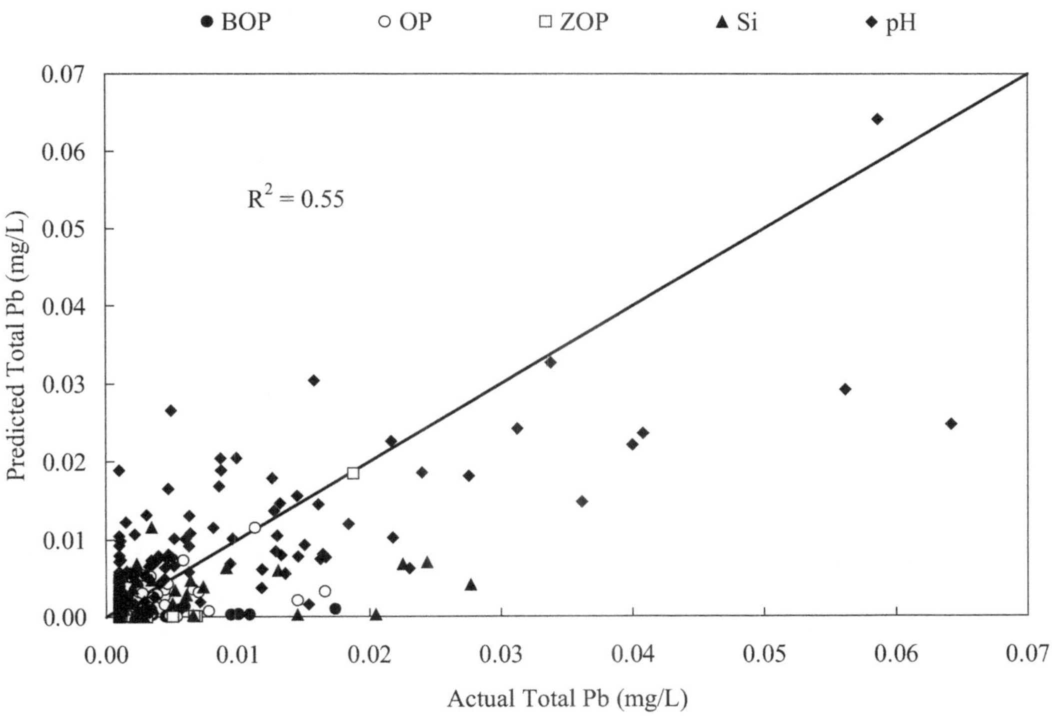

Figure 5.20 Model predicted versus actual total lead concentration

ZOP = zinc orthophosphate inhibitor (0, 1)
Si = silica inhibitor (0, 1)
pH_s = pH control (0, 1)
TP = total phosphorus, mg/L
Zn = zinc, mg/L
SiO_2 = silica, mg/L as SiO_2
pH = $-\log [H^+]$
alk = alkalinity, mg/L as $CaCO_3$
Cl = chloride, mg/L
SO_4 = sulfate, mg/L
temp = temperature, °C

The TP, Zn, and Si terms with their associated dummy variables for each inhibitor type are also similar to those shown in the total lead model. However, the exponent on the TP term for the BOP and ZOP inhibitors is positive. This suggests that addition of BOP would elevate the dissolved lead concentrations. The ZOP term has a negative exponent on the associated zinc term that will likely compensate for the minimal effects of the total phosphorus term with a small positive exponent. The *p*-values for the independent variables that were retained in the model are TP (BOP) – 0.6383, TP (OP) – 0.3062, TP (ZOP) – 0.9212, Zn – 0.8056, SiO_2 – 0.0038, pH – 0.0075, alk – 0.0207, Cl – <0.0001, SO_4 – <0.0001, and temp – <0.0001.

Figure 5.21 shows the actual dissolved lead concentrations versus the predictions by the dissolved lead model. The straight line represents a perfect agreement between the data set and

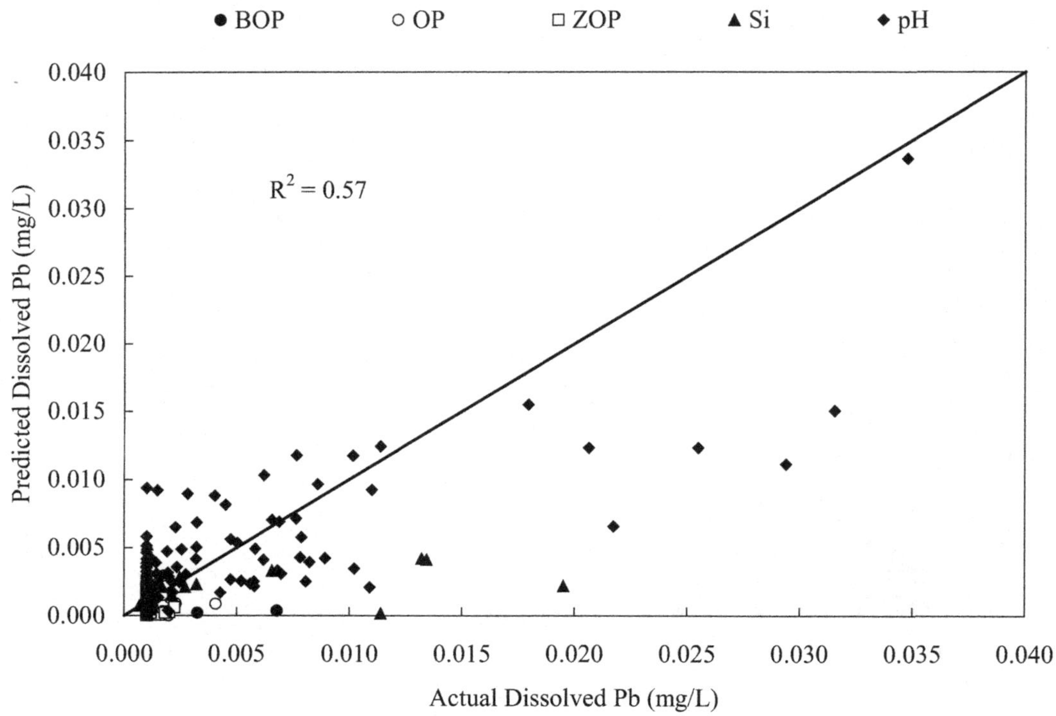

Figure 5.21 Model predicted versus actual dissolved lead concentration

the model predictions. The model achieves a reasonable fit between the predictions and the observations with an R^2 of 0.57.

Evaluation of Lead Concentration Probability Distribution

An analysis was conducted to define the probability distributions for the lead concentration. The approach is identical to the previously discussed procedure for definition of a probability distribution for copper. The empirical models developed in this study would be appropriate for definition of average concentrations. Federal regulations mandate compliance with a 90th percentile lead action level of 0.015 mg/L. Thus, identification of a suitable adjustment of the mean concentration would be necessary to relate the model output to the regulatory standard.

The normal distribution was accepted as an adequate description of the random variation in lead concentration. A normal distribution is appealing based on ease of application. Parameter estimates (mean and standard deviation) for the probability distribution are neither limited by software accessibility nor an advanced knowledge of statistics. Ratios defining the 90th percentile to the mean were calculated for all 56 populations (14 PDSs and four phases). The average ratio of the 90th percentile concentration to the mean concentration was 2.5 for total lead release. For the 90th percentile lead action level of 0.015 mg/L, the comparable required average concentration from the empirical model would be 0.006 mg/L in order to remain in compliance.

Table 5.31
Lead release summary for BOP PDSs

			Dissolved Pb			Total Pb		
Phase	PDS	Dose	Average	Maximum	Minimum	Average	Maximum	Minimum
I	01	0.57	0.000	0.003	0.000	0.003	0.017	0.000
	02	1.08	0.000	0.001	0.000	0.001	0.009	0.000
	03	1.82	0.001	0.007	0.000	0.001	0.011	0.000
II	01	0.57	0.000	0.001	0.000	0.002	0.009	0.000
	02	1.00	0.000	0.001	0.000	0.001	0.003	0.000
	03	1.95	0.000	0.000	0.000	0.001	0.006	0.000
III	01	0.49	0.000	0.001	0.000	0.001	0.003	0.000
	02	0.90	0.000	0.001	0.000	0.001	0.001	0.000
	03	1.73	0.000	0.000	0.000	0.000	0.001	0.000
IV	01	0.68	0.000	0.001	0.000	0.001	0.002	0.000
	02	1.31	0.000	0.001	0.000	0.001	0.001	0.000
	03	2.66	0.000	0.001	0.000	0.000	0.001	0.000
All	01	0.58	0.000	0.003	0.000	0.002	0.017	0.000
	02	1.07	0.000	0.001	0.000	0.001	0.009	0.000
	03	2.03	0.000	0.007	0.000	0.001	0.011	0.000

Control of Lead Release With BOP Inhibitor Addition

BOP Data Summary (PDS01 to PDS03)—Lead. Table 5.31 provides summary data for the corrosion loop total and dissolved lead concentrations for PDS01 to PDS03 that receive the BOP. This data is segregated by phase and by PDS and lists the average, maximum, and minimum values observed for the lead concentration.

These results suggest that there is a decrease in total lead release with increasing BOP dose. The majority of the lead is often present in a particulate form, especially for those cases with the greatest lead concentration. Also, many of the dissolved lead measurements were below the detection limit of analysis. Water quality data for the independent variables in the empirical models for metal release were presented previously in Table 5.14 for the BOP PDSs.

Analysis of BOP Inhibitor Performance—Lead. Application of the developed total lead model (Equation 5.25) to BOP PDSs results in the simplified model without dummy variables in Equation 5.27. Total lead release is influenced by total phosphate through BOP dosing, pH, chloride, sulfate, and temperature.

The exponents on terms in the model indicate the influence of each term on total lead release. The addition of BOP inhibitor is responsible for phosphorus concentrations in PDSs, represented by the TP term in Equation 5.27. The model suggests that BOP addition decreases lead release due to the negative exponent on the TP term. However, the magnitude of the exponent is the smallest in the model (–0.283).

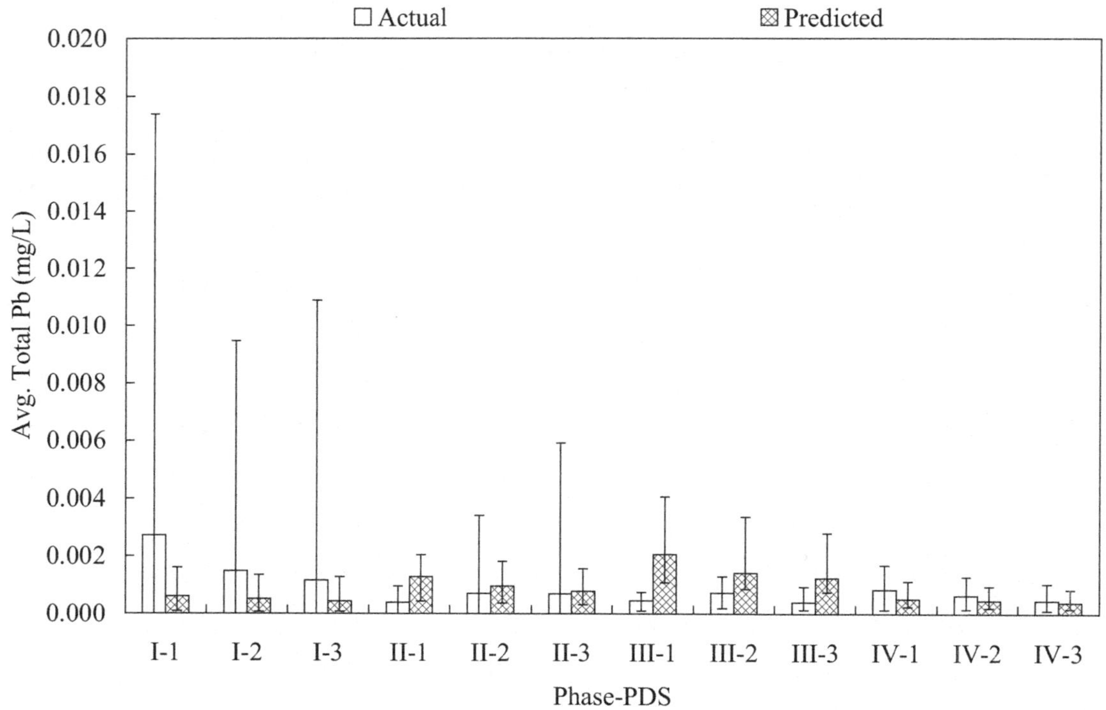

Figure 5.22 Total lead release model summary for BOP PDSs by phase

$$TPb = 6.64 \times 10^{-6}\, TP^{-0.283} \times pH^{-2.75} \times Cl^{1.98} \times SO_4^{-1.54} \times temp^{2.94} \quad (5.27)$$

Figure 5.22 shows separate bars for the average predicted and average actual concentrations of total lead release in each of the BOP PDSs for each phase. Whiskers are also shown to define the maximum and minimum measured and predicted concentrations. Actual total lead release decreases with increasing BOP dose during phases I and IV. In phases II and III, the medium BOP dose (PDS02) seems to result in higher actual total lead than the other two doses. The actual data from phase I showed the greatest variability for all BOP PDSs.

Model prediction by Equation 5.27 follows a decreasing trend with increasing dose for phases II and III, as shown in Figure 5.22. The beneficial effect of increasing BOP inhibitor dose is less noticeable for the model predictions in phases I and IV despite a clear declining trend for actual total lead data with increasing BOP dose. The empirical model overpredicts actual total lead release in phases II and III, while it underpredicts the actual data in phases I and IV. The prediction accuracy of the model in Equation 5.27 may be hindered by the abundance of below-detection-limit observations in the actual data.

The remaining terms in the model (pH, chloride, sulfate, and temperature) are related to the water quality of the source water blend. Phase II utilized a blend with a greater surface water composition (62%) than was used in other phases. The elevated surface water component resulted in increased sulfate (105 mg SO_4/L) for phase II. However, actual average total lead release in phase II was not noticeably different from phases II and IV, yet it was lower than lead releases in phase I. Sulfate in the empirical model has a negative exponent, indicting that higher sulfate would lead to lower total lead release.

Phases I and III received the same blend of source water, and, therefore, both phases exhibit comparable water quality. However, seasonal variability resulted in different average temperatures (21.3°C and 25.7°C for phases I and III, respectively). The higher predicted lead release observed in phase III compared to phase I is consistent with the inclusion of temperature in the empirical model in Equation 5.27. The temperature exponent in the empirical model is positive and has the largest magnitude of all exponents in the model (2.94). The anticipated lead concentrations would be quite sensitive to temperature, and any increase in temperature yields an increase in predicted total lead release. However, the model underpredicts actual total lead release in phase I and overpredicts actual data in phase III, as shown in Figure 5.22.

The empirical model is also sensitive to pH with an exponent of –2.75. However, the average variation in pH between phases for all three BOP PDSs was limited to a range of 7.88 in phase IV to 8.04 in phase I (a pH range of 0.16). Consequently, the predicted impact of pH on lead release is expected to be less significant than the temperature impact for the actual study conditions.

The model is also sensitive to chloride, where it has a positive exponent in the model with a significant magnitude (1.98). This suggests that an increase in chloride results in an increase in total lead release. Average chloride concentrations for phases II and III were higher than phases I and IV. However, the actual lead release did not vary greatly between phases as would be predicted based on chloride variability alone.

Even though Figure 5.22 shows a slight variability between predicted and actual total lead release, the model did not grossly overpredict actual releases beyond the lead action level of 0.015 mg Pb/L.

Given that BOP is a blend of orthophosphate and polyphosphate, a term accounting for the difference in phosphorus speciation in BOP was introduced during model development. However, this term was not found to be statistically significant. Therefore, the phosphate blend in BOP was accounted for in the total lead model by the dummy variable coefficient and the associated TP term. Equation 5.25 shows that the coefficient of dummy variables for BOP and OP are different (6.64×10^{-6} and 6.94×10^{-6}, respectively), and the exponent on the TP term associated with BOP is larger (less negative) than the exponent for OP (–0.283 and –1.47, respectively). These differences in model parameter estimates are an indication that the form of phosphorus in the BOP inhibitor affects lead release somewhat differently than with the OP inhibitor, even though actual measurements of delivered TP dose of both inhibitors are relatively equal.

Control of Lead Release With OP Inhibitor Addition

OP Data Summary (PDS04 to PDS06)—Lead. A summary of the total and dissolved lead release data for each of the corrosion loops receiving OP inhibitor is shown in Table 5.32 by phase and PDS. The majority of the lead released in the PDSs is in the particulate form. The data generally exhibit a decrease in lead release with increasing OP dose, suggesting that addition of OP inhibitor has a beneficial effect on control of lead release. Water quality data for the independent variables in the empirical models for metal release were presented previously in Table 5.16 for the OP PDSs.

Analysis of OP Inhibitor Performance—Lead. The total lead model for this study was presented previously in Equation 5.25. When the OP inhibitor is applied, the model is in the form shown in Equation 5.28, where all terms have been previously defined. The OP inhibitor adds a significant amount of phosphorus to the system. The model suggests that increasing TP

Table 5.32
Lead release summary for OP PDSs

			Dissolved Pb			Total Pb		
Phase	PDS	Dose	Average	Maximum	Minimum	Average	Maximum	Minimum
I	04	0.49	0.000	0.002	0.000	0.003	0.015	0.000
	05	1.04	0.000	0.002	0.000	0.001	0.003	0.000
	06	1.84	0.000	0.001	0.000	0.001	0.005	0.000
II	04	0.55	0.001	0.001	0.000	0.005	0.009	0.000
	05	0.90	0.000	0.000	0.000	0.000	0.000	0.000
	06	1.91	0.000	0.001	0.000	0.000	0.004	0.000
III	04	0.54	0.001	0.002	0.001	0.004	0.009	0.001
	05	0.99	0.000	0.001	0.000	0.000	0.001	0.000
	06	1.87	0.000	0.001	0.000	0.000	0.001	0.000
IV	04	0.47	0.001	0.004	0.000	0.004	0.017	0.000
	05	0.82	0.000	0.001	0.000	0.000	0.001	0.000
	06	1.69	0.000	0.001	0.000	0.003	0.026	0.000
All	04	0.51	0.001	0.004	0.000	0.004	0.017	0.000
	05	0.94	0.000	0.002	0.000	0.000	0.003	0.000
	06	1.83	0.000	0.001	0.000	0.001	0.026	0.000

concentrations reduces total lead release, as seen by the negative exponent on the TP term. This prediction is consistent with the data for the OP PDSs reported in Table 5.32.

$$TPb = 6.94 \times 10^{-6} \times TP^{-1.47} \times pH^{-2.75} \times Cl^{1.98} \times SO_4^{-1.54} \times temp^{2.94} \quad (5.28)$$

Figure 5.23 shows the average total lead release in the OP PDSs by phase compared to the predictions by the total lead model. The maximum and minimum observed and predicted concentrations are also shown. For the observed and predicted values, the low OP dose consistently exhibits the highest average total lead concentration. This observation is consistent with the model prediction of improved lead release control with increased OP inhibitor dose.

The average lead concentration in each of the OP PDSs never exceeded the lead action level of 0.015 mg/L. However, in phase IV, maximum observations in PDS04 and PDS06 both exceeded the action level with concentrations of 0.017 and 0.026 mg/L, respectively. These extreme values were individual occurrences with all other observations well below the action level. The data indicate that maintaining lead concentrations below the action level may be achieved by addition of OP inhibitor at all doses observed in this study. The results also indicate that periodic cases with elevated lead concentrations are experienced. It is believed that these episodes of high lead concentration are associated with release of lead in a particulate form.

Phase III is predicted to have the highest lead release in all OP PDSs by the model due to the high chloride concentration and high temperature observed in this phase. Sulfate concentrations were also low in phase III because of the low percentage of surface water. The positive exponents on the chloride and temperature terms indicate that increases in those variables would

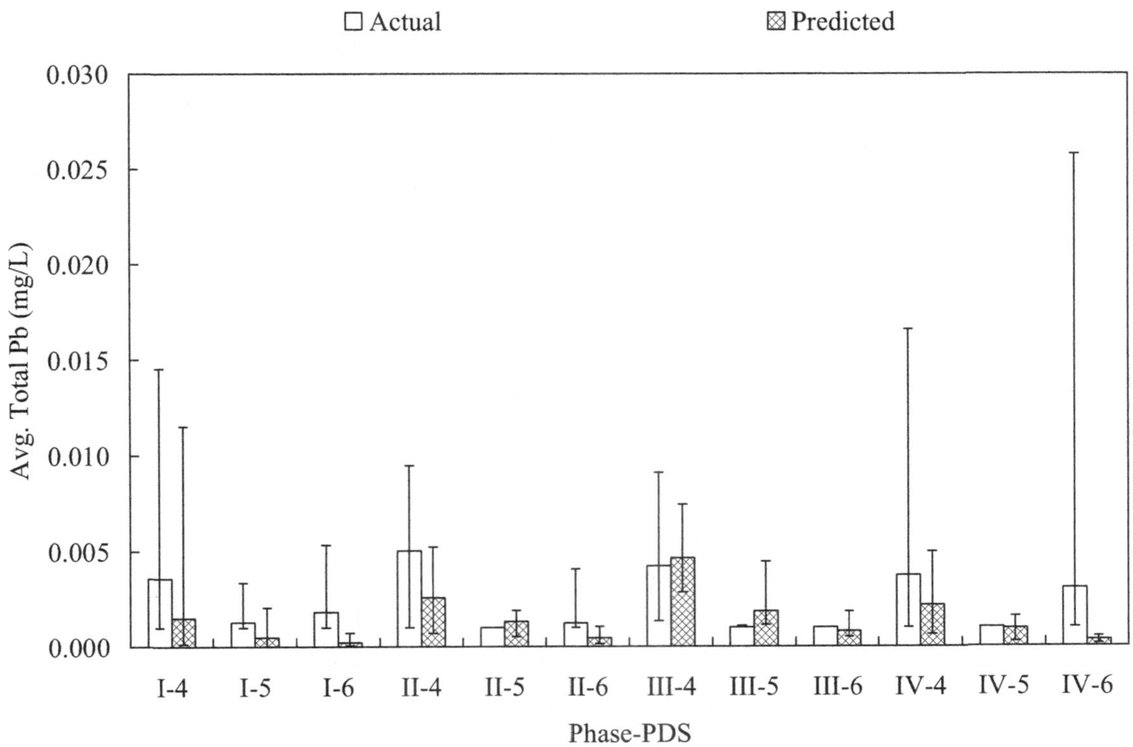

Figure 5.23 Lead release model summary for OP PDSs

contribute to lead release. In contrast, sulfate has a negative exponent, indicating that the reduced sulfate concentration in phase III would also aggravate lead concentrations. Based on these water quality trends, phase I is predicted to yield the lowest lead release, as the chloride and temperature were both lowest during phase I. The pH was relatively constant through all phases and all OP PDSs for this study, so the effect of changes in pH is not as obvious in the model predictions. The pH term does have a negative exponent, so increasing pH would be a beneficial control measure for reducing lead release.

Control of Lead Release With ZOP Inhibitor Addition

ZOP Data Summary (PDS07 to PDS09)—Lead. A summary of the total and dissolved lead release data for each of the PDSs receiving ZOP inhibitor is shown in Table 5.33. This data is segregated by phase and by PDS. Water quality data for the independent variables in the empirical models for metal release were presented previously in Table 5.18 for the ZOP PDSs. The majority of the lead released in the PDSs is in the particulate form. There is not any apparent improved inhibition of lead release with the increase in ZOP dose.

Analysis of ZOP Inhibitor Performance—Lead. The total lead model developed for this study was presented previously in Equation 5.25. When the ZOP inhibitor is applied, the model is of the form shown in Equation 5.29, where all terms have been previously defined.

$$TPb = 2.08 \times 10^{-7} \times TP^{-0.892} \times Zn^{-0.824} \times pH^{-2.75} \times Cl^{1.98} \times SO_4^{-1.54} \times temp^{2.94} \quad (5.29)$$

Table 5.33
Lead release summary for ZOP PDSs

			Dissolved Pb			Total Pb		
Phase	PDS	Dose	Average	Maximum	Minimum	Average	Maximum	Minimum
I	07	0.64	0.000	0.001	0.000	0.000	0.002	0.000
	08	0.99	0.000	0.002	0.000	0.000	0.003	0.000
	09	1.78	0.000	0.002	0.000	0.002	0.019	0.000
II	07	0.54	0.000	0.002	0.000	0.000	0.001	0.000
	08	0.89	0.000	0.001	0.000	0.000	0.002	0.000
	09	1.77	0.001	0.002	0.000	0.001	0.007	0.000
III	07	0.49	0.000	0.001	0.000	0.000	0.001	0.000
	08	0.88	0.000	0.000	0.000	0.001	0.001	0.000
	09	1.82	0.000	0.001	0.000	0.001	0.001	0.000
IV	07	0.46	0.000	0.001	0.000	0.001	0.001	0.000
	08	0.88	0.000	0.001	0.000	0.001	0.001	0.000
	09	1.58	0.001	0.002	0.000	0.001	0.005	0.000
All	07	0.54	0.000	0.002	0.000	0.000	0.002	0.000
	08	0.91	0.000	0.002	0.000	0.001	0.003	0.000
	09	1.74	0.000	0.002	0.000	0.001	0.019	0.000

Significant amounts of phosphorus and zinc were added to PDS07 to PDS09 because of the addition of ZOP inhibitor. The negative exponents on the TP term and zinc term in the model indicate that both total phosphorus and zinc were beneficial for reducing lead release. The data do not always appear to be consistent with the model, however, as it is seen from the overall average concentration data in Table 5.33 that increasing the dose of ZOP inhibitor does not offer improved lead control.

Figure 5.24 shows the average total lead release in each of the ZOP PDSs for each phase compared to the predicted lead release by the model. The maximum and minimum concentrations are also shown. PDS07, PDS08, and PDS09 were treated with a low, medium, and high dose of ZOP inhibitor, respectively. As shown by the observed data in the graph, increasing the ZOP inhibitor dose actually produced an adverse impact for lead release control for most of the cases. The model suggests a different trend. However, except for PDS09 in phase I, the lead concentrations through all phases of this study in all ZOP PDSs were well below the lead action level (0.015 mg/L). The predicted values from the empirical model confirm that compliance with the action level would be expected.

Control of Lead Release With Silicon Inhibitor Addition

Si Data Summary (PDS10 to PDS12)—Lead. The benefit of silica inhibitor treatment can be evaluated from the observed average lead concentrations shown in Table 5.34 between PDSs. Comparison of the lowest and highest levels of treatment suggests that silica was effective in reducing lead release for all phases. Analysis of the range for lead release indicates that silica limited the occurrence of an extreme observation. Although not as conspicuous, the effectiveness

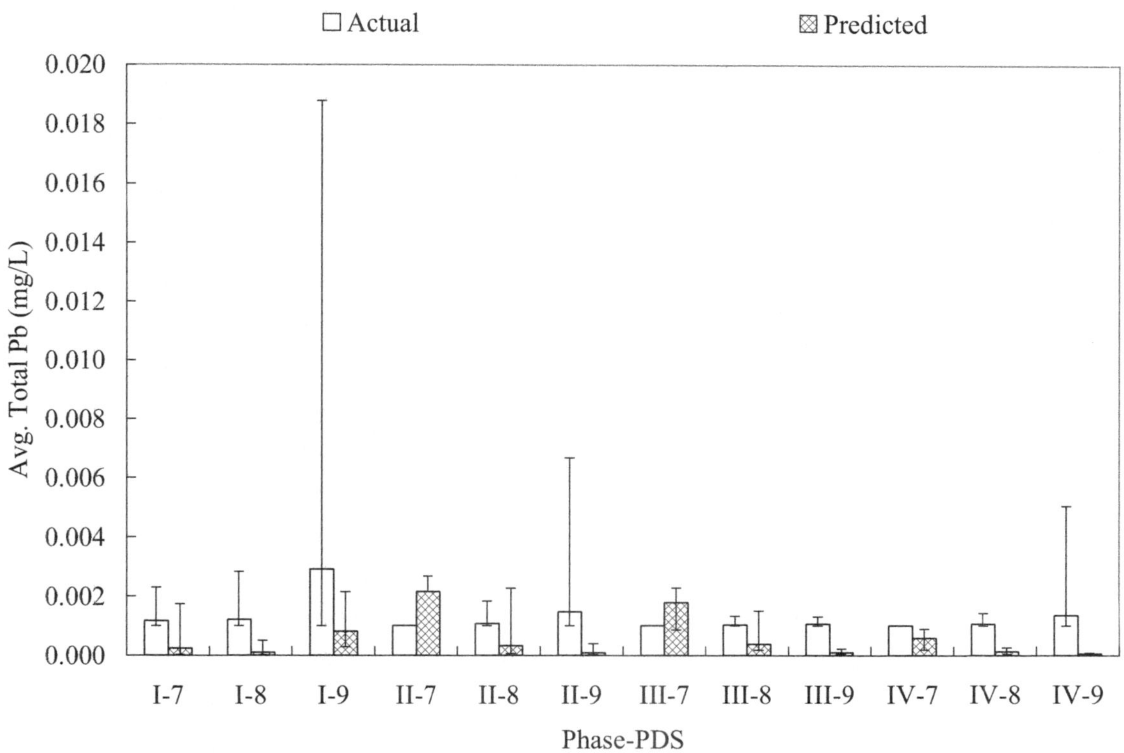

Figure 5.24 Total lead release model summary for ZOP PDSs by phase

of silica is also apparent from the evaluation of the dissolved fraction of lead. Further assessment of the factors controlling lead release is discussed with reference to an empirical model. This approach will both identify and quantify sources of variation attributed to the corrosion inhibitor and will consider water quality as well. The effect of variant water quality may be seen from Table 5.34, where phase III and phase IV concentrations appear significantly greater than the other phases. Water quality data for the independent variables in the empirical models for metal release were presented previously in Table 5.20 for the Si PDSs.

Analysis of Silicon Inhibitor Performance—Lead. The empirical model defining lead release during the study is presented as Equation 5.30, developed from the general model in Equation 5.25. Equation 5.30 is only appropriate for the prediction of total lead release when silica is applied as the corrosion inhibitor. From the observed lead release data shown in Figure 5.25, it is apparent that an increase in silica dose suppressed lead release. The model development procedure that defined significant water quality parameters determined that pH, chlorides, sulfate, and temperature are also necessary to accurately describe the data.

$$TPb = 3.80 \times 10^{-4} \times SiO_2^{-1.16} \times pH^{-2.75} \times Cl^{1.98} \times SO_4^{-1.54} \times temp^{2.94} \quad (5.30)$$

Equation 5.30 agrees with the observed relationship between silica dose and total lead release. This relationship is inferred from the negative exponent for the silica variable. The magnitude of the exponent suggests that an increase in dose has a dramatic effect on the expected value of the lead release. This effect is demonstrated from Figure 5.25 in which the average lead release

Table 5.34
Lead release summary for Si PDSs

			Dissolved Pb			Total Pb		
Phase	PDS	Dose	Average	Maximum	Minimum	Average	Maximum	Minimum
I	10	15.2	0.000	0.001	0.000	0.002	0.020	0.000
	11	21.5	0.001	0.011	0.000	0.002	0.015	0.000
	12	31.5	0.000	0.000	0.000	0.000	0.003	0.000
II	10	7.8	0.001	0.004	0.000	0.004	0.011	0.000
	11	11.7	0.000	0.000	0.000	0.000	0.002	0.000
	12	17.2	0.000	0.000	0.000	0.000	0.003	0.000
III	10	13.3	0.005	0.013	0.000	0.010	0.024	0.001
	11	17.0	0.000	0.001	0.000	0.001	0.002	0.000
	12	22.6	0.000	0.001	0.000	0.001	0.003	0.000
IV	10	9.2	0.005	0.020	0.000	0.009	0.028	0.001
	11	13.1	0.001	0.001	0.000	0.001	0.002	0.000
	12	17.7	0.001	0.002	0.000	0.001	0.002	0.000
All	10	11.5	0.003	0.020	0.000	0.006	0.028	0.000
	11	16.0	0.001	0.011	0.000	0.001	0.015	0.000
	12	22.5	0.000	0.002	0.000	0.001	0.003	0.000

from the high silica dose (PDS12) was nearly undetectable (0.001 mg/L), while that of the low silica dose (PDS10) was clearly much greater (0.009 mg/L). Phase I may appear anomalous regarding the effect of silica dose; however, the differences are more likely attributable to the variation in Si inhibitor dose that occurred in phase I.

Equation 5.30 identifies pH, chlorides, sulfate, and temperature as the water quality parameters affecting lead release. The effect of the water quality parameters may be validated from the evaluation of the observed data presented within Figure 5.25. Effects of selected factors such as chlorides, sulfate, and temperature may be isolated between phases that had limited variation in individual water quality parameters. A similar approach to evaluate the main effect of pH is not as simple. Because the silica inhibitor was highly alkaline, the inhibitor dose significantly affected the pH and/or alkalinity of the blend when applied. Thus, pH was not simply dependent on the characteristics of the blend alone, but on both the blend and the inhibitor treatment. It is possible that the singular effect of pH would be more apparent from contrasts in lead release between PDS13 and PDS14.

From Figure 5.25, lead release during phases I and II was much less than during phases III and IV. This observation becomes more apparent when comparing lead release from the low-dose treatments. However, all water quality parameters differed between phases I and II, suggesting that more than one main effect was responsible for the effective control of lead in phases I and II. Successful control of lead release during phase I may be attributed to a combination of low temperatures and low chlorides during operation. From Equation 5.30, both temperature and chlorides are positively correlated with lead concentration. Comparing phases I and III, in which each phase exhibited similar water quality excluding temperature and chlorides, the main effects of temperature and chlorides are clearly demonstrated. Phase III occurred during the later summer

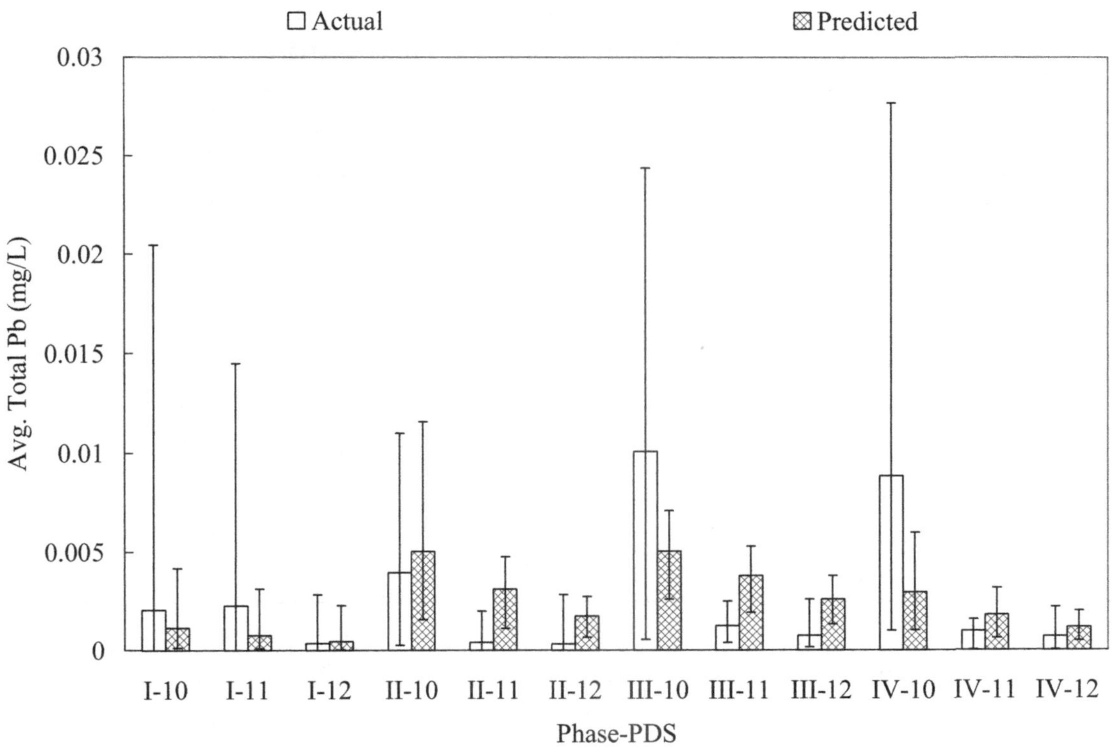

Figure 5.25 Total lead release summary for Si PDSs by phase

months of the study with temperatures greater than phase I. Also, phase I represented the lowest observed chlorides during the study.

Equation 5.30 also suggests that a source water blend that is high in sulfate would be beneficial for the control of lead within a distribution system. The ability for sulfate to deter lead release is evident from comparison of phase II and phase III results in Figure 5.25. Phase II experienced lower lead release because of its higher concentration of sulfate associated with a predominantly SW blend.

Control of Lead Release by pH Elevation (pH Controls)

pH Control Data Summary (PDS13 to PDS14)—Lead. A summary of the total and dissolved lead release data for the corrosion loops from the pH control PDSs is shown in Table 5.35 by phase and PDS. About 50% of the lead released in the corrosion loops is in the dissolved form for the pH control PDSs. Also, the operation at an elevated pH (PDS14) consistently achieved a significant improvement relative to operation at pH_s (PDS13) for control of lead concentration. This suggests that increasing the pH is beneficial for reducing lead release. The highest lead release for both PDSs occurred in phase III, which had high temperatures and high chloride concentrations. Water quality data for the independent variables in the empirical models for metal release were presented previously in Table 5.22 for the pH control PDSs.

Analysis of pH Control Performance—Lead. The total lead model developed for this study was presented previously in Equation 5.25. When the data for the pH control PDSs are selected for development of an empirical model for prediction of lead concentration, the general

Table 5.35
Lead release summary for pH control PDSs

Phase	PDS	Dissolved Pb			Total Pb		
		Average	Maximum	Minimum	Average	Maximum	Minimum
I	13	0.003	0.008	0.000	0.006	0.016	0.000
	14	0.001	0.006	0.000	0.003	0.015	0.000
II	13	0.005	0.010	0.001	0.012	0.024	0.000
	14	0.004	0.011	0.000	0.008	0.023	0.000
III	13	0.017	0.035	0.003	0.032	0.064	0.005
	14	0.005	0.022	0.001	0.010	0.036	0.001
IV	13	0.002	0.007	0.000	0.004	0.009	0.000
	14	0.001	0.005	0.000	0.003	0.014	0.001
All	13	0.007	0.035	0.000	0.013	0.064	0.000
	14	0.003	0.022	0.000	0.006	0.036	0.000

lead model takes on the form shown in Equation 5.31, where all terms have been previously defined. In the model, the increase in pH from pH_s to pH_s+0.3 is reflected in the pH term. The negative exponent on the pH variable suggests that increasing the pH will decrease the total lead release. This effect is consistent with the observed data as previously discussed.

$$TPb = 4.14 \times 10^{-5} \times pH^{-2.75} \times Cl^{1.98} \times SO_4^{-1.54} \times temp^{2.94} \quad (5.31)$$

Figure 5.26 shows the average total lead release by PDS and phase for the observations from the pH control PDSs compared with the predictions by the total lead model. The maximum and minimum concentrations for each phase are also indicated. Figure 5.26 indicates that the observations and predictions by the model consistently follow the trend of decreasing total lead with increasing pH.

Phases II and III were both characterized by high temperatures and high chloride concentrations. The positive exponents on both these terms in the model suggest that increasing these water quality parameters would increase the total lead release. However, as seen in that data summary and in Figure 5.26, the lead release in phase III is much higher. This may be the result of the high sulfates seen in phase II from the high percentage of surface water. The sulfate term in the model has a negative exponent, suggesting that increases will help to control total lead release. Phases I and IV are characterized by low temperatures and low chloride concentrations. The low lead concentrations for phases I and IV may be explained by the reduced temperature and chloride levels.

Comparisons between PDS13 and PDS14 within a single phase should focus on effects associated with variable chloride as well as variable pH. Chlorides were consistently higher in PDS13 because of the addition of HCl to maintain pH at the target value of pH_s. The HCl added significant levels of chloride to the PDS. The boost in chloride provides an additional explanation, along with the lowered pH, for the higher lead concentrations reported in PDS13 than PDS14. The lead action level of 0.015 mg/L was exceeded on average for PDS13 in phase III. In addition, the maximum observed lead concentration exceeded the action level in phases I, II, and III for PDS13. For PDS14, the average total lead concentration never exceeded the action level. These

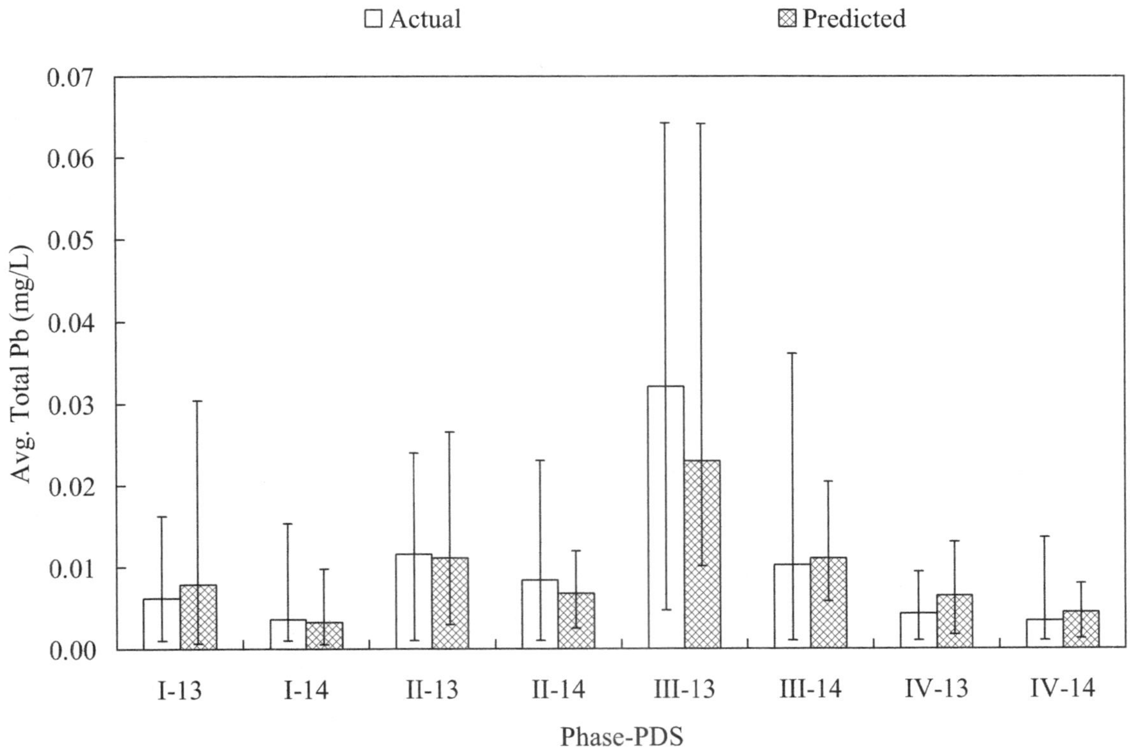

Figure 5.26 Total lead release model summary for pH control PDSs

results suggest that compliance with the action level for lead may be difficult without pH elevation above pH_s or application of an inhibitor.

Lead Release Summary

Review of Empirical Models. The empirical models that were developed for total and dissolved lead concentration (Equations 5.25 and 5.26) generally confirmed the relations that were defined in TBW I. Blends that exhibit elevated chloride would be expected to result in greater lead concentrations. Blends that maintain an elevated pH and/or elevated sulfate would be expected to exhibit muted lead concentrations. Seasonal differences in lead concentration would be expected, with greater violations of lead target concentrations expected during warmer periods. The alkalinity was demonstrated to significantly influence lead release for some, but not all, data sets. For total lead in TBW I and dissolved lead in the current study, greater alkalinities were associated with elevated lead levels. The relation between alkalinity and total lead was not shown to be significant in the current study.

Data Summary. A presentation of average output total lead concentrations for the 12-month study duration is shown in Figure 5.27. These data represent an average of the weekly determinations of corrosion loop lead concentrations. The bar chart presentation follows the same organization as described previously for iron in Figure 5.11.

The graphical summary of the data indicates that the corrosion control strategies achieved a similar benefit for lead, as discussed previously for copper. Elevation of pH is shown to substantially reduce the lead concentration relative to operation at pH_s, with a reduction in the annual

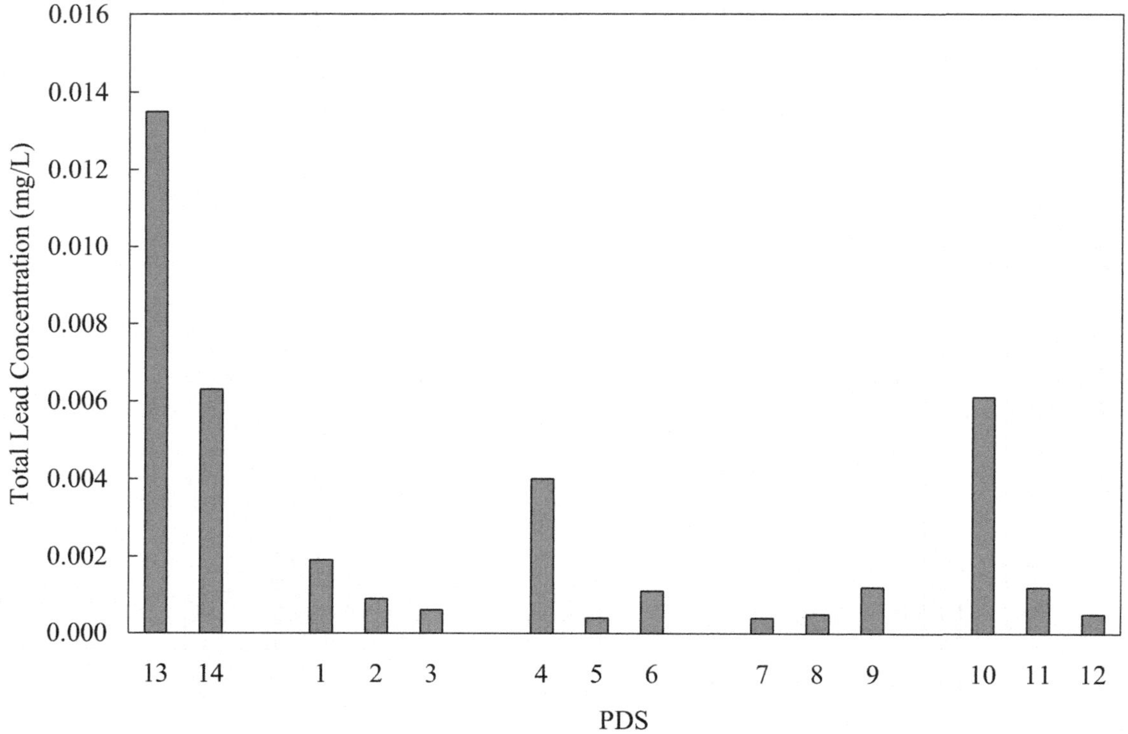

Figure 5.27 Data summary for total lead concentration

average corrosion loop concentrations from 0.013 mg/L in PDS13 to 0.006 mg/L in PDS14. As noted for copper, compliance with the lead action level (90th percentile concentration) of 0.015 mg/L would be a potential concern without implementation of a corrosion control strategy. The average (approximately 50th percentile) concentration for the pH_s control exceeded the lead action level for phase III. Elevation of the pH by 0.3 units was demonstrated to reduce the lead concentration by more than 50%.

The corrosion inhibitors were even more effective for control of lead concentrations. Relative to operation at pH_s, the annual average corrosion loop lead concentrations were reduced by 55% to 95% with inhibitors. All of the inhibitors were effective at all doses that were investigated. The data in Figure 5.27 generally suggest a benefit associated with increasing the inhibitor dose for BOP, OP, and Si. The trend for ZOP is difficult to discern, as much of the reported data were below the detection limit for lead analysis.

A statistical comparison of the mean lead concentrations (one-tailed, paired-data t-test for equality of means) between the control (PDS13) and the corrosion control treatments (PDS14, PDS01 to PDS12) is reported in Table 5.36. One data entry (phase IV, PDS10) is noted in boldface. For this single population, the average concentration in the pH_s control was exceeded by the average in the PDS that received a corrosion control measure.

A reduction in lead concentration was documented by elevation of pH in all study phases. This reduction was demonstrated to be statistically significant for three of five comparisons (phase II, phase III, and composite). Inhibitor addition was shown to reduce the average lead concentration in all but the one comparison noted in boldface. Thus, for all inhibitors at all doses for all study phases except one (59 of 60 comparisons), the inhibitors achieved a reduction in

Table 5.36
Comparison of inhibitors to pH_s control for total lead

	PDS13	PDS14	PDS01	PDS02	PDS03	PDS04	PDS05	PDS06	PDS07	PDS08	PDS09	PDS10	PDS11	PDS12
Treatment	pH	pH	BOP	BOP	BOP	OP	OP	OP	ZOP	ZOP	ZOP	Silicate	Silicate	Silicate
Dose	pH_s	pH_s+0.3	Low	Medium	High	Low	Medium	High	Low	Medium	High	Low	Medium	High
Phase I														
Average	0.006	0.003	0.003	0.001	0.001	0.003	0.001	0.001	0.000	0.001	0.002	0.002	0.002	0.000
Std. dev.	0.006	0.005	0.005	0.003	0.003	0.005	0.001	0.002	0.001	0.001	0.005	0.005	0.004	0.001
N	14	14	14	14	14	14	14	14	14	14	14	14	14	14
Significant		No	No	Yes	Yes	No	Yes	Yes	Yes	Yes	No	No	No	Yes
p-value		0.129	0.089	0.012	0.011	0.111	0.003	0.011	0.002	0.002	0.074	0.060	0.058	0.002
Phase II														
Average	0.012	0.008	0.002	0.001	0.001	0.005	0.000	0.000	0.000	0.000	0.001	0.004	0.000	0.000
Std. dev.	0.007	0.007	0.003	0.001	0.002	0.002	0.000	0.001	0.000	0.000	0.002	0.003	0.001	0.000
N	13	13	13	13	13	13	13	13	13	13	13	13	13	13
Significant		Yes	Yes	Yes	Yes	Yes	Yes	Yes	Yes	Yes	Yes	Yes	Yes	Yes
p-value		0.048	<0.001	<0.001	<0.001	0.003	<0.001	<0.001	<0.001	<0.001	<0.001	<0.001	<0.001	<0.001
Phase III														
Average	0.032	0.010	0.001	0.001	0.000	0.004	0.000	0.000	0.000	0.001	0.001	0.010	0.001	0.001
Std. dev.	0.019	0.009	0.001	0.000	0.000	0.002	0.000	0.000	0.000	0.000	0.000	0.008	0.001	0.001
N	13	13	13	13	13	13	13	13	13	13	13	13	13	13
Significant		Yes	Yes	Yes	Yes	Yes	Yes	Yes	Yes	Yes	Yes	Yes	Yes	Yes
p-value		<0.001	<0.001	<0.001	<0.001	<0.001	<0.001	<0.001	<0.001	<0.001	<0.001	<0.001	<0.001	<0.001

(continued)

Table 5.36 (Continued)

	PDS13	PDS14	PDS01	PDS02	PDS03	PDS04	PDS05	PDS06	PDS07	PDS08	PDS09	PDS10	PDS11	PDS12
Treatment	pH	pH	BOP	BOP	BOP	OP	OP	OP	ZOP	ZOP	ZOP	Silicate	Silicate	Silicate
Dose	pH_s	pH_s+0.3	Low	Medium	High	Low	Medium	High	Low	Medium	High	Low	Medium	High
Phase IV														
Average	0.004	0.003	0.001	0.001	0.000	0.004	0.000	0.002	0.000	0.001	0.001	**0.009**	0.001	0.001
Std. dev.	0.003	0.003	0.000	0.000	0.000	0.004	0.000	0.007	0.000	0.001	0.001	**0.008**	0.001	0.001
N	12	12	12	12	12	12	12	12	12	12	12	**12**	12	12
Significant		No	Yes	Yes	Yes	No	Yes	No	Yes	Yes	Yes	**Yes**	Yes	Yes
p-value		0.153	0.001	<0.001	<0.001	0.305	<0.001	0.224	<0.001	0.001	0.006	**0.015**	0.001	0.001
Composite														
Average	0.013	0.006	0.002	0.001	0.001	0.004	0.000	0.001	0.000	0.001	0.001	0.006	0.001	0.001
Std. dev.	0.015	0.007	0.003	0.001	0.002	0.003	0.001	0.004	0.000	0.001	0.003	0.007	0.002	0.001
N	52	52	52	52	52	52	52	52	52	52	52	52	52	52
Significant		Yes	Yes	Yes	Yes	Yes	Yes	Yes	Yes	Yes	Yes	Yes	Yes	Yes
p-value		<0.001	<0.001	<0.001	<0.001	<0.001	<0.001	<0.001	<0.001	<0.001	<0.001	<0.001	<0.001	<0.001

NOTE: The boldface population indicates the occurrence of an average concentration in the pH_s control that was exceeded by the average in the PDS that received a corrosion control measure.

average lead concentration. This benefit was demonstrated to be statistically significant for the vast majority of treatments (52 of 60 comparisons showed a significant benefit). The inhibitors proved to be effective for all blends and water qualities that were examined, including blends that were enriched in groundwater with an associated elevated alkalinity (phases I and III).

Corrosion Control Strategy Recommendations

Simulation Objectives

The subsequent discussion illustrates application of the empirical equations that have been developed for total iron, total copper, and total lead for evaluation of inhibitors and/or adjustment of pH. The analysis presumes a series of background water qualities that mimic the conditions present during phases I to IV of this project. It was intended that this variation in water quality would be representative of experienced blends in the TBW distribution network. Of course, the predictive models could also be applied to conduct a more general sensitivity analysis for other hypothetical water quality conditions. Caution is advised for extrapolation of the models to water quality conditions not representative of the empirical database developed in this study. The presentation is intended to demonstrate the utility of the empirical models as a tool for proactive selection of inhibitor dose and/or pH manipulation to achieve water quality criteria for metals in response to changes in blend composition.

Simulations are completed to predict metal concentrations (iron, copper, and lead) associated with a specific inhibitor dose and/or pH adjustment. The experimental data demonstrate success in compliance with the copper and lead action levels at the minimum inhibitor dose that was examined. In general, all of the corrosion control strategies also achieved compliance with the secondary standard for iron for the specific hydraulic conditions maintained in the PDS. In practice, manipulation of the HRT by distribution system flushing may be preferable to water conditioning for control of iron concentration (red water issues). Because the regulatory standards could be achieved with the minimum inhibitor dose, all simulations were conducted assuming the minimum dose: 0.5 mg/L TP or an increment of 3 mg/L SiO_2.

In general, water production for this research included stabilization to achieve a pH that was elevated relative to pH_s. Although operation at an elevated pH (or elevated alkalinity) was demonstrated to substantially mitigate release of iron, copper, and lead, it is recognized that associated precipitation of calcium carbonate may reduce the attractiveness of a high-pH operation. Therefore, a series of simulations are also presented for pH values that are lowered relative to the values that were present in this study with inhibitors (PDS01 to PDS12) or elevated pH (PDS14). These simulations would demonstrate potential trade-offs between a corrosion control strategy that relied exclusively on pH elevation in comparison to a strategy that incorporated pH elevation and/or inhibitor addition.

A final set of simulations was undertaken to examine the impact of variant blends of groundwater and surface water. It was noted in the historical research that operation at elevated alkalinities associated with GW-enriched blends may cause difficulties in compliance with the copper standard. Conversely, it was also noted that operation at reduced alkalinities associated with GW-deficient blends may cause violations of the secondary standard for iron or apparent color. These simulations illustrate the potential for inhibitors to expand the feasible operating range of blends that would simultaneously satisfy treatment objectives for both iron and copper.

Treatment Objectives

Iron. The dominant form of iron was demonstrated to be particulate. The presence of elevated iron levels resulted in excessive color. Secondary drinking water standards for color (15 cpu) or iron (0.3 mg/L) would be relevant considerations for establishment of maximum allowable iron concentrations. A correlation between iron and color was developed for the PDS data and is reported in Equation 5.32.

$$TFe = 0.011 \times color \tag{5.32}$$

where TFe = total iron concentration, mg/L
color = apparent color, cpu

The relationship between iron and color suggests that a violation of the color standard would be expected if the total iron concentration exceeded 0.16 mg/L. Because the maximum iron concentration associated with the color standard is lower than the secondary standard for iron, the color-related iron standard of 0.16 mg/L was adopted as a maximum allowable concentration for purposes of these simulations.

It was determined in the previous study that the increase in apparent color or total iron concentration was approximately proportional to the HRT in the pipe network. The empirical model developed in this study (Equation 5.6) would be specific for the 2-day HRT employed for data collection. Operation at these hydraulic conditions would correspond with HRTs of 10.2 hours and 3.8 hours in the UCI and GS segments of the hybrid PDS, respectively. The simulations of iron concentration were conducted using the empirical model without adjustment, with the understanding that increases or decreases in concentration in proportion to the HRT would be appropriate for alternate hydraulic conditions. Although these HRTs may exceed average values in a typical distribution system, they were selected for purposes of this experimental program as representative of localized portions (dead zones or dead ends) of the system with potential to generate adverse water quality conditions.

Copper. The federal action level for copper is defined by the 90th percentile concentration of 1.3 mg/L. Because the empirical model for total copper would be appropriate to define the mean concentration, it is necessary to establish a relationship between the mean value predicted by the model and the action level defined by the 90th percentile concentration. As discussed in a previous section of this report, examination of the total copper concentration data suggests that a normal probability distribution is valid. The ratio of the 90th percentile to the mean concentration is 1.3 for the project data. An average concentration of 1.0 mg/L would therefore correspond to the action level of 1.3 mg/L for total copper. For purposes of these simulations, a maximum average allowable total copper concentration of 1.0 mg/L was established.

Lead. The action level for lead is also defined by a 90th percentile concentration of 0.015 mg/L. As reviewed previously, the ratio of the 90th percentile concentration to the mean concentration for total lead was determined to be 2.5. In order to maintain compliance with the action level of 0.015 mg/L, it would be necessary to maintain an average concentration below 0.006 mg/L total lead. The maximum average allowable total lead concentration was defined as 0.006 mg/L for these simulations.

Calibration of the Total Lead Release Model. For an actual distribution system, it is likely that the lead concentrations would be subject to additional sources of variation not experienced in the experimental facilities. Specifically, all of the corrosion loops in the PDS were supplied with a lead coupon of identical surface area. In practice, there may be a considerable variation in exposed lead surface area, depending on characteristics of the plumbing fixtures, number of soldered joints, quantity of lead solder per joint, and other factors. To the extent that the surface area provided in the experimental system was not representative of a particular distribution system, the empirical model may under- or overpredict mean concentrations. Calibration of the model to adjust for any discrepancy in average surface area would be prudent. A procedure was developed in TBW I (Taylor et al. 2005) for calibration of the model to a particular data set, as outlined in the following paragraph.

The empirical model, in combination with the defined ratio of the 90th percentile to the average concentration (ratio = 2.5), would be used to predict a 90th percentile lead concentration for a specific water quality associated with a specific MG data set. This predicted total lead concentration ($Pb_{90,Predicted}$) would be representative of the lead coupon surface area used in the corrosion loops for this experimental study and the water quality associated with a different data set. This predicted lead concentration must be compared with the historical 90th percentile value ($Pb_{90,Historical}$) from that specific MG data set. Any discrepancy in these two values may be partially attributable to differences in the lead surface area in the pilot study and actual households for the specific data set. Calibration of the predictive model (Equation 5.25) to the specific data set is achieved by multiplication of the predicted value by the ratio of the measured 90th percentile value to the predicted 90th percentile value. The procedure is illustrated below, assuming a measured 90th percentile concentration of 0.005 mg/L. For this particular MG distribution system, a correction factor of 0.333 should be applied to predictions obtained with the empirical model (Equation 5.25):

$$Pb_{90,Historical} = 0.005 \text{ mg/L}$$

$$Pb_{90,Predicted} = 0.015 \text{ mg/L}$$

$$\text{correction factor} = 0.005/0.015 = 0.333$$

This calibration issue does not apply for copper predictions because copper pipe is present in both the corrosion loops in this study and residential construction. The corresponding copper surface area would be identical in the experimental and prototype systems.

Control of Iron by Inhibitor Addition and/or pH Adjustment

Inhibitor performance for mitigation of iron release is reviewed in Figures 5.28–5.31 for the water quality conditions that correspond with phases I to IV, respectively. All of the simulations presume use of the minimum inhibitor dose employed in the experimental study. Iron concentrations were predicted for various values of pH, recognizing that alteration of pH may also have merit to address metal corrosion and release. For purposes of these simulations, pH was varied from 7.0 to 8.2. These values are believed to encompass the probable operating range for pH. Stabilization of the water to pH_s would produce a finished water with a minimum pH of 7.10 or greater, depending on the blend. The average pH_s values of the three principal sources for the study duration were 7.14 (GW), 7.60 (SW), and 8.14 (RO).

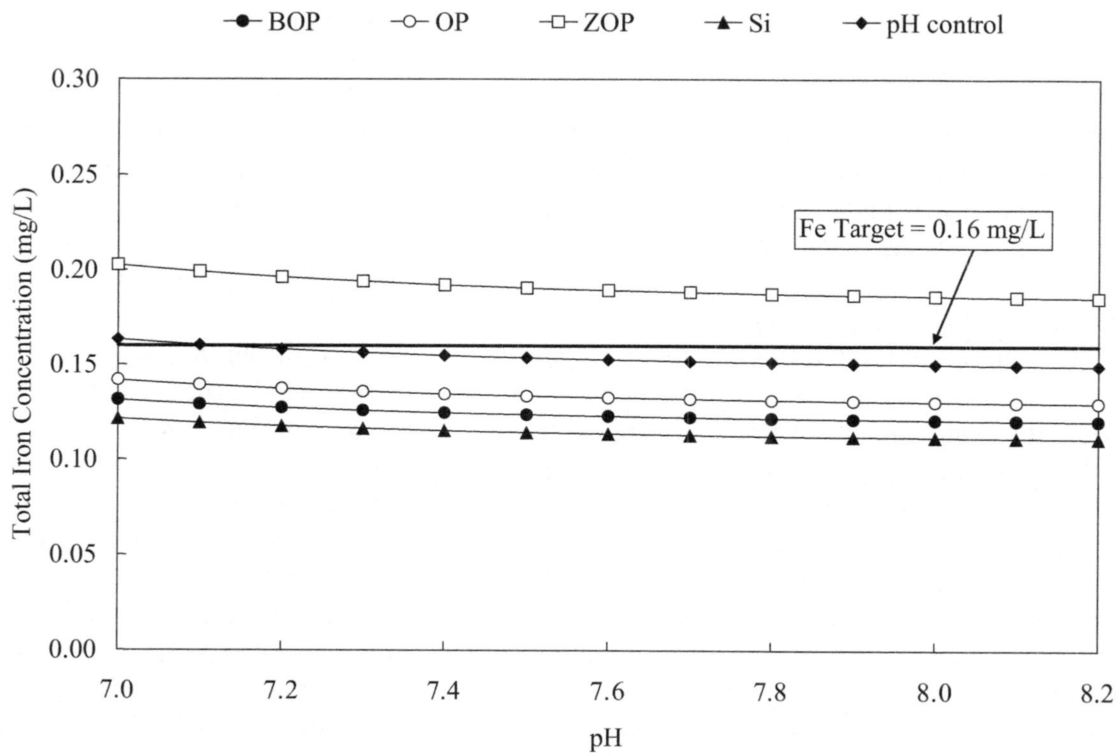

Figure 5.28 Iron simulations for phase I water quality specifications

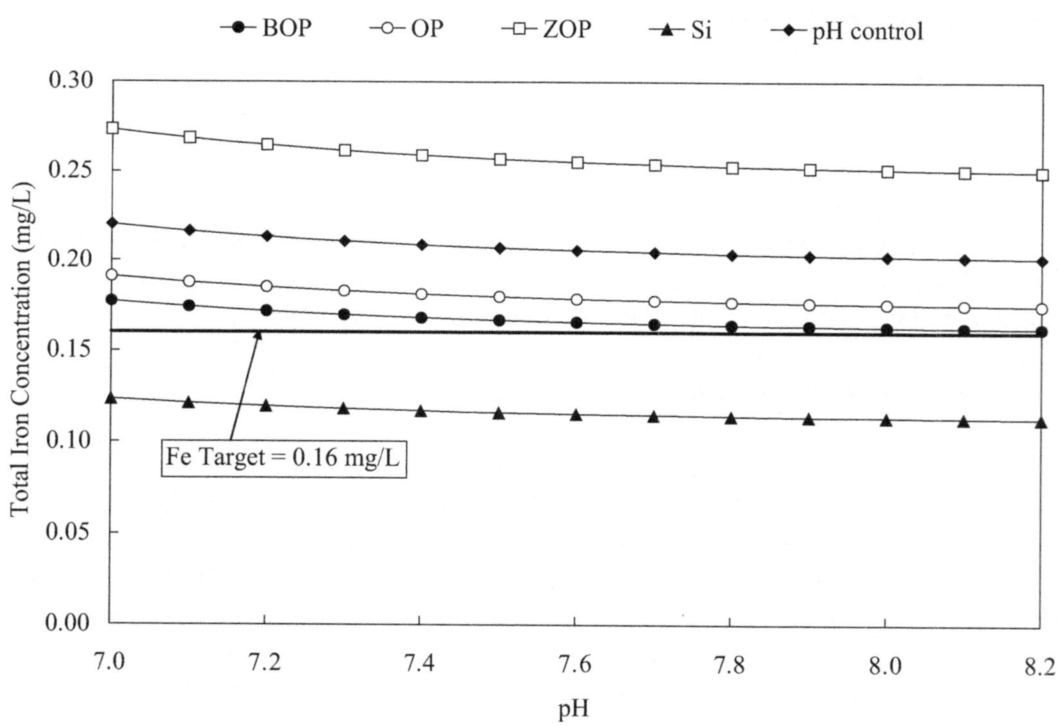

Figure 5.29 Iron simulations for phase II water quality specifications

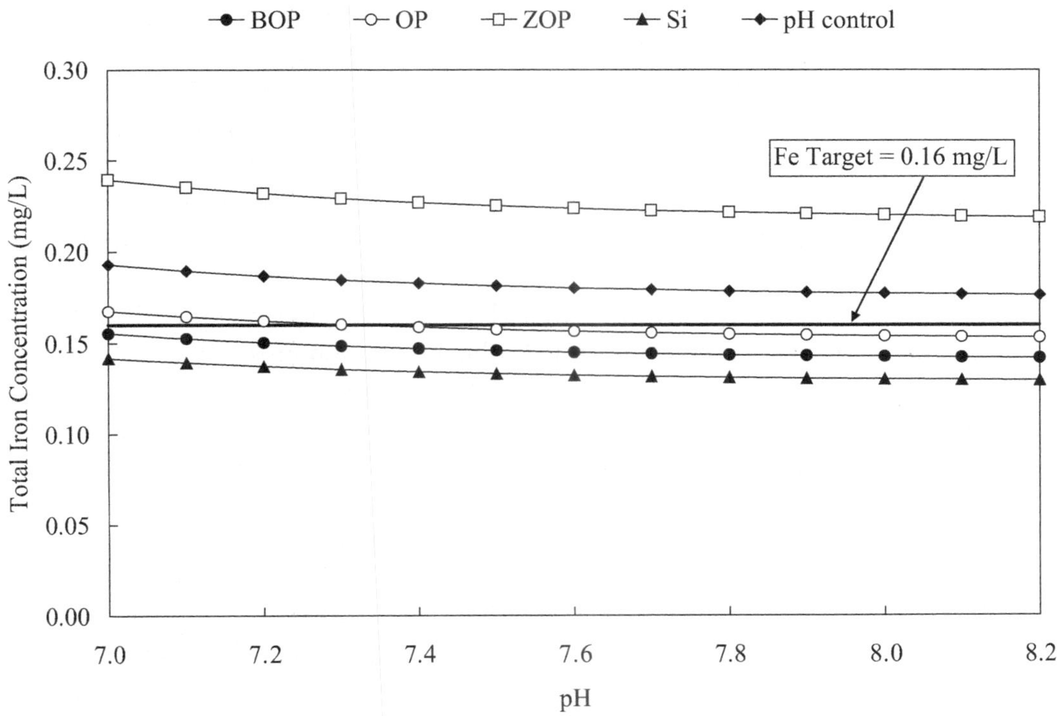

Figure 5.30 Iron simulations for phase III water quality specifications

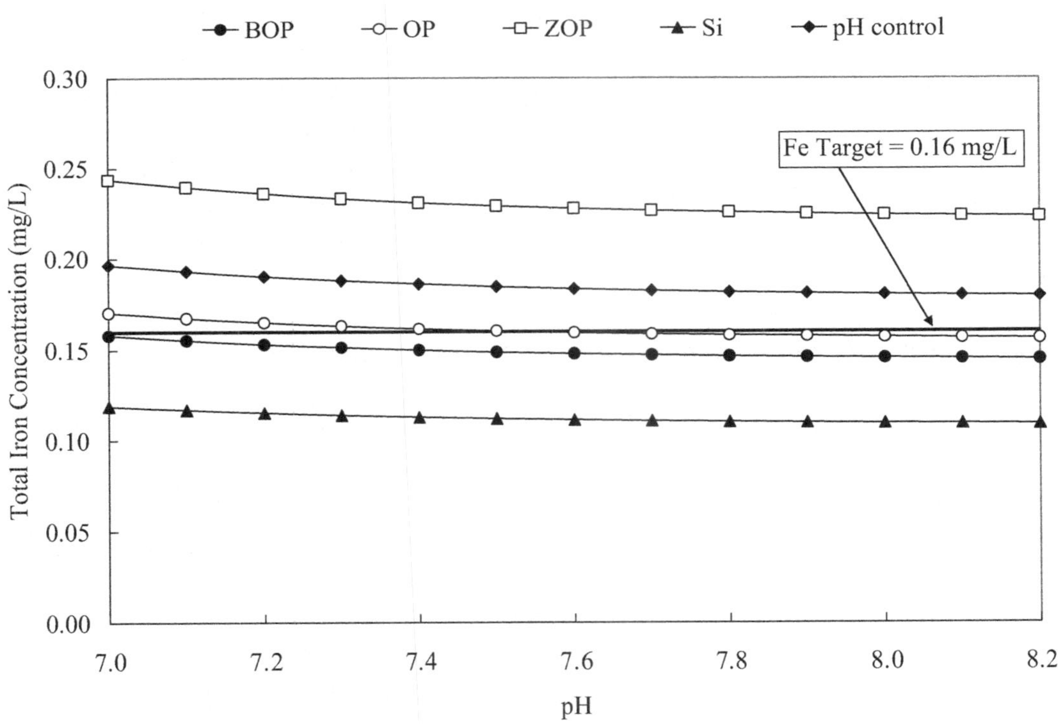

Figure 5.31 Iron simulations for phase IV water quality specifications

The simulations are consistent with the prior data summary (Figure 5.11). The data summary for an elevated pH (PDS14) indicated an improvement in comparison to operation at pH_s (PDS13). The simulation results confirm a slight benefit of an elevated pH, expressed by the alkalinity term in the total iron empirical model (Equation 5.6). The experimental results from the current study and the simulations continue to provide support for supplementation of alkalinity in the SW and RO sources to address red water concerns. Target alkalinities for the blend in excess of 100 mg/L as $CaCO_3$ would be beneficial to mitigate release of iron corrosion products for the UCI and GS pipe materials. Addition of the various inhibitors has a modest impact on iron concentrations. The ZOP inhibitor is not predicted to be beneficial in comparison to a corrosion control strategy that depends exclusively on pH adjustment, supporting the data summary from Figure 5.11. The other inhibitors achieve marginal improvements in iron concentrations, with the Si inhibitor projected to perform slightly better than the phosphate-based inhibitors. These same trends are evident for all water quality specifications that were examined.

The simulated iron concentrations during conditions representative of phase II indicated iron concentrations routinely in excess of the target concentration (0.16 mg/L) established to maintain compliance with the secondary standard for color. Compliance with the secondary iron standard of 0.30 mg/L was generally achieved in the simulations. The system was operated in phase II with an elevated blend percentage (62%) of surface water. The associated reduction in alkalinity in phase II would be expected to result in greater iron release per the empirical model. Concentrations in excess of the target concentration were noted in simulations for all four phases. Improvements in iron concentration could be obtained by increasing the pH, supplementing alkalinity, increasing the inhibitor dose (for the phosphate-based inhibitors), and/or reducing the HRT. The benefit associated with inhibitor addition appears to be limited. As noted previously (Table 5.3), an increase in PDS output iron concentration (relative to PDS input iron concentration) was demonstrated to be statistically significant for all inhibitors at all doses that were studied. Furthermore, the degradation of water quality was shown in TBW I to be proportional to the HRT. Consequently, operation at an excessive HRT would produce iron concentrations (or apparent color) in excess of water quality criteria. It does not appear that chemical treatment of the finished water by inhibitor addition and/or pH adjustment can be expected to maintain acceptable iron concentrations in the absence of maintenance of suitable hydraulic conditions in the distribution system.

Control of Copper by Inhibitor Addition and/or pH Adjustment

Inhibitor performance for control of copper is reviewed in Figures 5.32–5.35 for the water quality conditions that correspond with phases I to IV, respectively. All of the simulations presume use of the minimum inhibitor dose employed in the experimental study. The copper concentrations were predicted for various values of pH, recognizing that alteration of pH may also have merit to address metal corrosion and release. Simultaneous manipulation of pH and inhibitor addition may be a viable option for mitigation of copper release.

The simulations are consistent with the prior data summary (Figure 5.19 and Table 5.30). The data summary provided evidence that pH elevation or addition of any of the inhibitors at the doses employed in this study would significantly lower copper concentrations in comparison to the pH_s control. The improvement in performance was demonstrated to be statistically significant for all blends. The data summary also indicated that the addition of an inhibitor provided improved performance in comparison to the pH_s control. The simulations reflect these results for

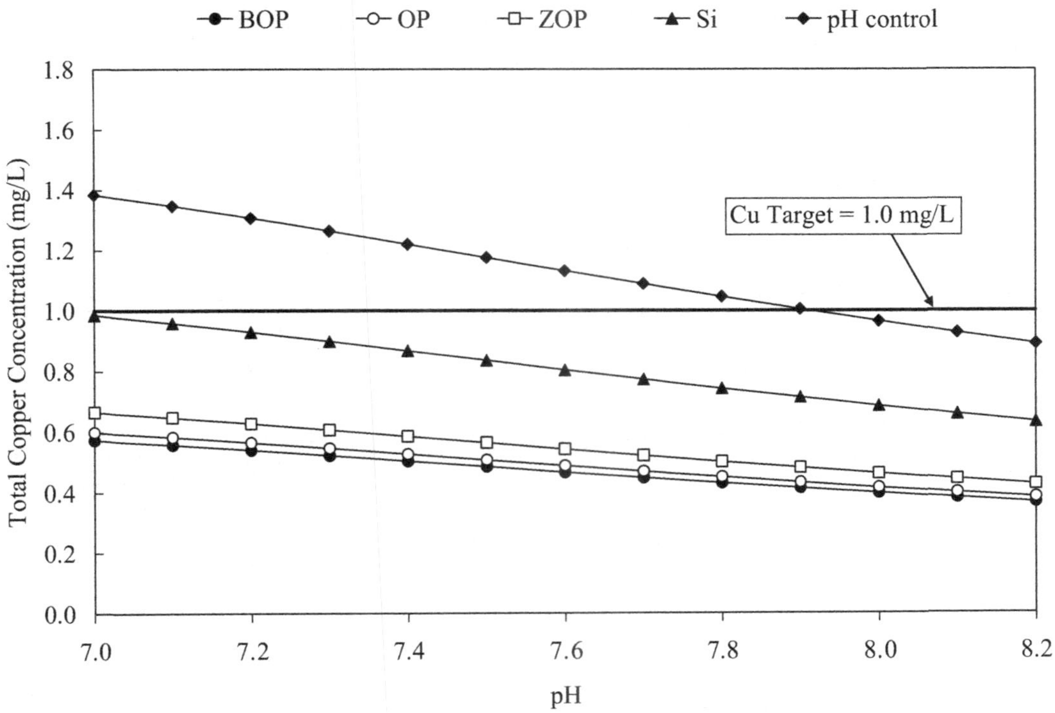

Figure 5.32 Copper simulations for phase I water quality specifications

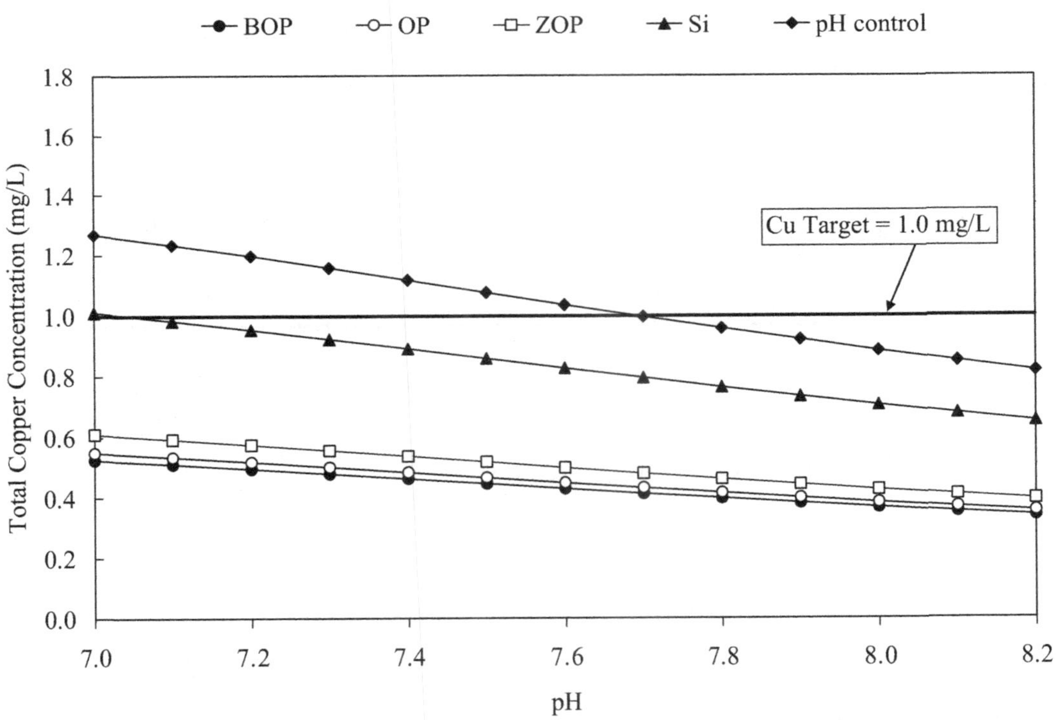

Figure 5.33 Copper simulations for phase II water quality specifications

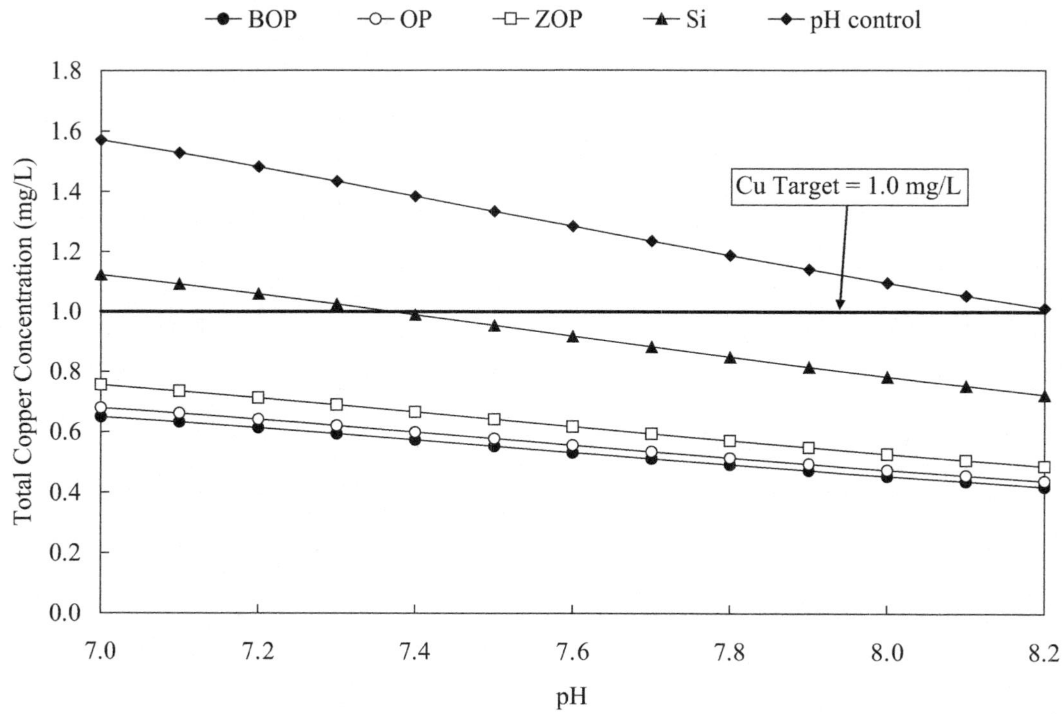

Figure 5.34 Copper simulations for phase III water quality specifications

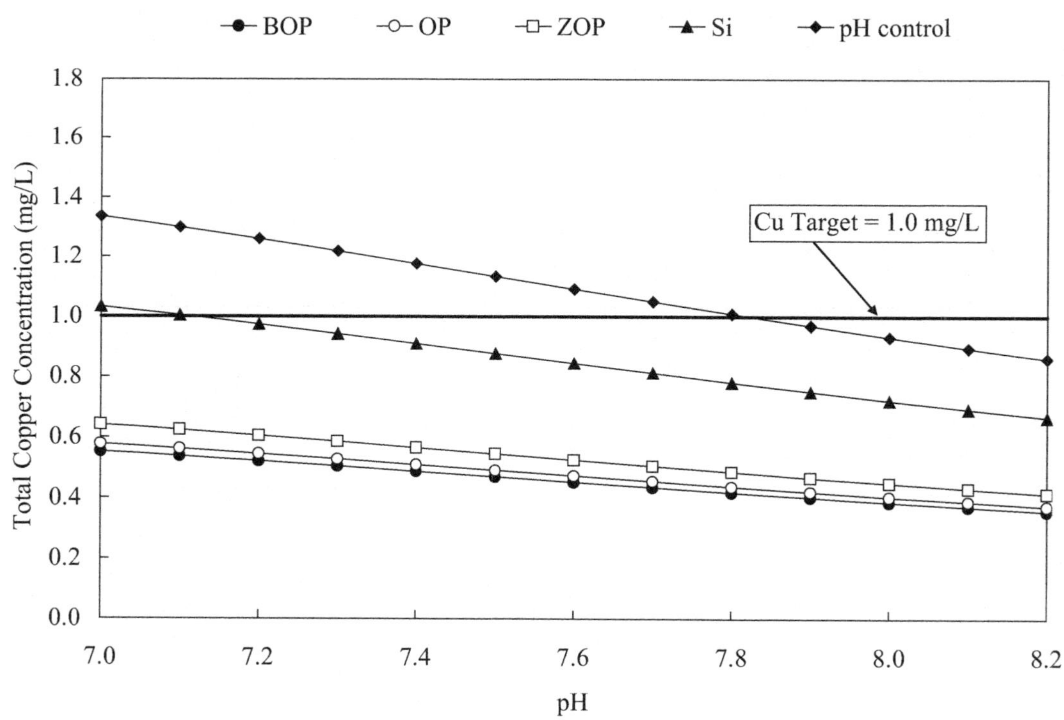

Figure 5.35 Copper simulations for phase IV water quality specifications

the different water quality in the four study phases. For the minimum inhibitor doses, the simulations indicate that the phosphate-based inhibitors achieve lower copper concentrations than the Si inhibitor. The simulated performance of all three of the phosphate-based inhibitors was virtually identical when compared with an equivalent TP dose.

The simulated copper concentrations experienced without inhibitor addition were in excess of the target average concentration (1.0 mg/L) established for compliance with the copper action level for most pH values and all water quality conditions. The largest copper concentrations were predicted for the phase III water quality specification, corresponding with a GW-enriched blend and associated elevated alkalinity. The adverse effect of alkalinity, as well as the benefit of elevated pH, on copper release is evident from the empirical model (Equation 5.18). Addition of any of the phosphate-based inhibitors successfully addressed copper concentrations for all water quality specifications (all phases) over the range of pH that was examined. Although the simulations for the Si inhibitor in Figures 5.32–5.35 did not indicate equivalent performance to the phosphate-based inhibitors, it is recognized that increase of the Si dose would be expected to achieve compliance with the copper target concentrations.

All of the inhibitors could be applied with success for compliance with the copper action level for the various blends and associated water quality addressed by the simulations. Operation at a pH in the vicinity of pH_S to avoid calcium carbonate precipitation would be feasible with an inhibitor. In contrast, without addition of an inhibitor, compliance would be problematic for most operating conditions. Although pH elevation without inhibitor addition was shown by simulation and experimentation to achieve mitigation of copper release, the required pH for copper control may exceed 8.2. Scaling problems at these high-pH conditions may preclude exclusive reliance on pH elevation for control of copper concentrations.

Control of Lead by Inhibitor Addition and/or pH Adjustment

Inhibitor performance for control of lead at varying pH is reviewed in Figures 5.36–5.39 for the water quality conditions that correspond with phases I to IV, respectively. All of the simulations presume use of the minimum inhibitor dose employed in the experimental study.

The simulations are consistent with the prior data summary (Figure 5.27 and Table 5.36). Examination of the water quality data indicated that pH elevation or addition of any of the inhibitors at the doses employed in this study would significantly lower lead concentrations in comparison to the pH_S control. A statistical analysis confirmed that the improvement in performance was significant relative to operation at pH_S. For the minimum inhibitor doses, the simulations indicate that the ZOP and BOP inhibitors achieve lower lead concentrations than the OP or Si inhibitors.

For phase II (Si and control) and phase III (OP, Si, and control) conditions, the simulated lead concentrations experienced without inhibitor addition did not meet the target average concentration (0.006 mg/L) established for compliance with the lead action level. Maximum temperatures were experienced during the two phases, contributing to the elevated lead concentrations per the empirical model (Equation 5.25). Compliance with the target concentration was achieved with the minimum BOP or ZOP dose of 0.5 mg TP/L for all water quality and pH conditions employed in the simulations. An increase in the inhibitor dose for OP or Si would also be expected to achieve compliance with the average lead concentration target.

All of the inhibitors could be applied with success for compliance with the lead action level for the various blends and associated water quality addressed by the simulations. As noted in the previous discussion for copper control, operation at a pH in the vicinity of pH_S to avoid

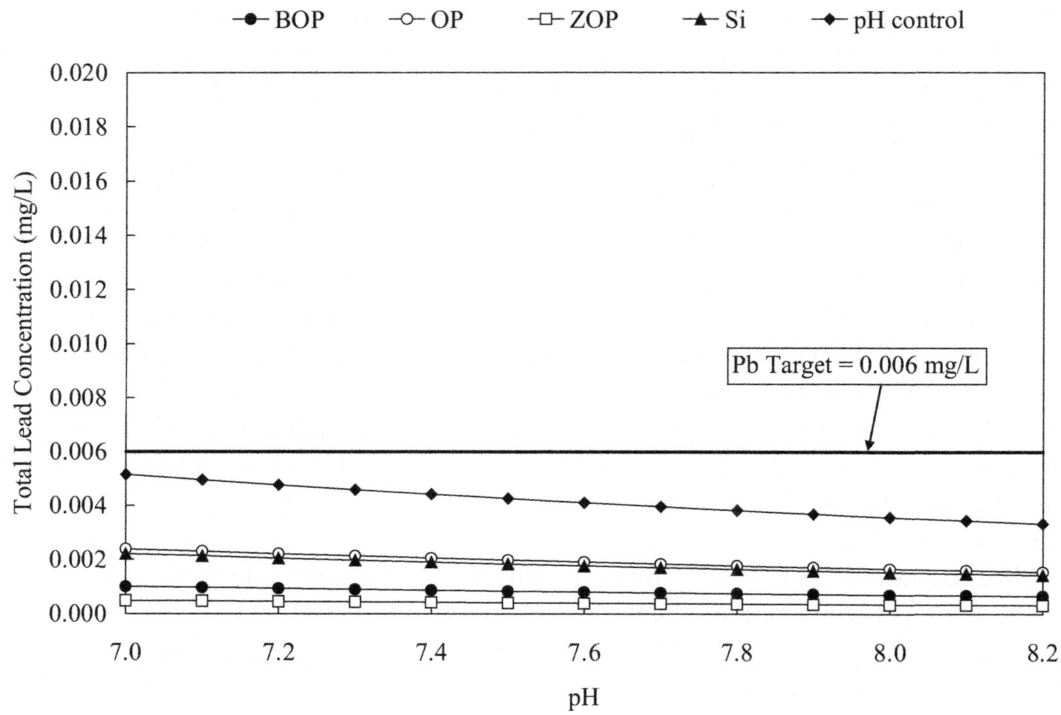

Figure 5.36 Lead simulations for phase I water quality specifications

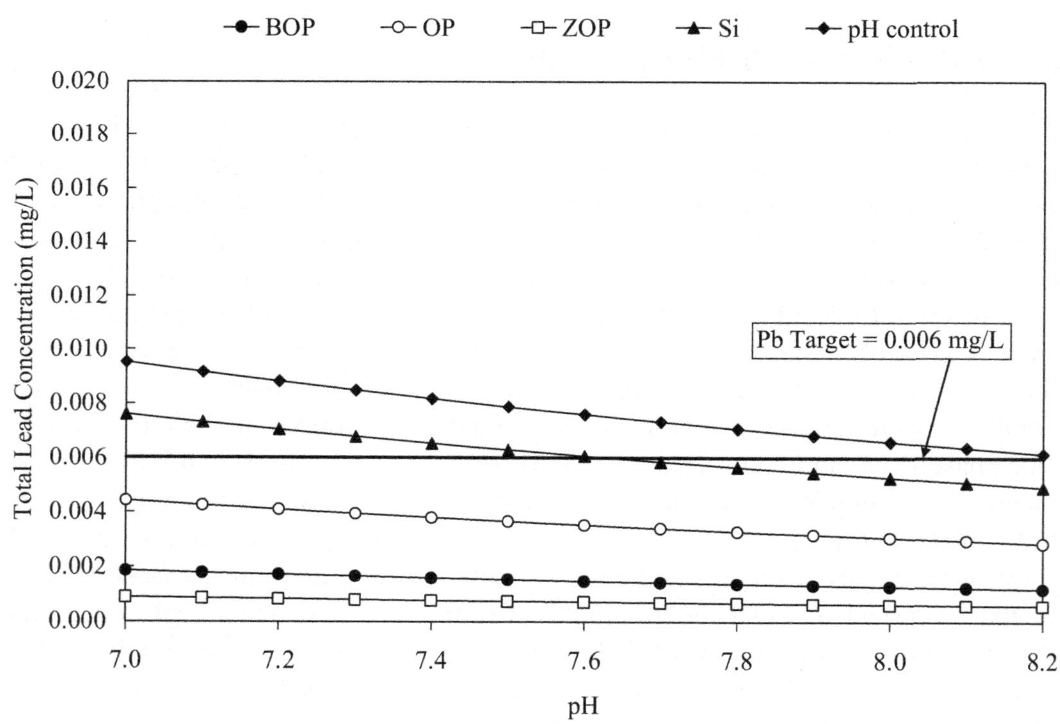

Figure 5.37 Lead simulations for phase II water quality specifications

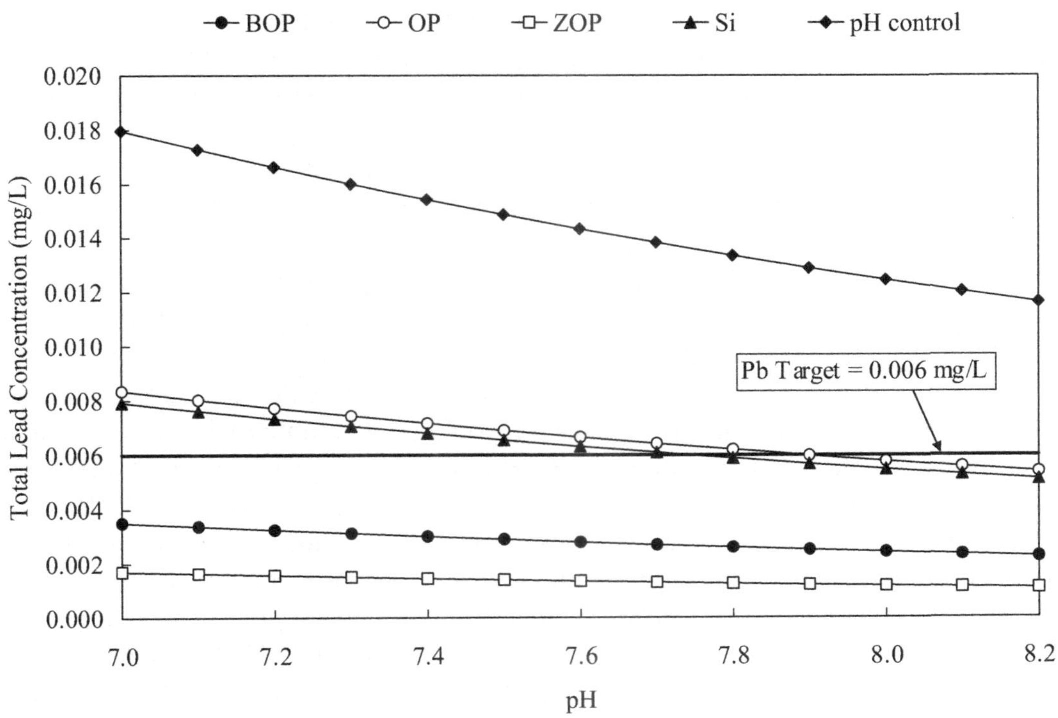

Figure 5.38 Lead simulations for phase III water quality specifications

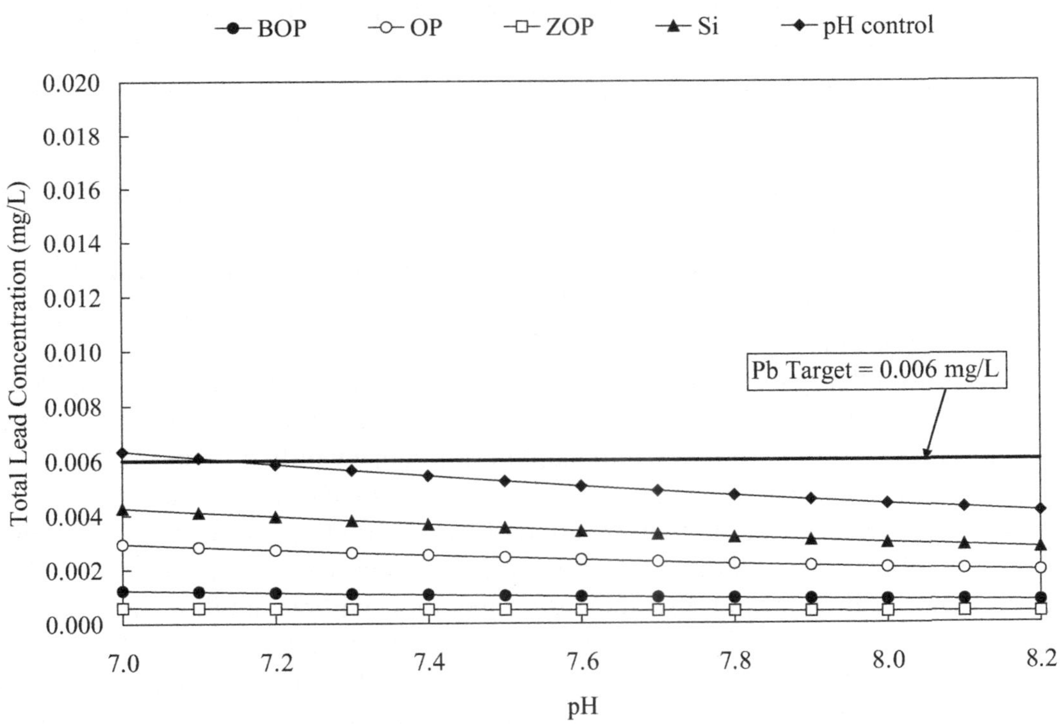

Figure 5.39 Lead simulations for phase IV water quality specifications

calcium carbonate precipitation would be feasible with an inhibitor. For the most adverse water quality specifications (phase III), the simulation results indicate that without an inhibitor, operation at a pH near pH_s to minimize scaling potential would not be feasible because of excessive lead concentrations.

Effect of Groundwater/Surface Water Blend Ratio on Metal Concentration

Water supply within the TBW distribution system relies on three principal water sources: GW, SW, and RO. The relative composition of the water that is delivered to users may vary with location in the distribution system and time. Results from the current and prior study have identified conflicting objectives for target alkalinity. Specifically, provision of an alkalinity that is characteristic of a GW source (>200 mg/L as $CaCO_3$) would be desirable to address red water issues. In contrast, in the absence of pH elevation or inhibitor addition, provision of an alkalinity that is characteristic of an SW source (<100 mg/L as $CaCO_3$) would be desirable to address elevated copper concentrations. It is recognized that these objectives are mutually exclusive. Furthermore, as the blend composition changes with time and location, it is probable that compliance concerns may cycle back and forth between iron and copper. Simulations are presented to assess the potential benefit of inhibitors to expand the feasible operating range for alkalinity that could be used while maintaining compliance with iron and copper treatment objectives. These simulations are structured to mimic water quality variation associated with variant fractions of groundwater and surface water in the product blend, as described in Table 5.37.

These simulations address the fundamental dilemma that was identified during TBW I regarding simultaneous compliance with iron and copper standards. The data for operation at pH_s (PDS13) in Table 5.29 suggest that compliance with the copper action level would be unlikely for the blends utilized in the current study (phase average alkalinity range of 92 to 149 mg/L as $CaCO_3$). Operation at an elevated pH to achieve copper control may be undesirable because of calcium carbonate scale formation. Results from TBW I with operation at an alkalinity less than 100 mg/L as $CaCO_3$ demonstrated a potential for elevated color. The simulations are intended to illustrate the ability of the inhibitors to expand the feasible operating range for alkalinity, and thus the feasible blend composition. All simulations were conducted at pH_s with the water quality defined in Table 5.37 using the minimum inhibitor doses employed in the experimental study (TP = 0.5 mg/L P, Si = incremental addition of 3 mg/L SiO_2). The simulations are presented in Figures 5.40–5.42 for iron, copper, and lead, respectively.

Iron. The simulated iron concentrations reveal general violations of the target concentration (0.16 mg/L) for all scenarios with an alkalinity less than 100 mg/L as $CaCO_3$, except when the Si inhibitor is used. A modest improvement in performance would be expected with increasing inhibitor doses (phosphate-based inhibitors). The behavior of the Si inhibitor deviates from the phosphate-based products because of an elevation of background SiO_2 in GW-enriched blends. As an increase in the SiO_2 concentration is experienced, the empirical model (Equation 5.6) predicts that the iron concentration would also increase. Operation without inhibitor addition at pH_s is projected to result in excessive apparent color for all of the blends with an alkalinity less than 150 mg/L as $CaCO_3$. As discussed previously, management of the HRT would be critical to control of iron concentrations in dead zones in the distribution system. In conjunction with control of HRT, maintenance of a minimum alkalinity of 100 mg/L as $CaCO_3$ and a pH greater than pH_s would be prudent to mitigate release of iron corrosion products.

Table 5.37
Water quality for GW/SW blends to assess inhibitor performance as a function of alkalinity

Blend	GW (%)	SW (%)	RO (%)	pH_s	Alkalinity (mg/L $CaCO_3$)	Influent iron (mg/L)	Cl^- (mg/L)	SO_4 (mg/L)	SiO_2 (mg/L)
1	90	0	10	7.24	196	0.054	41	27	12.6
2	85	5	10	7.26	190	0.052	42	34	12.0
3	80	10	10	7.28	183	0.051	42	42	11.4
4	75	15	10	7.29	176	0.050	43	49	10.7
5	70	20	10	7.31	170	0.048	44	57	10.1
6	65	25	10	7.33	163	0.047	45	64	9.4
7	60	30	10	7.35	157	0.045	45	72	8.8
8	55	35	10	7.37	150	0.044	46	79	8.2
9	50	40	10	7.39	143	0.042	46	86	7.5
10	45	45	10	7.41	137	0.041	47	94	6.9
11	40	50	10	7.43	130	0.040	48	101	6.2
12	35	55	10	7.46	124	0.038	49	109	5.6
13	30	60	10	7.48	117	0.037	50	116	5.0
14	25	65	10	7.51	110	0.036	50	124	4.3
15	20	70	10	7.54	104	0.034	51	131	3.7
16	15	75	10	7.57	97	0.033	52	139	3.0
17	10	80	10	7.60	91	0.031	53	146	2.4
18	5	85	10	7.63	84	0.030	53	153	1.8
19	0	90	10	7.67	77	0.029	54	161	1.1

Copper. The simulations for total copper concentration demonstrate a clear benefit associated with all of the inhibitors. In the absence of an inhibitor or pH elevation above pH_S, the projected copper concentration exceeds the allowable target (1 mg/L Cu) whenever the alkalinity exceeds 110 mg/L as $CaCO_3$. For the blend compositions employed in these simulations, the region of copper noncompliance would correspond with any blend that contains more than 25% groundwater. Imposition of a 25% constraint on the maximum allowable percentage of groundwater would cripple TBW operations, based on current capacities of alternate sources and existing water demand.

Addition of any of the inhibitors at the minimum dose is simulated to eliminate violations of the copper standard for all of the water quality blends in Table 5.37. This mitigation of copper release is projected for operation at pH_s so that potential for calcium carbonate precipitation is minimized. At their respective minimum doses, the phosphate-based inhibitors perform better than the Si product, although an increase in the Si inhibitor dose would achieve a corresponding reduction in copper concentration. The simulations do not indicate any substantial difference in performance among the three phosphate-based inhibitors.

Lead. The simulations for lead concentration also suggest a universal benefit associated with use of any of the inhibitors. The ZOP and BOP products are projected to achieve the greatest

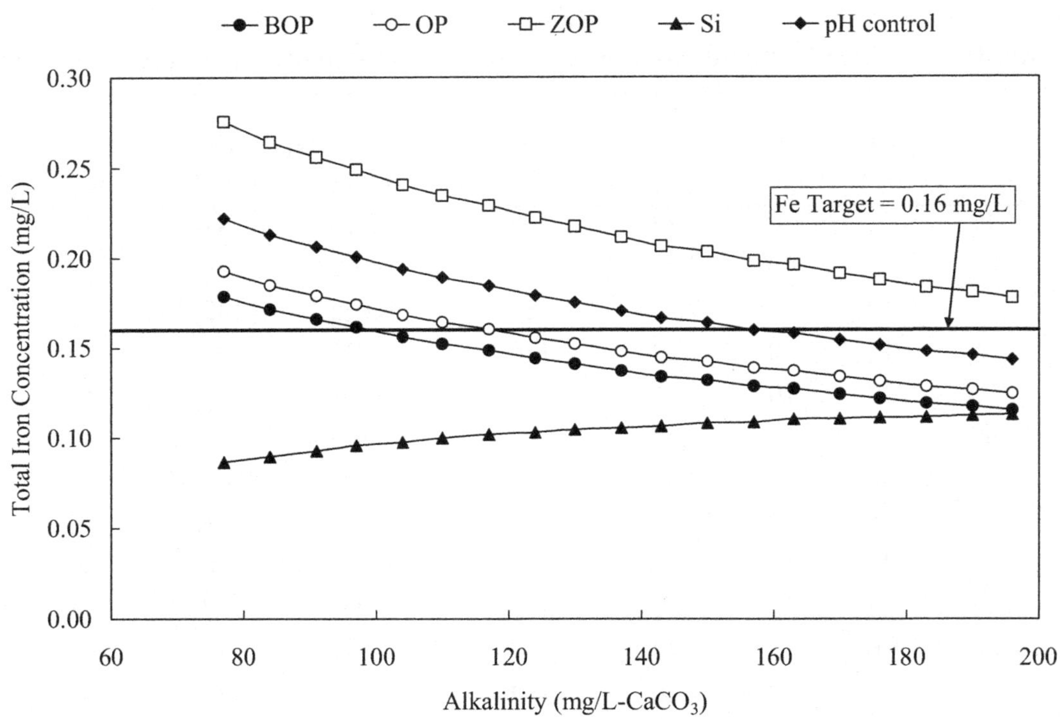

Figure 5.40 Iron as a function of alkalinity for minimum inhibitor dose

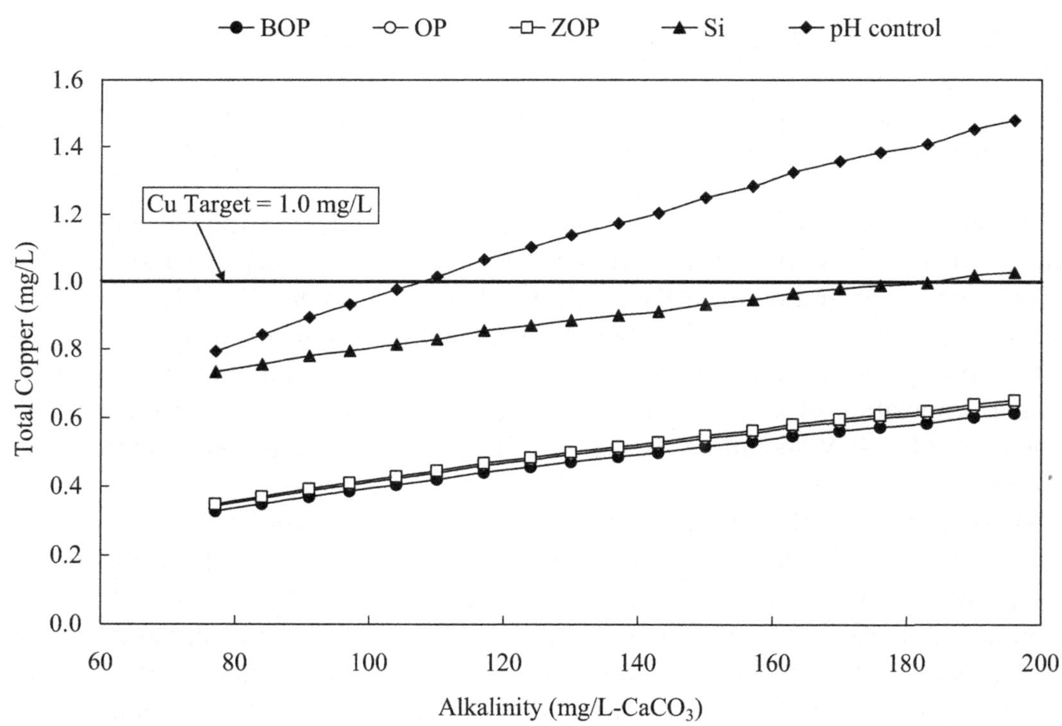

Figure 5.41 Copper as a function of alkalinity for minimum inhibitor dose

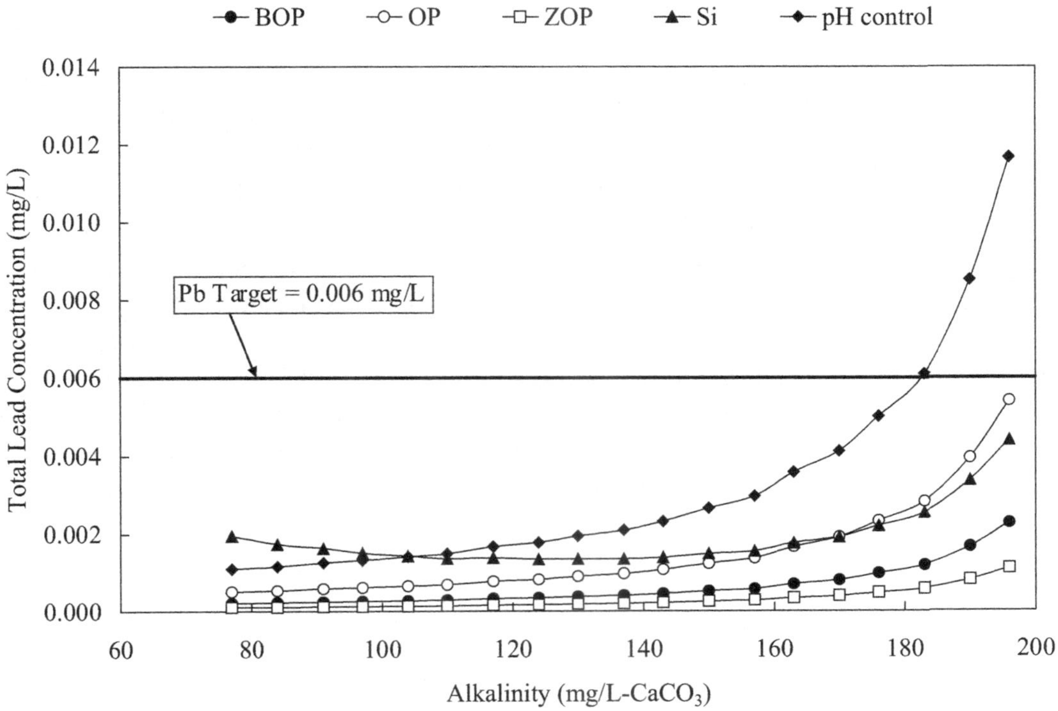

Figure 5.42 Lead as a function of alkalinity for minimum inhibitor dose

degree of lead control. Elevation of lead concentrations at high alkalinities is forecast by these models to coincide with GW-dominant blends. For the conditions of these simulations, these high-alkalinity blends would be deficient in sulfate (derived from surface water). The empirical lead concentration model (Equation 5.25) identifies that a lead mitigation benefit is realized for elevated sulfate concentrations. Finally, it should be noted that violation of the lead action level is not forecast for the range of blends (maximum alkalinity = 150 mg/L as $CaCO_3$) employed in this study. The simulations were conducted for an assumed temperature of 20°C. Operation at higher temperatures would result in increased lead concentrations per Equation 5.25.

ELECTROCHEMICAL CORROSION MONITORING

Throughout the remainder of this chapter, the term *electronoise (EN)* refers to the electrochemical corrosion monitoring process used in this study. Three corrosion rates—linear polarization resistance (LPR), harmonic distortion analysis (HDA), and electrochemical noise (ECN)—and twelve EN variables were monitored. Some of the EN variables were used to calculate the corrosion rates. These parameters were used to develop models predicting iron, copper, and lead release. The following sections cover the measurement of EN parameters, predictive model development, and the model results and discussion. The chapter concludes with a discussion of how utilities can use the real-time data from electrochemical corrosion monitoring to proactively manage metal release for regulatory compliance within the distribution system.

Measurement of EN Parameters

The corrosion rates from LPR, HDA, and ECN are abbreviated as LPRCR, HMCR, and ECNCR, respectively, and are reported in units of mils per year, or mpy (1 mpy = 0.0254 mm/yr). All three methods employ Ohm's law as the basis of corrosion rate determination. Ohm's law is shown in Equation 5.33. LPR applies a small direct current (DC) voltage across the electrodes, measures the current passing through the electrodes, and evaluates the resistance (R), as shown in Equation 5.33. HDA applies a small alternating current (AC) voltage over a range of frequencies, measures the current response at each frequency, and evaluates the impedance (Z), as shown in Equation 5.34. With LPR, the resistance, R, is independent of the applied voltage. However, with HDA, the impedance, Z, depends on the frequency of applied voltage. The relationship between impedance, Z, and the applied voltage frequency within HDA may be analyzed to determine both the anodic Tafel slope Ba and the cathodic Tafel slope Bc. The Stearn–Geary coefficient (B) is computed from the anodic and cathodic Tafel slopes. This coefficient, B, is then used to compute the HMCR, as shown in Equation 5.35.

$$V = I \times R \tag{5.33}$$

$$V = I \times Z \tag{5.34}$$

$$I_{corr} = \frac{B}{R_p} \tag{5.35}$$

where V = potential across the electrodes, volts
I = current through the electrodes, amperes
R = resistance of the electrodes, ohms

In electrochemical measurements, the total measured resistance for an applied voltage is the sum of resistance in the solution and at the electrode surfaces. The solution resistance (RSOL) relates to the conductivity of solution, whereas the polarization resistance (R_p) relates to the resistance caused by polarization at the electrode surfaces. The true electrode resistance, R_p, is calculated by subtracting RSOL form the total measured resistance. RSOL is determined during measurement of HDA by evaluating the changes in impedance with the changes in the frequency and amplitude of the voltage.

Unlike LPR and HDA, the ECN technique does not apply a DC or AC voltage to the electrodes, but rather, it monitors the variability of the naturally occurring voltage and current fluctuations caused by reactions on the electrode surfaces. This information is used to calculate the noise resistance R_N, as shown in Equation 5.36. The ECN technique uses R_N to calculate the corrosion current, I_{corr}, using Equation 5.37. In contrast, LPR and HDA use R_p to calculate the corrosion current, as shown previously in Equation 5.35. For all three techniques, after I_{corr} is calculated, Equation 5.38 is used to convert the corrosion current to the corrosion rate.

$$R_N = \frac{\sigma_E}{\sigma_i} \tag{5.36}$$

$$I_{corr} = \frac{B}{R_N} \tag{5.37}$$

$$CR = \frac{A \times I_{corr}}{z \times F \times \rho} = \frac{K \times I_{corr} \times A}{z \times \rho} \tag{5.38}$$

where σ_E = standard deviation of potential
σ_i = standard deviation of current
CR = corrosion rate of LPRCR or HMCR or ECNCR, mpy
A = atomic weight of the metal
I_{corr} = corrosion current density, amperes/cm^2
z = number of electrons transferred per atom of metal corroded
F = Faraday's constant = 96,500 coulombs/equivalent
ρ = density of the metal, g/cm^3
K = 1.288×10^5 constant for conversion of units

In addition, the standard deviation, skew, and kurtosis were calculated for both voltage and current. Equations for each of these statistical parameters are shown in Equations 5.39, 5.40, and 5.41 respectively. The pitting factor (PF) is shown in Equation 5.42.

$$\sigma = \frac{\sum_{1}^{n}(Y_i - \bar{Y})^2}{(n-1)} \tag{5.39}$$

$$\text{skew} = \frac{\sum_{1}^{n}(Y_i - \bar{Y})^3}{(n-1)\sigma^3} \tag{5.40}$$

$$\text{kurtosis} = \frac{\sum_{1}^{n}(Y_i - \bar{Y})^4}{(n-1)\sigma^4} \tag{5.41}$$

$$PF = \frac{\sigma_i}{(I_{corrHDA})} \tag{5.42}$$

where σ = standard deviation
$\bar{Y}$ = mean
σ_i = standard deviation of corrosion current density from ECNCR
I_{corr} = mean corrosion current density from HMCR

Model Development

Regression models were developed using EN parameters to predict metal release for both total and dissolved forms of iron, copper, and lead. The models used dummy variables to segregate the effect of inhibitor dose by inhibitor type. The effects of the EN parameters and temperature were included in the model without regard to inhibitor type. Models using all 12 EN parameters did not converge. Subsequent modeling attempts using reduced sets of EN parameters eventually produced a model based on LPRCR, HMCR, HMCR, and PF.

The general form of the EN power model with dummy variables for inhibitor type is shown in Equation 5.43. The remaining eight EN parameters were added one at a time and are represented by EN^{ai} in Equation 5.43, where EN represents either B, Ba, Bc, skew of the potential (PSKW), kurtosis of the potential (PKRT), skew of the current (ISKW), kurtosis of the current (IKRT), or RSOL. The same approach was used to develop models for both total and dissolved forms of iron, copper, and lead. The subscripts of EN parameters represent the metal for which the parameter was measured (i.e., in Equation 5.43, iron release (Me = Fe) uses LPRCR for iron, which is denoted as $LPRCR_{Fe}$).

$$\begin{aligned} Me = {} & (LPRCR_{Me}{}^{a1} \times HMCR_{Me}{}^{a2} \times ECNCR_{Me}{}^{a3} \times PF_{Me}{}^{a4} \times EN_{Me}{}^{ai}) \\ & \times (a \times BOP \times TP^{f} + b \times OP \times TP^{g} + c \times ZOP \times TP^{h} \times Zn^{i} + d \times Si \times SiO_2{}^{j} \\ & + e \times pH_s) \times temp^{o} \end{aligned} \tag{5.43}$$

where
- Me = Fe or Cu or Pb
- a1 = $LPRCR_{Me}$ exponent
- a2 = $HMCR_{Me}$ exponent
- a3 = $ECNCR_{Me}$ exponent
- a4 = PF_{Me} exponent
- $EN_{Me}{}^{ai}$ = one of the terms associated with exponents a5, a6, a7, a8, a9, a10, a11, or a12
- a5 = B_{Me} exponent
- a6 = Ba_{Me} exponent
- a7 = Bc_{Me} exponent
- a8 = $PSKW_{Me}$ exponent
- a9 = $PKRT_{Me}$ exponent
- a10 = $ISKW_{Me}$ exponent
- a11 = $IKRT_{Me}$ exponent
- a12 = $RSOL_{Me}$ exponent
- BOP = blended ortho-polyphosphate inhibitor dummy variable (0, 1)
- OP = orthophosphate inhibitor dummy variable (0, 1)
- ZOP = zinc orthophosphateinhibitor dummy variable (0, 1)
- Si = silica inhibitor dummy variable (0, 1)
- pH_s = pH control dummy variable (0, 1)
- TP = total phosphorus, mg/L
- Zn = zinc, mg/L
- SiO_2 = silica, mg/L as SiO_2
- temp = temperature, °C
- a = BOP dummy variable coefficient

b = OP dummy variable coefficient
c = ZOP dummy variable coefficient
d = silica dummy variable coefficient
e = pH control dummy variable coefficient
f = TP exponent associated with BOP dummy variable
g = TP exponent associated with OP dummy variable
h = TP exponent associated with ZOP dummy variable
i = Zn exponent associated with ZOP dummy variable
j = SiO_2 exponent associated with silica dummy variable
o = temperature exponent

The dummy variables shown in Equation 5.43 were used for all three metals and represent the effects of influent concentrations of total phosphate or silicate. The temperature used for all EN models was the average temperature in the flush tanks that supplied water to the EN trailer. The resulting models for total and dissolved iron, copper, and lead show the dummy variables abbreviated as DV for convenience. The abbreviated dummy variables as DV are shown in Equation 5.44.

$$\begin{aligned} DV = (&a \times BOP \times TP^f + b \times OP \times TP^g + c \times ZOP \times TP^h \times Zn^i \\ &+ d \times Si \times SiO_2{}^j + e \times pH_s) \times temp^o \end{aligned} \tag{5.44}$$

Model Results and Discussion

Iron Release

Water Quality Correlation With EN for Iron. Correlations between iron EN parameters were assessed to determine whether the parameters were independent. Statistically significant ($p < 0.05$) Pearson's linear correlation coefficients for the iron EN parameters are shown in Table 5.38. The only correlation coefficient above 0.5 is 0.67 for the correlation of $LPRCR_{Fe}$ and $HMCR_{Fe}$, which indicates that confounding of EN parameters for iron is not an issue. Correlation coefficients greater than 0.9 are cause for concern.

Pearson's linear correlation coefficients for the iron EN parameters and water quality are shown in Table 5.39. The average temperature from the EN trailer, T_{16}, is taken from temperature data during a 16-hour period prior to sampling and is shown in Table 5.39. This time corresponds to the combined HRT in UCI and GS sections of each PDS. The data indicate that alkalinity is modestly correlated with $LPRCR_{Fe}$ and PF_{Fe} and possibly with $HMCR_{Fe}$. EN parameters would correlate with water quality, as both are global measures of the same environment.

By examining the correlations in Table 5.39, it is possible to deduce which iron EN parameters may correspond with iron release. Both $LPRCR_{Fe}$ and $HMCR_{Fe}$ measure a general corrosion rate and are positively correlated with alkalinity. In contrast, the PF, which has been proposed as a measure of localized corrosion, is negatively correlated with alkalinity. Given that a higher alkalinity mitigates iron release, a negative correlation between PF_{Fe} and alkalinity is theoretically sound. PF_{Fe} is positively correlated with chlorides, sulfates, and temperature, which are historically known to induce pitting. Unlike PF_{Fe}, $LPRCR_{Fe}$ and $HMCR_{Fe}$ were poorly correlated with temperature, as shown by the low correlation coefficients in Table 5.39. Thus, PF_{Fe}

Table 5.38
Correlation coefficients for EN parameters for iron

	$LPRCR_{Fe}$	$HMCR_{Fe}$	$ECNCR_{Fe}$	PF_{Fe}	B_{Fe}	Ba_{Fe}	Bc_{Fe}	$PSKW_{Fe}$	$PKRT_{Fe}$	$ISKW_{Fe}$	$IKRT_{Fe}$	$RSOL_{Fe}$
$LPRCR_{Fe}$	1.00											
$HMCR_{Fe}$	0.67	1.00										
$ECNCR_{Fe}$	0.13	0.07	1.00									
PF_{Fe}	–0.39	–0.22	N/A	1.00								
B_{Fe}	0.110	0.07	N/A	N/A	1.00							
Ba_{Fe}	N/A	–0.11	N/A	N/A	N/A	1.00						
Bc_{Fe}	N/A	N/A	N/A	N/A	N/A	N/A	1.00					
$PSKW_{Fe}$	0.11	0.09	–0.43	–0.16	N/A	N/A	N/A	1.00				
$PKRT_{Fe}$	N/A	N/A	0.11	N/A	N/A	N/A	N/A	0.65	1.00			
$ISKW_{Fe}$	–0.10	N/A	N/A	0.26	N/A	N/A	N/A	0.07	N/A	1.00		
$IKRT_{Fe}$	–0.12	N/A	N/A	N/A	N/A	N/A	N/A	0.16	0.24	–0.18	1.00	
$RSOL_{Fe}$	–0.21	–0.12	N/A	–0.09	N/A	N/A	N/A	N/A	N/A	N/A	N/A	1.00

N/A = not applicable.

Table 5.39
Pearson's correlation coefficient for EN versus water quality

Pearson's R	Alkalinity	Ca	Cl	SO_4	Cl_2 Influent	T_{16}
$LPRCR_{Fe}$	0.688	0.494	–0.558	–0.504	–0.366	–0.300
$HMCR_{Fe}$	0.442	0.284	–0.374	–0.286	–0.314	–0.178
PF_{Fe}	–0.517	—	0.382	0.604	0.321	0.407

NOTE: All correlation coefficients were significant at 5% significance.

correlated with iron release, and water quality in that PF_{Fe} increased as alkalinity decreased and chlorides and sulfates increased.

At very low general corrosion rates, the value of the pitting factor can be very large, even though aggressive localized corrosion is not occurring. Examining Equation 5.42, when the average current, I_{corr}, is near zero, it becomes clear that misleadingly high values of PF_{Fe} would be generated. PF_{Fe} values over 0.5 are strong indicators of localized corrosion. Plots of PF_{Fe} versus alkalinity, chlorides, and sulfates are shown in Figures 5.43–5.45, respectively. The plots show that there is no direct linear correlation between these variables and PF_{Fe}; instead, there appear to be threshold values of alkalinity, chlorides, and sulfates that correspond with sudden increases in PF_{Fe}. The relatively high PF_{Fe} shown in Figure 5.43 for alkalinities less than 120 mg/L (as $CaCO_3$) correspond to the same range of alkalinity that was noted for high release of total iron during the TBW I project. Similarly, relatively high PF_{Fe} are shown in Figures 5.44 and 5.45 for chlorides and sulfates concentrations above 60 and 90 mg/L, respectively. Consequently, the data presented in Figures 5.43–5.45 support the use of PF_{Fe} for monitoring water quality and predicting high iron release.

The average iron EN corrosion rates for all phases are shown in Table 5.40. There are few general EN parameters that can be taken as indicative of corrosion, but $PF_{Fe} > 0.5$ indicates localized

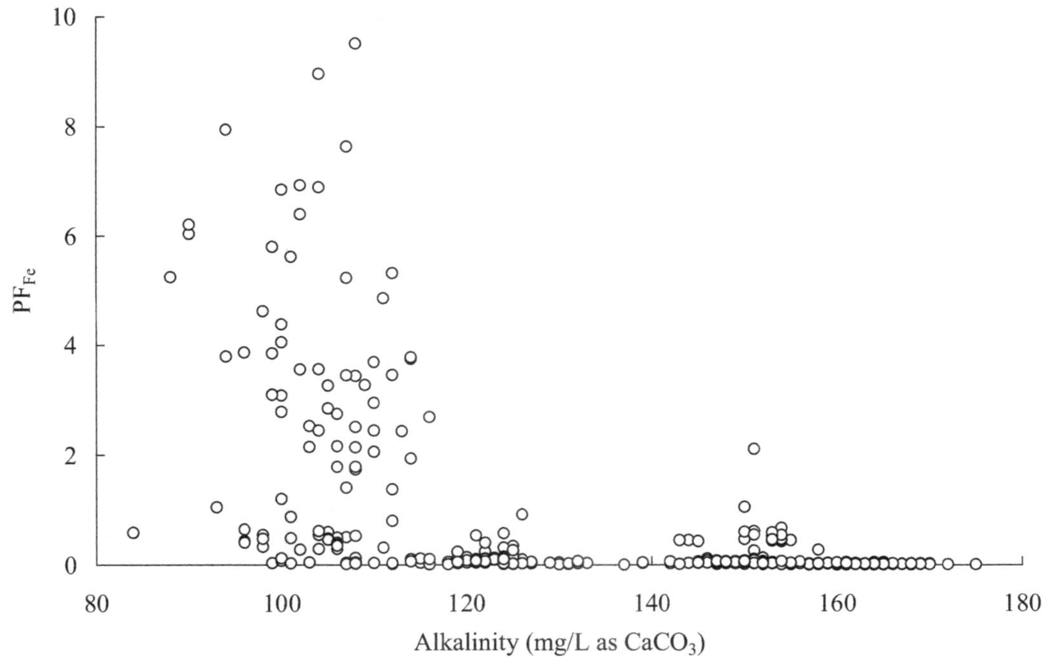

Figure 5.43 Iron pitting factor correlation with alkalinity

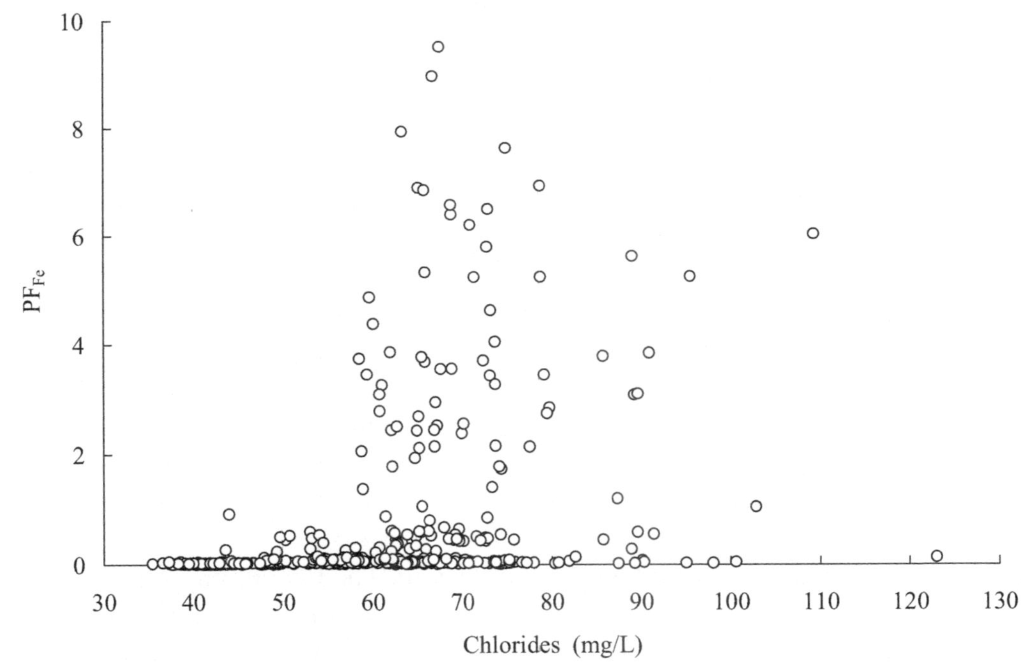

Figure 5.44 Iron pitting factor correlation with chlorides

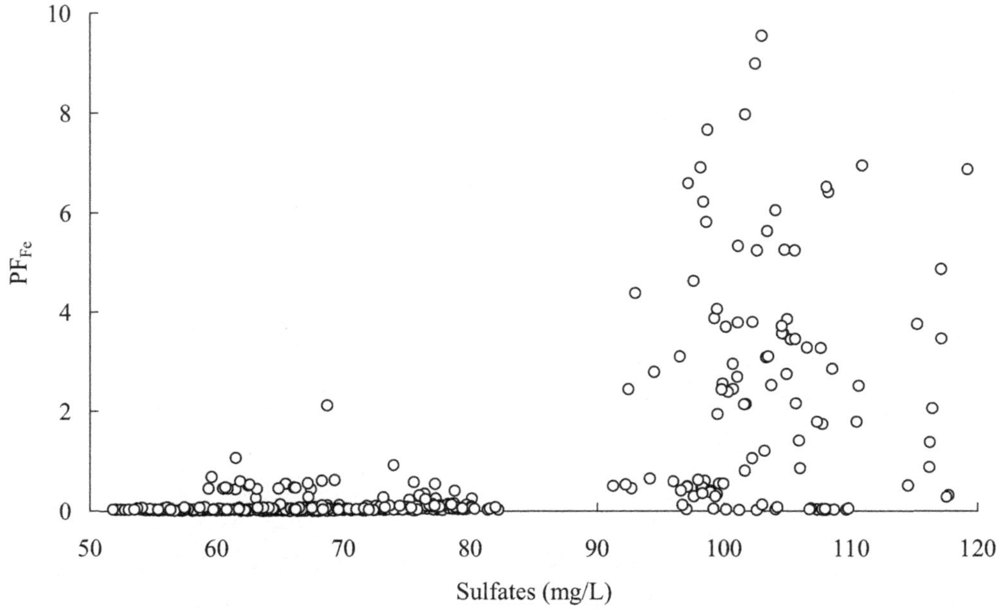

Figure 5.45 Iron pitting factor correlation with sulfates

Table 5.40
Iron corrosion rates and pitting factor

Statistic	$LPRCR_{Fe}$ (mpy)	$HMCR_{Fe}$ (mpy)	$ECNCR_{Fe}$ (mpy)	PF_{Fe}
Mean	1.52	1.53	21.56	0.71
Std. dev.	1.62	2.19	113.95	1.58
N	716	716	716	716

NOTE: 1 mpy = 0.0254 mm/yr.

corrosion and $LPRCR_{Fe} < 2$ indicates low to moderate general and localized corrosion. Consequently, the PF_{Fe} and $LPRCR_{Fe}$ means reported in Table 5.40 indicate iron corrosion during this project. Other $ECNCR_{Fe}$ and associated water quality are shown in Table 5.41 (Duranceau, Townley, and Bell 2004), and the $ECNCR_{Fe}$ differ significantly within Table 5.41 and relative to Table 5.40. No $ECNCR_{Fe}$ correlations with water quality were found among the data shown in Tables 5.40 and 5.41, which may have been due to the lack of standardization for measurement of EN parameters.

EN Empirical Model for Total Iron Concentration. The EN dummy variable models developed for total iron concentration in the hybrid PDSs are shown in Table 5.42. These models were reduced by discarding any independent variables with $p > 0.05$, which generated the model shown in the last row of Table 5.42, which is the method of model development that was used for all EN models. R^2 is 0.27, which is comparable to the R^2 of 0.29 from the water quality model for total

Table 5.41
ECNCR and water quality from other EN studies*

Water quality parameter	Sarasota	IRWD 1	IRWD 2	LVVWD 2	PCU 2
pH	7.82	7.92	7.92	7.61	7.74
Hardness (mg/L as $CaCO_3$)	129	204	204	255	226
Calcium (mg/L as $CaCO_3$)	44	121	121	141	199
Alkalinity (mg/L as $CaCO_3$)	117	94	94	162	202
Total dissolved solids (mg/L)	593	455	455	293	378
Sodium (mg/L)	150	65	65	6	14
Sulfate (mg/L)	288	154	154	44	21
Chloride (mg/L)	28	69	69	8	12
Dissolved oxygen (mg/L)	7.24	7.82	7.82	7.32	0.66
Temperature (°C)	27.1	22.6	22.6	23.5	23.9
Influent P (mg/L)	0.33	0.33	0.33	0.33	0.33
Langelier saturation index	–0.24	0.26	0.26	0.26	0.54
Fe corrosion—control (mpy)	14.7	8.0	4.0	6.5	3.5
Fe corrosion—inhibitor (mpy)	14.9	7.5	6.0	1.7	2.3
Cu corrosion—control (mpy)	1.4	—	3.0	1.7	2.5
Cu corrosion—inhibitor (mpy)	0.7	—	1.0	1.6	2.3
Pb solder corrosion—control (mpy)	0.4	0.4	—	0.5	2.3
Pb solder corrosion—inhibitor (mpy)	0.7	0.3	—	25.9	1.7

* Sarasota = City of Sarasota, Fla.
IRWD = Irvine Ranch Water District, Calif.
LVVWD = Las Vegas Valley Water District, Nev.
PCU = Pinellas County Utilities, Fla.
Inhibitor = 1 ppm ZOP as PO_4
1 mpy = 0.0254 mm/yr

Table 5.42
Total iron release EN models

Model	R^2	Standard error (mg/L)	Model *p*-value	Parameter with highest *p*-value	Highest *p*-value
$TFe = (ECNCR_{Fe}^{a3} * PF_{Fe}^{a5}) * DV$	0.27	0.069	<1e–4	$ECNCR_{Fe}$	0.054
$TFe = (PF_{Fe}^{a5}) * DV$	0.27	0.069	<1e–4	PF_{Fe}	<1e–4

NOTE: DV = dummy variable terms, as shown previously in Equation 5.44.

iron release presented previously, which infers that EN can be used as well as water quality to predict total iron release in this work.

The PF_{Fe} model shown in Table 5.42 is presented with all variables that appear in Equation 5.45. The exponent for PF_{Fe} is positive, which means that total iron release will increase as PF_{Fe} increases. The exponent on temperature is negative, which means that increasing temperature decreases iron release in this work, which is also consistent with the water quality model for total iron release. Each corrosion control strategy (inhibitor or pH elevation) has different contributing terms, multiplied by the dummy variables, which adjust the predicted iron concentration as a function of the corrosion control strategy. Figure 5.46 shows the fit of the total iron model predictions with the actual observations for all phases. The diagonal line on the plot represents perfect agreement between the model and the observed data. The model tends to underpredict the higher concentrations of iron, and a number of observations that exceed the secondary iron standard of 0.3 mg/L are not predicted.

$$\begin{aligned} TFe = {} & (PF_{Fe}{}^{0.0694}) \times (0.240 \times BOP \times TP^{-0.071} + 0.277 \times OP \times TP^{0.047} \\ & + 0.239 \times ZOP \times TP^{-0.088} \times Zn^{-0.157} + 0.071 \times Si \times SiO_2{}^{0.456} \\ & + 0.334 \times pH_s\,) \times temp^{-0.144} \end{aligned} \tag{5.45}$$

where
TFe = total iron, mg/L
PF_{Fe} = pitting factor for iron, unitless
BOP = blended ortho-polyphosphate inhibitor (0, 1)
OP = orthophosphate inhibitor (0, 1)
ZOP = zinc orthophosphate inhibitor (0, 1)
Si = silica inhibitor (0, 1)
pH_s = pH control (0, 1)
TP = total phosphorus, mg/L as P
Zn = zinc, mg/L
SiO_2 = silica, mg/L as SiO_2
$temp$ = temperature, °C

The negative exponent for temperature with a value of –0.144 suggests that total iron will decrease with temperature. A change in temperature from 10°C to 30°C would reduce the temperature effect by the ratio $(30/10)^{-0.144}$, that is, by 0.85, and decrease the model predicted total iron release by 15%. The study with inhibitor addition was started in February 2006, and the total iron concentration decreased with time after inhibitor addition began. This is seen in Figure 5.47, which shows total iron release for the entire study with symbols for each phase. The solid black squares were the data for phase I, wherein the bulk of the data had a low temperature and resulted in total iron similar to other phases. All the data below 15°C was from phase I, and the highest iron release occurred during phase I at 18°C and ranged from 0.4 to 0.6 mg/L, as shown in Figure 5.47.

Phase I iron release versus temperature by inhibitor type is shown in Figure 5.48. The data shows that all total iron concentrations exceeding 0.2 mg/L in phase I were for inhibitors and not pH controls. The few spikes of iron exceeding 0.3 mg/L (i.e., the secondary standard) were for ZOP and Si inhibitors only. Phase I had a blend with high alkalinity, which mitigates iron release. Hence, the phase I total iron release data indicates that the inhibitors caused an initial release of total iron, which did not continue. Release of materials from distribution systems is not uncommon during an initial period following inhibitor addition. However, the initial release of

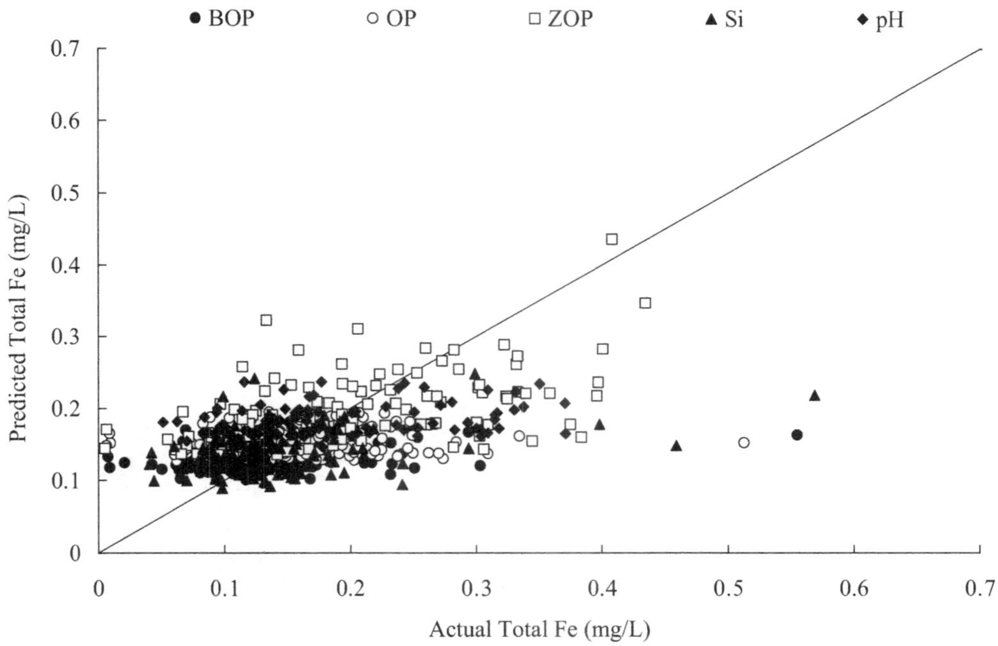

Figure 5.46 PF model predicted versus actual total iron concentration

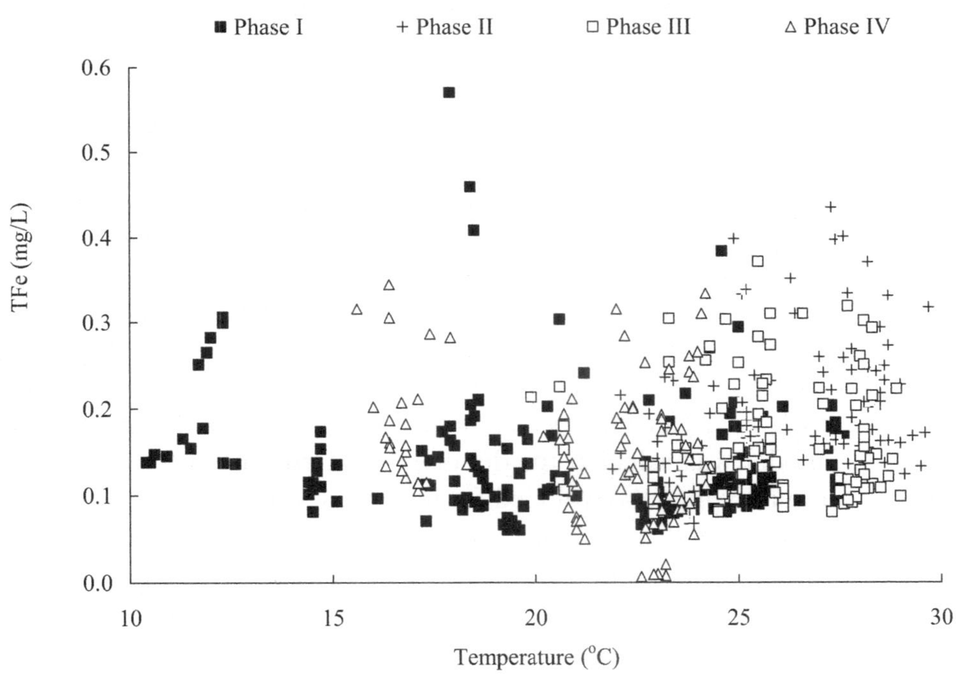

Figure 5.47 Temperature versus total iron concentration by phase

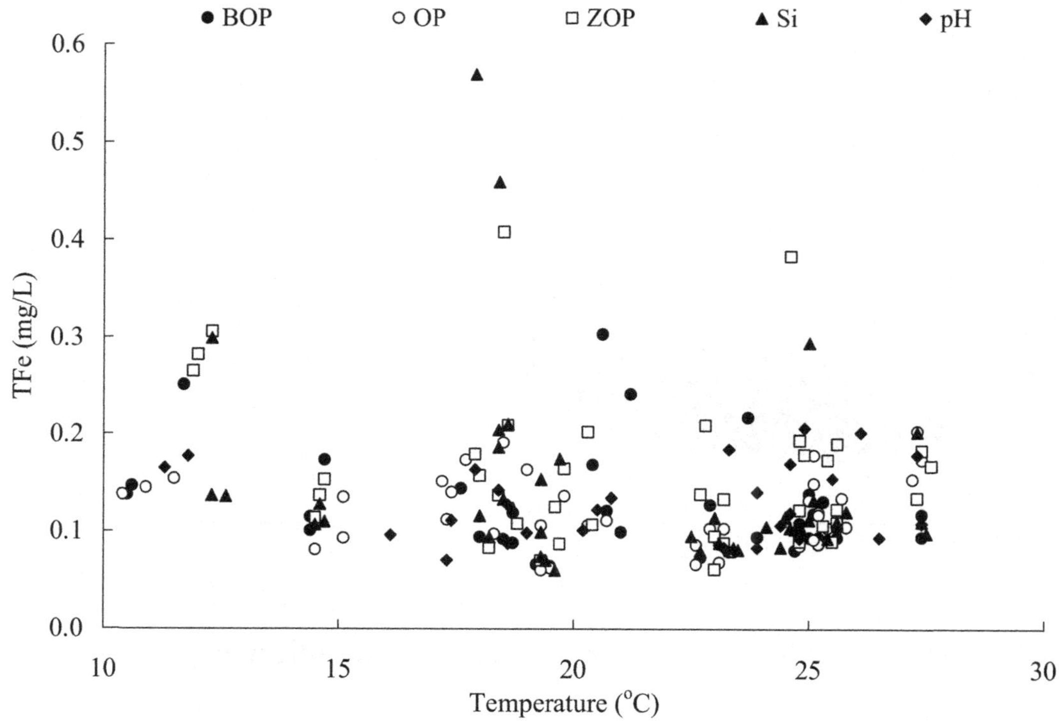

Figure 5.48 Temperature versus total iron concentration for phase I

total iron may have biased the relationship between iron release and temperature as iron release was not expected to decrease with increasing temperature.

EN Empirical Model for Dissolved Iron Concentration. Empirical models for dissolved iron in the hybrid PDSs are shown in Table 5.43. After elimination of statistically insignificant variables with $p > 0.05$, a dissolved iron model remained that was based on either $LPRCR_{Fe}$, $HMCR_{Fe}$, $ECNCR_{Fe}$, or $RSOL_{Fe}$. As shown in Table 5.43, all of these models were statistically significant. The dummy variable exponents from the $HMCR_{Fe}$ model were positive for OP and SiO_2 and negative for BOP and ZOP. This is consistent with the dummy variable model for dissolved iron developed from water quality parameters in the "Metal Release" section of this chapter. The $HMCR_{Fe}$ model was selected for dissolved iron and has a 0.43 R^2, which is also comparable to the water quality model developed for dissolved iron release. The $HMCR_{Fe}$ model is shown in Equation 5.46 and indicates that dissolved iron release will increase as $HMCR_{Fe}$ increases and will decrease as temperature decreases.

$$DFe = (HMCR_{Fe}{}^{0.486}) \times (246.71 \times BOP \times TP^{-0.128} + 157.63 \times OP \times TP^{0.275} + 342.92 \times ZOP \times TP^{-0.534} \times Zn^{0.109} + 74.79 \times Si \times SiO_2{}^{0.188} + 172.21 \times pH_s) \times temp^{-3.22} \quad (5.46)$$

where DFe = dissolved iron, mg/L
$HMCR_{Fe}$ = corrosion rate from harmonic distortion for iron, mpy
BOP = blended ortho-polyphosphate inhibitor (0, 1)
OP = orthophosphate inhibitor (0, 1)

Table 5.43
Dissolved iron release EN models

Model	R^2	Standard error (mg/L)	Model *p*-value	Parameter with highest *p*-value	Highest *p*-value
$DFe = (LPRCR_{Fe}{}^{a1} * HMCR_{Fe}{}^{a2} * ECNCR_{Fe}{}^{a3} * RSOL_{Fe}{}^{a12}) * DV$	0.46	0.025	<1e–4	$ECNCR_{Fe}$	0.043
$DFe = (LPRCR_{Fe}{}^{a1}) * DV$	0.43	0.026	<1e–4	$LPRCR_{Fe}$	<1e–4
$DFe = (HMCR_{Fe}{}^{a2}) * DV$	0.43	0.026	<1e–4	$HMCR_{Fe}$	<1e–4
$DFe = (ECNCR_{Fe}{}^{a3}) * DV$	0.42	0.026	<1e–4	$ECNCR_{Fe}$	<1e–4
$DFe = (RSOL_{Fe}{}^{a12}) * DV$	0.43	0.026	<1e–4	$RSOL_{Fe}$	<1e–4

NOTE: DV = dummy variable terms as shown previously in Equation 5.44.

ZOP = zinc orthophosphate inhibitor (0, 1)
Si = silica inhibitor (0, 1)
pH_s = pH control (0, 1)
TP = total phosphorus, mg/L as P
Zn = zinc, mg/L
SiO_2 = silica, mg/L as SiO_2
temp = temperature, °C

The actual versus predicted dissolved iron release using the $HMCR_{Fe}$-DFe model is shown in Figure 5.49 for all phases. A change in temperature from 10°C to 30°C would reduce the temperature effect by the ratio $(30/10)^{-3.22}$ (i.e., by a ratio of 0.03) and decrease the model predicted dissolved iron release by 97%. This seems counterintuitive but is consistent with Figure 5.50, which shows dissolved iron release by phase. A similar effect was seen with total iron. This effect is demonstrated for dissolved iron in Figure 5.50 where phase I data indicate lower temperature and higher dissolved iron release than other phases. The effect of inhibitors on dissolved iron release by inhibitor for phase I data are shown in Figure 5.51. The data show that dissolved iron release increased in the presence of BOP, ZOP, and Si inhibition relative to pH control on several occasions.

Copper Release

Water Quality Correlation With EN for Copper. The correlation coefficients for copper EN parameters are shown in Table 5.44. As with iron, there is not significant confounding of copper EN parameters. The greatest correlation coefficient for the copper EN parameters is 0.54 for the correlation of $LPRCR_{Cu}$ and $HMCR_{Cu}$.

All statistically significant correlations at $p < 0.05$ of copper EN parameters and PDS water quality are shown in Table 5.45. Only the PF_{Cu} was significant with alkalinity, chlorides, sulfate, and temperature; however, the correlation coefficients indicate that PF_{Cu} is not highly confounded with any of these water quality coefficients. PF_{Cu} is a ratio of standard deviation of the ECN instantaneous current over the HDA mean current. As noted in the EN iron discussion, a low general corrosion rate is associated with a low mean current, which can generate a misleadingly high PF_{Cu}. The

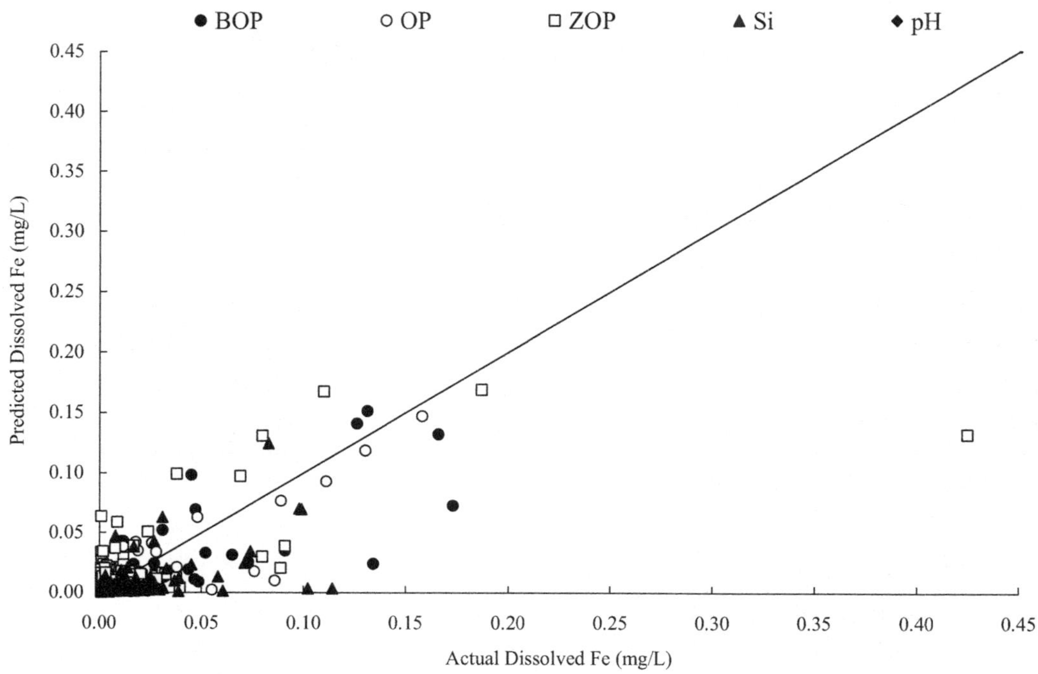

Figure 5.49 HMCR model predicted versus actual dissolved iron concentration

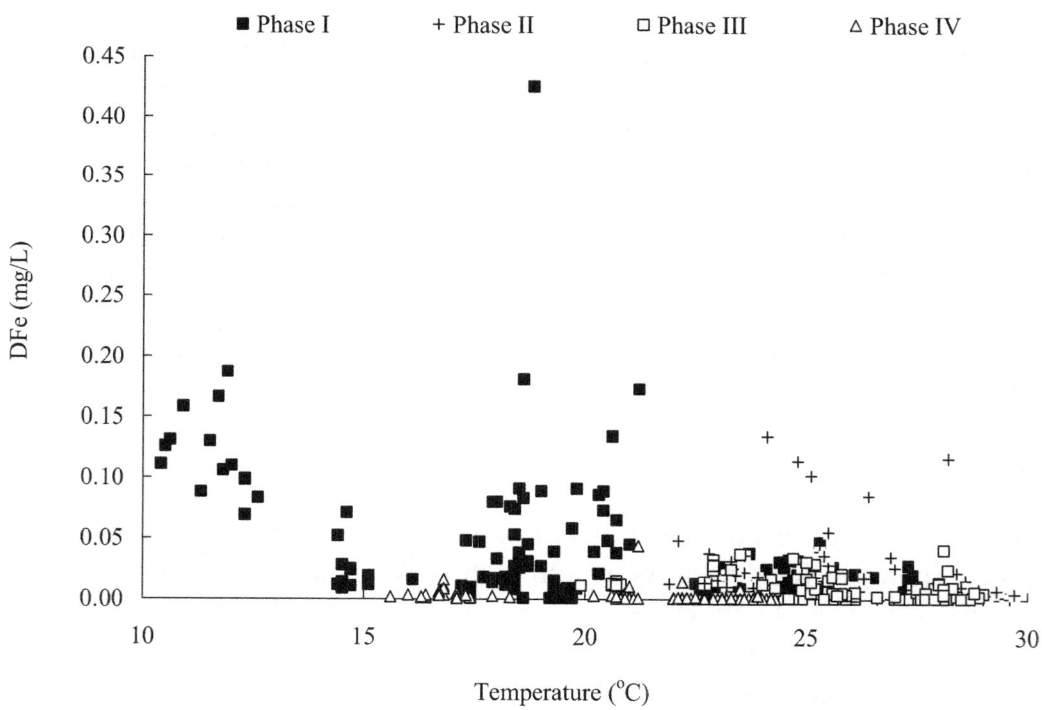

Figure 5.50 Temperature versus dissolved iron concentration by phase

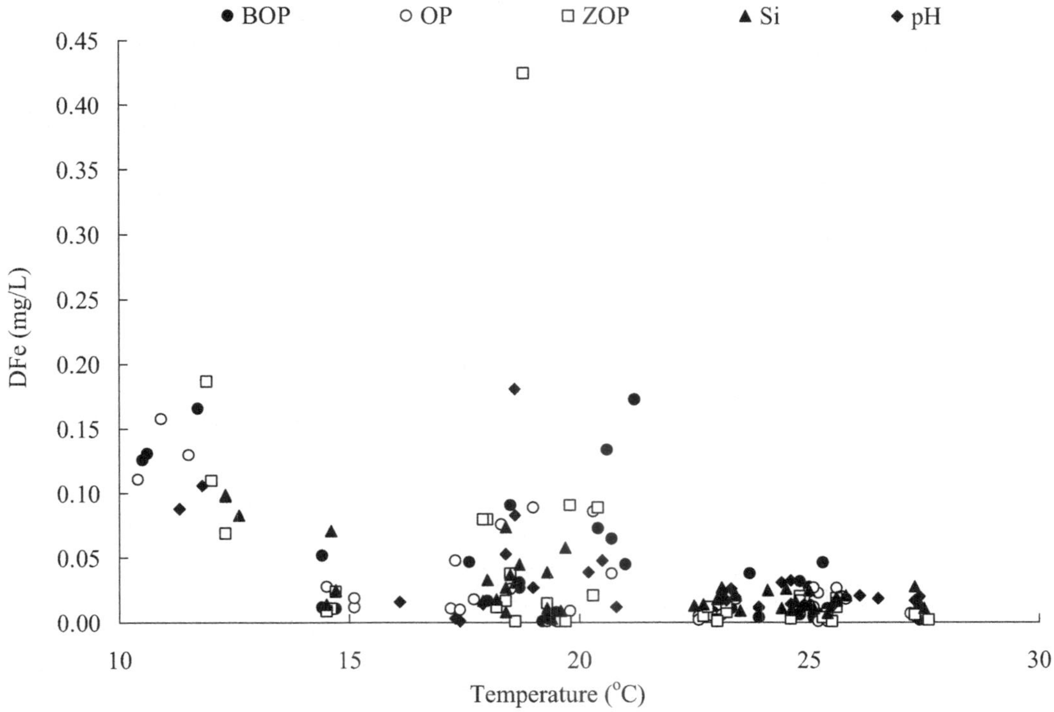

Figure 5.51 Temperature versus dissolved iron concentration for phase I

Table 5.44
Correlation coefficients for copper EN parameters

	$LPRCR_{Cu}$	$HMCR_{Cu}$	$ECNCR_{Cu}$	PF_{Cu}	B_{Cu}	Ba_{Cu}	Bc_{Cu}	$PSKW_{Cu}$	$PKRT_{Cu}$	$ISKW_{Cu}$	$IKRT_{Cu}$	$RSOL_{Cu}$
$LPRCR_{Cu}$	1.00											
$HMCR_{Cu}$	0.54	1.00										
$ECNCR_{Cu}$	0.37	0.15	1.00									
PF_{Cu}	–0.13	0.21	N/A	1.00								
B_{Cu}	N/A	N/A	N/A	N/A	1.00							
Ba_{Cu}	N/A	N/A	N/A	–0.08	N/A	1.00						
Bc_{Cu}	N/A	N/A	N/A	N/A	N/A	N/A	1.00					
$PSKW_{Cu}$	0.28	N/A	–0.10	–0.17	N/A	N/A	N/A	1.00				
$PKRT_{Cu}$	0.23	N/A	0.18	N/A	N/A	N/A	N/A	0.51	1.00			
$ISKW_{Cu}$	N/A	N/A	N/A	N/A	N/A	N/A	N/A	0.10	N/A	1.00		
$IKRT_{Cu}$	N/A	N/A	N/A	N/A	N/A	N/A	N/A	N/A	N/A	–0.25	1.00	
$RSOL_{Cu}$	N/A	0.14	N/A	–0.09	N/A	N/A	N/A	N/A	N/A	–0.09	N/A	1.00

N/A = not applicable.

Table 5.45
Pearson's correlation coefficient for copper PF_{Cu} versus selected water quality

Pearson's R	Alkalinity	Cl	SO_4	Cl_2 Influent	T_{16}
PF_{Cu}	–0.487	0.387	0.532	0.351	0.342

NOTE: All correlation coefficients were significant at 5% significance.

Table 5.46
Copper corrosion rates and PF_{Cu}

Statistic	$LPRCR_{Cu}$ (mpy)	$HMCR_{Cu}$ (mpy)	$ECNCR_{Cu}$ (mpy)	PF_{Cu}
Mean	0.12	0.28	15.93	1.06
Std. dev.	0.25	0.29	165.60	1.90
N	717	717	717	717

NOTE: 1 mpy = 0.0254 mm/yr.

average $LPRCR_{Cu}$, $HMCR_{Cu}$, $ECNCR_{Cu}$, and PF_{Cu} for all phases is shown in Table 5.46 and is similar to the EN iron data shown previously.

$LPRCR_{Cu}$ and $HMCR_{Cu}$ corrosion rates are similar, whereas $ECNCR_{Cu}$ is significantly higher than either $LPRCR_{Cu}$ or $HMCR_{Cu}$. The standard deviations of $LPRCR_{Cu}$, $HMCR_{Cu}$, and PF_{Cu} were all approximately twice their means. Hence, the corrosion rates were highly variable, and both general corrosion and localized corrosion of copper occurred.

EN Empirical Model for Total Copper Concentration. The empirical EN models developed for total copper release in this study are shown in Table 5.47. A three-parameter model based on $LPRCR_{Cu}$, $ECNCR_{Cu}$, and $PKRT_{Cu}$ for total copper release was developed, which has a 0.71 R^2 and was similar to the water quality model for copper release.

The EN model for total copper release is given in Equation 5.47. The model indicates that total copper release will increase as $LPRCR_{Cu}$ and $ECNCR_{Cu}$ increase. This model also suggests that an increase in phosphate, zinc, or silicate from any inhibitor will decrease total copper release as the model exponents for TP, Zn, and SiO_2 are negative, which is the same for the water quality copper release model. The temperature exponent (–0.063) indicates that copper release will vary slightly with temperature but that higher temperature will decrease total copper release. A change in temperature from 10°C to 30°C would reduce the temperature effect by the ratio $(30/10)^{-0.063}$ (i.e., by 0.93) and decrease the model predicted total copper release by 7%, which is not very significant.

$$\begin{aligned} TCu = {} & (LPRCR_{Cu}^{0.028} \times ECNCR_{Cu}^{0.031}) \times (0.510 \times BOP \times TP^{-0.116} \\ & + 0.476 \times OP \times TP^{-0.281} + 0.471 \times ZOP \times TP^{-0.232} \times Zn^{-0.048} \\ & + 1.960 \times Si \times SiO_2^{-0.324} + 1.505 \times pH_s) \times temp^{-0.063} \end{aligned} \tag{5.47}$$

where TCu = total copper, mg/L
$LPRCR_{Cu}$ = corrosion rate from linear polarization for copper, mpy
$ECNCR_{Cu}$ = corrosion rate from electrochemical noise for copper, mpy

Table 5.47
Total copper release EN models

Model	R^2	Standard error (mg/L)	Model *p*-value	Parameter with highest *p*-value	Highest *p*-value
$TCu = (LPRCR_{Cu}{}^{a1} * ECNCR_{Cu}{}^{a3} * PKRT_{Cu}{}^{a9}) * DV$	0.71	0.18	<1e–4	$PKRT_{Cu}$	0.002
$TCu = (LPRCR_{Cu}{}^{a1}) * DV$	0.71	0.18	<1e–4	$LPRCR_{Cu}$	<0.0001
$TCu = (ECNCR_{Cu}{}^{a3}) * DV$	0.71	0.18	<1e–4	$ECNCR_{Cu}$	<0.0001
$TCu = (PKRT_{Cu}{}^{a9}) * DV$	0.69	0.19	<1e–4	$PKRT_{Cu}$	<0.0001
$TCu = (LPRCR_{Cu}{}^{a1} * ECNCR_{Cu}{}^{a3}) * DV$	0.72	0.18	<1e–4	All	<0.0001

NOTE: DV = dummy variable terms, as shown previously in Equation 5.44.

BOP = blended ortho-polyphosphate inhibitor (0, 1)
OP = orthophosphate inhibitor (0, 1)
ZOP = zinc orthophosphate inhibitor (0, 1)
Si = silica inhibitor (0, 1)
pH_s = pH control (0, 1)
TP = total phosphorus, mg/L as P
Zn = zinc, mg/L
SiO_2 = silica, mg/L as SiO_2
temp = temperature, °C

Figure 5.52 shows the fit of the total copper model predictions with the actual observations for all phases. A line is indicated on the plot that would represent perfect agreement between the model predictions and the actual data. The model tends to have a good fit for low copper concentrations and fits modestly at the higher copper concentrations. The model also shows that copper release is highest for pH control and all inhibitors reduced total copper release.

EN Empirical Model for Dissolved Copper Concentration. The empirical EN models developed for dissolved copper release in this study are shown in Table 5.48. These models are the EN dummy variable models that converged using nonlinear least-squares regression. An EN model was developed for the release of dissolved copper based on $LPRCR_{Cu}$ and $ECNCR_{Cu}$. The model selected for total copper has the same form as the model selected for dissolved copper release and is shown in boldface in Table 5.48.

The EN model for dissolved copper release is given in Equation 5.48. The model indicates that an increase in corrosion rates ($LPRCR_{Cu}$ and $ECNCR_{Cu}$) will increase copper release. This model also suggests that an increase in phosphate, zinc, or silicate from any inhibitor will decrease copper release because of the negative exponent for TP, Zn, and SiO_2. The temperature exponent indicates that higher temperature will lower total copper release; however, copper release is not practically sensitive to temperature. A change in temperature from 10°C to 30°C would decrease the model predicted dissolved copper release by 3%, which is practically insignificant.

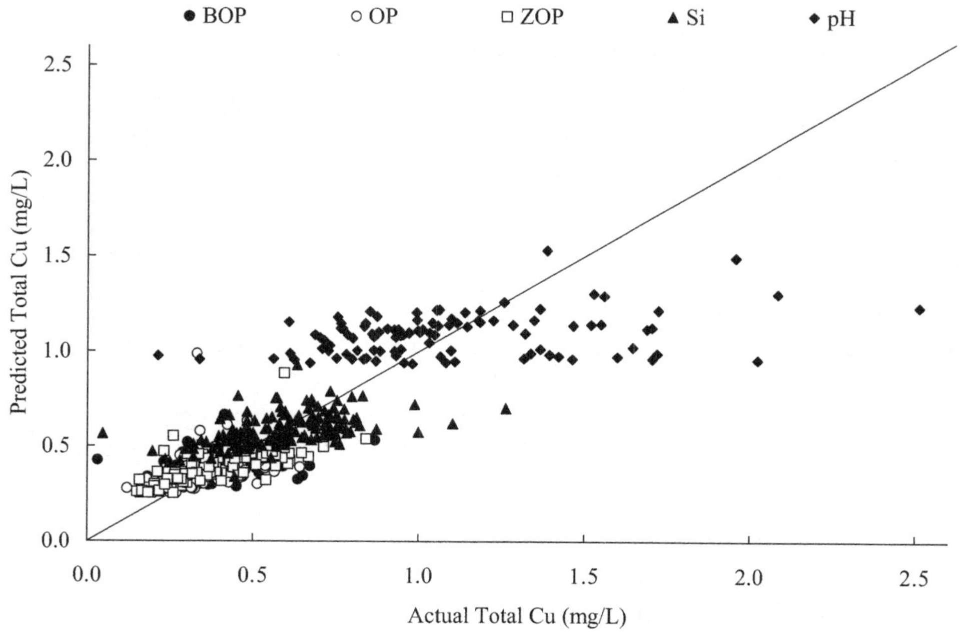

Figure 5.52 EN model predicted versus actual total copper concentration

Table 5.48
Dissolved copper release EN models

Model	R^2	Standard error (mg/L)	Model *p*-value	Parameter with highest *p*-value	Highest *p*-value
$DCu = (LPRCR_{Cu}^{a1} * ECNCR_{Cu}^{a3}) * DV$	**0.68**	**0.17**	**<1e–4**	**$ECNCR_{Cu}$**	**0.002**
$DCu = (LPRCR_{Cu}^{a1}) * DV$	0.68	0.17	<1e–4	$LPRCR_{Cu}$	<0.0001
$DCu = (ECNCR_{Cu}^{a3}) * DV$	0.67	0.18	<1e–4	$ECNCR_{Cu}$	<0.0001

NOTE: DV = dummy variable terms as shown previously in Equation 5.44.

$$\begin{aligned} DCu = {} & (LPRCR_{Cu}^{0.0259} \times ECNCR_{Cu}^{0.027}) \times (0.406 \times BOP \times TP^{-0.107} \\ & + 0.372 \times OP \times TP^{-0.293} + 0.386 \times ZOP \times TP^{-0.242} \times Zn^{-0.021} \\ & + 1.528 \times Si \times SiO_2^{-0.320} + 1.181 \times pH_s) \times temp^{-0.028} \end{aligned} \tag{5.48}$$

where DCu = dissolved copper, mg/L
$LPRCR_{Cu}$ = corrosion rate from linear polarization for copper, mpy
$ECNCR_{Cu}$ = corrosion rate from electrochemical noise for copper, mpy
BOP = blended ortho-polyphosphate inhibitor (0, 1)
OP = orthophosphate inhibitor (0, 1)
ZOP = zinc orthophosphate inhibitor (0, 1)
Si = silica inhibitor (0, 1)
pH_s = pH control (0, 1)
TP = total phosphorus, mg/L as P
Zn = zinc, mg/L

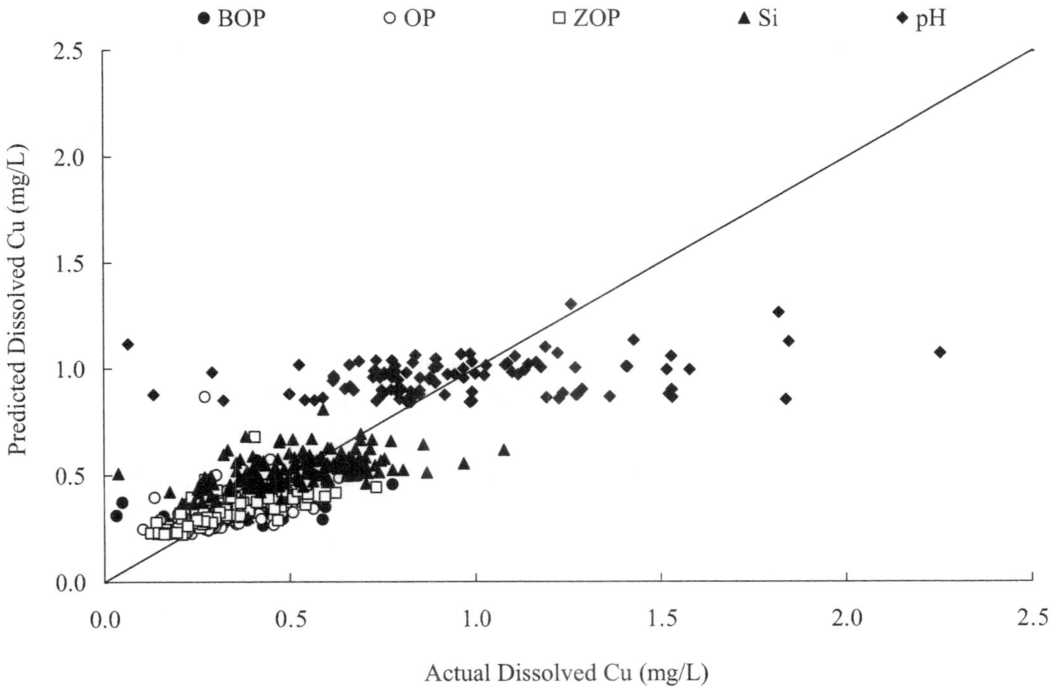

Figure 5.53 EN model predicted versus actual dissolved copper concentration

SiO_2 = silica, mg/L as SiO_2
temp = temperature, °C

Figure 5.53 shows the fit of the dissolved copper model predictions with the actual observations for all phases. The model achieves a reasonable fit between the predictions and the observations. The R^2 and standard error are 0.68 and 0.17 mg/L, which are comparable to the water quality model.

Lead Release

Project field operations began on Jan. 2, 2006, with inhibitor feed started to the blended waters in the PDSs. Downstream of the PDSs, copper loops house lead/tin (50:50) coupons which simulate lead release from solders and fixtures in this project. Lead/tin electrodes compatible to these coupons were not initially available; hence, pure lead electrodes were used for the first 27 days of February 2006. Pure lead electrodes were replaced by lead/tin electrodes on Feb. 27, 2006.

Water Quality Correlation With EN for Lead. The correlation of lead EN parameters is shown in Table 5.49. As with iron and copper, the greatest correlation is for $LPRCR_{Pb}$ and $HMCR_{Pb}$, and it is less than 0.60. As with iron and copper, the correlation of lead EN parameters found no confounding effects among lead EN parameters.

Only PF_{Pb} correlated with water quality, as shown in Table 5.50. Summary statistics for lead corrosion rates and PF_{Pb} are shown in Table 5.51 and indicate significant differences between each parameter. As for iron and copper, the EN data shown in Table 5.51 indicates that both general and localized corrosion mechanisms affected lead release in this work.

Table 5.49
Correlation coefficients for EN parameters for lead

	$LPRCR_{Pb}$	$HMCR_{Pb}$	$ECNCR_{Pb}$	PF_{Pb}	B_{Pb}	Ba_{Pb}	Bc_{Pb}	$PSKW_{Pb}$	$PKRT_{Pb}$	$ISKW_{Pb}$	$IKRT_{Pb}$	$RSOL_{Pb}$
$LPRCR_{Pb}$	1.00											
$HMCR_{Pb}$	0.55	1.00										
$ECNCR_{Pb}$	0.34	N/A	1.00									
PF_{Pb}	N/A	0.19	N/A	1.00								
B_{Pb}	N/A	N/A	N/A	N/A	1.00							
Ba_{Pb}	–0.37	–0.50	N/A	N/A	N/A	1.00						
Bc_{Pb}	N/A	N/A	N/A	N/A	N/A	N/A	1.00					
$PSKW_{Pb}$	0.10	N/A	–0.35	–0.12	N/A	N/A	N/A	1.00				
$PKRT_{Pb}$	0.40	N/A	0.13	N/A	N/A	N/A	N/A	0.58	1.000			
$ISKW_{Pb}$	N/A	N/A	N/A	N/A	N/A	N/A	N/A	0.38	N/A	1.00		
$IKRT_{Pb}$	N/A	N/A	N/A	N/A	N/A	N/A	N/A	0.30	N/A	0.71	1.00	
$RSOL_{Pb}$	N/A	0.10	N/A	–0.10	N/A	N/A	N/A	N/A	N/A	N/A	N/A	1.00

N/A = not applicable.

Table 5.50
Pearson's correlation coefficient for lead PF versus water quality

Pearson's R	Alkalinity	Ca	Cl	SO_4	Cl_2 Influent	T_{16}
PF_{Pb}	–0.45	—	0.37	0.56	0.26	0.29

NOTE: All correlation coefficients were significant at 5% significance.

Table 5.51
Lead corrosion rates and PF

Statistic	$LPRCR_{Pb}$ (mpy)	$HMCR_{Pb}$ (mpy)	$ECNCR_{Pb}$ (mpy)	PF_{Pb}
Mean	0.06	0.86	6.25	0.54
Std. dev.	0.30	1.08	101.83	1.07
N	652	652	652	652

NOTE: 1 mpy = 0.0254 mm/yr.

EN Empirical Model for Total Lead Concentration. The empirical EN models developed for total lead release in this study are shown in Table 5.52. Sequential removal of single EN variables, which was statistically insignificant ($p > 0.05$), resulted in a total lead release model based on $LPRCR_{Pb}$, $HMCR_{Pb}$, $ECNCR_{Pb}$, PF_{Pb}, and $PKRT_{Pb}$. Table 5.52 shows the model having these five EN parameters with an R^2 of 0.57. Other models were developed and a model based on $LPRCR_{Pb}$, $HMCR_{Pb}$, and PF_{Pb} was selected and is shown in boldface in Table 5.52. The number of significant EN parameters in this model indicates that lead release was caused by of general and localized corrosion.

Table 5.52
Total lead release EN models

Model	R^2	Standard error (mg/L)	Model *p*-value	Parameter with highest *p*-value	Highest *p*-value
$TPb = (LPRCR_{Pb}^{a1} * HMCR_{Pb}^{a2} * ECNCR_{Pb}^{a3} * PF_{Pb}^{a4} * PKRT_{Pb}^{a9}) * DV$	0.57	0.005	<1e–4	$PKRT_{Pb}$	0.001
$TPb = (LPRCR_{Pb}^{a1} * HMCR_{Pb}^{a2} * ECNCR_{Pb}^{a3} * PF_{Pb}^{a4}) * DV$	0.49	0.005	<1e–4	$ECNCR_{Pb}$	0.040
$TPb = (LPRCR_{Pb}^{a1} * HMCR_{Pb}^{a2} * PF_{Pb}^{a4}) * DV$	**0.55**	**0.005**	**<1e–4**	**All**	**<1e–4**

NOTE: DV = dummy variable terms, as shown previously in Equation 5.44.

The best EN model for total lead release shown in Table 5.52 is presented with all variables in Equation 5.49. This model suggests that an increase in phosphate or silicate from any inhibitor will decrease lead release because of the negative exponent for TP and SiO_2. This is in agreement with the water quality lead release model presented in the "Metal Release" section in this chapter. The model suggests that an increase in corrosion rates ($LPRCR_{Pb}$ and $HMCR_{Pb}$) will increase lead release.

$$TPb = (LPRCR_{Pb}^{0.217} \times HMCR_{Pb}^{0.337} \times PF_{Pb}^{-0.347}) \times (1.58 \times 10^{-8} \times BOP \times TP^{-0.805} + 1.53 \times 10^{-9} \times OP \times TP^{-4.188} + 5.49 \times 10^{-8} \times ZOP \times TP^{-0.784} \times Zn^{0.735} + 5.56 \times 10^{-8} \times Si \times SiO_2^{-0.326} + 1.83 \times 10^{-7} \times pH_s) \times temp^{3.67} \quad (5.49)$$

where
TPb = total lead, mg/L
$LPRCR_{Pb}$ = corrosion rate from linear polarization for lead, mpy
$HMCR_{Pb}$ = corrosion rate from harmonic distortion for lead, mpy
PF_{Pb} = pitting factor for lead (unit less)
BOP = blended ortho-polyphosphate inhibitor (0, 1)
OP = orthophosphate inhibitor (0, 1)
ZOP = zinc orthophosphate inhibitor (0, 1)
Si = silica inhibitor (0, 1)
pH_s = pH control (0, 1)
TP = total phosphorus, mg/L as P
Zn = zinc, mg/L
SiO_2 = silica, mg/L as SiO_2
temp = temperature, °C
1 mpy = 0.0254 mm/yr

This model indicates that increasing the pitting factor (PF_{Pb}) will decrease lead release because of the negative PF_{Pb} exponent. This seems counterintuitive but is also shown in Figure 5.54. Figure 5.54 shows that most of the low PF_{Pb} values associated with high total lead release from this study were for pH controls.

The temperature exponent (3.67) for the total lead release EN model (shown in Equation 5.49) suggests that lead release is very sensitive to temperature and higher temperature will increase total lead release. A change in temperature from 10°C to 30°C would increase the

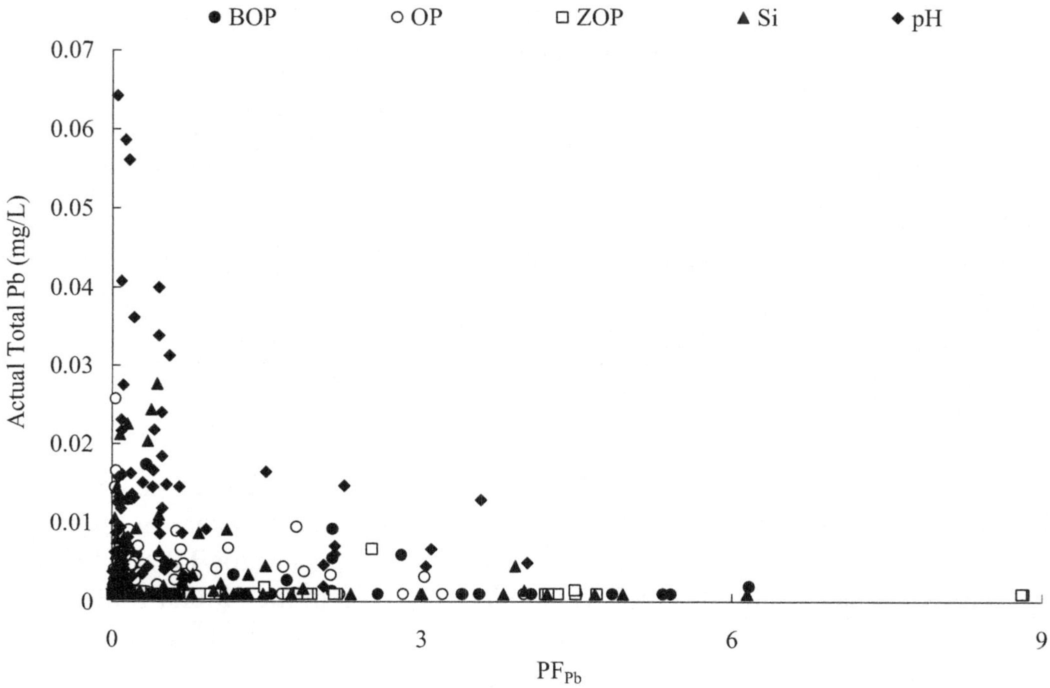

Figure 5.54 Pitting factor versus actual total lead concentration

temperature effect by the ratio $(30/10)^{3.67}$ (i.e., by 56 times), all other factors being equal. Figure 5.55 shows the total lead release for the entire study with symbols for each phase. All the highest lead release data occurred in phase III. Though both phase II and phase III had the highest temperatures in this study, phase II had the highest sulfates because of higher surface water in the blend. The water quality lead release model presented in the "Metal Release" section of this chapter has a negative exponent for sulfate and suggests that for similar temperatures with all other factors being the same, phase II with higher sulfate should release less lead than phase III.

Figure 5.56 shows the fit of the total lead model predictions with the actual observations for all phases. The model tends to have a scatter in the fit for low lead concentrations (TPb < 0.01 mg/L) but fits decently at the higher concentrations of lead. The model also shows that lead release is highest for pH control and that all inhibitors reduced total lead release.

EN Empirical Model for Dissolved Lead Concentration. The empirical EN models developed for dissolved lead release are shown in Table 5.53. Sequential removal of statistically insignificant EN variables resulted in models based on several EN variables, which indicated that general and localized corrosion mechanisms were significant for dissolved lead release.

The best EN model for total lead release shown in Table 5.53 is presented with all variables in Equation 5.50. This model suggests that an increase in phosphate or silicate from any inhibitor will decrease lead release because of the negative exponent for TP and SiO_2. This is in agreement with the water quality lead release model presented in the "Metal Release" section of this chapter. The model suggests that an increase in corrosion rates ($LPRCR_{Pb}$ and $HMCR_{Pb}$) will increase lead release.

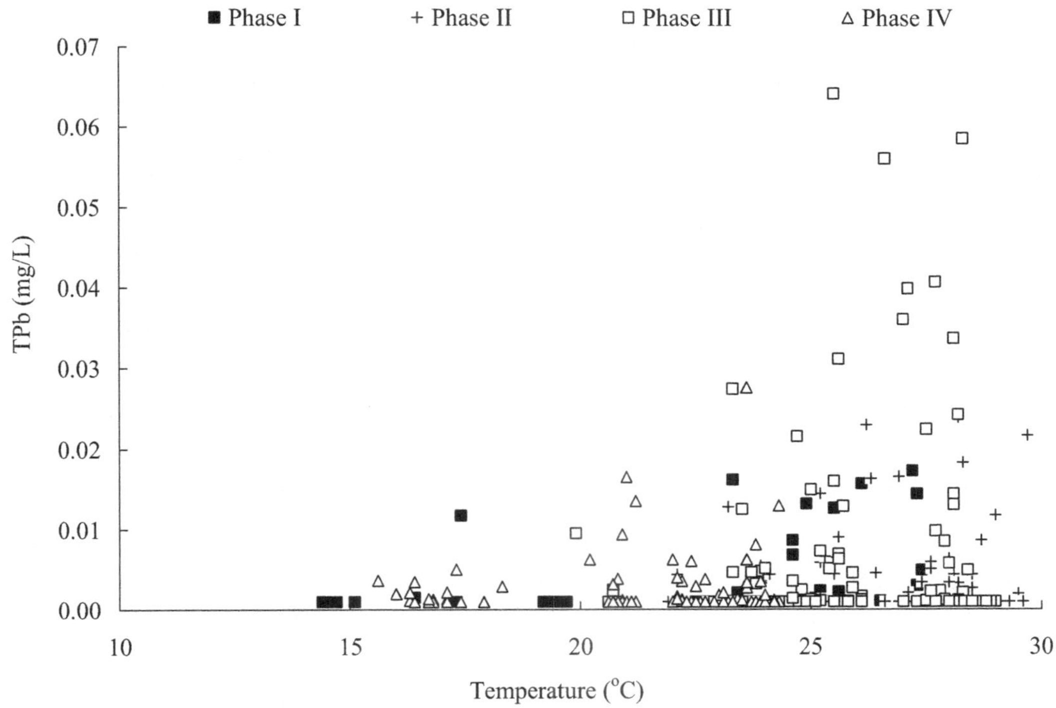

Figure 5.55 Temperature versus total lead concentration by phase

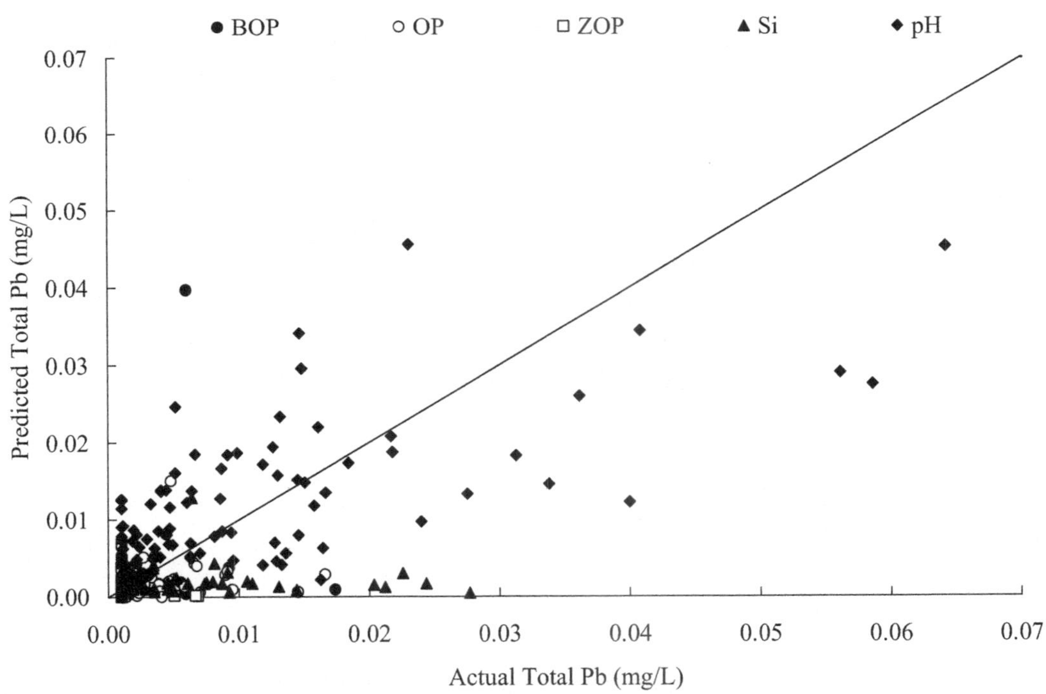

Figure 5.56 EN model predicted versus actual total lead concentration

Table 5.53
Dissolved lead release EN models

Model	R^2	Standard error (mg/L)	Model *p*-value	Parameter with highest *p*-value	Highest *p*-value
$DPb = (LPRCR_{Pb}^{a1} * HMCR_{Pb}^{a2} * ECNCR_{Pb}^{a3} * PF_{Pb}^{a4} * PKRT_{Pb}^{a9}) * DV$	0.44	0.003	<1e–4	$PKRT_{Pb}$	0.045
$DPb = (LPRCR_{Pb}^{a1} * HMCR_{Pb}^{a2} * ECNCR_{Pb}^{a3} * PF_{Pb}^{a4} * IKRT_{Pb}^{a11}) * DV$	0.48	0.003	<1e–4	$ECNCR_{Pb}$	0.043
$DPb = (LPRCR_{Pb}^{a1} * HMCR_{Pb}^{a2} * PF_{Pb}^{a4} * IKRT_{Pb}^{a11}) * DV$	0.47	0.003	<1e–4	$IKRT_{Pb}$	0.004
$DPb = (LPRCR_{Pb}^{a1} * HMCR_{Pb}^{a2} * PF_{Pb}^{a4}) * DV$	0.44	0.003	<1e–4	All	<1e–4
$DPb = (LPRCR_{Pb}^{a1} * HMCR_{Pb}^{a2}) * DV$	0.37	0.003	<1e–4	All	<1e–4
$DPb = (LPRCR_{Pb}^{a1} * IKRT_{Pb}^{a11}) * DV$	0.34	0.003	<1e–4	$IKRT_{Pb}$	0.006
$DPb = (LPRCR_{Pb}^{a1}) * DV$	0.33	0.003	<1e–4	$LPRCR_{Pb}$	<1e–4
$DPb = (HMCR_{Pb}^{a2}) * DV$	0.31	0.003	<1e–4	$HMCR_{Pb}$	0.013
$DPb = (ECNCR_{Pb}^{a3}) * DV$	0.27	0.003	<1e–4	$ECNCR_{Pb}$	0.013
$DPb = (PF_{Pb}^{a4}) * DV$	0.33	0.003	<1e–4	PF_{Pb}	0.006
$DPb = (IKRT_{Pb}^{a11}) * DV$	0.31	0.003	<1e–4	$IKRT_{Pb}$	0.015

NOTE: DV = dummy variable terms, as shown previously in Equation 5.44.

$$DPb = (LPRCR_{Pb}^{0.189} \times HMCR_{Pb}^{0.281} \times PF_{Pb}^{-0.297}) \times (1.09 \times 10^{-7} \times BOP \times TP^{-0.115} + 2.78 \times 10^{-8} \times OP \times TP^{2.550} + 3.72 \times 10^{-7} \times ZOP \times TP^{-0.690} \times Zn^{0.580} + 1.92 \times 10^{-7} \times Si \times SiO_2^{-0.088} + 8.3 \times 10^{-7} \times pH_s) \times temp^{2.96} \quad (5.50)$$

where DPb = dissolved lead, mg/L
$LPRCR_{Pb}$ = corrosion rate from linear polarization for lead, mpy
$HMCR_{Pb}$ = corrosion rate from harmonic distortion for lead, mpy
PF_{Pb} = pitting factor for lead, unitless
BOP = blended ortho-polyphosphate inhibitor (0, 1)
OP = orthophosphate inhibitor (0, 1)
ZOP = zinc orthophosphate inhibitor (0, 1)
Si = silica inhibitor (0, 1)
pH_s = pH control (0, 1)
TP = total phosphorus, mg/L as P
Zn = zinc, mg/L
SiO_2 = silica, mg/L as SiO_2
$temp$ = temperature, °C

The model shown in Equation 5.50 indicates that dissolved lead release will decrease as PF_{Pb} increases because of the negative PF_{Pb} exponent. It is inferred that PF_{Pb} may not represent true pitting behavior but that high PF_{Pb} represents the inflating effect of a low average corrosion rate in the denominator of Equation 5.42, shown previously. A change in temperature from 10°C to 30°C would increase the temperature effect by $(30/10)^{2.96}$ or almost 29 times if all other factors

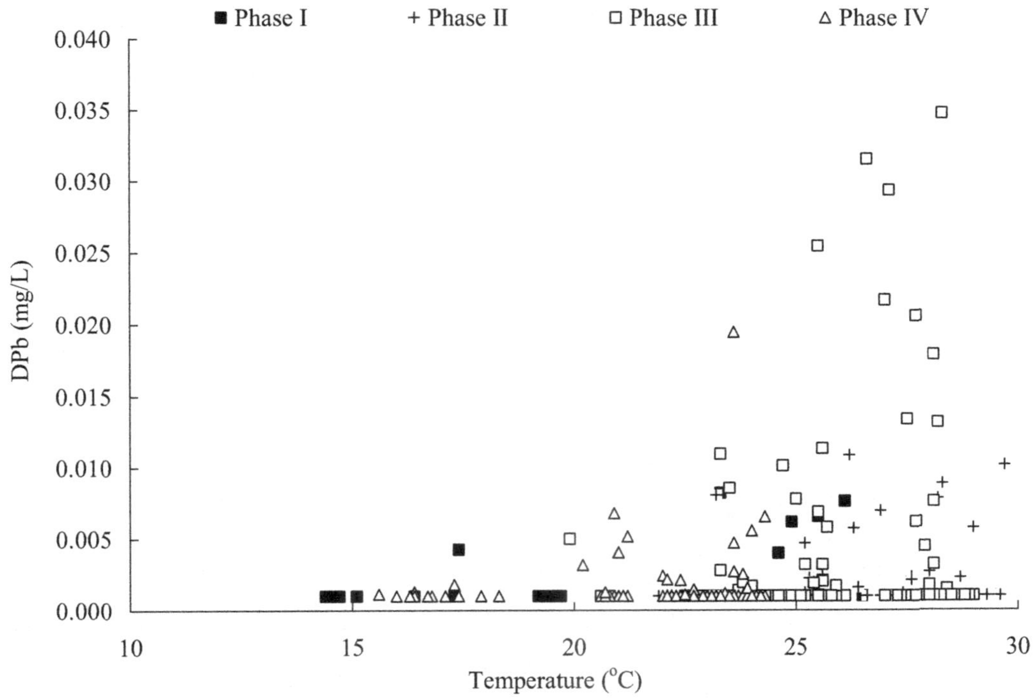

Figure 5.57 Temperature versus dissolved lead concentration by phase

are equal. Figure 5.57 shows the dissolved lead release for the entire study with symbols for each phase. All the highest lead releases occurred in phase III. Though both phase II and phase III had the highest temperatures in this study, phase II had the highest sulfates because of higher surface water in the blend. The water quality lead release model presented in the "Metal Release" section of this chapter has a negative exponent for sulfate and suggests that for similar temperatures with all other factors being the same, phase II with higher sulfate should release less lead than phase III, which is shown in Figure 5.57.

Figure 5.58 shows the fit of the dissolved lead model predictions with the actual observations for all phases. The model achieves a reasonable fit between the predictions and the observations. R^2 and standard error are 0.44 and 0.003 mg/L, which are comparable to the water quality model.

Water Quality Implications of EN Parameters

A study by Boyle Engineering and M.J. Schiff & Associates (Duranceau, Townley, and Bell 2004) investigated the application of electrochemical corrosion monitoring using the ECN method. ECN monitoring was conducted at four utilities, along with some water quality sampling. The study concluded that ECN can be used to derive practical information for assessment of corrosion susceptibility of metals and the type of corrosion process. A quote from the report reads, "the usefulness of this type of testing is demonstrated in this project by the conclusion that for some conditions, 'inhibition' increases the corrosion rate." Inhibition herein was referred to as addition of the corrosion inhibitor ZOP at 1 mg/L as P. The average water quality and corrosion rates from the referenced study are already presented in Table 5.41.

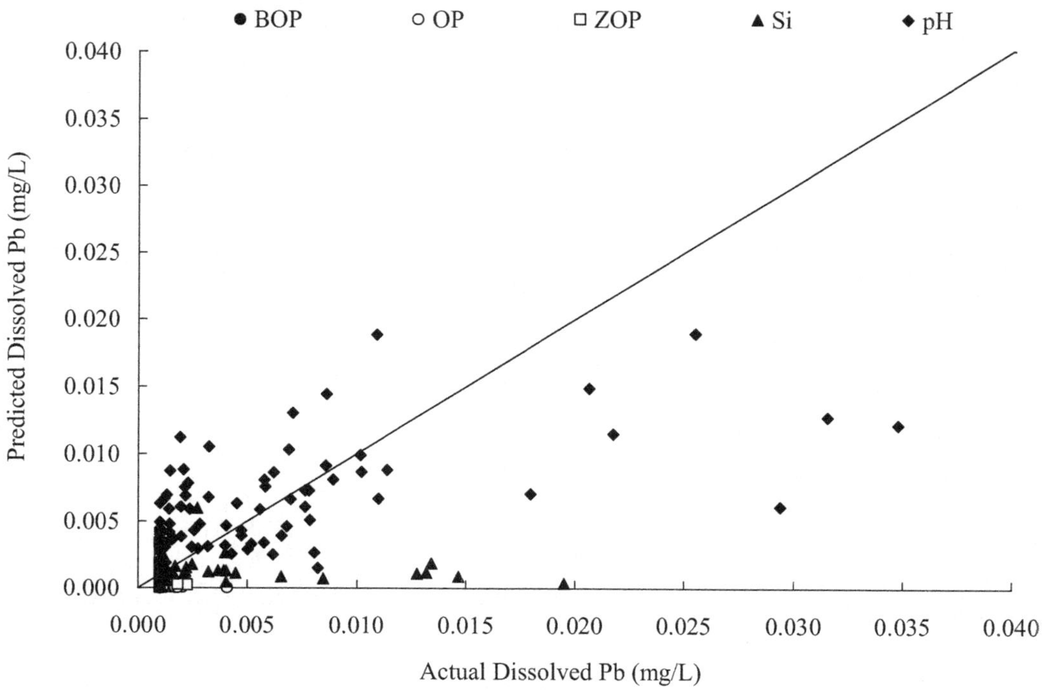

Figure 5.58 EN model predicted versus actual dissolved lead concentration

The TBW II data was averaged for each PDS by phase in order to incorporate that data with data from the Boyle report. TBW II EN averaged water quality and EN data were generated for 14 PDSs and four phases. The created data set with and without inhibitors is shown in Table 5.54 for iron, copper, and lead. Larson's ratio 1 (LR1), Larson's ratio 2 (LR2), modified Larson's ratio (LRM), and Larson's ratio with temperature (LR1-T) were determined from this data set as defined in Equations 5.51–5.54, respectively. The LRM was developed by Imran et al. (2005b) from the TBW I study, and LR1-T is a combination of temperature and Larson's ratio. EN measurements can be viewed as electrical water quality measurements, and because all metal release models utilize water quality parameters as independent variables for prediction, correlation of water quality and EN parameters provide additional insight into the utilization of EN parameters for prediction of any redox product that is typically predicted from water quality parameters.

$$LR1 = \frac{[SO_4^-] + [Cl^-]}{[HCO_3^-]} \tag{5.51}$$

$$LR2 = \frac{[Cl^-]}{[HCO_3^-]} \tag{5.52}$$

Table 5.54
TBW II EN averaged water quality and EN data

Water quality parameter	Phase I	Phase II	Phase III	Phase IV
pH	8.04	7.89	8.03	7.88
Hardness (mg/L as $CaCO_3$)	108	105	110	86
Calcium (mg/L as $CaCO_3$)	94	90	90	72
Alkalinity (mg/L as $CaCO_3$)	161	104	150	123
Total dissolved solids (mg/L)	417	428	461	396
Sodium (mg/L)	7	38	40	32
Sulfate (mg/L)	62	103	66	76
Chloride (mg/L)	45	67	67	58
Dissolved oxygen (mg/L)	8.77	7.54	7.76	8.85
Temperature (°C)	21.4	27.7	25.1	20.0
Influent P (mg/L)	0.823	0.731	0.709	0.776
Langelier saturation index	0.47	0.12	0.43	0.08
Fe corrosion—control (mpy)	27.2	2.8	9.8	2.7
Fe corrosion—inhibitor (mpy)	32.3	23.3	22.1	7.1
Cu corrosion—control (mpy)	0.7	0.9	1.2	0.5
Cu corrosion—inhibitor (mpy)	4.4	23.2	23.7	12.3
Pb solder corrosion—control (mpy)	0.8	1.8	2.3	0.4
Pb solder corrosion—inhibitor (mpy)	1.2	18.2	3.3	0.3

NOTE: 1 mpy = 0.0254 mm/yr.

$$LRM = \frac{([SO_4^-] + [Cl^-] + [Na^+])^{0.5}}{[HCO_3^-]} \times \frac{temp}{25} \times HRT \qquad (5.53)$$

$$LR1\text{-}T = \frac{[SO_4^-] + [Cl^-]}{[HCO_3^-]} \times \frac{temp}{25} \qquad (5.54)$$

where [ ] = concentration, meq/L
HCO^-_3 = bicarbonate, meq/L

The correlations between water quality and ECNCR from the combined data and PF for TBW II data for iron, copper, and lead are shown in Table 5.55 and were significant for alkalinity, temperature, sulfates, chlorides, DO, and Larson's ratio. The PF was not available from the Boyle report. As with the EN models for metal release shown previously, PF correlated with most water quality parameters, more than ECNCR. Interestingly, the PF for all iron, copper, and lead were highly correlated with each other. The ECNCR for iron and lead for inhibitors did not show any strong correlation ($R > 0.5$) with any water quality parameters or inhibitor dose.

Table 5.55
Water quality correlations of ECNCR and PF-combined data

ECNCR (mpy)	Alkalinity (mg/L as $CaCO_3$)	Sulfate (mg/L)	Chloride (mg/L)	Temperature (°C)	Dissolved oxygen (mg/L)	LR1	LR1-T	Hardness (mg/L as $CaCO_3$)	PF_{Fe}
Fe control	0.70					–0.60	–0.56		
Fe inhibitor									
Cu control				0.56	–0.86			0.80	
Cu inhibitor	–0.75								
Pb control				0.79	–0.55				
Pb inhibitor									
PF_{Fe}	–0.72	0.87		0.74	–0.61	0.79	0.85		
PF_{Cu}	–0.75	0.88	0.47	0.72	–0.63	0.82	0.86		0.97
PF_{Pb}	–0.66	0.82		0.75	–0.64	0.76	0.84		0.98

NOTE: Correlations are shown for R > 0.5 and are statistically significant (p-value < 0.05).

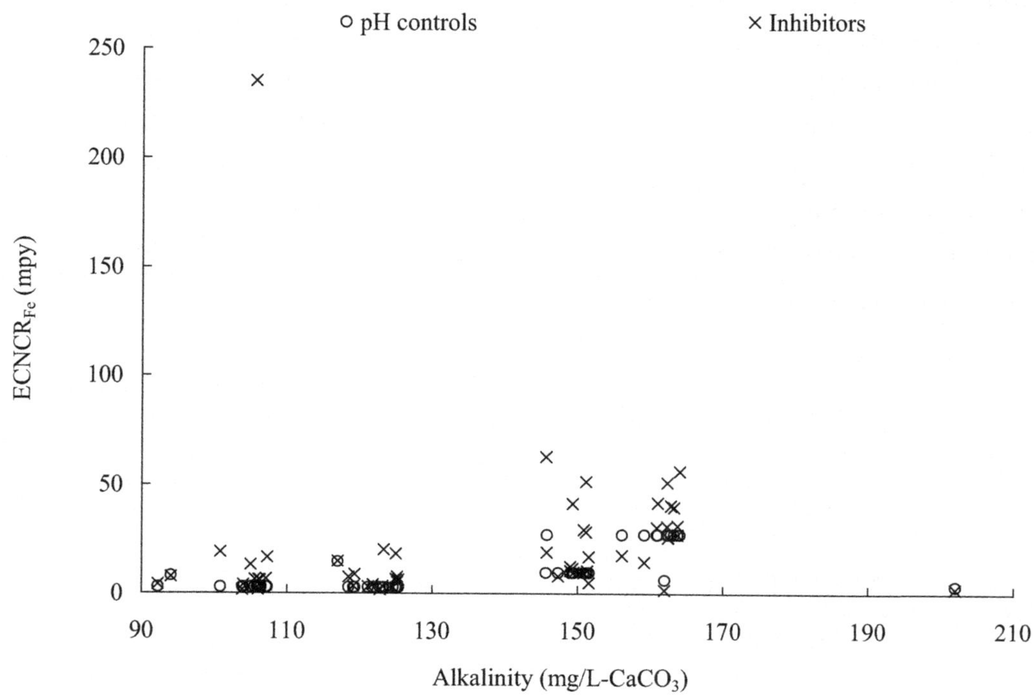

Figure 5.59 Alkalinity versus ECNCR of iron-combined data

Figure 5.59 shows the plot of alkalinity and $ECNCR_{Fe}$ for the combined data set with different symbols for pH controls and inhibitors. The graph shows that the iron corrosion rate ($ECNCR_{Fe}$) increases with alkalinity for both pH controls and inhibitors. $ECNCR_{Fe}$ was found to be negatively correlated with alkalinity in TBW II; and alkalinity was found to mitigate iron release by Taylor et al. (2005) during the TBW I study. TBW II data is consistent with the findings of Duranceau, Townley, and Bell (2004) that corrosion rates ($ECNCR_{Fe}$) may increase when inhibitors are used.

Table 5.56
Average ECNCR from TBW II by phase

Method for corrosion rate (mpy)	Phase I	Phase II	Phase III	Phase IV
$ECNCR_{Fe}$ control	27.18	2.82	9.78	2.69
$ECNCR_{Fe}$ inhibitors	32.34	23.32	22.12	7.12
$ECNCR_{Cu}$ control	0.69	0.89	1.16	0.47
$ECNCR_{Cu}$ inhibitors	4.40	23.17	23.71	12.32
$ECNCR_{Pb}$ control	0.83	1.75	2.27	0.36
$ECNCR_{Pb}$ inhibitors	1.23	18.16	3.30	0.28
$LPRCR_{Fe}$ control	3.45	0.44	1.58	0.56
$LPRCR_{Fe}$ inhibitors	3.51	0.31	1.09	0.37
$LPRCR_{Cu}$ control	0.13	0.11	0.13	0.10
$LPRCR_{Cu}$ inhibitors	0.16	0.07	0.15	0.06
$LPRCR_{Pb}$ control	0.02	0.19	0.01	0.01
$LPRCR_{Pb}$ inhibitors	0.02	0.09	0.03	2.5E–03
$HMCR_{Fe}$ control	3.62	0.71	1.21	0.55
$HMCR_{Fe}$ inhibitors	3.34	0.44	0.75	0.58
$HMCR_{Cu}$ control	0.15	0.32	0.27	0.38
$HMCR_{Cu}$ inhibitors	0.24	0.32	0.27	0.38
$HMCR_{Pb}$ control	0.07	1.12	1.22	0.82
$HMCR_{Pb}$ inhibitors	0.06	0.93	1.01	0.96

NOTE: 1 mpy = 0.0254 mm/yr.

Numerous reactions are expected to occur on a metal surface in a drinking water environment. Metal scale formation, oxygen consumption, hydrogen generation, chlorine reduction, biological oxidation, and oxidation of metals other than iron, copper, or lead are examples of reactions that change surface potential (voltage) and result in current flow, which generates ECN and are not direct measurements of iron, copper, or lead release. Hence, ECNCR can overestimate the actual corrosion rate of metal dissolution because ECNCR represents the total current activity on a metal surface. Table 5.56 shows the corrosion rates by all three techniques averaged by phase for this study. From Table 5.56 it is observed that ECNCR for all metals was higher in the presence of inhibitors relative to pH controls. Hence, EN data shows evidence that corrosion inhibitors caused an increase in redox reactions occurring on the metal surface for iron, copper, and lead. LPRCR and HMCR values did not show a trend between inhibitors and pH controls.

Figure 5.60 shows that $ECNCR_{Fe}$ was negatively correlated with both total iron release and LR1-T in this study. This is consistent with theory in that alkalinity mitigated the iron release and a higher LR1-T was associated with higher iron release. Hence, the $ECNCR_{Fe}$ correlation with mitigation by alkalinity and aggression by both sulfate and chlorides is supported by theory and other research (Taylor et al. 2005; Duranceau, Townley, and Bell 2004).

As for all models developed for total iron release, R^2 for the TBW II PF_{Fe} model for total iron release was low (0.27) but significantly showed that total iron release increased as PF_{Fe}

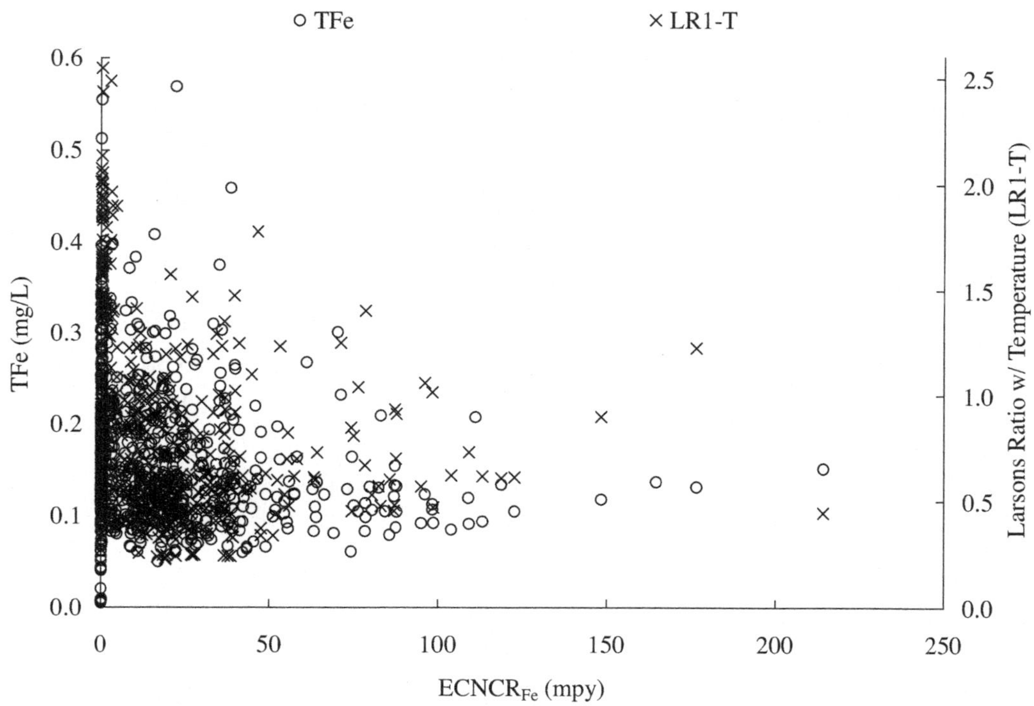

Figure 5.60 ECNCR of iron versus total iron and Larson's ratio TBW II data

increased. The dotted lines in Figure 5.61 separate the data into three groups, wherein PF_{Fe} was less than 0.1, between 0.1 and 1, and greater than 1.0.

Figure 5.62 shows that total iron release increases with LR1-T for all groups of inhibitors. From Figure 5.62 it is observed that most of the data with an LR1-T < 1 shows a high variability in total iron release but most of the data has TFe < 0.3 mg/L, the secondary standard for iron. LR1-T > 1 shows an increasing trend with total iron release.

Figure 5.63 shows that higher PF_{Fe} is associated with a higher LR1-T. Thus PF_{Fe} is consistent with the water quality (LR1-T) that correlates with iron release; hence, it may be used quantitatively to determine corrosivity towards iron release. From Figure 5.63 it is observed that most of the data with LR1-T < 1 has a $PF_{Fe} < 0.10$, and data with $PF_{Fe} > 0.10$ shows a consistent increasing trend with LR1-T. The correlation (R^2) of LR1-T for all inhibitors and PF_{Fe} is shown in Figure 5.63 and indicates that PF_{Fe} can be used for predicting iron release.

PF was the only EN parameter for both copper and lead that correlated with water quality from this study, but it did not correlate with either metal release. Figure 5.64 shows that PF_{Cu} tends to increase with LR1-T. Similarly, Figure 5.65 shows that PF_{Pb} tends to increase with LR1-T. Only iron release correlated with LR1-T.

The advantages of EN are rapid response and less labor relative to traditional water quality monitoring in distribution systems. PF ideally represents pitting, but, as previously shown, PF is directly correlated with LR1-T. PF is the dimensionless ratio of standard deviation of corrosion current from ECNCR to the average current from HMCR. ECNCR quantifies the natural current-voltage response from all redox reactions, whereas the HMCR quantifies the current-voltage response from an applied alternating current. Hence, PF values obtained from this study correspond to the variation in naturally occurring corrosion current to the current obtained from the

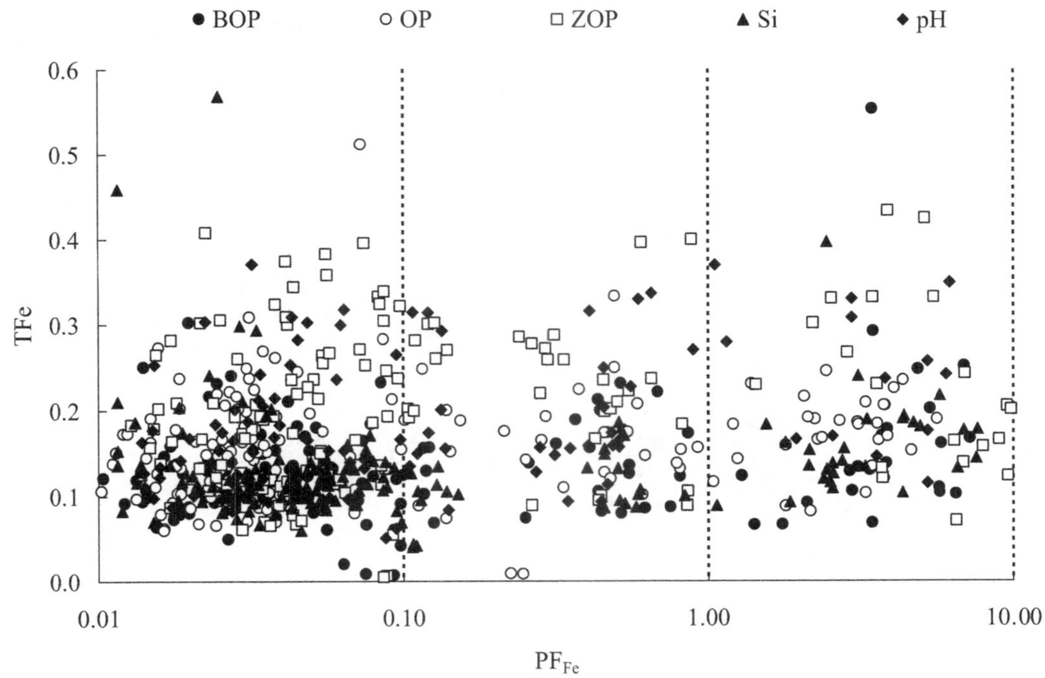

Figure 5.61 Total iron release versus PF of iron for TBW II data

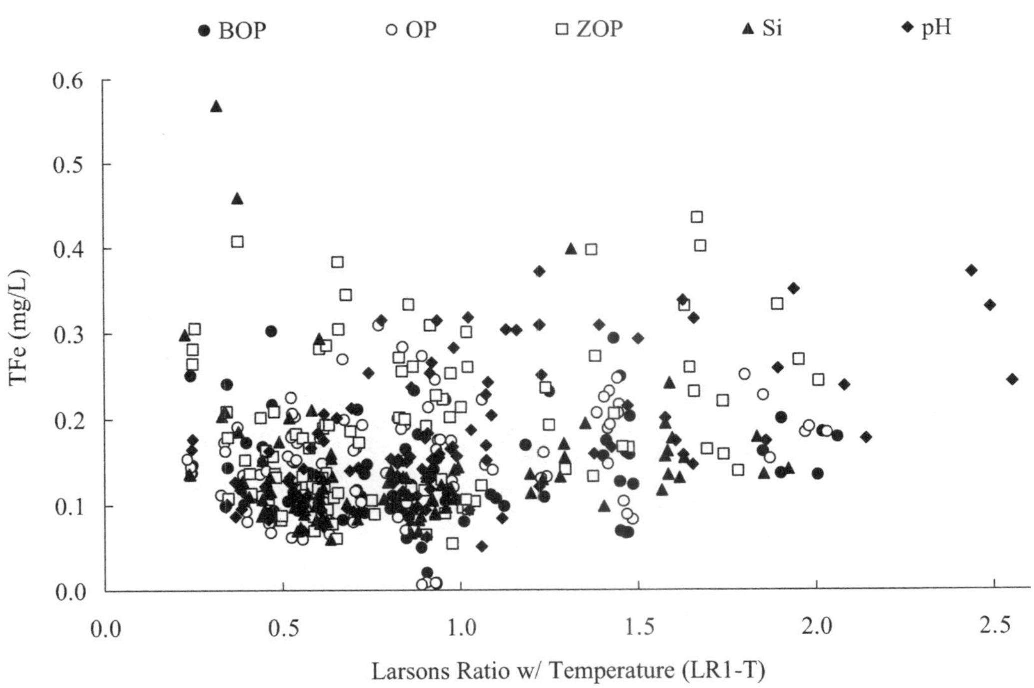

Figure 5.62 Total iron release versus LR1-T for TBW II data

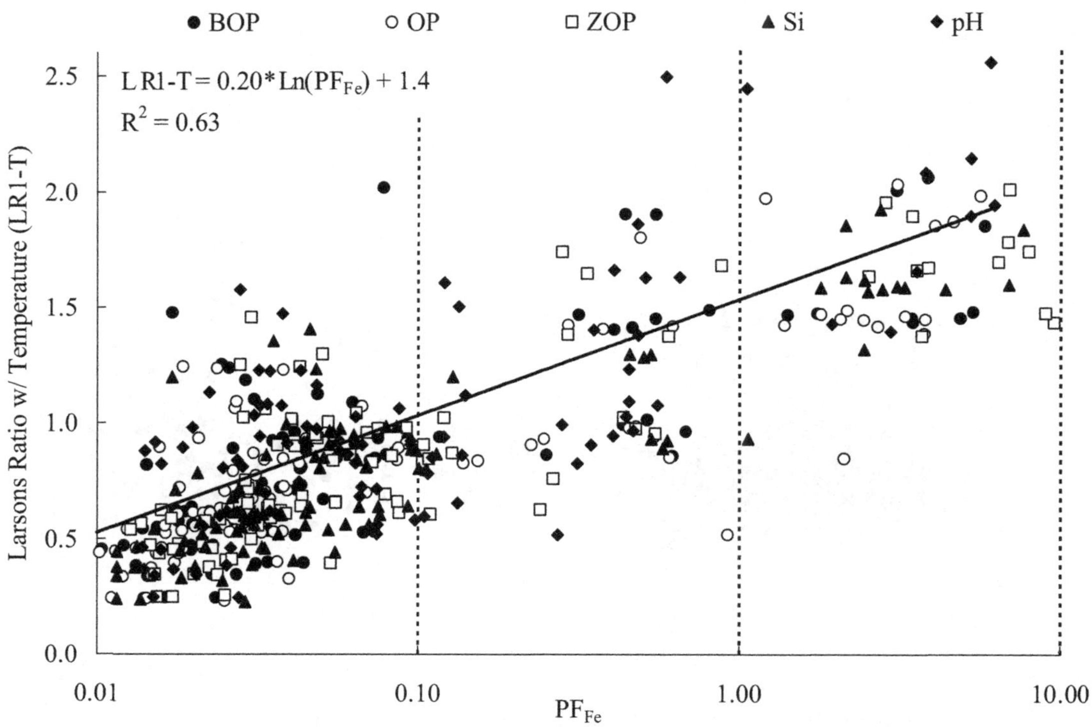

Figure 5.63 LR1-T versus PF_{Fe} for TBW II data

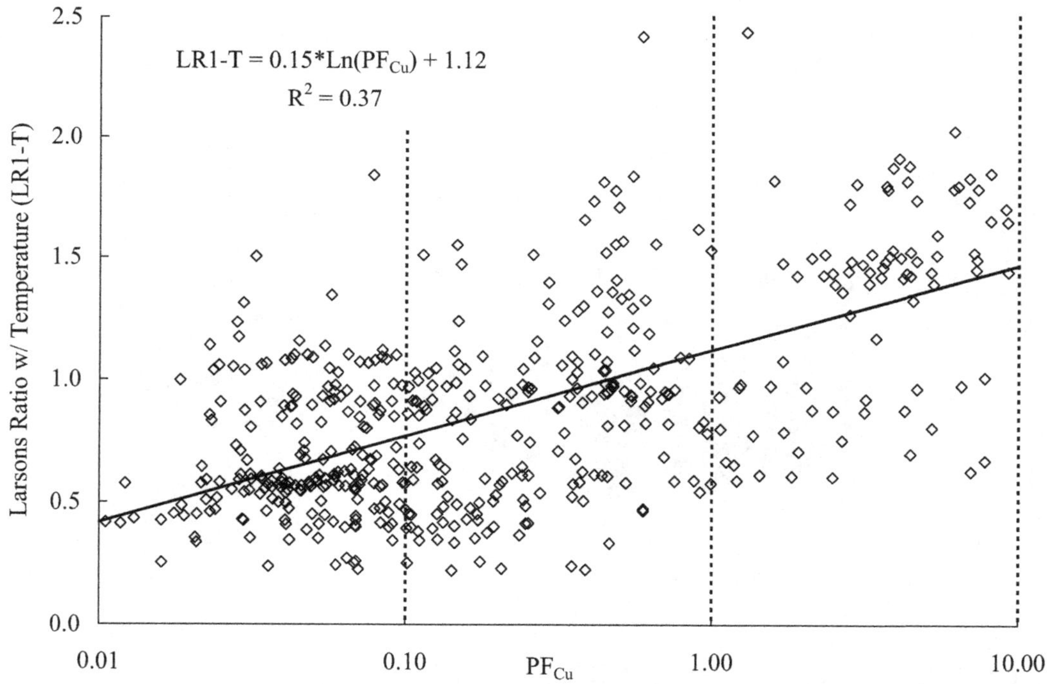

Figure 5.64 LR1-T versus PF_{Cu} for TBW II data

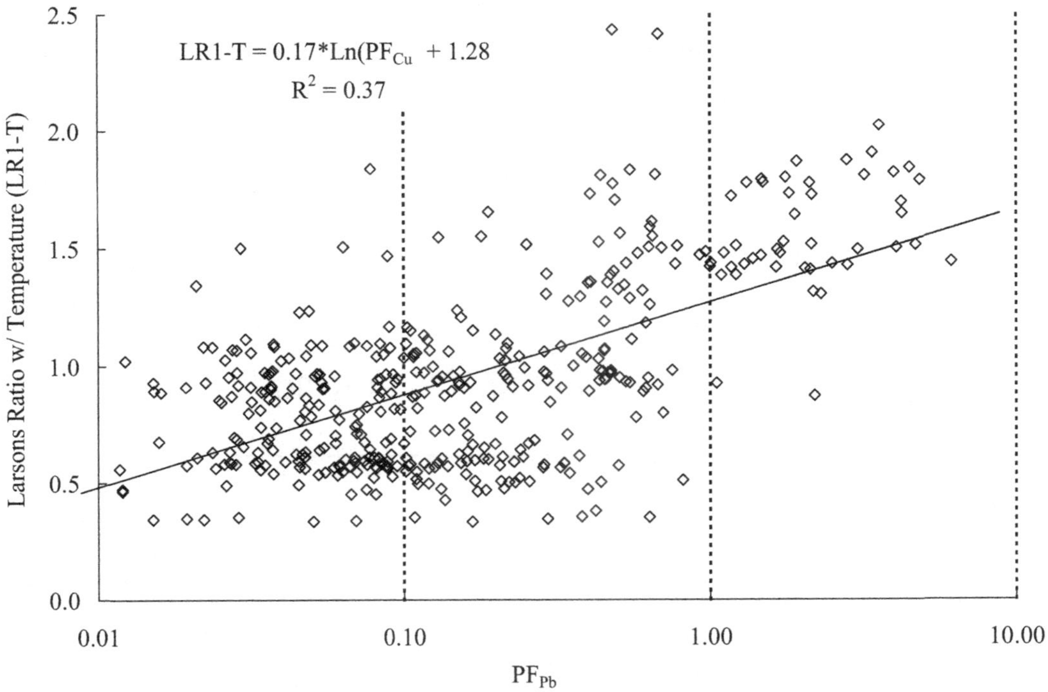

Figure 5.65 LR1-T versus PF_{Pb} for TBW II data

system if it was corroding at a uniform steady rate. PF represents the likely variation in mean corrosion current for this study and is correlated with water quality through LR1-T.

User guidelines for corrosion control of iron, copper, or lead release can be developed from the three EN models for total metal release that were presented previously. An example of possible user guidelines are shown in Table 5.57 for iron, copper, or lead release, assuming 1 mg/L P BOP inhibition. The metal concentrations shown in Table 5.57 were developed with regard to existing goals and regulations. An iron concentration in excess of 0.16 mg/L corresponds to the violation of the secondary standard for color (<15 cpu), and 0.3 mg/L Fe corresponds to the secondary standard for iron; which was previously discussed in the metal release discussion.

Only PF for control of iron release can be determined, as shown in Table 5.57, as the EN models for copper and lead release contain multiple terms. Many combinations of $LPRCR_{Cu}$ and $ECNCR_{Cu}$ values for copper and $LPRCR_{Pb}$, $HMCR_{Pb}$, and PF_{Pb} values for lead are possible within the associated ranges. However, the point of Table 5.57 is that EN monitoring of strategically placed iron, copper, and lead electrodes could provide nearly instantaneous monitoring and rapid response to proactively control unwanted metal release. Typical recommended actions for such control are shown in Table 5.58 and range from no action to enhanced general and specific flushing, inhibitor addition, and alkalinity addition. EN parameters can be used in the same way as water quality parameters in that they are descriptive of general redox water quality. The results of this work have demonstrated that EN parameters can be used to describe metal release similar to modeling based on conventional water quality. Utilization of EN monitoring for gaining insight and improving water treatment and distribution system water quality seems fruitful and worthy of pursuit by the water community.

Table 5.57
EN recommendations for total iron, copper, and lead release using BOP inhibition at 1 mg/L as P

Metal	Level	Metal concentration	EN model	Parameter value(s)
Fe (mg/L)	I	<0.06	$PF_{Fe}^{0.0694}$	<3E–9
	II	0.06 to 0.16		3E–9 to 0.003
	III	>0.16		>0.006
Cu (mg/L)	I	<1.0	$LPRCR_{Cu}^{0.0280} \times ECNCR_{Cu}^{0.031}$	Multiple values
	II	1.0 to 1.3		
	III	>1.3		
Pb (μg/L)	I	<5	$LPRCR_{Pb}^{0.217} \times HMCR_{Pb}^{0.337} \times PF_{Pb}^{-0.347}$	Multiple values
	II	5 to 15		
	III	>15		

NOTES: Fe = Mean iron concentration for a level.
Cu = 90th percentile copper concentration for a level.
Pb = 90th percentile lead concentration for a level.

Table 5.58
Recommended action by metal release level

Action	Level I	Level II	Level III
Flushing	No action	Increased flushing at specific location	Increased general and specific flushing
Inhibitor addition	No action	Add inhibitor to distribution system; monitoring and regulation as needed	Inhibitor addition
Alkalinity addition	No action	Add inhibitor to distribution system; monitoring and regulation as needed	Decrease alkalinity for Cu; increase alkalinity for Fe
pH adjustment	No action	Add inhibitor to distribution system; monitoring and regulation as needed	Increase pH for Fe, Cu, and Pb

Summary

EN measures the corrosion current that represents electron transfers due to all oxidation reactions on a metal surface, without regard to whether metal is released to the water or incorporated in an oxidized scale. Hence, a direct correlation between EN and metal release in drinking waters may not be possible and does not make sense mechanistically, given that all corrosion does not lead to metal release. Electrochemical corrosion monitoring (EN) was employed to measure corrosion rates by three methods, which were LPR, HDA, and ECN. The EN output yields

12 variables, of which some are used to derive the corrosion rate based on one of the above-mentioned three techniques.

Iron corrosion was found to be dependent on both general and localized corrosion. However, localized corrosion was more important for particulate iron, which comprised most of the total iron release in this study. None of the regression models were found to provide a reasonable fit to total iron release, and the best EN model, which used PF_{Fe} alone, had a low fit with an R^2 of 0.27. However, the EN model did equally well as the water quality model developed for total iron release, which was presented in the metal release discussion, with a similar R^2 of 0.29. PF_{Fe} was found to be correlated with alkalinity, chlorides, sulfate, and temperature; however, PF_{Fe} did not correlate with iron release. PF_{Fe} may not represent pitting or localized corrosion but represents the variation in general corrosion rate. EN regression models were found to provide a reasonable fit to dissolved iron release. The best model used $HMCR_{Fe}$ alone and had a modest fit with an R^2 of 0.43. The EN model did better than the water quality model developed in the metal release discussion for dissolved iron release, which had a lower R^2 of 0.37. The effect of temperature on iron release was negligible for this study. The data indicate that there may be a concern for ZOP and SiO_2 inhibitors, which increased iron release during the first few weeks of inhibitor dosing.

Copper corrosion was found to be both general and localized. However, general corrosion was more important for dissolved copper release, which was the predominant form in this study. There was a significant variability in corrosion rates for copper, suggesting that localized corrosion events may also be important for copper release. EN regression models using $LPRCR_{Cu}$ and $ECNCR_{Cu}$ provided a reasonable fit to total copper release with an R^2 of 0.72. The EN model had a similar R^2 to the water quality model (0.76) for copper that was presented in the metal release discussion. Temperature effect on total copper release was negligible for this study. Similar observations were made for dissolved copper release. Regression models for dissolved copper, which incorporated $LPRCR_{Cu}$ and $ECNCR_{Cu}$, provide a reasonable fit with an R^2 of 0.68. The EN model and water quality model for dissolved copper release had a similar R^2 of 0.72.

Lead release was found to be mostly in particulate form and depends on both general and localized corrosion. There was significant variability in corrosion rates for lead, indicating that localized corrosion is important for lead release. EN regression models provided a reasonable fit to total lead release. The best model has an R^2 of 0.55 and includes $LPRCR_{Pb}$, $HMCR_{Pb}$, and PF_{Pb}. The EN model was found to be comparable to the water quality model for total lead release presented in the metal release discussion; both models have an R^2 of 0.55. EN regression models, containing $LPRCR_{Pb}$, $HMCR_{Pb}$, and PF_{Pb}, provided a reasonable fit to dissolved lead release with an R^2 of 0.44. The EN model did not fit dissolved lead release as well as the water quality model developed in the metal release discussion, which had an R^2 of 0.57. Temperature effect on both total and dissolved lead release was similar and significant for this study, with increased temperature resulting in increased lead release.

PF was the only EN parameter that consistently correlated with water quality for all three metals. The water quality parameters considered include alkalinity, chlorides, sulfate, and temperature. Correlations of PF_{Cu} with metal release suggested that PF_{Cu} only represents localized corrosion of copper. Because PF is a dimensionless ratio of the standard deviation of instantaneous current to the mean current, it represents the variability in general corrosion rate and not localized corrosion. The PF should be used with caution when the HMCR indicates low general corrosion rates. The three corrosion rates, HMCR, LPRCR, and ECNCR, varied significantly for all metals, and a combination of these was needed to account for both general and localized corrosion in empirical metal release models. Iron release was not modeled well by EN because most iron release was in particulate form. Lead release was modestly modeled by EN, although most lead release was in particulate form. Copper release was reasonably modeled by EN with most copper release in dissolved form. EN models fitted the actual data as well as the water quality models for metal release for all metals studied. PF correlated with the modified Larson's ratio for temperature, LR1-T. While LR1-T correlated with total iron release, it did not correlate with copper or lead release. PF_{Cu} and PF_{Pb} were modestly correlated with LR1-T. The PF for all metals (iron, copper, and lead) were highly correlated with each other. LPRCR, HMCR, ECNCR (corrosion rates), and PF demonstrated the capability of EN to predict iron, copper, and lead release significantly more rapidly and with the same degree of confidence as chemical models.

IDENTIFICATION AND MODELING OF CORROSION SCALE PRODUCTS

Surface composition of scales on new iron, old galvanized steel, new copper, and new lead/tin coupons was analyzed by XPS to gain insight into the controlling solid phase and mechanism of iron, zinc, copper, or lead release with and without phosphorus, silica, or pH-based inhibition. Each material is considered in sequence, with sections on thermodynamic modeling of metal release and on surface composition by XPS.

Thermodynamic equilibrium models for metal release were developed by considering the metal salts most likely to form for a given water quality. Metal phosphate and silicate compounds were considered where thermodynamic data were available. The predicted metal release from each solid is compared to actual metal release, and the solid that most accurately predicts actual metal release is discussed.

After each section on thermodynamic modeling, the results from XPS analysis of surface composition are presented. Following each phase of operations, iron, galvanized steel, copper, and lead/tin coupons were analyzed for surface composition using XPS. In XPS, a magnesium anode (1,254 eV kα) bombards the sample surface with high energy X-rays that interact with atoms in the top 5–6 nm of the surface (Seal and Barr 2001). Then, according to the photoelectric effect, these atoms emit high-energy electrons. An electron analyzer counts the number of electrons emitted over a range of energies, producing scan spectra that may be used for surface composition analysis.

Coupons were inserted into the PDSs at the beginning of each phase of pilot-plant operation and then were retrieved at the end of the phase. The XPS scanning consisted of a two-step process where an initial survey scan was performed to identify all elements present on the coupon. A survey scan, conducted over a broad range of energy levels, is useful for confirming the presence or absence of elements on the surface. In contrast, a high-resolution scan, conducted over a narrow range of energy levels, is useful for establishing the chemical states present for a given element. High-resolution spectra were shifted along the energy axis to conform to a 284.6 eV adventitious carbon (C 1s) standard.

The elements are indicated on the survey by pronounced peaks above the background of the survey, and these peaks are then selected individually for the high-resolution scan for each element. Elements of interest, such as those associated with inhibitors, coupon material, or water quality, were analyzed with high-resolution scans. Possible compounds associated with a given element were determined through deconvolution of the high-resolution scans.

Iron Surface Characterization

Thermodynamic Model Development for Iron Release

Iron release from pipe surfaces may be predicted by the solubility of solid phases formed within the corrosion scales. This thermodynamic approach is limited to prediction of dissolved iron and will underpredict total iron release because most iron is present in particulate form. The XPS analysis of iron coupons indicated the possible controlling solid forms that may be responsible for iron release. Thermodynamic models were constructed by assuming that an iron solid acts as the controlling solid phase for iron release into water. Dissolved iron concentrations were calculated by assuming that free dissolved iron is in equilibrium with the controlling solid phase. Concentrations of other dissolved iron complexes were determined using the appropriate equilibrium constants.

The equilibrium modeling approach was similar to that of TBW I (Taylor et al. 2005) with an expansion of iron complexes, as shown in Equation 5.55. The use of phosphate and silica complexes, which represented significantly higher concentrations than that of TBW I, resulted in the inclusion of $FeH_2PO_4^+$ and $FeHPO_4^0$ into the equilibrium expression. Thermodynamic literature was inconclusive in identifying a stable complex between iron and silica.

$$\begin{aligned} Fe = {} & [Fe^{2+}] + [FeOH^+] + [Fe(OH)_2^0] + [Fe(OH)_3^-] + [FeHCO_3^+] + [FeHCO_3^0] \\ & + [FeSO_4^0] + [FeCl^+] + [FeH_2PO_4^+] + [FeHPO_4^0] \end{aligned} \tag{5.55}$$

Model development was based on the assumption that the oxidation–reduction potential was 0.43 V (the average of water quality in all phases) at a temperature of 25°C. Under these conditions, Fe^{2+} concentrations are much more abundant than Fe^{3+} and, therefore, ferric species are ignored in model development.

The complexation of iron, as described by Equation 5.55, suggests that water quality influences, such as that of pH, alkalinity, sulfate, chloride, and phosphate, affect the total solubility of iron. Inclusion of hydroxyl ligands suggests that an increasingly alkaline solution would promote iron release. Increasing the pH would also shift the distribution of carbonate constituents, favoring the formation of the more stable $FeCO_3^0$. Such complexes as $FeHCO_3^+$ and $FeCO_3^0$ are dependent on source water alkalinity, implying that increasing alkalinity may complicate control of iron release. However, alkalinity was found to be beneficial to mitigating iron, as demonstrated by the empirical model for iron release.

The empirical model describing iron release also identified chloride as a significant water quality parameter that promotes iron release. This relationship agrees well with Equation 5.55. However, equilibrium modeling suggests that the effect of chloride is not entirely explained from complexation with iron, as chloride complexes represented a small portion of soluble iron. Sulfate was modeled even though it was not found to be significant from empirical modeling. The following sections consider dissolved iron solubility and controlling solid phases without inhibitor addition, with phosphate inhibitor addition, and with silicate inhibitor addition.

Inhibitor Control of Iron Release

Average iron release is shown in Table 5.59 by phase for each inhibitor and relative to the pH_s+0.3 control. On average, all inhibitors reduced dissolved iron release relative to the pH_s+0.3

Table 5.59
Comparison of actual iron release from inhibitor PDSs to pH_s+0.3

Phase	Inhibitor	Average release (mg/L)		% Ratio to pH_s+0.3	
		Dissolved Fe	Total Fe	Dissolved Fe	Total Fe
I	pH_s+0.3	0.04	0.11		
	BOP	0.04	0.12	104	109
	OP	0.03	0.13	79	112
	ZOP	0.04	0.16	98	144
	Si	0.03	0.14	74	127
II	pH_s+0.3	0.03	0.19		
	BOP	0.01	0.17	22	87
	OP	0.01	0.18	20	95
	ZOP	0.01	0.25	28	129
	Si	0.02	0.17	52	86
III	pH_s+0.3	0.01	0.14		
	BOP	0.01	0.12	96	84
	OP	0.01	0.17	94	123
	ZOP	0.01	0.18	107	129
	Si	0.01	0.11	95	78
IV	pH_s+0.3	0.00	0.14		
	BOP	0.01	0.11	357	78
	OP	0.00	0.16	93	111
	ZOP	0.00	0.21	89	148
	Si	0.00	0.11	84	78

control. However, only BOP and Si inhibitors reduced total iron release relative to the pH_s+0.3 control. OP showed no better control of iron release, whereas ZOP caused higher release than pH_s+0.3, as shown in Table 5.59. Total or dissolved iron release was not always more or less in the presence of phosphate or silica relative to pH_s+0.3 inhibition.

Expected Iron Release From Thermodynamic Models

pH+0.3. In the absence of inhibitors, the identified iron compounds from the XPS analysis of iron coupons were the same as in TBW I, which concluded that siderite was the controlling solid phase in iron release. Although siderite ($FeCO_3$) was not detected in the XPS analysis, considerable quantities of ferric oxide (Fe_2O_3) were detected. This is significant because in contact with air, $FeCO_{3(s)}$ decomposes to Fe_2O_3 (Taylor et al. 2005). Given the XPS results and the beneficial effect of alkalinity on iron release, it is reasonable to assume that $FeCO_3$ acts at the controlling solid phase for iron release in the absence of phosphate or silicate inhibitors. This model is shown in Equation 5.56, where all units are in moles per liter.

Table 5.60
Iron thermodynamic modeling of pH_s+0.3

Inhibitor	Phase	Actual iron release (mg/L) Dissolved Fe	Total Fe	Modeled iron release (mg/L) $FeCO_3$
pH_s+0.3	I	0.04	0.11	0.10
	II	0.03	0.19	0.17
	III	0.01	0.14	0.09
	IV	0.00	0.14	0.14

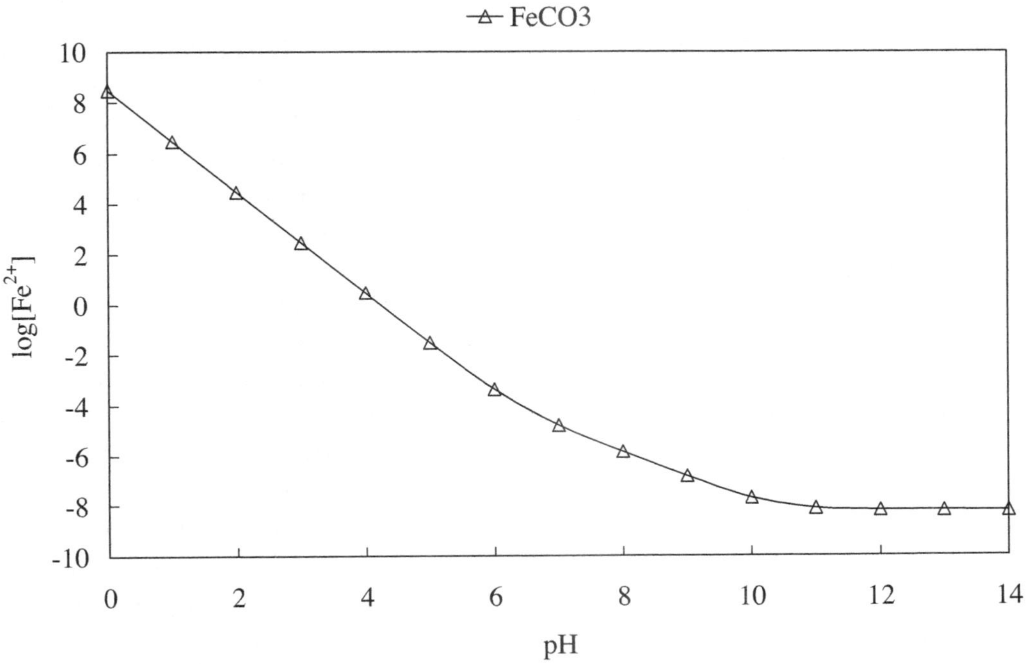

Figure 5.66 Solubility limit of iron for pH_s and pH_s+0.3

$$Fe = [Fe^{2+}] + [FeOH^{+}] + [Fe(OH)_2^{0}] + [Fe(OH)_3^{-}] + [FeHCO_3^{+}] + [FeCO_3^{0}] + [FeSO_4^{0}] + [FeCl^{+}] \quad (5.56)$$

Average actual and predicted dissolved iron concentrations using Equation 5.56 are presented in Table 5.60. Data shows that the siderite model overpredicted actual dissolved iron concentrations in the study. This observation is consistent with results of the siderite model in TBW I. Figure 5.66 shows the solubility diagram of siderite, $FeCO_3$, without complexation, with respect to pH. Solubility of siderite increases with decreasing pH.

Blended Ortho-polyphosphate. The XPS analysis of iron coupons exposed to phosphate-based inhibitors revealed the presence of ferric phosphate ($FePO_4$) as a solid compound on the surface. Investigation of ferric phosphate as the possible controlling solid phase of dissolved iron

Table 5.61
Iron thermodynamic modeling of BOP

Inhibitor	Phase	Actual iron release (mg/L)		Modeled iron release (mg/L)	
		Dissolved Fe	Total Fe	$FePO_4$	$FeCO_3$
BOP	I	0.04	0.10	1.0E–05	0.11
	II	0.01	0.15	1.2E–05	0.20
	III	0.01	0.11	5.9E–07	0.10
	IV	0.01	0.13	2.4E–07	0.17

showed that the calculated concentrations from ferric phosphate are much lower (10,000 times) than those calculated for siderite, as shown in Table 5.61. Thus, ferric phosphate was eliminated as the controlling solid phase. Calculated iron concentrations from both $FePO_4$ and $FeCO_3$ models are compared to actual dissolved iron in Table 5.61. Predicted concentrations by the siderite model slightly overpredicted actual release. This suggests that dissolved iron release is not solely dependent on $FeCO_3$ but rather may depend on $FePO_4$ or other iron phosphate solids that form and reduce iron release with BOP inhibitor addition. Given that ferric phosphate ($FePO_4$) was detected in XPS analysis on three of four coupons exposed to BOP, it is reasonable that BOP addition resulted in the formation of a solid film ($FePO_4$) that reduced iron release.

Solubility limits of iron for both solids ($FePO_4$ and $FeCO_3$) without complexation, are illustrated in Figure 5.67. The solubility diagram suggests that under equilibrium conditions, ferric phosphate ($FePO_4$) was the limiting solid in iron release, which agrees with predicted iron concentrations from the $FePO_4$ thermodynamic model for all BOP-, OP-, and ZOP-treated PDSs (Tables 5.61, 5.62, and 5.63, respectively). This confirms that the presence of a phosphate-based solid contributes to the reduction of iron release. Although phosphate inhibitor addition did not change the controlling solid for iron release, it may promote the formation of iron phosphate solids that reduce the rate of iron release, while the equilibrium iron concentration remains unchanged. A mechanism of kinetic inhibition of iron release would explain why inhibitors are least effective in reducing iron concentrations in low-flow dead-end conditions.

Orthophosphate. The thermodynamic model for iron solubility with OP inhibitor addition followed the same development as for the BOP inhibitor. Thermodynamic models demonstrated that $FeCO_3$ was the controlling solid for iron release with OP inhibitor addition; however, the siderite model tended to slightly overpredict actual dissolved iron (Table 5.62). Also, prediction of the $FePO_4$ model grossly underpredicted actual release. Similar to BOP, PDSs receiving OP inhibitor had lower dissolved iron release than the pH_s+0.3 treatment (Table 5.59). However, for total iron release, OP inhibitor had almost the same release as the pH_s+0.3 treatment. This difference could be related to the absence of $FePO_4$ solid in XPS analysis for most coupons receiving OP inhibitor. This effect suggests that the presence of $FePO_4$ is beneficial in reducing iron release.

Zinc Orthophosphate. Thermodynamic models for dissolved iron release provided results, shown in Table 5.63, that were similar to those for BOP and OP inhibitors. Unlike BOP and OP, total iron release with ZOP addition was higher than the pH_s+0.3 control, as shown in Table 5.59 (137%). Although $FePO_4$ was detected by XPS analysis, the presence of zinc may have reduced the effectiveness of the $FePO_4$ scale in inhibiting iron release. Also, ZOP PDSs experienced an average lower pH than other phosphates PDSs through all phases (between 0.05 and 0.18 pH units), which may have contributed to increased iron release.

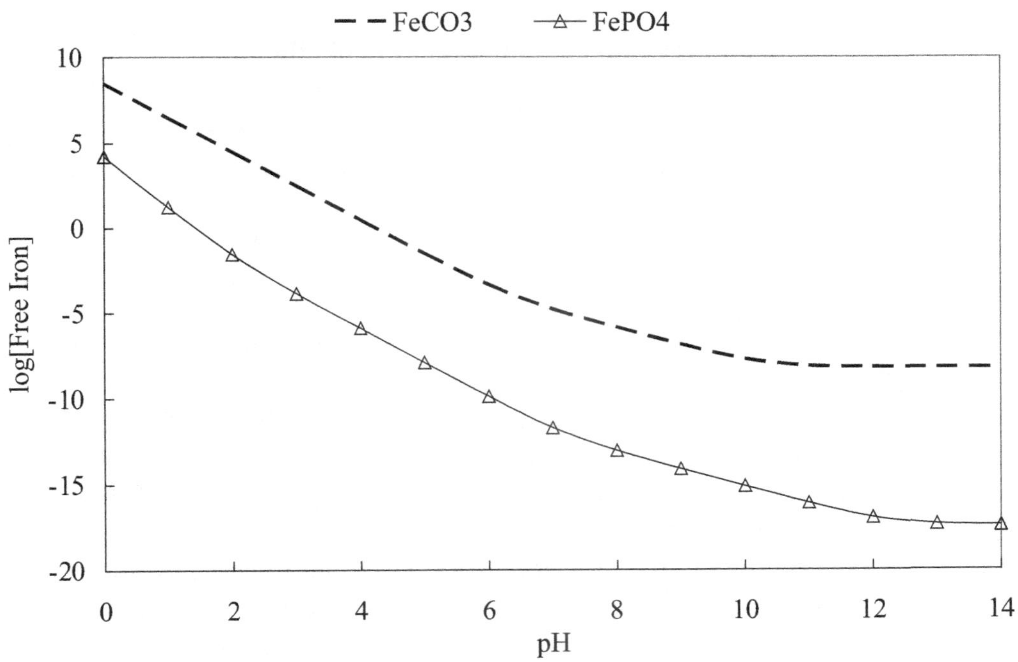

Figure 5.67 Solubility limits of iron for phosphate-based inhibitors

Table 5.62
Iron thermodynamic modeling of OP

		Actual iron release (mg/L)		Modeled iron release (mg/L)	
Inhibitor	Phase	Dissolved Fe	Total Fe	$FePO_4$	$FeCO_3$
OP	I	0.03	0.14	9.7E–06	0.10
	II	0.01	0.21	1.2E–05	0.20
	III	0.01	0.23	6.3E–07	0.11
	IV	0.00	0.13	2.6E–07	0.18

Table 5.63
Iron thermodynamic modeling of ZOP

		Actual iron release (mg/L)		Modeled iron release (mg/L)	
Inhibitor	Phase	Dissolved Fe	Total Fe	$FePO_4$	$FeCO_3$
ZOP	I	0.04	0.14	1.1E–05	0.12
	II	0.01	0.23	1.7E–05	0.27
	III	0.01	0.22	7.6E–07	0.13
	IV	0.00	0.24	3.0E–07	0.21

Table 5.64
Iron thermodynamic modeling of silicon

Inhibitor	Phase	Actual iron release (mg/L)		Modeled iron release (mg/L)	
		Dissolved Fe	Total Fe	$FeSiO_3$	$FeCO_3$
Si	I	0.03	0.12	0.004	0.06
	II	0.02	0.14	0.005	0.10
	III	0.01	0.10	0.005	0.07
	IV	0.00	0.10	0.005	0.08

Silicon. The XPS analysis for iron coupons exposed to Si inhibitor showed the presence of a silicate compound on the surface. $FeSiO_3$ was considered as a controlling solid phase with silicate inhibitor addition. No other iron–silicate solids or complexes were found in the literature for consideration in model development (Latimer 1952, Sillen and Martell 1964).

Table 5.64 shows predicted iron concentrations based on iron silicate and siderite models. The iron silicate model underpredicted actual dissolved iron release by more than 90%. Table 5.64 shows that prediction of iron by the $FeCO_3$ model slightly overpredicted actual release but was closer to the actual data than the $FeSiO_3$ model. Therefore, it seems reasonable that $FeCO_3$ is the controlling solid phase for dissolved iron release with Si inhibitor addition, as was the case with the control and phosphate inhibitor addition.

Actual average dissolved iron concentrations from the Si PDSs are shown in Table 5.59 and are about 30% less than concentrations for pH_s+0.3. This suggests that dissolved iron release is not solely dependent on $FeCO_3$ but that the presence of silicate solids, supported by XPS analysis, helps reduce iron release with Si inhibitor addition. Therefore, it is reasonable that the addition of Si inhibitor resulted in the formation of a solid silicate film that inhibited iron release more effectively than pH elevation alone.

Solubility limits of iron for both solids ($FeSiO_3$ and $FeCO_3$) are illustrated in Figure 5.68. The solubility diagram suggests that under equilibrium conditions and for pH values above 7, $FeSiO_3$ was the limiting solid in iron release, which is in agreement with predicted iron concentrations from the $FeSiO_3$ thermodynamic model, as shown in Table 5.64.

This suggests that $FeCO_3$ is the controlling solid phase in iron release; however, actual dissolved iron release was lower than what the $FeCO_3$ model predicted. This is likely attributed to the presence of $FeSiO_3$ from the Si inhibitor addition. As was discussed with phosphate inhibitors, although addition of silicate inhibitor may not change the equilibrium iron concentration, a protective silicate film could slow the rate of iron release, resulting in reduced iron concentrations. A mechanism of kinetic inhibition of iron release would explain why inhibitors are least effective in reducing iron concentrations in low-flow dead-end conditions.

Iron Scale Elemental Analysis

An example iron XPS survey scan is shown in Figure 5.69. Zinc, iron, oxygen (O), calcium, carbon (C), phosphorus, and silicon are indicated on the survey by pronounced peaks. Specific compounds for each of these elements were identified using high-resolution XPS scans.

Table 5.65 shows the number of coupons where each element was detected. Calcium, silica, and zinc were detected on 19 of the 24 iron coupons, while carbon, iron, and oxygen were

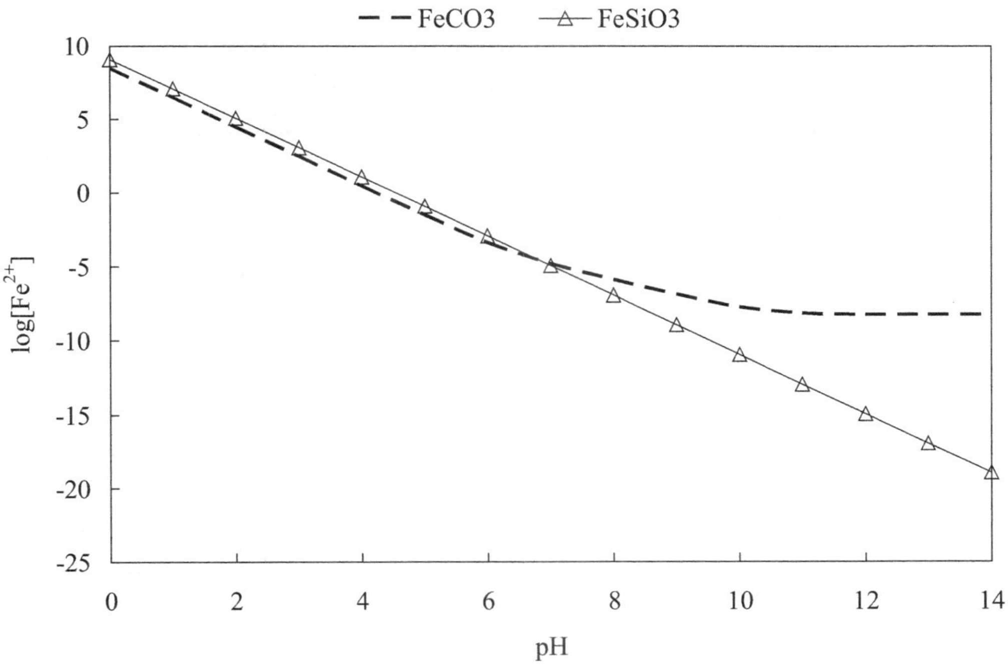

Figure 5.68 Solubility limits of iron for Si inhibitor

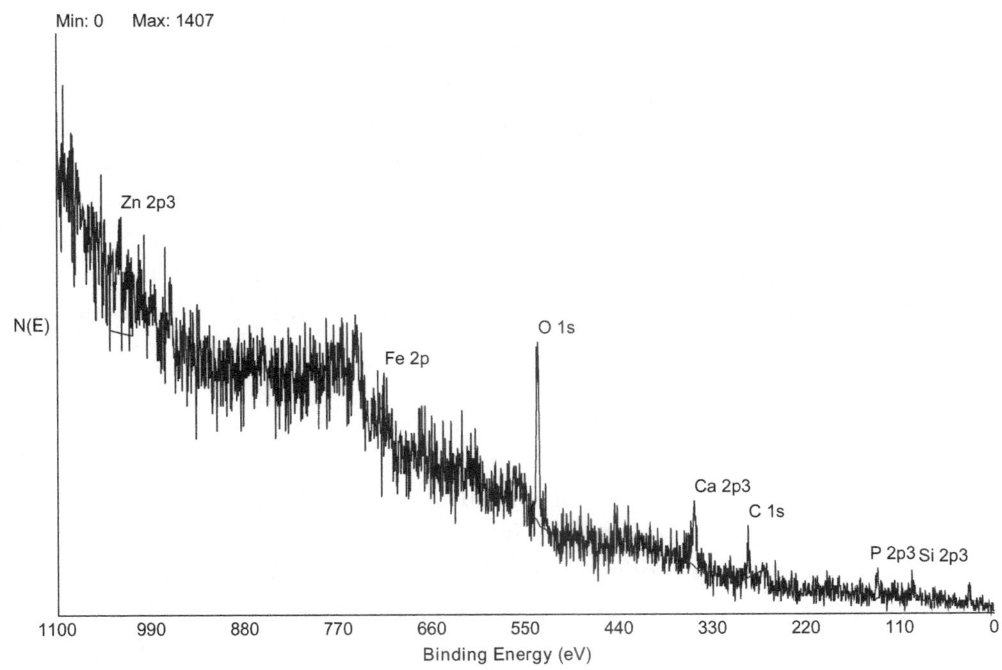

Figure 5.69 Representative XPS survey scan for iron coupon

Table 5.65
Number of coupons containing detected elements from XPS scan out of 24 iron coupons

Detected element	No. of coupons (from 24 total)
Carbon (C)	24
Calcium (Ca)	19
Iron (Fe)	24
Oxygen (O)	24
Phosphorus (P)	9
Silicon (Si)	19
Zinc (Zn)	19

detected on all the iron coupons. This is reasonable since the raw water always contained carbonate alkalinity (carbon), calcium hardness, dissolved oxygen, trace phosphorus (0.05 mg/L as P), and silica. Twelve of the 24 iron coupons were exposed to phosphate-based inhibitors. Four of the iron coupons were exposed to the silica-based inhibitor. Eight of the coupons were exposed to the pH control inhibitor.

Phosphorus was found in the scale of nine of the twelve iron coupons exposed to phosphate inhibition. Hence, XPS analysis found a solid phosphate film on the iron coupon for 75% of the observations from this work. Carbon, iron, and oxygen were found in all scale samples, which suggests that iron oxides were uniformly distributed throughout the iron coupon scales. Calcium carbonate is a common form of calcium and carbon found on iron scale; however, calcium was not found in 5 of 24 samples, which suggests that $CaCO_3$ was not uniformly distributed throughout the iron scale. Given the nature of corrosion in distribution systems, it is not unusual to find nonuniform scales. Hence, the partial or complete presence of carbon, calcium, iron, oxygen, phosphorus, silicon, and zinc on iron coupons is not surprising. The presence of silicon and zinc in the scales of 19 of 24 coupons indicates that the raw water silicon was adequate to form scale on the iron coupons. The iron, lead, and copper coupons were housed in the EN trailer and received PDS effluent, which contained metal release from GS and UCI pipe, which could contain zinc, iron, and calcium solids.

Identification of Iron Scale Corrosion Products

Identification of coupon corrosion products were done by deconvolution of high-resolution iron scans (Fe_{2p} or Fe_{2p3}). Findings from previous work in TBW I (Taylor et al. 2005), suggest that major iron corrosion products included hydrated ferric oxide (FeOOH), ferric oxide (Fe_2O_3), and magnetite (Fe_3O_4). Scans of coupons exposed to phosphate-based inhibitors revealed the presence of ferric phosphate ($FePO_4$). A high-resolution scan of identified corrosion products on an iron coupon exposed to a phosphate inhibitor is shown in Figure 5.70.

Figure 5.71 shows an iron curve fit with identified corrosion products when ferric phosphate was not detected. Ferric phosphate was detected in 8 of the 12 iron coupons exposed to phosphate-based inhibitors. This suggests that phosphate has an influence on the formation of iron corrosion products. Fe_2O_3, Fe_3O_4, and FeOOH were detected in all iron coupons except one. Hence, this data suggests that the same oxygen-based corrosion products form on the surface of

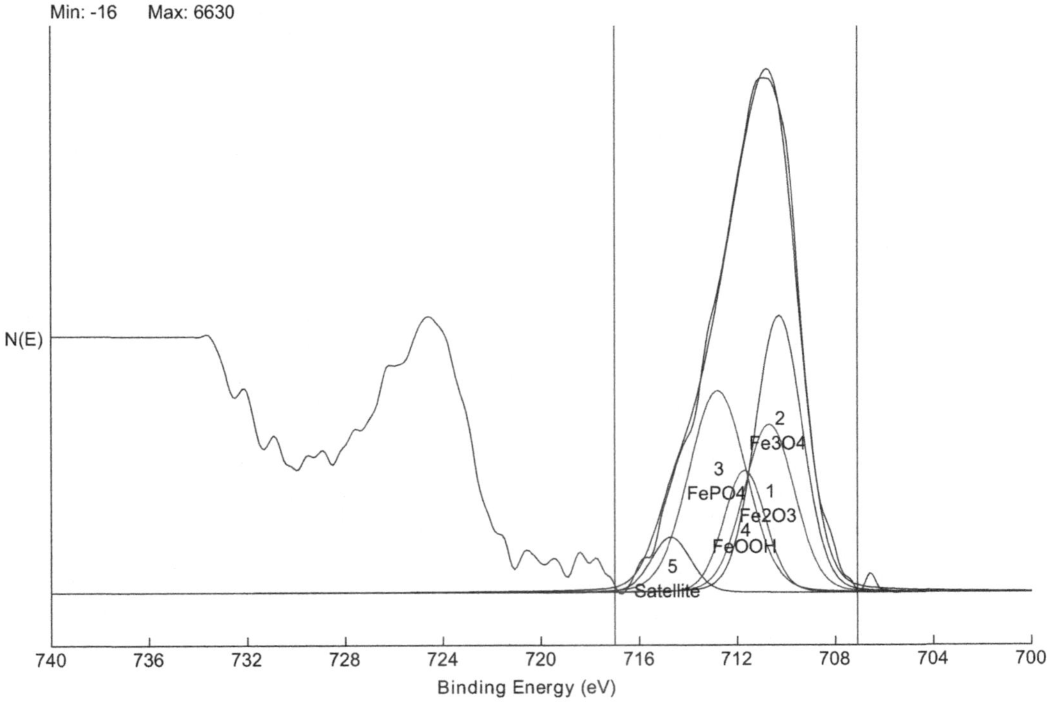

Figure 5.70 High-resolution XPS scan and deconvolution of iron peak of corrosion products on iron coupon from PDS02 during phase I

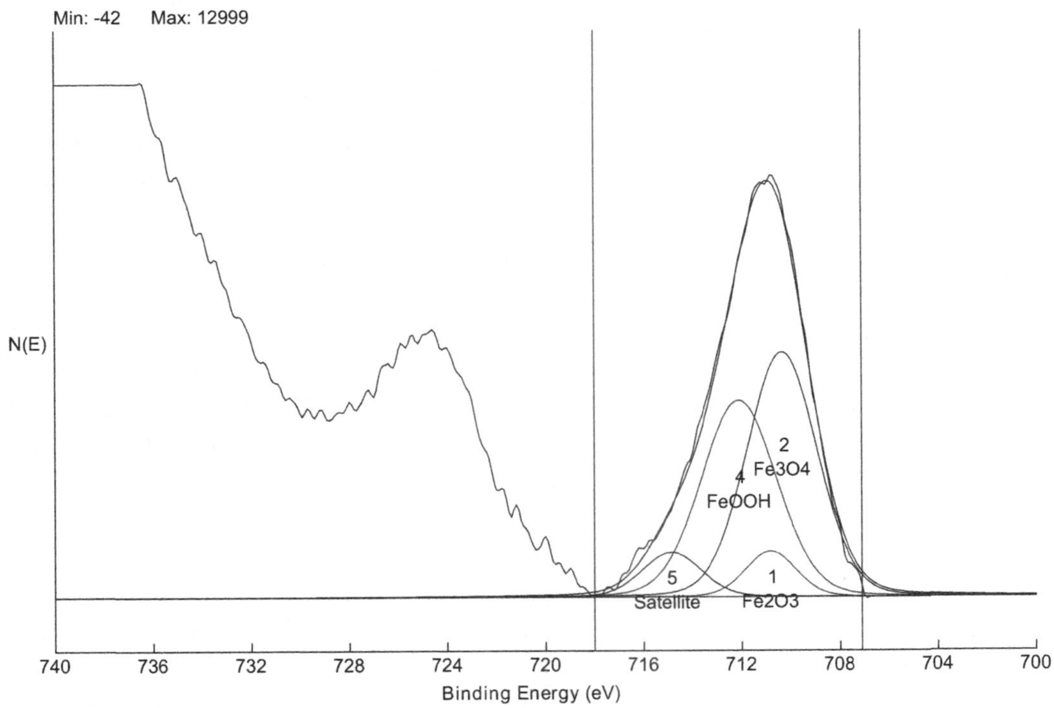

Figure 5.71 High-resolution XPS scan and deconvolution of iron peak of corrosion products on iron coupon from PDS11 during phase II

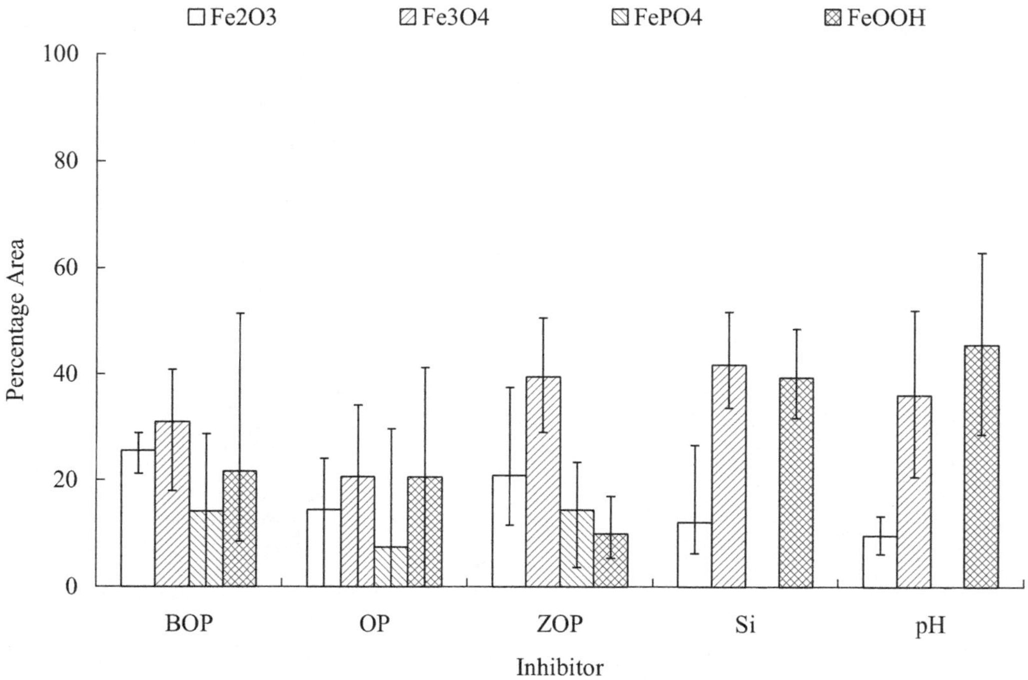

Figure 5.72 Distribution of iron compounds by inhibitor for all phases

iron pipe in the presence of phosphate-based inhibitors and that phosphate compounds will also form on the surface of iron pipe in the presence of phosphate inhibitors. Furthermore, it is rational that the presence of phosphate solids could affect metal release from metal surfaces.

The relative distribution of iron corrosion products by inhibitor type is presented in Figure 5.72. This plot shows average composition of the iron surface scale on iron coupons from all four phases. In each phase, one iron coupon was exposed to one of six treatments: BOP, OP, ZOP, Si, pH_s, and pH_s+0.3. Doses of the inhibitors applied to the iron coupons did not vary between phases.

The percentage of Fe_2O_3 in iron scale did not vary with inhibitor type; Fe_3O_4 was the dominant iron compound for coupons receiving ZOP inhibitor. The fraction of FeOOH increased in the scales for nonphosphate inhibitors, which did not have any $FePO_4$. The percentage of FeOOH in the iron scales ranged from 5% to 17% when $FePO_4$ was detected in the same coupon, while the percentage of FeOOH ranged from 20% to 64%, increasing when no $FePO_4$ was present. Ferric phosphate was not detected in coupons exposed to nonphosphate-based inhibitors.

Effect of Phase on Iron Corrosion Scale Products

The effect of phase composition of the scale on iron coupons was evaluated for the identified iron compounds, Fe_2O_3, Fe_3O_4, $FePO_4$, and FeOOH. Figure 5.73 shows variation of Fe_2O_3 percentages by phase across all inhibitors. Fe_2O_3 varied between 4% and 28% for phases I, II, and IV. Phase III coupons had similar Fe_2O_3 composition to other phases, except for the OP and ZOP coupons, where the OP coupon had no Fe_2O_3 (0%) and the ZOP coupon had the highest of Fe_2O_3

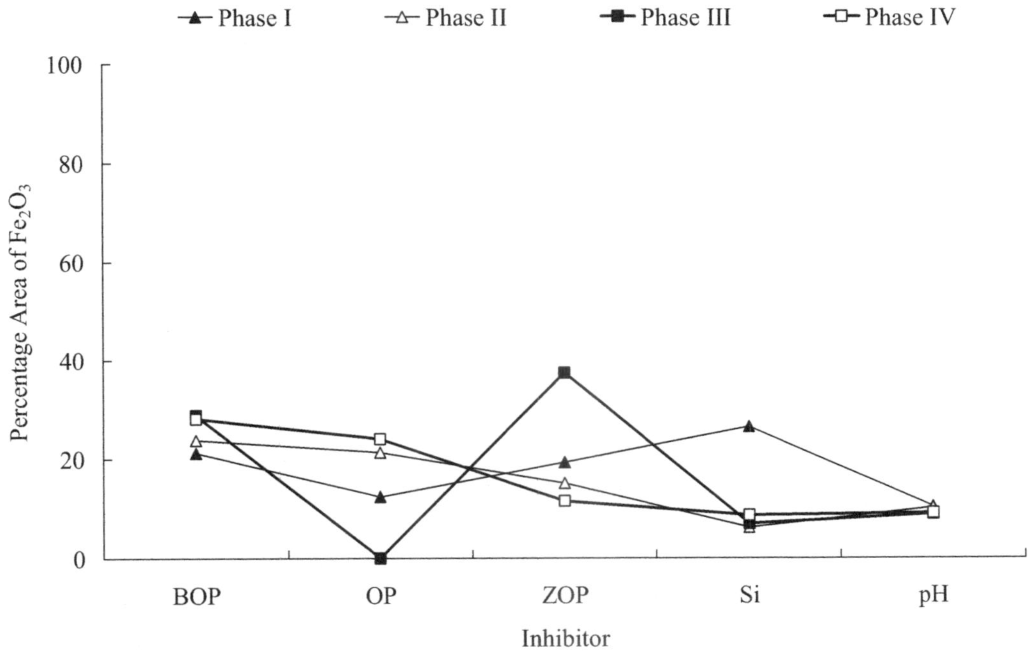

Figure 5.73 Variation of composition of the iron scale on iron coupons of FeO by phase and across all inhibitors

composition (37%), as shown in Figure 5.73. Fe_2O_3 composition in phase III coupons ranged from 5% to 29%, which is similar to phases I, II, and IV. As shown in Figure 5.73, there is no relationship between phase and fraction of Fe_2O_3 in the iron scale. Hence, variation in water quality by phase has limited influence on Fe_2O_3 composition on iron coupons.

Phase effect on Fe_3O_4 was similar to the effect on Fe_2O_3, where composition of the iron scale on iron coupons in phases I, II, and IV were in the range of 18% to 57%. Phase III had similar influence on Fe_3O_4 as other phases, except that no iron was detected in the iron scan of the OP iron coupon.

Detection of ferric phosphate was highly dependent on the addition of phosphate-based inhibitors because no $FePO_4$ was detected in coupons exposed to Si or pH treatment, as shown in Figure 5.74. Inhibitor type influenced the percentage of $FePO_4$ more than phase change. Variations of $FePO_4$ composition in the iron scale on iron coupons exposed to BOP or ZOP ranged from 4% to 29%, whereas no $FePO_4$ was found in the iron scale on OP iron coupons except in phase IV.

Figure 5.75 shows variations of percentages for hydrated ferric oxide by phase. There seems to be no clear difference in scale composition of iron coupons by phase except for coupons exposed to OP inhibitor, where percentages of FeOOH in iron scales formed during phases I and II were higher than in phases III and IV. However, interaction between inhibitor types and FeOOH composition is more influential than phase change. The composition of FeOOH on Si and pH coupons was between 29% and 64%, whereas for most coupons exposed to phosphate-based inhibitors (8 of 12 coupons), FeOOH composition was less than 20%, as shown in Figure 5.75.

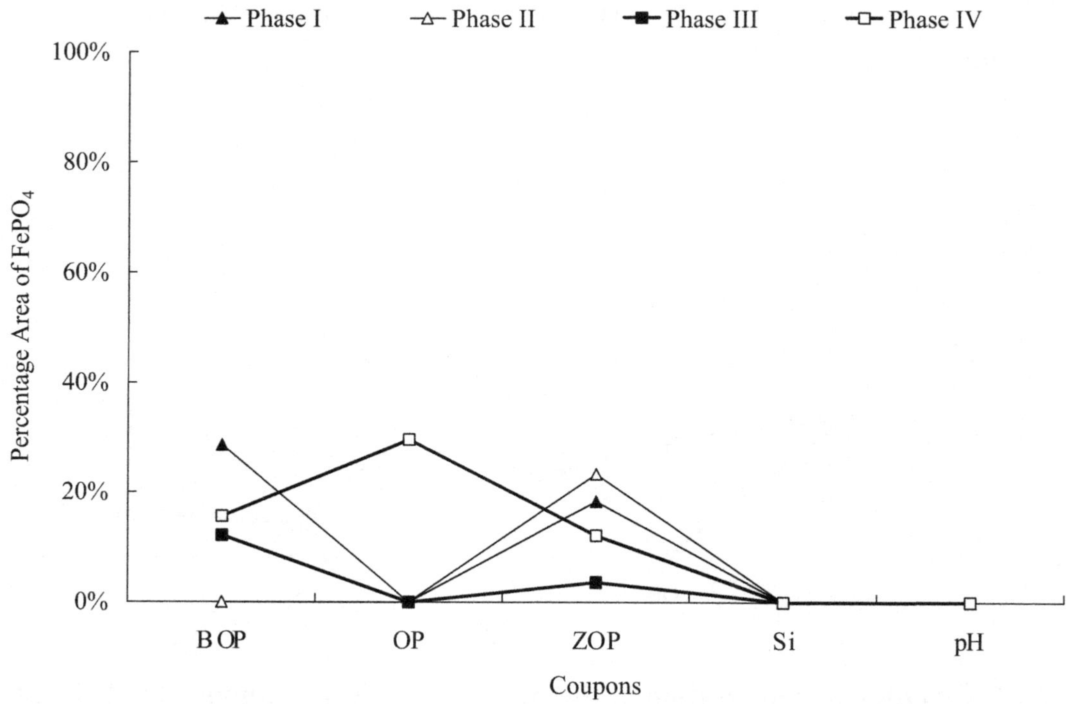

Figure 5.74 Variation of composition of the iron scale on iron coupons of $FePO_4$ by phase and across all inhibitors

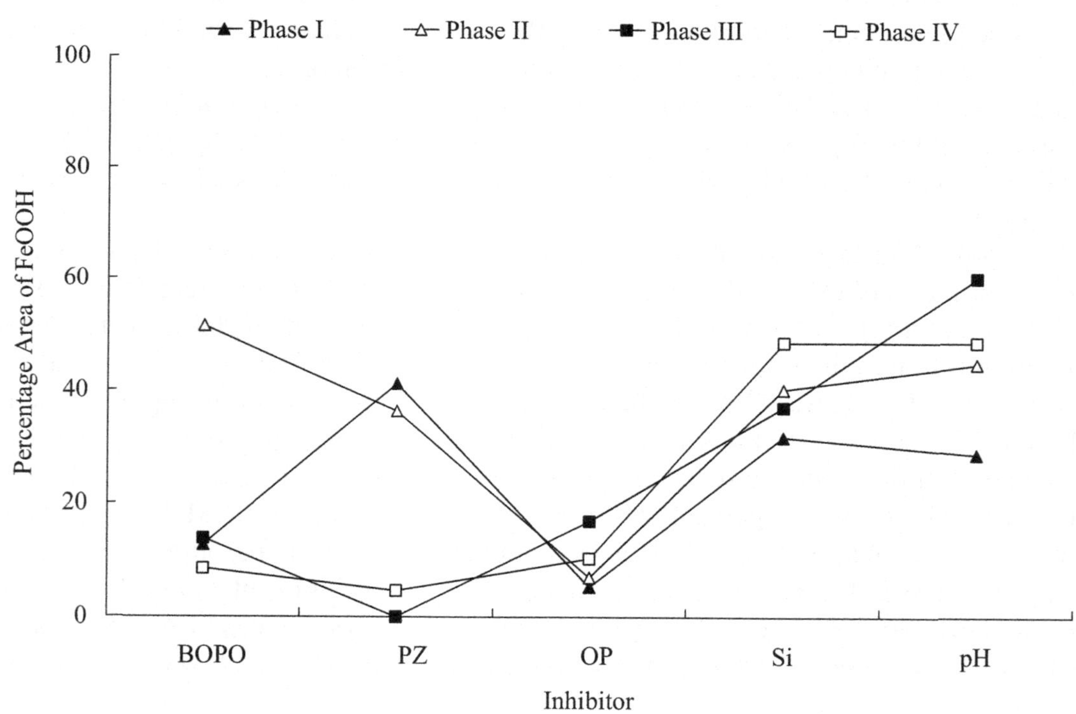

Figure 5.75 Variation of composition of the iron scale on iron coupons of FeOOH by phase and across all inhibitors

Table 5.66
Compounds identification for BOP coupons

Element	Compound	Phase I	Phase II	Phase III	Phase IV
C	CO_3	X	X	X	X
Ca	$CaCO_3$	X	X	X	X
Fe	Fe_2O_3	X	X	X	X
	Fe_3O_4	X	X	X	X
	$FePO_4$	X		X	X
	FeOOH	X	X	X	X
O	O^{2-}	X	X	X	X
	CO_3^{2-}	X	X	X	X
	OH^-	X	X	X	X
	PO_4^{3-}	X	X	X	X
P	PO_4^{3-}	X			X
	Pyro P	X	X		X
	Meta P			X	
Si	SiO_x	X	X	X	X
Zn	ZnO	X		X	X
	Zn^{2+}		X		

Effect of Inhibitor on Iron Corrosion Scale Products

Blended Ortho-polyphosphate. Four iron coupons were exposed to the BOP inhibitor, with one coupon in each phase. Detected elements and compounds from XPS deconvolution of iron coupon scales in the presence of BOP are shown in Table 5.66. Carbon and calcium were found in all phases, for BOP, as calcium carbonate. For iron, Fe_2O_3, Fe_3O_4, and FeOOH were found in every phase in the BOP iron coupon scale (four coupons); however, $FePO_4$ was only found in three of the four coupons. Oxygen was found in the form of O^{2-}, CO_3^{2-}, OH^-, and PO_4^{3-}. These forms are consistent with the solid forms of Ca, Fe, Zn, and Si shown in Table 5.66. Phosphorus in coupons exposed to BOP was found in the phosphate (PO_4^{3-}), pyrophosphate, or meta-phosphate forms, or in a combination of these forms. Phosphorus was found by the survey scan in every phase but not phosphate, pyrophosphate, or meta-phosphate in all phases. However, the form of phosphorus present did not hinder the formation of ferric phosphate, as $FePO_4$ was detected in phase III in the presence of meta-phosphate, as shown in Table 5.66. This suggests that other forms of phosphorus in the BOP inhibitor, besides OP, can influence scale formation on iron.

Si as SiO_x was present in the iron coupon scale on all samples, suggesting deposition of silica from the raw water. Zn as ZnO (zinc oxide) in the iron scale probably came from the GS pipes in the PDS. The BOP iron scale data presented in Table 5.66 does show the presence of $FePO_4$, which may affect metal release. Complexation of calcium by polyphosphates from phosphate-based inhibitors is noted as a possible mechanism of corrosion inhibition, as is precipitation of a calcium phosphate film (Moriarty 1990); however, no calcium phosphate precipitate was detected. Iron complexation by phosphates could explain why BOP PDSs tended to release more dissolved and less total iron than was released in the OP PDSs, which is shown in Table 5.64.

Table 5.67
Compounds identification for OP coupons

Element	Compound	Phase I	Phase II	Phase III	Phase IV
C	CO_3^{2-}		X		X
Ca	$CaCO_3$		X		X
Fe	Fe_2O_3	X	X		X
	Fe_3O_4	X	X		X
	$FePO_4$				X
	FeOOH	X	X		X
O	O^{2-}	X	X	X	X
	CO_3^{2-}		X		X
	OH^-	X	X	X	X
	PO_4^{3-}		X		X
P	PO_4^{3-}		X		X
	Pyro P				
	Meta P				
Si	SiO_x	X		X	X
Zn	ZnO	X	X		X
	Zn^{2+}			X	

Orthophosphate. The compounds identified on the iron coupons receiving inhibitor are given in Table 5.67, with one iron coupon per phase. These results are different relative to the BOP iron scans in that fewer solid species were detected. CO_3^{2-} in the carbon scan was only detected in phases II and IV (two of four iron coupons). Similarly, calcium was found in phases II and IV only, which indicates that the presence of calcium or CO_3^{2-} in the scale is in the $CaCO_3$ form for the OP coupons.

No iron oxide products were found in phase III, but all iron oxide products were detected in the remaining phases. $FePO_4$ was only detected in the phase IV iron scale, which was consistent with the detection of phosphorus and PO_4^{3-} in the phase IV iron coupon scale. Oxygen was found in O^{2-}, CO_3^{2-}, OH^-, and PO_4^{3-} forms. These forms are consistent with the solid forms of calcium, iron, phosphorus, zinc, and silicon. Silicon was detected in all OP scales except those from phase II. Zinc was detected in all OP scales as either ZnO or Zn^{2+}.

Zinc Orthophosphate. Table 5.68 contains identified elements and their associated compounds from the four coupons exposed to ZOP inhibitor, where one coupon was used per phase. Compounds identified from the scans of carbon, calcium, iron, oxygen, phosphorus, and zinc were similar for all phases, with the exception of phase II in which calcium, phosphorus, and zinc were not detected.

In phase II, CO_3^{2-} was detected; however, no calcium was detected on the scale. Hence, the carbonate may be the remnants of a $FeCO_3$ compound. Previous discussion in the thermodynamic modeling section, "Identification and Modeling of Corrosion Scale Products," found $FeCO_3$ to be the controlling solid phase for iron release; however, upon exposure to air, siderite, $FeCO_{3(s)}$, decomposes to Fe_2O_3 (Taylor et al. 2005).

Table 5.68
Compounds identification for ZOP coupons

Element	Compound	Phase I	Phase II	Phase III	Phase IV
C	CO_3^{2-}	X	X	X	X
Ca	$CaCO_3$	X		X	X
Fe	Fe_2O_3	X	X	X	X
	Fe_3O_4	X	X	X	X
	$FePO_4$	X	X	X	X
	FeOOH	X	X	X	X
O	O^{2-}	X	X	X	X
	CO_3^{2-}	X	X	X	X
	OH^-	X	X	X	X
	PO_4^{3-}	X		X	X
P	PO_4^{3-}	X		X	X
	Pyro P				
	Meta P				
Si	SiO_x		X	X	
Zn	ZnO	X		X	X
	Zn^{2+}				

Phosphorus as PO_4^{3-} was detected in three of the four iron coupons exposed to ZOP. The differences between coupons exposed to the phosphate-based inhibitors are in the phosphorus compounds, where phosphate, pyrophosphate, or meta-phosphate was found in the iron scale composition on BOP iron coupons while only orthophosphate was found in the ZOP iron scale. In contrast, phosphorus was detected infrequently on OP iron scales. ZnO was detected in all scales except the sample from phase II. No zinc phosphate was detected in the iron scale in the presence of ZOP, which indicates that zinc phosphate precipitation was not a factor for inhibiting iron release by ZOP inhibition. Silicon was only detected in the ZOP iron scales from phases II and III.

Silicon. For the four iron coupons exposed to Si inhibitor, the carbon scan showed CO_3^{-2} to be present in all phases except phase III. Similarly, the calcium scan also showed CO_3^{2-} to be present in all phases except phase III, as shown in Table 5.69. This indicates that CO_3^{2-} was present on the scale as $CaCO_3$. The iron scan detected Fe_2O_3, Fe_3O_4, and FeOOH. $FePO_4$ was not found, which is consistent with addition of silicate inhibitor. The absence of PO_4^{3-} in the scale is supported by the oxygen scan, which found O^{2-}, CO_3^{2-}, and OH^-, but not PO_4^{3-}.

Silica, as silicate, was detected in all phases for coupons exposed to the silicate inhibitor. However, the analysis did not reveal the exact silicate compound because no specific iron–silicate compounds were found corresponding to the energy level of silicate in the scanned coupons. The consistent presence of silicate throughout all phases suggests that Si has a role in scale formation and may help inhibit iron release. In the four coupons exposed to Si inhibitor, zinc in the form of ZnO was detected in phase III only.

pH. Tables 5.70 and 5.71 show the compounds identified on iron coupons for pH_s and pH_s+0.3, respectively. Carbonate was detected in both carbon and calcium scans for all phases but phase III. This suggests that CO_3^{2-} found in the scale is in $CaCO_3$ form. The absence of

Table 5.69
Compounds identification for Si coupons

Element	Compound	Phase I	Phase II	Phase III	Phase IV
C	CO_3	X	X		X
Ca	$CaCO_3$	X	X		X
Fe	Fe_2O_3	X	X	X	X
	Fe_3O_4	X	X	X	X
	$FePO_4$				
	FeOOH	X	X	X	X
O	O^{2-}	X	X	X	X
	CO_3	X	X		X
	OH	X	X	X	X
	PO_4				
P	PO_4				
	Pyro P				
	Meta P				
Si	SiO_x	X	X	X	X
Zn	ZnO			X	
	Zn^{2+}				

phosphorus from scanned coupons exposed to pH or Si treatment indicates that the phosphorus in the iron scale on iron coupons came from the phosphate-based inhibitors and not from trace phosphorus in the source water. Hence, any differences in iron release between pH control and phosphate-based inhibitors may be a result of the presence of phosphate, especially as iron phosphate solids.

Analysis of Fe scans showed the presence of corrosion products Fe_2O_3, Fe_3O_4, and FeOOH on every coupon for pH_s and pH_s+0.3 for all four phases. These compounds are consistent with those identified in the TBW I XPS analysis, where no phosphate or silicate inhibitors were used. Oxygen was found as O^{2-}, CO_3^{2-}, and OH^- in all iron coupons in both pH_s and pH_s+0.3 control PDSs and for all phases. PO_4^{3-} was not detected in the oxygen scan, which is consistent with the absence of phosphate-based inhibitors. Silicate and ZnO were detected in scans of the pH_s and pH_s+0.3 coupons. In pH_s coupons, silicate and ZnO were found in all phases except phase III. For pH_s+0.3, silicate was detected in all phases except phase II, while ZnO was found in phases III and IV. The consistent presence of $CaCO_3$ scale in pH_s+0.3 coupons is in agreement with the expected $CaCO_3$ precipitation at higher pH conditions, compared to pH_s coupons, which showed $CaCO_3$ in three phases only.

Summary for Iron Scale Surface Characterization

Surface composition of iron scale on iron coupons exposed to varying inhibitors was analyzed using survey and high-resolution XPS spectra scans. Carbon, calcium, iron, oxygen, phosphorus, silicon, and zinc were identified from surface scans and selected for high-resolution scans. Fe_2O_3, Fe_3O_4, FeOOH, and $FePO_4$ were detected in the surface composition of the iron

Table 5.70
Compounds identification for pH_s coupons

Element	Compound	Phase I	Phase II	Phase III	Phase IV
C	CO_3	X	X		X
Ca	$CaCO_3$	X	X		X
Fe	Fe_2O_3	X	X	X	X
	Fe_3O_4	X	X	X	X
	$FePO_4$				
	FeOOH	X	X	X	X
O	O^{2-}	X	X	X	X
	CO_3^{2-}	X	X	X	X
	OH^-	X	X	X	X
	PO_4^{3-}				
P	PO_4^{3-}				
	Pyro P				
	Meta P				
Si	SiO_x	X	X		X
Zn	ZnO	X	X		X
	Zn^{2+}				

Table 5.71
Compounds identification for pH_s+0.3 coupons

Element	Compound	Phase I	Phase II	Phase III	Phase IV
C	CO_3	X	X	X	X
Ca	$CaCO_3$	X	X	X	X
Fe	Fe_2O_3	X	X	X	X
	Fe_3O_4	X	X	X	X
	$FePO_4$				
	FeOOH	X	X	X	X
O	O^{2-}	X	X	X	X
	CO_3^{2-}	X	X	X	X
	OH^-	X	X	X	X
	PO_4^{3-}				
P	PO_4^{3-}				
	Pyro P				
	Meta P				
Si	SiO_x	X		X	X
Zn	ZnO			X	X
	Zn^{2+}				

scale on 8 of 12 coupons exposed to phosphate inhibition, which was attributed to inhibitor addition as no phosphate scale was detected in the Si- or pH-inhibited iron coupons, although trace phosphates were discovered in all finished waters. The variation in water quality by phase did not affect the composition of the surface scales on iron coupons; however, inhibitor type did affect the scale composition. Silicate scale was detected on all coupons exposed to Si inhibitor, but was also detected to a lesser extent in iron scales of coupons exposed to other inhibitors and was assumed to come from the raw water. The lack of zinc phosphate scale in the presence of ZOP indicates that zinc phosphate precipitation was not a factor in inhibiting iron release by ZOP inhibition. Scales on coupons exposed to inhibitors other than ZOP showed traces of zinc compounds on the iron scales. This zinc presence is likely attributed to the GS pipe portion in every PDS.

All phosphorus- and silicon-based inhibitors tended to reduce the average of dissolved iron release relative to pH_s+0.3. For total iron release, the overall average release was reduced by both BOP and Si, whereas OP showed no improvement relative to pH_s+0.3. ZOP, however, resulted in a higher overall average total iron release relative to pH_s+0.3. $FePO_4$ was confirmed by XPS on 8 of 12 iron coupons in the presence of phosphate inhibition and is likely the cause of reduced iron release by phosphate inhibition. Zinc enhanced iron release more frequently than not, suggesting that zinc may interfere with $FePO_4$ scale formation. Silica enhanced the inhibition of iron release and tended to be more effective than any other inhibitor in reducing iron release. Although absent from the XPS reference scan library and not confirmed by XPS, silica is found in many solids in combination with other metals, which could occur with iron and could enhance iron inhibition by formation of a solid that reduces iron release.

Galvanized Steel Surface Characterization

Surface compositions of scales on GS coupons were analyzed by XPS to gain insight into the controlling solid phase and mechanism of iron release in the presence of inhibition. Thermodynamic equilibrium models for iron release and zinc release are developed and discussed relative to the actual and predicted metal release to aid the interpretation of the XPS results.

Thermodynamic Model Development for Zinc Release

Iron release from GS was described using the same model that was developed to describe iron release from iron coupons. Therefore, the iron equilibrium modeling is not repeated here; however, a thermodynamic model for zinc solubility is considered in the following sections.

The zinc coating in galvanized iron pipes acts as a sacrificial anode to protect iron and prevent iron release. In this study, despite the age of the GS coupons, the zinc coating was still present, mostly as ZnO and zinc hydroxide ($Zn(OH)_2$). Zinc release from GS pipe surfaces can be predicted by the solubility of solid phases formed within the corrosion scales as well. The possible controlling solid forms that may be responsible for zinc release were selected according to the XPS analysis of GS coupons. Thermodynamic models were constructed by assuming that each of the detected solid forms was the controlling solid phase for zinc release into water. Zinc ion concentrations were calculated by assuming that free zinc ions are in equilibrium with the controlling solid phase, and subsequently, concentrations of dissolved zinc complexes were determined using the appropriate equilibrium constants. The zinc complexes that were incorporated in equilibrium modeling of dissolved zinc are shown in Equation 5.57.

Table 5.72
Comparison of observed zinc release from inhibitor PDSs to pH_s+0.3

Phase	Inhibitor	Average release (mg/L) total Zn	% Ratio to pH_s+0.3 total Zn
I	pH_s+0.3	0.006	
	BOP	0.004	60
	OP	0.004	57
	ZOP	0.132	2,092
	Si	0.003	46
II	pH_s+0.3	0.002	
	BOP	0.003	126
	OP	0.002	104
	ZOP	0.099	4,392
	Si	0.004	176
III	pH_s+0.3	0.003	
	BOP	0.004	119
	OP	0.003	98
	ZOP	0.161	5,403
	Si	0.006	218
IV	pH_s+0.3	0.002	
	BOP	0.002	108
	OP	0.002	107
	ZOP	0.159	8,492
	Si	0.023	1,235

$$Zn = [Zn^{2+}] + [Zn(OH)^{+}] + [Zn(OH)_2^{0}] + [Zn(OH)_3^{-}] + [Zn(OH)_4^{2-}] + [ZnCO_3^{0}] + [ZnCl^{+}] + [ZnCl_2^{0}] + [ZnSO_4^{0}] \quad (5.57)$$

The following sections consider dissolved zinc solubility and controlling solid phases without inhibitor addition, with phosphate inhibitor addition, and with silicate inhibitor addition.

Inhibitor Control of Zinc Release

The average zinc release observed during the study is shown in Table 5.72 by phase for each inhibitor and relative to the pH_s+0.3 control. For all PDSs, only total zinc data was collected. The effect of a phosphate-based inhibitor on zinc release was not significant. Significant zinc was added to PDSs receiving ZOP inhibitor. Silicate inhibitors tended to correspond with higher zinc concentrations than phosphate inhibitors.

Table 5.73
Actual and thermodynamically predicted zinc concentrations for pH_s+0.3 treatment

Inhibitor	Phase	Actual zinc release (mg/L) total Zn	Modeled zinc release (mg/L) $ZnCO_3$	Modeled zinc release (mg/L) $Zn(OH)_2$
	I	0.006	0.09	4.09
	II	0.002	0.15	5.51
pH_s+0.3	III	0.003	0.08	3.11
	IV	0.002	0.13	5.52
	All	0.004	0.11	4.44

Thermodynamic Modeling of Zinc Release

pH+0.3. Zinc oxide and zinc hydroxide were identified on seven of eight GS coupons from PDSs not receiving inhibitors. Freshly precipitated zinc hydroxide undergoes an aging process that eventually leads to the formation of zinc oxide (ZnO), which is a stable solid phase. Thus, in the absence of inhibitors, zinc hydroxide ($Zn(OH)_2$) was first assumed to be the solid phase controlling zinc ion concentrations in water for pH_s and pH_s+0.3. However, model-predicted zinc concentrations are much higher (1,000 times) than the actual values, as shown in Table 5.73; therefore, zinc hydroxide was eliminated as the controlling solid phase.

Instead, another thermodynamic model was developed through the same approach, assuming zinc carbonate ($ZnCO_3$) was the controlling solid phase. The model includes the major species affecting zinc release and equilibrium between free Zn(II) ion and the controlling solid phase, as shown in Equation 5.58.

$$Zn = [Zn^{2+}] + [Zn(OH)^{+}] + [Zn(OH)_2^{0}] + [Zn(OH)_3^{-}] + [Zn(OH)_4^{2-}] + [ZnCO_3^{0}] + [ZnCl^{+}] + [ZnCl_2^{0}] + [ZnSO_4^{0}] \quad (5.58)$$

A plot of solubility for $ZnCO_3$ and $Zn(OH)_2$, without consideration of zinc complexation, is shown in Figure 5.76. Figure 5.76 suggests that zinc carbonate, $ZnCO_3$, limits the solubility of zinc within the pH range of the study.

Actual and predicted total zinc concentrations from the $ZnCO_3$ model and the $Zn(OH)_2$ model are presented in Table 5.73. Both models tended to overpredict the actual zinc concentrations in the study. This is not unreasonable given that the HRT for the GS pipe was only 4 hours, which suggests that actual zinc concentrations were less than the equilibrium concentrations.

XPS analysis of galvanized coupons exposed to phosphate-based inhibitors did not reveal the presence of zinc phosphate, $Zn_3(PO_4)_2$, as a surface solid. Investigation of zinc phosphate as the possible controlling solid phase of dissolved zinc also showed that the calculated zinc concentration from zinc phosphate was higher than that of $ZnCO_3$ although less than that of $Zn(OH)_2$.

A plot of solubility for $ZnCO_3$, $Zn(OH)_2$, and $Zn_3(PO_4)_2$, without consideration of zinc complexation, is shown in Figure 5.77. Figure 5.77 suggests that for PDSs under phosphate-based inhibitors treatment, zinc carbonate ($ZnCO_3$) should control the solubility of zinc within the pH range of the study.

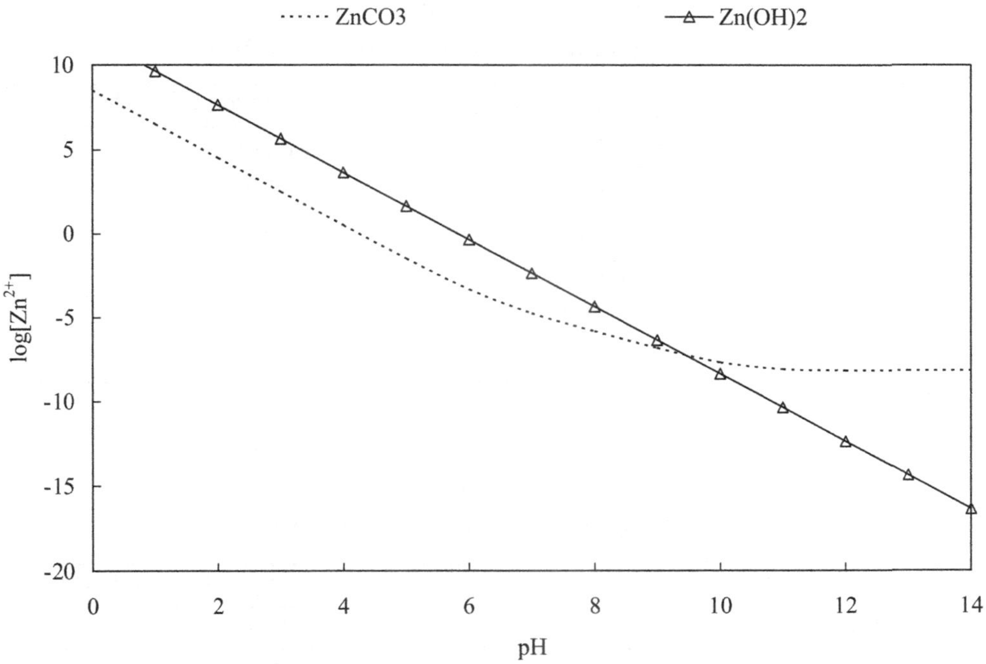

Figure 5.76 Solubility of zinc solids, without complexation, showing relative predominance without inhibitor addition

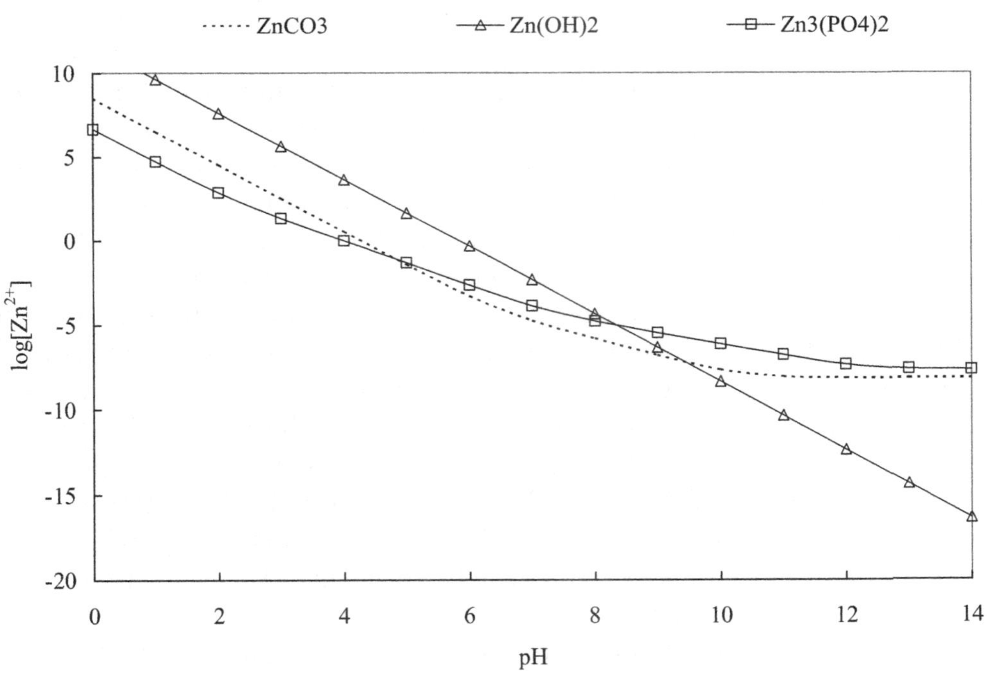

Figure 5.77 Solubility of zinc solids, without complexation, showing relative predominance with phosphate inhibitor addition

Table 5.74
Zinc thermodynamic modeling of BOP

Inhibitor	Phase	Actual zinc release (mg/L) total Zn	Modeled zinc release (mg/L)	
			$ZnCO_3$	$Zn_3(PO_4)_2$
	I	0.004	0.09	1.33
	II	0.003	0.17	1.59
BOP	III	0.004	0.09	1.27
	IV	0.002	0.14	1.62
	All	0.003	0.11	1.44

Table 5.75
Zinc thermodynamic modeling of OP

Inhibitor	Phase	Actual zinc release (mg/L) total Zn	Modeled zinc release (mg/L)	
			$ZnCO_3$	$Zn_3(PO_4)_2$
	I	0.004	0.08	1.31
	II	0.002	0.17	1.61
OP	III	0.003	0.09	1.31
	IV	0.002	0.15	1.67
	All	0.003	0.11	1.46

The consideration of zinc–phosphate complexes in the model will cause the calculated dissolved zinc to increase because of the additional complexed zinc. However, as shown in Table 5.72, the impact of phosphate-based inhibitor on zinc release is insignificant. Therefore, if the GS coupons are fully protected by the zinc coating that remains on the surface, corrosion products on coupons exposed to phosphate-based inhibitor and on coupons in the control PDSs should be very similar, and iron corrosion products should not be identified.

The presence of iron corrosion products in all GS coupons and identification of ferric phosphate ($FePO_4$) on 9 of 12 coupons exposed to phosphate-based inhibitor indicates the zinc coating was not uniform and had corroded over time. Under such circumstances, GS pipes will behave like UCI pipes, and the mechanism of iron release inhibition is also likely due to the formation of a solid film, as discussed in the "Iron Surface Characterization" section.

Blended Ortho-polyphosphate. For PDSs under BOP inhibitor treatment, actual and predicted total zinc from the $ZnCO_3$ model and the $Zn_3(PO_4)_2$ model are presented in Table 5.74. These data show that both models overpredicted actual zinc. This may be because the zinc concentrations in the BOP PDSs did not reach equilibrium.

Orthophosphate. Actual and predicted total zinc concentrations for OP are presented in Table 5.75. Similar to BOP PDSs, both thermodynamic models shown in Table 5.75 overpredicted actual zinc.

Zinc Orthophosphate. Significant amounts of zinc were added to ZOP PDSs because of the addition of ZOP inhibitor. Therefore, total zinc in the ZOP PDSs are much higher than other PDSs. Actual and predicted total zinc for ZOP PDSs are presented in Table 5.76. The predictions

Table 5.76
Zinc thermodynamic modeling of ZOP

Inhibitor	Phase	Actual zinc release (mg/L) total Zn	Modeled zinc release (mg/L)	
			$ZnCO_3$	$Zn_3(PO_4)_2$
	I	0.132	0.10	1.43
	II	0.099	0.23	1.98
ZOP	III	0.161	0.11	1.47
	IV	0.159	0.18	1.88
	All	0.140	0.13	1.67

Table 5.77
Comparison of predicted zinc concentrations for Si inhibitor

Inhibitor	Phase	Actual zinc release (mg/L) total Zn	Modeled zinc release (mg/L)	
			$ZnCO_3$	$ZnSiO_3$
	I	0.003	0.058	0.00002
	II	0.004	0.096	0.00002
Si	III	0.006	0.063	0.00002
	IV	0.023	0.082	0.00002
	All	0.007	0.069	0.00002

from the zinc carbonate model are close to the actual total zinc concentrations, indicating possible equilibrium in the ZOP PDSs.

Silicon. The XPS analysis for GS coupons exposed to Si inhibitor indicated that silicate compounds were present on the coupon surface. $ZnSiO_3$ was considered as a controlling solid phase for zinc release under Si inhibitor treatment. However, model-predicted zinc concentrations are much lower than the actual values, as shown in Table 5.77. Therefore, another thermodynamic model was developed assuming that zinc carbonate ($ZnCO_3$) was the controlling solid phase.

Table 5.77 shows predicted zinc concentrations based on the developed models for zinc carbonate and zinc silicate. The $ZnCO_3$ model overpredicted actual zinc release for the Si inhibitor. This result suggests that zinc concentrations did not reach equilibrium in the Si PDSs.

Elemental Analysis

An example XPS survey spectrum for GS is shown in Figure 5.78. Peaks were typically pronounced for zinc, iron, calcium, carbon, phosphorus, and silicon.

The occurrence of elements on 24 GS coupons is summarized in Table 5.78. Carbon, iron, and oxygen were detected on all coupons. The frequent detection of these elements suggests the deposition of zinc-, silica-, iron- and calcium-based carbonate and oxide compounds on the surface of the GS coupons. Zinc was observed in 23 of 24 GS coupons. Phosphorus was identified in 9 of the 12 coupons that were exposed to phosphate-based inhibitors. This is reasonable given

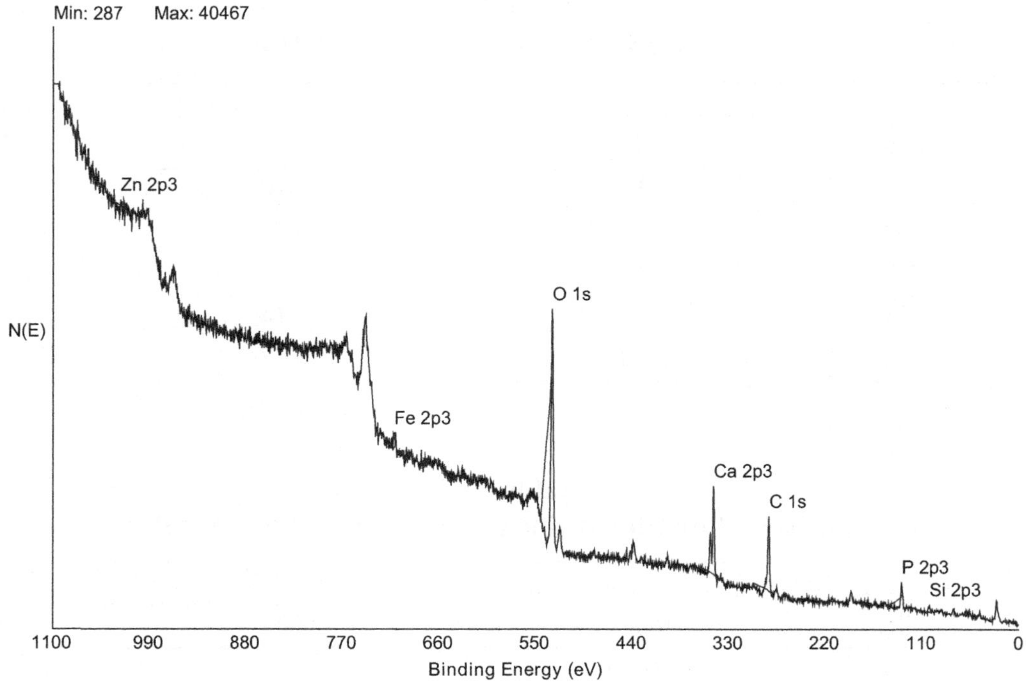

Figure 5.78 Representative XPS survey for GS coupons

Table 5.78
Element occurrence on 24 GS coupons

Element	No. of coupons (from 24 total)
Carbon (C)	24
Calcium (Ca)	20
Iron (Fe)	24
Oxygen (O)	24
Phosphorus (P)	9
Silicon (Si)	18
Zinc (Zn)	23

that zinc is initially present in GS coupons but phosphorus is not. Phosphorus deposition on 75% of the coupons indicates solid phosphate films are present and are likely a primary means of corrosion inhibition for phosphate-based inhibitors, but an adsorbed liquid solid film involving some form of phosphate could also inhibit metal release. Phosphorus was detected with the same frequency, 75% for the iron coupons receiving phosphate inhibitor, showing a consistent effect. In the following results, if zinc was not found on a GS coupon, this may be due to a precipitated corrosion product, complete leaching of zinc from the scanned areas, or an analytical error.

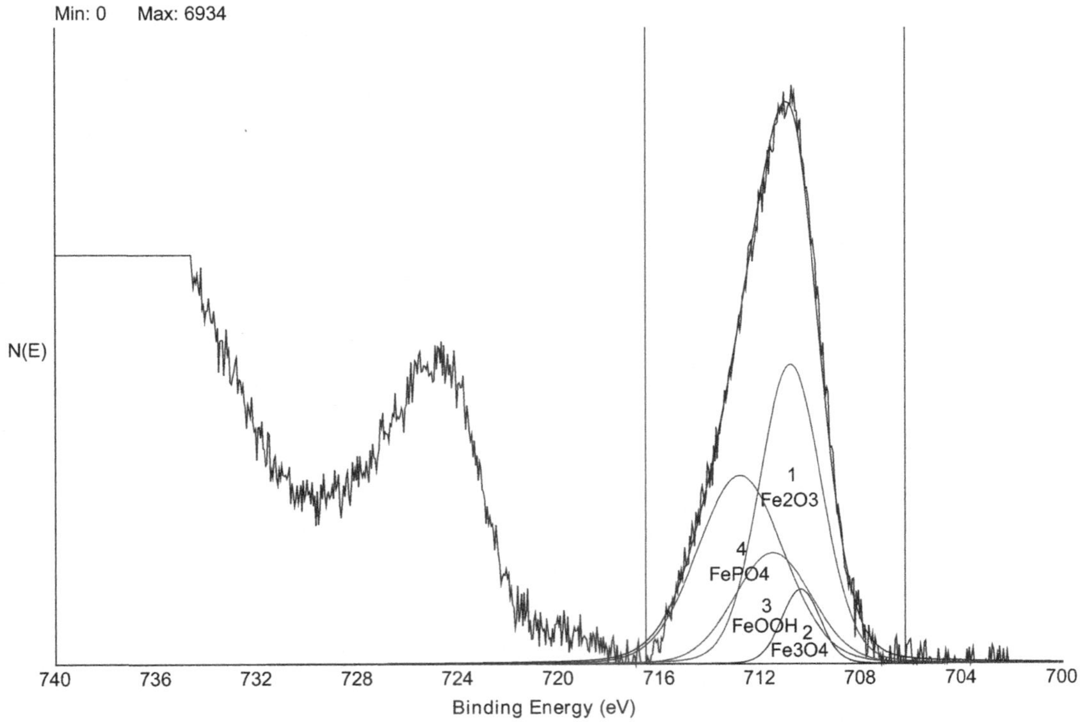

Figure 5.79 Representative high-resolution XPS scan and deconvolution of iron peak of corrosion products on a GS coupon from PDS05 (OP PDS) during phase III

Identification of Corrosion Products

High-resolution spectra for iron were deconvoluted to identify the corrosion products of iron on the surface and to determine their relative abundance. Figures 5.79 and 5.80 give example curve fits for iron with and without phosphate inhibitor addition. Ferric oxide (Fe_2O_3), magnetite (Fe_3O_4), hydrated ferric oxide (FeOOH), and ferric phosphate ($FePO_4$) were detected in the scale of the GS coupon exposed to the phosphate-based inhibitor, while only ferric oxide, magnetite, and hydrated ferric oxide were found without phosphate inhibitor addition. Ferric phosphate was only found on coupons exposed to phosphate inhibitors. Except for ferric phosphate, all these iron corrosion products were identified during TBW I by X-ray diffraction and then verified by XPS (Taylor et al. 2005). Binding energies for each iron compound were consistent with those used during TBW I. Energies for Fe_2O_3, Fe_3O_4, and FeOOH were 710.8, 710.4, and 711.5 eV, respectively.

Relative distribution of identified iron corrosion products is illustrated in Figure 5.81 by inhibitor using a bar-and-whiskers format. The bars represent the average and the whiskers represent the minimum and maximum of the four phases. The percentages represent the relative fraction of each corrosion product in the scale of the GS coupons. These data indicate that the surfaces of GS coupons contained a mixture of several iron corrosion products, with no dominant iron compound. Such results are consistent with a corrosion mechanism that consists of the oxidation of zero valence iron to ferrous and ferric iron, respectively. Ferrous carbonate has been shown to form in anaerobic conditions (Taylor et al. 2005); however, ferrous carbonate tends to degrade to Fe_2O_3 following exposure to air before XPS analysis. Corrosion inhibitors do not appear to influence the proportion of each iron oxide present. However, ferric phosphate is only present among the phosphorus inhibitors.

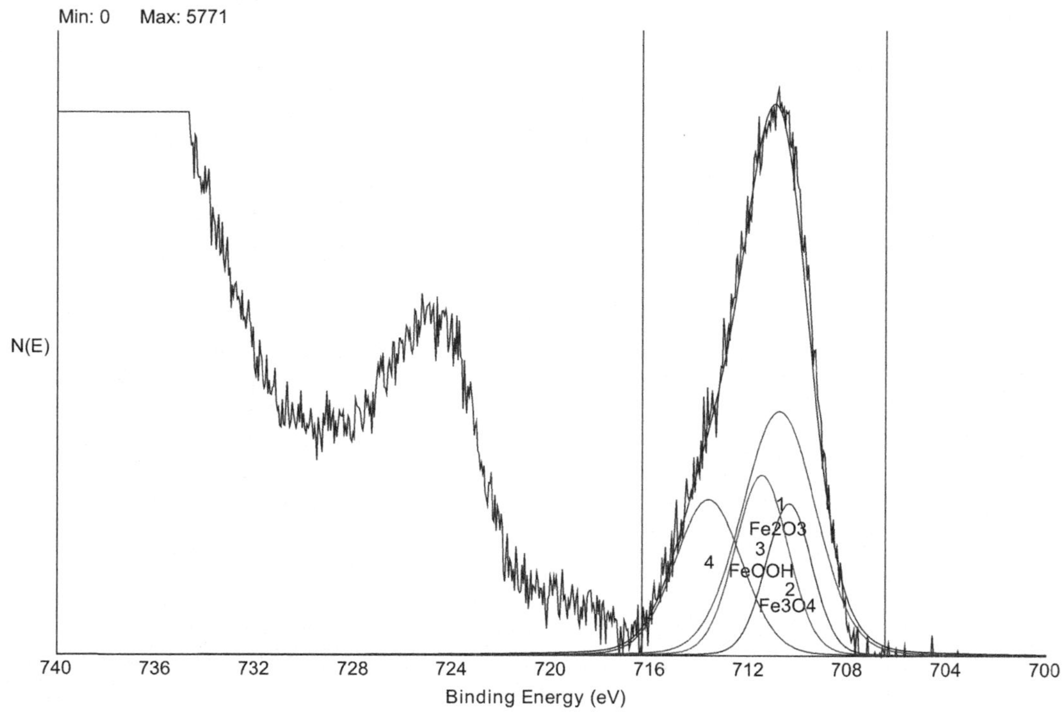

Figure 5.80 Representative high-resolution XPS scan and deconvolution of iron peak of corrosion products on a GS coupon from PDS11 (Si PDS) during phase II

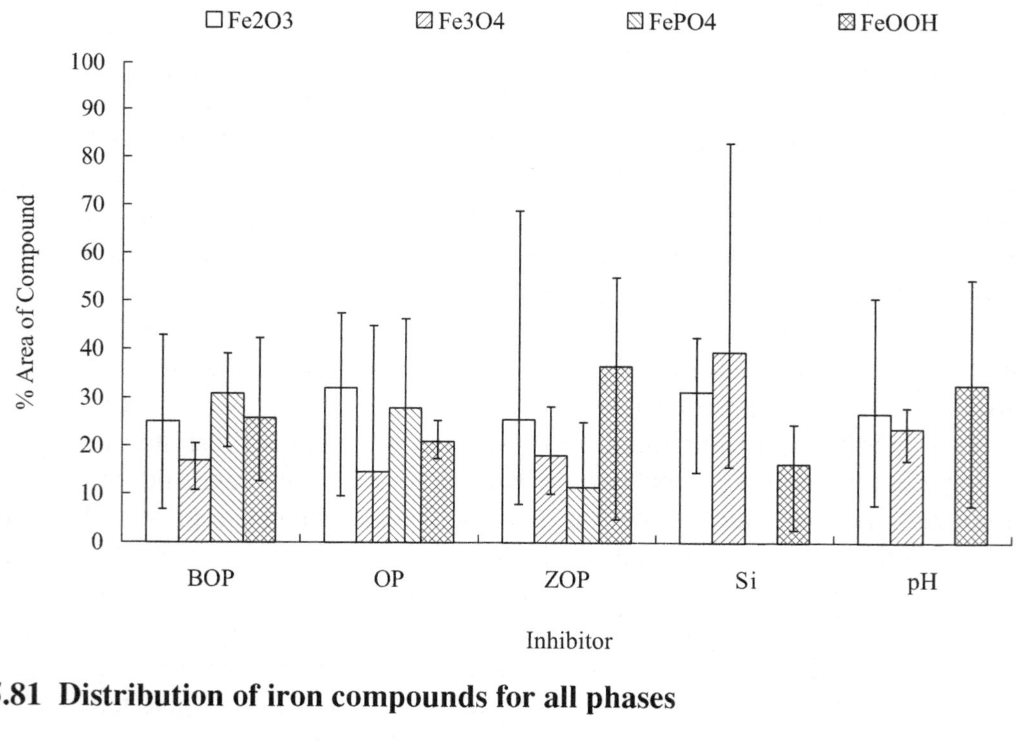

Figure 5.81 Distribution of iron compounds for all phases

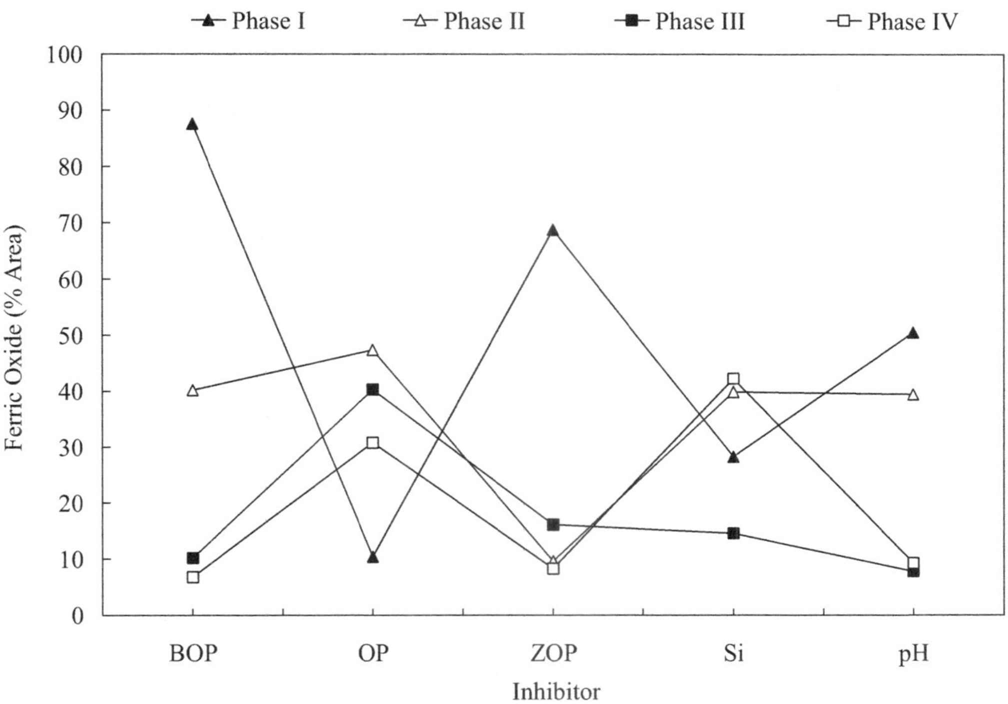

Figure 5.82 Percent area of ferric oxide versus inhibitors for all phases

Effect of Phase

The relationship of phase water quality and the relative composition of the iron compounds in the scale on the GS coupons was evaluated for each iron compound. A plot of the relative scale composition of ferric oxide (Fe_2O_3) in the scale on the GS coupons exposed to the midlevel inhibitor dose and both pH controls is presented in Figure 5.82. Similar analysis was conducted for magnetite and hydrated ferric oxide. No significant relationship was found between phase water quality and the percentage of the three iron compounds found in the surface scale of the GS coupons. Hence, Figure 5.82 indicates that variation in phase water quality (84 to 175 mg/L as $CaCO_3$ alkalinity, 35.4 to 122.9 mg/L chlorides, and 136 to 252 mg/L as $CaCO_3$ total hardness) did not affect the relative distribution of the solid iron oxides found in the scale of the GS coupons. The relative iron release for any of the inhibitors relative to pH_s+0.3 did not vary as much between phases as between the inhibitors, as shown previously in Table 5.59.

Effect of Inhibitor

Blended Ortho-polyphosphate. Table 5.79 gives a summary of the compounds detected on four GS coupons exposed to BOP inhibitor. One GS coupon was tested for each phase. The coupons exposed to BOP exhibited the same iron corrosion products discussed in the previous section. Ferric phosphate ($FePO_4$) was identified on all four coupons. Deconvolution of oxygen demonstrated the presence of CO_3^{2-}, O^{2-}, OH^-, and PO_4^{3-} for all phases. All these forms of oxygen correspond to various compounds of calcium, iron, zinc, and silicate. Calcium was present as solely $CaCO_3$ on all four coupons, implying that calcium is predominantly in the form of

Table 5.79
Compounds identified for BOP coupons

Element	Compound	Phase I	Phase II	Phase III	Phase IV
Fe	Fe_2O_3	X	X	X	X
	Fe_3O_4	X	X	X	X
	FeOOH	X	X	X	X
	$FePO_4$	X	X	X	X
O	CO_3^{2-}	X	X	X	X
	O^{2-}	X	X	X	X
	OH^-	X	X	X	X
	PO_4^{3-}	X	X	X	X
C	CO_3^{2-}	X	X	X	X
Ca	$CaCO_3$	X	X	X	X
P	PO_4^{3-}	X	X	X	X
	Pyro P	X		X	
Si	SiO	X		X	
	SiO_x	X	X		
Zn	ZnO	X	X	X	X
	$Zn(OH)_2$			X	

calcium carbonate when BOP is utilized within the PDS. Si was observed on three of four coupons as a mixture of SiO (silicon oxide) and SiO_x. The absence of Si in the phase IV coupon was likely due to the nonuniform nature of corrosion in distribution systems. There was no evidence to support the presence of Si in SiO_2 form on the coupon.

Zinc deposits were present on the scale on all four coupons, with a majority of the zinc existing as zinc oxide (ZnO). Also, zinc hydroxide ($Zn(OH)_2$) was found on one of four coupons. The single detection was on the phase III coupon. Phosphorus was observed for all four phases with a majority of the phosphorus existing as phosphate and pyrophosphate. Hence, these data indicate that solid forms of phosphates, such as ferric phosphate ($FePO_4$), may be inhibiting corrosion of galvanized steel when phosphate-based inhibitors are present.

Orthophosphate. Table 5.80 gives a summary of the compounds detected on four GS coupons exposed to the OP inhibitor. One GS coupon was tested for each phase. The coupons exposed to OP exhibited the same iron corrosion products discussed in the previous section. Ferric phosphate ($FePO_4$) was identified on three of four coupons. The absence of $FePO_4$ on the phase IV coupon is likely due to the nonuniform nature of corrosion in distribution systems.

Deconvolution of oxygen demonstrates the presence of CO_3^{2-} and O^{2-} on all four coupons and OH^- and PO_4^{3-} on three of four coupons. All these forms of oxygen correspond to various compounds of calcium, iron, zinc, and silicate. Phosphorus was observed on three of four coupons, existing mostly as phosphate and pyrophosphate. The single exception was on the phase IV coupon. Calcium was present as $CaCO_3$ on the phase I and III coupons, implying that calcium is predominantly in the form of calcium carbonate when OP is utilized. Si was found for the first three phases as a mixture of SiO and SiO_x. Zinc deposits were also present on the scale for all four coupons with a majority of the zinc existing as zinc oxide (ZnO). Also, zinc hydroxide ($Zn(OH)_2$)

Table 5.80
Compounds identified for OP coupons

Element	Compound	Phase I	Phase II	Phase III	Phase IV
Fe	Fe_2O_3	X	X	X	X
	Fe_3O_4	X		X	X
	FeOOH	X	X	X	X
	$FePO_4$	X	X	X	
O	CO_3^{2-}	X	X	X	X
	O^{2-}	X	X	X	X
	OH^-	X	X	X	
	PO_4^{3-}	X	X	X	
C	CO_3^{2-}	X	X	X	X
Ca	$CaCO_3$	X		X	
P	PO_4^{3-}	X	X	X	
	Pyro P		X		
Si	SiO	X			
	SiO_x	X	X	X	
Zn	ZnO	X	X	X	X
	$Zn(OH)_2$		X		X

was found on phase II and phase IV coupons. Likewise, with the exception of phase IV, these data indicate that solid forms of phosphates may be inhibiting corrosion of galvanized steel when phosphate-based inhibitors are present.

Zinc Orthophosphate. Table 5.81 gives a summary of the compounds detected on four GS coupons exposed to ZOP inhibitor. One GS coupon was tested for each phase. The coupons exposed to ZOP exhibited the same iron corrosion products as were found on GS coupons exposed to BOP and OP inhibitors. Ferric phosphate ($FePO_4$) was identified on two of four coupons. The absence of ferric phosphate on phase II and phase IV coupons is likely due to the nonuniform nature of corrosion in distribution systems. Phosphorus as phosphate was observed on all four coupons. Deconvolution of oxygen demonstrates the presence of CO_3^{2-}, O^{2-}, OH^-, and PO_4^{3-} on all four coupons. All these forms of oxygen correspond to various compounds of calcium, iron, zinc, and silicate. Calcium was deconvoluted as solely $CaCO_3$ for all phases, implying that calcium is predominantly in the form of calcium carbonate when ZOP is utilized within the PDS. Si was observed on the phase I and phase III coupons as a mixture of SiO and SiO_x. Zinc deposits were also present on the scale for all phases, and zinc oxide (ZnO) and zinc hydroxide ($Zn(OH)_2$) were found on all four coupons. Zinc phosphate ($Zn_3(PO_4)_2$) was not detected in the scale in the presence of ZOP, which indicates zinc phosphate precipitation was not a factor for inhibiting iron release by ZOP inhibitor. With the exception of phases II and IV, this data indicates that solid forms of phosphates like ferric phosphate ($FePO_4$) may be inhibiting corrosion of galvanized steel when phosphate-based inhibitors are present.

Silicon. Table 5.82 gives a summary of the compounds detected on four GS coupons exposed to Si inhibitor. One GS coupon was tested for each phase. The same iron oxides were found in the scale on the GS coupons in the presence of the silica inhibitor as were found in the

Table 5.81
Compounds identified for ZOP coupons

Element	Compound	Phase I	Phase II	Phase III	Phase IV
Fe	Fe_2O_3	X	X	X	X
	Fe_3O_4	X	X	X	X
	FeOOH	X	X	X	X
	$FePO_4$		X		X
O	CO_3^{2-}	X	X	X	X
	O^{2-}	X	X	X	X
	OH^-	X	X	X	X
	PO_4^{3-}	X	X	X	X
C	CO_3^{2-}	X	X	X	X
Ca	$CaCO_3$	X	X	X	X
P	PO_4^{3-}	X	X	X	X
	Pyro P				
Si	SiO	X		X	
	SiO_x	X			
Zn	ZnO	X	X	X	X
	$Zn(OH)_2$	X	X	X	X

Table 5.82
Compounds identified for Si coupons

Element	Compound	Phase I	Phase II	Phase III	Phase IV
Fe	Fe_2O_3	X	X	X	X
	Fe_3O_4	X	X	X	X
	FeOOH	X	X	X	X
	$FePO_4$				
O	CO_3^{2-}	X	X	X	X
	O^{2-}	X	X	X	X
	OH–	X	X	X	X
	PO_4^{3-}				
C	CO_3^{2-}	X	X	X	X
Ca	$CaCO_3$	X	X	X	X
P	PO_4^{3-}				
	Pyro P				
Si	SiO	X	X	X	
	SiO_x	X	X	X	X
Zn	ZnO	X	X	X	X
	$Zn(OH)_2$	X	X		

presence of the BOP, OP, and ZOP inhibitors; however, ferric phosphate ($FePO_4$) was not found in the scale on the GS coupons in the presence of the silica inhibitor. Calcium was deconvoluted as solely $CaCO_3$ on all four coupons, implying that calcium is predominantly in the form of calcium carbonate when Si is utilized. Si was observed on all four coupons as a mixture of SiO and SiO_x. There was no evidence to support the presence of Si in SiO_2 form on the coupon. Zinc deposits were also present on the scale for all phases with a majority of the zinc existing as zinc oxide (ZnO). Also, zinc hydroxide ($Zn(OH)_2$) was found on the phase I and phase II coupons.

No phosphate solids were detected in the surface scale of the GS coupons in the presence of the silicate inhibitor, which demonstrates that addition of a phosphate inhibitor was necessary to form ferric phosphate ($FePO_4$) in the scale on the GS coupons. Deconvolution of oxygen further supports the absence of phosphate compounds. The result demonstrates the presence of CO_3^{2-}, O^{2-}, and OH^- on all four coupons. In contrast, silicates were present in all raw waters and were detected on all GS surface scales, which indicates that adequate silica was present in the raw water to form solid silica forms on the GS coupons. Hence, silica could impact metal release in distribution systems.

pH. Tables 5.83 and 5.84 give a summary of the compounds detected on eight GS coupons from the pH control PDSs. One GS coupon was tested for each treatment for each phase. These coupons exhibited the same iron oxide corrosion products as found in the GS scales for the BOP, OP, ZOP, and Si inhibitors. Ferric phosphate ($FePO_4$) was not found in the surface scale of the pH-only GS coupons. Deconvolution of oxygen indicated the presence of CO_3^{2-}, O^{2-}, and OH^- on all eight pH-only GS coupons, which was consistent with the phosphorus scan, where phosphate compounds were not found. All these forms of oxygen correspond to various compounds of calcium, iron, zinc, and silicate. Calcium was deconvoluted as solely $CaCO_3$ on seven of eight coupons, implying that calcium is predominantly in the form of calcium carbonate. It appears that calcium carbonate accounts for most of the surface scale. Si was observed on all four coupons for pH_s and on the phase II and phase III coupons for pH_s+0.3 as a mixture of SiO and SiO_x. Zinc was present for both treatments. Zinc oxide (ZnO) was found on seven of eight coupons. Zinc hydroxide ($Zn(OH)_2$) was found on the phase IV coupon for pH_s and on the phase II coupon for pH_s+0.3.

Summary of XPS Surface Characterization of Galvanized Steel Coupons

Surface characterization of 24 GS coupons found that ferric oxide (Fe_2O_3), magnetite (Fe_3O_4), and hydrated ferric oxide (FeOOH) are the major corrosion products in the surface scale of the GS coupons. Nine of the 12 coupons incubated in phosphate-based inhibitors had ferric phosphate ($FePO_4$); in contrast, none of the other 12 coupons exposed to nonphosphate treatments had $FePO_4$. Therefore, addition of corrosion inhibitor changed the relative distribution of corrosion products. The surface of GS coupons was a mixture of several iron corrosion products, with no predominant iron compound. There was no evidence of correlation between Fe_2O_3, Fe_3O_4, and FeOOH surface compounds and iron release in this work; however, iron oxides were, by far, the greatest solid form of iron in the GS films and are involved in the release of iron from galvanized steel.

Variation of phase water quality did not significantly affect the distribution of solid iron forms in the GS films, which suggests that variations in water quality in this work would not adversely affect iron release from galvanized steel.

Table 5.83
Compounds identified for pH_s coupons

Element	Compound	Phase I	Phase II	Phase III	Phase IV
Fe	Fe_2O_3	X	X	X	
	Fe_3O_4	X	X	X	X
	FeOOH	X	X	X	X
	$FePO_4$				
O	CO_3^{2-}	X	X	X	X
	O^{2-}	X	X	X	X
	OH^-	X	X	X	X
	PO_4^{3-}				
C	CO_3^{2-}	X	X	X	X
Ca	$CaCO_3$	X		X	X
P	PO_4^{3-}				
	Pyro P				
Si	SiO			X	
	SiO_x	X	X		X
Zn	ZnO	X	X	X	X
	$Zn(OH)_2$				X

Table 5.84
Compounds identified for pH_s+0.3 coupons

Element	Compound	Phase I	Phase II	Phase III	Phase IV
Fe	Fe_2O_3	X	X	X	X
	Fe_3O_4	X	X	X	X
	FeOOH	X	X	X	X
	$FePO_4$				
O	CO_3^{2-}	X	X	X	X
	O^{2-}	X	X	X	X
	OH^-	X	X	X	X
	PO_4^{3-}				
C	CO_3^{2-}	X	X	X	X
Ca	$CaCO_3$	X	X	X	X
P	PO_4^{3-}				
	Pyro P				
Si	SiO			X	
	SiO_x		X		
Zn	ZnO		X	X	X
	$Zn(OH)_2$		X		

Phosphorus was detected in 9 of the 12 GS coupons that were exposed to phosphate-based inhibitors, which indicates that solid forms of compounds containing solid phosphorus could be barriers to metal release from galvanized steel. Coupons exposed to Si and pH treatments did not contain any form of phosphorus. Although silica forms were found in all GS scales, the addition of Si-based inhibitors was necessary to reduce iron release.

Zinc deposits were present in the GS scale for 23 of 24 observations. A majority of zinc existed as zinc oxide (ZnO). Zinc hydroxide ($Zn(OH)_2$) was identified in less than half (11/24) of all GS coupons.

The common presence and relative distribution of iron oxides in the surface scale of the GS coupons in all environments shows inhibitor addition does not eliminate the formation of iron oxide in the corrosion process. The presence of phosphate in the surface scale of the GS coupons shows the presence of a solid phosphate form that affects iron release.

Copper Surface Characterization

Thermodynamic Model Development for Copper Release

Copper corrosion scales were identified by XPS to determine any possible controlling solid phases. Thermodynamic modeling of the various blends was accomplished to evaluate the validity of a copper-controlling solid phase. The equilibrium model assumed the presence of Cu(II) species. The equilibrium model used to predict dissolved copper during TBW I (Taylor et al. 2005) was expanded to include phosphate, sulfate, chloride, and ammonia complexes and is presented in Equation 5.59. Presence of a silica ligand was also considered to contribute to the dissolved concentration of copper. However, review of the literature concerning coordination chemistry did not reveal any thermodynamic data corresponding to a copper–silica complex.

$$\begin{aligned} Cu_T = {} & [Cu^{2+}] + [CuOH^+] + [Cu(OH)_2^0] + [Cu(OH)_3^-] + [CuHCO_3^+] + [CuCO_3^0] \\ & + [Cu(CO_3)^{2-}] + [Cu(OH)CO_3^-] + [Cu(OH)_2CO_3^{2-}] + [CuH_2PO_4^+] \\ & + [CuHPO_4^0] + [CuSO_4^0] + [Cu(NH_3)^{2+}] + [Cu(NH_3)_2^{2+}] + [Cu(NH_3)_3^{2+}] \\ & + [Cu(NH_3)_4^{2+}] + [Cu(NH_3)_5^{2+}] + [CuCl^+] + [CuCl^0] \end{aligned} \qquad (5.59)$$

The complexes considered within Equation 5.59 demonstrate the various water quality parameters that theoretically govern copper release. The speciation of dissolved copper suggests that such water quality parameters as alkalinity, chloride, sulfate, ammonia, and phosphate may promote further copper release. From the empirical model defined in the previous section of the report describing copper release, pH, chloride, and alkalinity were established as the main effects attributed to water quality. Increasing the pH was beneficial for controlling copper release, whereas raising chlorides or alkalinity promoted further copper release.

As previously discussed, the empirical model for copper release suggests that chlorides complicate the control of copper. The equilibrium model shown in Equation 5.59 demonstrates the corrosive property of introducing chlorides with the presence of $CuCl^+$ and $CuCl^0$. Although chloride was found to be significant in the empirical model, equilibrium calculations suggest that the thermodynamic model could have ignored the effect of $CuCl^+$ and $CuCl^0$. With $CuCl^+$ and $CuCl^0$ accounting for approximately 1/1,000th of a percent of the dissolved copper, the effect of chloride seems to be poorly described by complexation with copper.

The consideration of alkalinity-dependent complexes defined in the equilibrium model agrees well with the empirical model for copper release. Copper complexes dependent of alkalinity included $CuHCO_3^-$, $CuCO_3^0$, $Cu(CO_3)_2^{2-}$, $Cu(OH)CO_3^-$, and $Cu(OH)_2CO_3^{2-}$. Equilibrium calculations demonstrated the profound effect of alkalinity on copper release. When combined, these complexes accounted for approximately 90% of all soluble copper species.

Equation 5.59 implies that as pH increases, hydroxide and carbonate complexes will increase as well. Development of the empirical model suggests the opposite trend between pH and copper release. Although the equilibrium and empirical models may appear to predict opposite trends with pH, the concentration of all copper complexes is dependent on the activity of Cu^{2+}. Thus, the effect of pH is more likely sensitive in the control of Cu^{2+} associated with the dissolution of the controlling solid rather than altering the distribution of hydroxide and carbonate ligands.

Copper phosphate complexes were considered for describing the expected impact of applying BOP, OP, and ZOP, assuming a common controlling solid. Both $CuH_2PO_4^+$ and $CuHPO_4^0$ were not considered during TBW I because of the relatively low OP concentrations of the untreated (exclusion of phosphate inhibitor) blend. However, with the addition of BOP, OP, and ZOP, the presence of these complexes may represent a significant portion of the dissolved lead species. Consideration of the phosphate ligands proved unnecessary, as equilibrium calculations revealed that the addition of copper phosphate complexes in the equilibrium model only accounted for approximately 0.5% of all soluble copper species. The inclusion of $CuSO_4^0$ also represented a relatively small portion of soluble copper in the equilibrium model (≈0.3%). Empirical modeling found sulfate to be insignificant for the prediction of copper release. Given that disinfection during the study was accomplished by providing a chloramine residual, ammonia was considered as a ligand. However, equilibrium calculations demonstrated a negligible effect on the ability of ammonia to stabilize a complex with copper.

Inhibitor Control of Copper Release

The average copper release observed during the study is shown in Table 5.85 by phase for each inhibitor and relative to the pH_s+0.3 control.

In all cases, inhibitors reduced copper release relative to the control. Phosphate inhibitors reduced copper release to about 40% of that observed with the control. Silicate inhibitors reduced copper release to about 70% of that observed with the control.

Expected Copper Release From Thermodynamic Models

pH+0.3. Table 5.85 compares copper release observed with inhibitor addition to copper release observed with the initial source water. A thermodynamic model was developed to consider copper release as a function of a copper-controlling solid phase. The prediction of copper release was determined from the previously defined thermodynamic model for copper release.

From the XPS analysis, it was determined that cupric hydroxide ($Cu(OH)_2$ was the predominant copper corrosion product present on the copper coupons. Tenorite (CuO), cuprous oxide (Cu_2O), and cupric salts (Cu(II)) were also present on the copper coupons.

Binding energy data from the NIST database (Wagner et al. 2003) support the inclusion of cupric chloride ($CuCl_2$), copper sulfate ($CuSO_4$), copper carbonate ($CuCO_3$), and a cupric silicate ($CuSiO_3$) under the Cu(II) curve given by Xiao (2004). In addition $Cu_3(PO_4)_2$ (cupric phosphate), $Cu_2(OH)_2CO_3$ (malachite), and $Cu_4SO_4(OH){\cdot}6H_2O$ (a cupric hydroxysulfate) are all feasible

Table 5.85
Comparison of observed copper release from inhibitor PDSs to pH_s+0.3

Phase	Inhibitor	Average release (mg/L)		% Ratio to pH_s+0.3	
		Dissolved Cu	Total Cu	Dissolved Cu	Total Cu
I	pH_s+0.3	0.89	1.00		
	BOP	0.34	0.36	38	36
	OP	0.38	0.41	42	41
	ZOP	0.34	0.36	38	36
	Si	0.61	0.68	68	68
II	pH_s+0.3	0.73	0.82		
	BOP	0.29	0.33	40	40
	OP	0.28	0.35	39	43
	ZOP	0.32	0.39	44	48
	Si	0.51	0.60	69	73
III	pH_s+0.3	0.79	0.85		
	BOP	0.30	0.32	38	38
	OP	0.29	0.35	37	41
	ZOP	0.30	0.34	37	40
	Si	0.58	0.65	74	77
IV	pH_s+0.3	0.83	0.93		
	BOP	0.31	0.37	38	39
	OP	0.32	0.36	38	39
	ZOP	0.32	0.38	38	41
	Si	0.51	0.58	62	63

cupric solids (Schock, Lytle, and Clement 1995); however, no published XPS binding energies were identified for these three compounds.

$Cu(OH)_2$ has been identified as a metastable intermediate that forms in young copper pipe, and has been used as the basis for thermodynamic models of copper (Schock, Lytle, and Clement 1995; Xiao 2004). Cupric hydroxide can age to its dehydrated form, tenorite (CuO), which is less soluble. In the presence of alkalinity, old copper pipe tends to form the more stable, less soluble cupric hydroxycarbonate $Cu_2(OH)_2CO_3$, malachite (Schock, Lytle, and Clement, 1995).

Both cupric hydroxide and tenorite were considered as the controlling solid phase during thermodynamic modeling. Table 5.86 shows that the cupric hydroxide thermodynamic model underpredicted the dissolved copper concentrations by about 0.3 mg/L Cu. Similarly, the tenorite thermodynamic model underpredicted the dissolved copper concentrations by about 0.7 mg/L Cu. Equation 5.60 describes the equilibrium of dissolved cupric ion with cupric hydroxide solids. A log equilibrium constant of 8.89 was used to predict copper dissolution from cupric hydroxide scales. This same value has been used in other studies modeling cupric hydroxide as a controlling solid phase for copper release (Shock, Lytle, and Clement 1995; Xiao 2004).

$$Cu(OH)_{2(s)} + 2H^{+} \leftrightarrow Cu^{2+} + H_2O \qquad (5.60)$$

Table 5.86
Copper thermodynamic modeling of pH_s+0.3

Inhibitor	Phase	Actual copper release (mg/L)		Modeled copper release (mg/L)	
		Dissolved Cu	Total Cu	$Cu(OH)_2$	CuO
	I	0.89	1.00	0.59	0.07
	II	0.73	0.82	0.46	0.06
pH_s+0.3	III	0.79	0.85	0.48	0.06
	IV	0.83	0.93	0.54	0.07
	All	0.81	0.90	0.54	0.07

The dissolved copper samples were collected 6 hours after flushing of the copper loops, and 6 hours is less than the time required for copper concentrations to reach equilibrium values (Schock, Lytle, and Clement 1995). Therefore, one would expect the dissolved copper measurements to be less than the predicted equilibrium concentration, not greater than it. In both this study and TBW I, which were conducted using the same PDS and copper loops, the cupric hydroxide model with log equilibrium constant 8.89 consistently underpredicted dissolved copper release (Xiao 2004). The variability of equilibrium constants for copper scales, including copper hydroxide, was discussed by Schock, Lytle, and Clement (1995). Using earlier work by Schindler et al. (Schindler et al. 1965, Schindler 1967), Schock et al. found that the range of log equilibrium constants for cupric hydroxide varied from 8.67 to 8.89, when corrected for ionic strength. Hidmi and Edwards (1999) reported a log equilibrium constant of 9.36 ± 0.02 for freshly precipitated $Cu(OH)_2$. The observed copper concentrations presented in these results are consistent with what would be expected using a greater log equilibrium constant. The log equilibrium constant for $Cu(OH)_2$ has been observed to decrease with aging of the cupric hydroxide solids (Hidmi and Edwards 1999). Other reported values for the $Cu(OH)_2$ log equilibrium constant include 8.3, 8.7, 9.2, and 9.4 (King 1959, Hahn and Welcher 1968, Hogness and Johnson 1957, Schenk 1996). A unit variation in log equilibrium constant would result in a log-order variation in predicted copper release.

Blended Ortho-polyphosphate. The XPS analysis for the five copper coupons exposed to BOP during four phases of operation indicated the presence of $Cu(OH)_2$, CuO, Cu_2O, and Cu(II) salts for phases I, II, and III. Only $Cu(OH)_2$ and Cu(II) salts were detected on the phase IV coupon. A duplicate copper coupon in phase III suggested the formation of a calcium film on the copper surface with no copper detected, but rather $CaCO_3$, hydroxides, and OP. Phosphorus was detected on four of five copper coupons in both ortho- and poly- forms. OP was detected on two of five coupons, pyrophosphate was detected on three of five coupons, and metaphosphate was detected on one of five coupons.

The dramatic reductions in copper release with the addition of phosphate inhibitor, as shown in Table 5.85, suggest that a less soluble copper phosphate solid may form; however, very little information on solid forms of copper OP solids is available (Schock, Lytle, and Clement 1995). A review of the literature confirmed the lack of information identifying copper OP solids in water distribution systems. Thermodynamic data was available for the two forms, $Cu_3(PO_4)_2 \cdot 2H_2O$ and $Cu_3(PO_4)_2$; hence, these were used to model cupric phosphate solubility (Shock, Lytle, and Clement 1995). A plot of solubility for these four copper solids, without

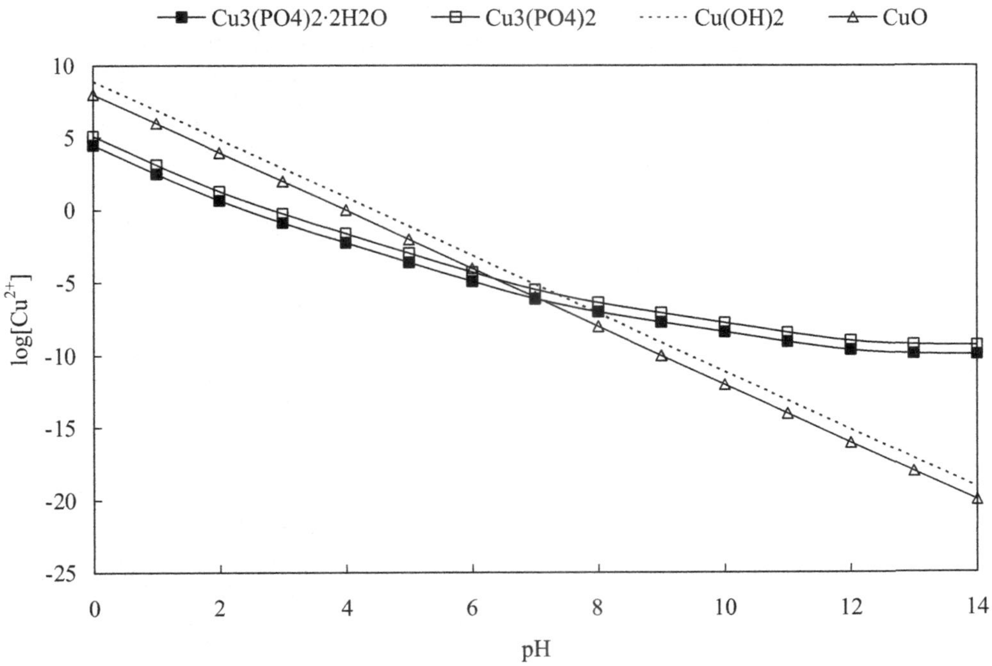

Figure 5.83 Solubility of copper solids, without complexation, showing relative predominance with phosphate inhibitor addition

consideration of copper complexation, is shown in Figure 5.83. The copper phosphate solids are plotted assuming 1.0 mg/L as P of phosphoric acid.

For the range of pH values considered in this study with phosphate inhibitors (7.7–8.0), tenorite (CuO), is thermodynamically favored; however, tenorite usually forms by aging of cupric hydroxide ($Cu(OH)_2$) solids. Tenorite was detected by XPS on copper coupons for all inhibitors but was not as abundant as cupric hydroxide. This indicates that cupric solids are slow to reach equilibrium form as CuO. Therefore, cupric hydroxide solids can persist as a metastable solid (Xiao 2004). Near pH 8, cupric phosphate dihydride ($Cu_3(PO_4)_2{\cdot}2H_2O$) and cupric hydroxide ($Cu(OH)_2$) have similar predicted copper solubility, indicating that neither compound is thermodynamically favored over the other.

If the effective log equilibrium constant for cupric hydroxide were elevated to be consistent with observed dissolved copper release for pH_s+0.3, the cupric phosphate dihydride model would explain the decrease in copper release with BOP addition. No data on the XPS binding energies for cupric phosphate dihydride were located in the literature; however, they may be represented as part of the Cu(II) salts curve.

Cupric hydroxide, tenorite, and the two forms of cupric phosphate were considered separately as controlling solid phases for copper release. Table 5.87 shows the actual copper release and modeled copper release for the four controlling solid forms. Predicted copper release from cupric hydroxide and the cupric phosphate dihydride provided the most reasonable prediction of copper release. However, as mentioned previously, OP was not consistently detected by XPS on the copper coupons receiving BOP inhibitor. Therefore, it is not clear whether a copper OP solid

Table 5.87
Copper thermodynamic modeling of BOP

Inhibitor	Phase	Actual copper release (mg/L)		Modeled copper release (mg/L)			
		Dissolved Cu	Total Cu	$Cu(OH)_2$	CuO	$Cu_3(PO_4)_2{\cdot}2H_2O$	$Cu_3(PO_4)_2$
BOP	I	0.34	0.36	0.57	0.07	0.66	2.84
	II	0.29	0.33	0.50	0.06	0.40	1.74
	III	0.30	0.32	0.50	0.06	0.64	2.74
	IV	0.31	0.37	0.61	0.08	0.47	2.03
	All	0.31	0.34	0.58	0.07	0.56	2.42

Table 5.88
Copper thermodynamic modeling of OP

Inhibitor	Phase	Actual copper release (mg/L)		Modeled copper release (mg/L)			
		Dissolved Cu	Total Cu	$Cu(OH)_2$	CuO	$Cu_3(PO_4)_2{\cdot}2H_2O$	$Cu_3(PO_4)_2$
OP	I	0.38	0.41	0.56	0.07	0.67	2.88
	II	0.28	0.35	0.52	0.06	0.41	1.75
	III	0.29	0.35	0.53	0.07	0.63	2.70
	IV	0.32	0.36	0.64	0.08	0.47	2.01
	All	0.32	0.37	0.60	0.07	0.56	2.42

is responsible for the reductions in copper release with BOP addition. There was no evidence in the XPS data that the addition of phosphate inhibitors increased the relative abundance of tenorite.

Orthophosphate. Copper corrosion products identified by the XPS analysis for the four copper coupons exposed to OP found $Cu(OH)_2$, CuO, Cu_2O, and Cu(II) copper scales. However, $Cu(OH)_2$ and CuO were the most predominant scales. OP was found on three of four copper coupons. Calcium phosphate was not detected on any of these coupons, although calcium carbonate was detected on two of the three coupons with OP present. As explained in the previous section describing thermodynamic modeling of copper coupons exposed to BOP, the formation of tenorite is thermodynamically favored, but this solid is not as abundant as cupric hydroxide because of kinetic limitations. Considering the other solids, cupric hydroxide and cupric phosphate dihydride are equally favorable thermodynamically.

Equilibrium calculations are provided in Table 5.88, considering cupric hydroxide, tenorite, cupric phosphate dihydride, and cupric phosphate as the possible controlling solid phases. From Table 5.85 there is a clear improvement with the control of copper release with the addition of OP when compared to pH_s+0.3. As with BOP inhibitor, both cupric hydroxide and cupric phosphate dihydride solids overpredict the observed dissolved copper release. If the effective log equilibrium constant for cupric hydroxide were elevated to be consistent with observed dissolved copper release for pH_s+0.3, the cupric phosphate dihydride model would explain the decrease in copper release with OP addition.

Table 5.89
Copper thermodynamic modeling of ZOP

Inhibitor	Phase	Actual copper release (mg/L)		Modeled copper release (mg/L)			
		Dissolved Cu	Total Cu	$Cu(OH)_2$	CuO	$Cu_3(PO_4)_2{\cdot}2H_2O$	$Cu_3(PO_4)_2$
ZOP	I	0.34	0.36	0.64	0.08	0.63	2.70
	II	0.32	0.39	0.68	0.08	0.37	1.60
	III	0.30	0.34	0.63	0.08	0.59	2.52
	IV	0.32	0.38	0.72	0.09	0.43	1.86
	All	0.32	0.37	0.72	0.09	0.52	2.24

Zinc Orthophosphate. Copper corrosion products identified by the XPS analysis for the four copper coupons exposed to ZOP include $Cu(OH)_2$, CuO, Cu_2O, and Cu(II) copper scales. However, $Cu(OH)_2$ and CuO were the most predominant scales. Cu(II) salts were not detected in phase III. OP was found on three of four copper coupons. Calcium phosphate was not detected on any of the copper coupons exposed to ZOP, although calcium carbonate was detected on two copper coupons. As explained in the section describing thermodynamic modeling of copper coupons exposed to BOP, the formation of tenorite is thermodynamically favored, but the formation of this solid is kinetically limited by transformation from cupric hydroxide. Considering the other solids, cupric hydroxide and cupric phosphate dihydride are equally favorable thermodynamically.

Equilibrium calculations are provided in Table 5.89, considering cupric hydroxide, tenorite, cupric phosphate dihydride, and cupric phosphate as the possible controlling solid phases. From Table 5.85 there is a clear improvement with the control of copper release with the addition of ZOP when compared to pH_s+0.3. As with BOP and OP inhibitors, both cupric hydroxide and cupric phosphate dihydride solids overpredict the observed dissolved copper release. If the effective log equilibrium constant for cupric hydroxide were elevated to be consistent with observed dissolved copper release for pH_s+0.3, the cupric phosphate dihydride model would explain the decrease in copper release with ZOP addition.

Silicon. The XPS analysis for the five copper coupons exposed to silicate inhibitor during four phases of operation indicated the presence of $Cu(OH)_2$, CuO, Cu_2O, and Cu(II) salts for all four phases. An additional copper coupon was placed in PDS12 during phase III. Of the five copper coupons analyzed by XPS, three of the coupons contained amorphous silica.

Following precipitation of silica, copper ions may be incorporated into the lattice of the amorphous structure, precipitated on the copper surface. Copper silicate hydride ($CuSiO_3{\cdot}H_2O$) was modeled to represent a possible bond between copper and silica. Cupric hydroxide and tenorite were also considered as possible controlling solids for copper release in the presence of silicate inhibitor.

A plot of solubility for $Cu(OH)_2$, CuO, and $CuSiO_3{\cdot}H_2O$ without consideration of copper complexation is shown in Figure 5.84. The copper silicate solid is plotted, assuming 6.0 mg/L as SiO_2 of silicate inhibitor.

At the pH of the lines receiving silicate inhibitor (8.2), tenorite (CuO), is thermodynamically favored; however, as mentioned earlier under the thermodynamic discussion in the BOP section, tenorite forms slowly by the aging of cupric hydroxide. Therefore, significant amounts of

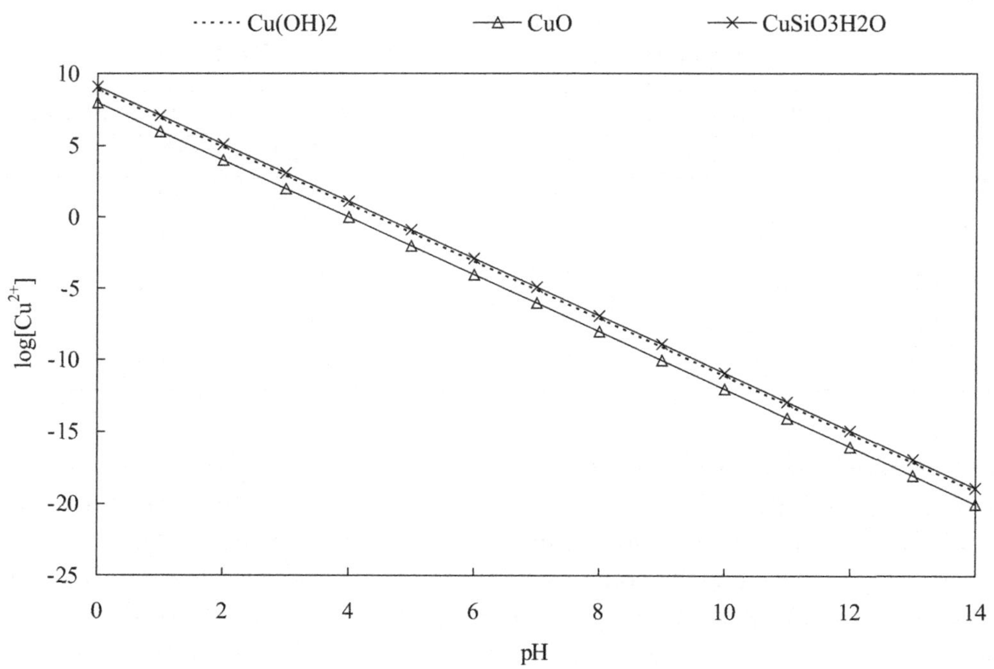

Figure 5.84 Solubility of copper solids, without complexation, showing relative predominance with silicate inhibitor addition

metastable cupric hydroxides remain on the copper surface. According to Figure 5.84, cupric silicate hydride ($CuSiO_3 \cdot H_2O$) is nearly as thermodynamically favored as $Cu(OH)_2$. Copper silicates were not identified during XPS deconvolution; however, they may be represented as part of the Cu(II) salts curve.

Cupric hydroxide, tenorite, and the cupric silicate hydride were considered separately as the controlling solid phases for copper release. Table 5.90 shows the actual copper release and modeled copper release for the three controlling solid forms. Predicted copper release from cupric hydroxide and the cupric silicate hydride provided the most reasonable prediction of copper release, with both solids slightly underpredicting dissolved copper release. Cupric silicate hydride more accurately predicted dissolved copper concentrations; however, the predicted solubility for the cupric hydroxide and the cupric silicate hydride model are very close.

If the effective log equilibrium constant for cupric hydroxide were elevated to be consistent with observed dissolved copper release for pH_s+0.3, the data in Table 5.90 would suggest that $CuSiO_3 \cdot H_2O$ acts as a controlling solid for copper release in the presence of silicate inhibitor.

Copper Scale Elemental Analysis

An example copper survey spectrum is shown in Figure 5.85. The elements found in the scale on the copper coupons exposed to all methods of corrosion inhibitor and controls are shown in Table 5.91. Peaks were dependent on the inhibitor to which the coupon had been exposed and were typically pronounced for copper, carbon, oxygen, calcium, phosphorus, silica, and zinc.

Table 5.90
Copper thermodynamic modeling of Si

Inhibitor	Phase	Actual copper release (mg/L)		Modeled copper release (mg/L)		
		Dissolved Cu	Total Cu	$Cu(OH)_2$	CuO	$CuSiO_3{\cdot}H_2O$
	I	0.61	0.68	0.32	0.04	0.51
	II	0.51	0.60	0.25	0.03	0.40
Si	III	0.58	0.65	0.31	0.04	0.51
	IV	0.51	0.58	0.29	0.04	0.47
	All	0.55	0.63	0.31	0.04	0.50

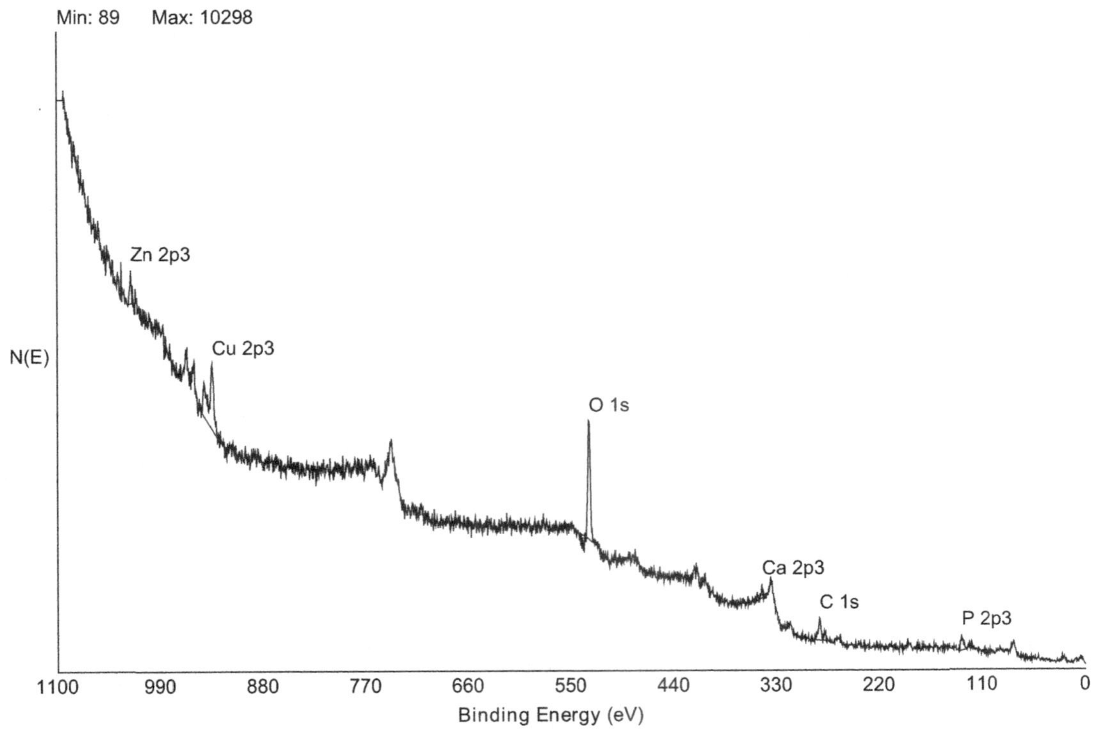

Figure 5.85 Survey spectrum of a copper coupon from 1.0 ZOP, phase III

Identification of Copper Corrosion Products

High-resolution spectra for copper were deconvoluted (Figure 5.86) to identify copper corrosion products and to determine their relative abundance. Copper compounds detected included Cu_2O, CuO, $Cu(OH)_2$; and cupric salts Cu(II) (Xiao 2004). Binding energy data from the NIST database (Wagner et al. 2003) support the inclusion of $CuCl_2$, $CuSO_4$, $CuCO_3$, and $CuSiO_3$ under the Cu(II) curve given by Xiao. In addition, $Cu_3(PO_4)_2$, $Cu_2(OH)_2CO_3$ (malachite), and $Cu_4SO_4(OH)_6H_2O$ (a cupric hydroxysulfate) are all feasible forms of Cu(II) (Schock, Lytle, and Clement 1995); however, no published XPS binding energies were identified for these compounds. The surface composition of copper coupons is not uniform. Failure to detect a compound by XPS in one scan does not imply that the compound was not in the scale.

Table 5.91
Number of coupons containing detected elements from XPS scan of the 26 copper coupons

Detected element	No. of coupons (from 26 total)
Carbon (C)	26
Calcium (Ca)	12
Copper (Cu)	26
Oxygen (O)	26
Phosphorus (P)	11
Silicon (Si)	4
Zinc (Zn)	5

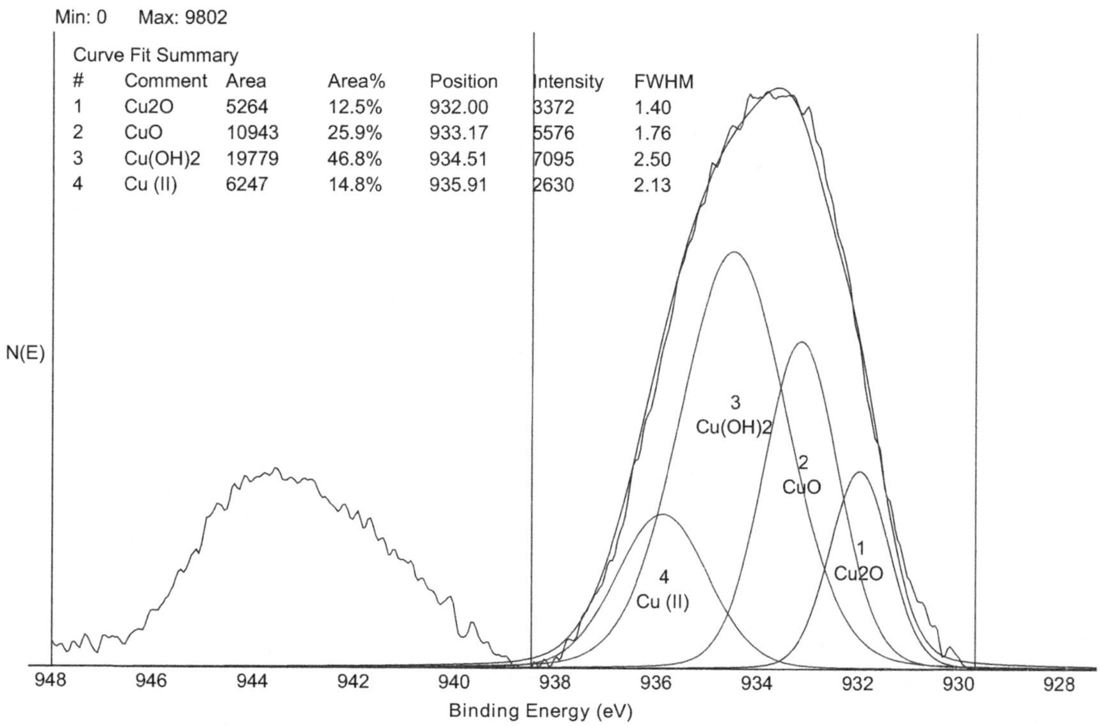

Figure 5.86 Deconvolution of copper for the coupon exposed to BOP during phase IV

A combination of these four compounds was sufficient to describe the high-resolution spectra for all copper coupons. Figure 5.87 depicts the percent area of the various compounds over all phases by inhibitor treatment. Cupric hydroxide was the most abundant copper surface compound. This may imply that $Cu(OH)_2$ was the controlling solid phase for copper solubility (Schock 1999, Xiao 2004). However, as shown in the thermodynamic models for copper, cupric phosphate dihydride ($Cu_3(PO_4)_2{\cdot}2H_2O$) and cupric silicate hydride ($CuSiO_3{\cdot}H_2O$), models provided reasonable predictions of dissolved copper release from copper loops receiving phosphate and silicate inhibitors. $Cu(OH)_2$ commonly forms in young copper pipe. Tenorite is more thermodynamically stable than cupric hydroxide; however, the rate of transformation of cupric hydroxide to tenorite is slow, resulting in cupric hydroxide being present as a metastable intermediate. Older

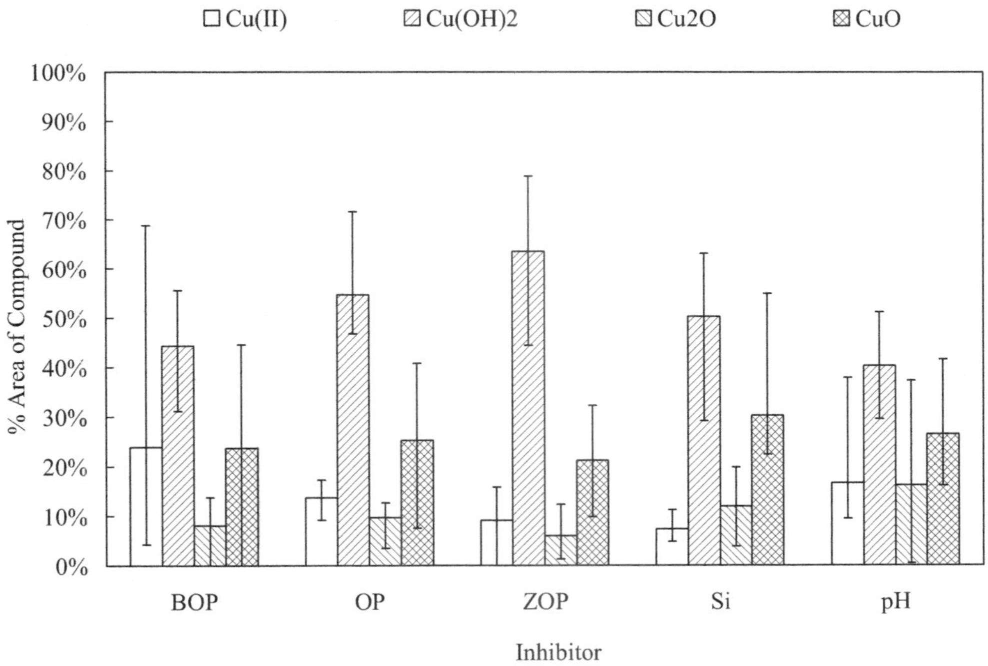

Figure 5.87 Distribution of copper compounds for all phases

copper pipe tends to form the more stable, less soluble cupric hydroxycarbonate $Cu_2(OH)_2CO_3$, malachite (Schock, Lytle, and Clement 1995); nonetheless, increased levels of alkalinity can aggravate copper release (Taylor et al. 2005), even in older copper pipe (Schock and Fox 2001).

Effect of Phase on Copper Corrosion Scale Products

The relationship between phase change and percent area of the copper compounds was evaluated for each copper compound. An interaction plot for cupric hydroxide is shown in Figure 5.88. Similar plots were constructed for CuO, Cu_2O, and Cu(II).

Cupric hydroxide varied by phase and treatment, as shown in Figure 5.88. Among the phosphorus inhibitors, $Cu(OH)_2$ was greatest in phase II, decreasing in phase III, phase I, and phase IV, respectively. In the silicate line, $Cu(OH)_2$ was greatest in phases II and III, decreasing in phases IV and I, respectively. For the pH lines, $Cu(OH)_2$ was greatest in phases III and IV, decreasing in phases II and I, respectively. Also, for the pH lines in phases II, III, and IV, the percent $Cu(OH)_2$ is greater for pH_s+0.3, which is reasonable, given that $Cu(OH)_2$ solids become less soluble as pH increases.

There appeared to be some interaction between phase water quality and inhibitor type. For phases I and II, $Cu(OH)_2$ was more predominant in the phosphate lines. However, for phase IV, $Cu(OH)_2$ was equally predominant for all treatments except BOP. $Cu(OH)_2$ surface films have been identified as important to copper release, especially in new pipe (Schock 1999, Xiao 2004).

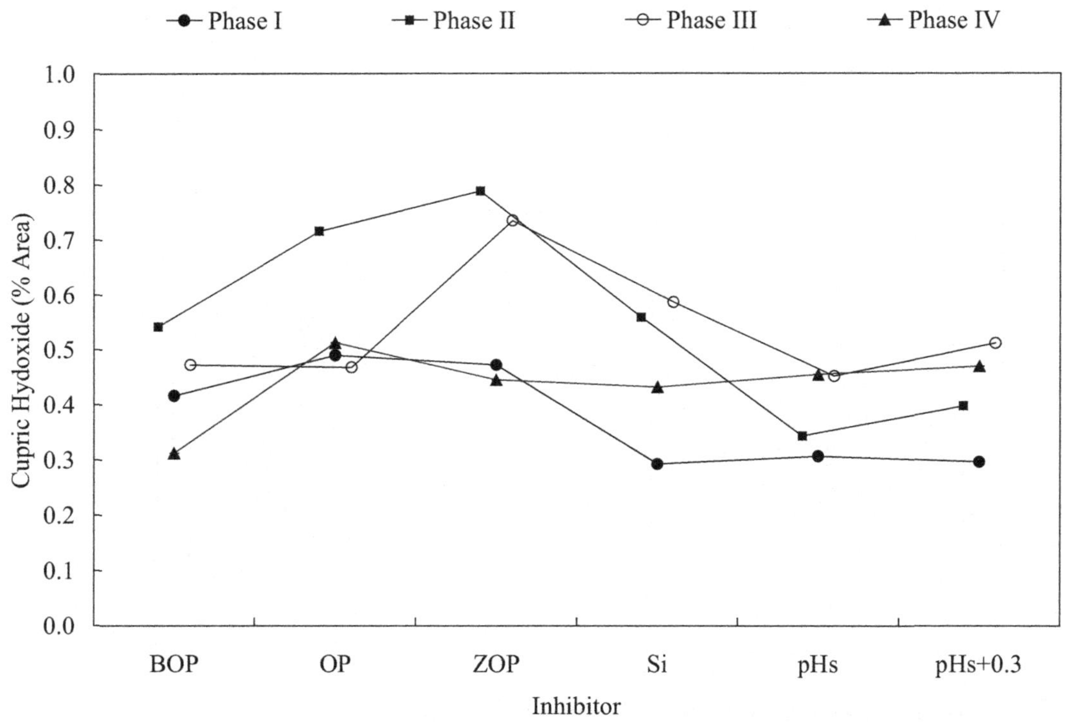

Figure 5.88 Percent area of cupric hydroxide versus treatments for all phases

Effect of Inhibitor on Copper Corrosion Scale Products

Blended Ortho-polyphosphate. Five copper coupons exposed to BOP were analyzed for the elements shown in Table 5.92. One copper coupon was tested for each phase, with a duplicate copper coupon for phase III. Species identified on the phase III duplicate copper coupon are indicated by an X enclosed in parentheses. Carbon was detected on all five coupons, appearing as carbonate in each case.

Calcium was detected on four of five coupons, appearing as calcium carbonate on three of five coupons, and calcium phosphate on one of five coupons. Phase II had the lowest alkalinity of all the phases, which may account for the absence of calcium carbonate on the coupon. However, pH was equal to or greater than pH_s in all PDSs, indicating that calcium carbonate precipitation would be expected. Calcium carbonate was detected on three of the five coupons, and calcium phosphate was detected on one of the five coupons. Phase II had the highest sulfates and lowest alkalinity of all four phases, which may account for the absence of $CaCO_3$ from the phase II BOP copper coupon. There was a difference in the calcium compound detected on the phase III coupon, $Ca_3(PO_4)_2$, and the phase III duplicate coupon, $CaCO_3$.

This discrepancy might be due to nonuniform composition of the surface scales. Also, for the phase III coupon, the presence of pyrophosphates and metaphosphates might have inhibited the precipitation of a $CaCO_3$ film. This effect of polyphosphates has been demonstrated in the literature (Hatch and Rice 1945).

Table 5.92
Compounds identified in scale on copper coupons in the presence of BOP inhibitor

Element	Compound	Phase I	Phase II	Phase III (and duplicate)	Phase IV
C	CO_3^{2-}	X	X	X(X)	X
Ca	$CaCO_3$	X		(X)	X
	$Ca_3(PO_4)_2$			X	
	CaO				
Cu	Cu_2O	X	X	X	
	CuO	X	X	X	
	$Cu(OH)_2$	X	X	X	X
	Cu(II)	X	X	X	X
O	O^{2-}	X	X	(X)	X
	CO_3^{2-}	X	X	X(X)	X
	OH^-	X	X	X(X)	X
	PO_4^{3-}	X		(X)	
P	PO_4^{3-}	X		(X)	
	Meta			X	
	Pyro	X	X	X	
Si	SiO_x				X
	SiO_2				X
Zn					

Copper was detected on four of five copper coupons. Forms of copper detected included Cu_2O, CuO, $Cu(OH)_2$, and Cu(II). No copper was detected in the phase III duplicate; however, $CaCO_3$, and OP were detected on the duplicate coupon, indicating that a precipitate may have obscured the metal surface for this coupon. This result is not unreasonable, given the nonuniform nature of surface scales. For the phase IV coupon, only $Cu(OH)_2$ and Cu(II) salts were detected on the surface, perhaps indicating that water quality conditions from phase IV did not correspond with the stability of Cu_2O surface scales; but rather promoted more complete oxidation to cupric compounds. Cu(II) was detected in all coupons except the phase III duplicate coupon. Carbonate ion was detected in phase II for carbon and oxygen; however, no calcium was detected on the phase II coupon. This indicates that copper existed in the form of a copper carbonate compound, such as $CuCO_3$ or, more likely, $Cu_2(OH)_2CO_3$. Similarly, phosphate was present in phases I and III (duplicate coupon), and it was not associated with $Ca_3(PO_4)_2$ for those coupons. This suggests that PO_4^{3-} may be present as $Cu_3(PO_4)_2$. However, it is also possible that phosphate was adsorbed to the scale surface as a liquid–solid film.

Oxygen was detected and appeared as carbonate and hydroxides on all five coupons. Metal oxide, O^{2-}, was detected on four of five coupons except the phase III coupon. Oxygen in OP was detected on two coupons, from phase I and from phase III, which both had high-alkalinity blends run at different seasons of the year.

Table 5.93
Compounds identified in scale on copper coupons in the presence of OP inhibitor

Element	Compound	Phase I	Phase II	Phase III	Phase IV
C	CO_3^{2-}	X	X	X	X
Ca	$CaCO_3$	X		X	
	$Ca_3(PO_4)_2$				
	CaO				
Cu	Cu_2O	X	X	X	X
	CuO	X	X	X	X
	$Cu(OH)_2$	X	X	X	X
	Cu(II)		X	X	X
O	O^{2-}	X	X	X	X
	CO_3^{2-}	X	X	X	X
	OH^-	X	X	X	X
	PO_4^{3-}	X	X	X	
P	PO_4^{3-}	X	X	X	
	Meta				
	Pyro				
Si					
Zn	ZnO			X	

For BOP, phosphorus was detected in various forms in the scale on four of the five copper coupons. OP was only detected on the phase I coupon and the phase III duplicate coupon. Pyrophosphate was detected for phases I, II, and III, but not for phase IV. Metaphosphate (polyphosphate) was also detected for phase III. Because OP was not detected on the phase III coupon, it is possible that the $Ca_3(PO_4)_2$ solid detected in calcium may actually represent a calcium pyrophosphate solid. As mentioned in the scale modeling section, phosphate may also be present as a copper phosphate solid. Although the actual form of such a solid in distribution systems has not been established in the literature, the thermodynamic model for $Cu_3(PO_4)_2{\cdot}2H_2O$ gave reasonable predictions for dissolved copper release for all phosphate inhibitors.

For BOP, silica was detected in the scale on only one of the five copper coupons, in the form of SiO_2 and SiO_x. However, the XPS scan indicated that SiO_2 and SiO_x were present in only trace amounts. No silica was detected on any other coupon for BOP inhibitor. No zinc was detected on any of the five copper coupons for BOP.

Orthophosphate. Four copper coupons exposed to OP were analyzed for the elements shown in Table 5.93. One copper coupon was tested for each phase. Carbon was detected in the copper scale on all four coupons, always appearing as carbonate. Calcium was detected on two of four coupons, appearing as $CaCO_3$ in phases I and III. This may indicate that in the presence of a similar predominantly GW blend receiving OP inhibitor, conditions may favor the formation of $CaCO_3$ films. No other calcium compounds were detected on copper coupons in the presence of OP inhibitor.

Copper was detected on all four coupons, appearing as a mixture of Cu_2O, CuO, $Cu(OH)_2$, and Cu(II) salts, except for phase I. No Cu(II) salts were detected on the copper coupon for phase I. In phases II, III, and IV, Cu(II) may be present as a copper carbonate compound, such as $CuCO_3$ or $Cu_2(OH)_2CO_3$, given that carbonate was detected for all these phases but calcium was detected only in phase III. Similarly, though it is not precipitated with calcium, phosphate may be present as $Cu_2(PO_4)_3$. However, it is also possible that phosphate was adsorbed to the scale surface as a liquid–solid film. Oxygen was detected on all four coupons, appearing as a mixture of O^{2-}, CO_3^{2-}, OH^-, and PO_4^{3-}, except for phase IV. No PO_4^{3-} was detected on the copper coupon for phase IV. This result is consistent with observations for BOP, where no PO_4^{3-} was detected on the copper coupon in phase IV.

Phosphorus was detected on three of four coupons, appearing as PO_4^{3-} in phases I, II, and III only. No phosphorus was detected on the phase IV coupon, which is consistent with the results for oxygen. As mentioned in the scale modeling section, phosphate may also be present as a copper phosphate solid. Although the actual form of such a solid in distribution systems has not been established in the literature, the thermodynamic model for $Cu_3(PO_4)_2 \cdot 2H_2O$ gave reasonable predictions for dissolved copper release for all phosphate inhibitors. No silica was detected on any of the four coupons. Zinc was detected on one of the four coupons, being found in the form of ZnO for phase III. This zinc might have come from the GS pipe upstream of the cradles.

Zinc Orthophosphate. Four copper coupons exposed to ZOP were analyzed for the elements shown in Table 5.94. One copper coupon was tested for each phase. Carbon was detected on all four coupons, always appearing as carbonate. Calcium was detected on three of four coupons, appearing as $CaCO_3$ in phases I and IV, while appearing as CaO in phase III. No calcium was detected on the phase II copper coupon. When compared to BOP and OP, these results indicate that the presence of zinc in ZOP may alter the conditions that will lead to the formation of $CaCO_3$ films. Copper was detected on all four coupons, appearing as a mixture of Cu_2O, CuO, $Cu(OH)_2$, and Cu(II) salts, except for phase III. No Cu(II) salts were detected on the copper coupon for phase III.

Oxygen was detected on all four coupons, appearing as a mixture of O^{2-}, CO_3^{2-}, OH^-, and PO_4^{3-}, except for phase III. No PO_4^{3-} was detected on the copper coupon for phase III. The absence of both Cu(II) and PO_4^{3-} from the phase III coupon for ZOP suggests that one form of the Cu(II) salts may be $Cu_3(PO_4)_2$. Cupric phosphate films have been shown to be favored over cupric hydroxide and malachite films at pH < 7.5 (Schock, Lytle, and Clement 1995). Similarly, as mentioned for BOP and OP inhibitors, no $Ca_3(PO_4)_2$ compound is present to account for the phosphate; hence, it seems reasonable that the phosphate was either in a copper compound or adsorbed to the copper scale in liquid–solid film. OP was found on the phase IV coupon for ZOP. This is notable given that no OP was detected on the phase IV copper coupons for BOP and OP. Thus, the presence of zinc may aid in the precipitation of phosphates in water quality conditions similar to that of phase IV.

Phosphorus was detected on three of four coupons, appearing as PO_4^{3-} in phases I, II, and IV only. No phosphorus was detected on the phase III coupon, which is consistent with the results for oxygen. As mentioned in the scale modeling section, phosphate may also be present as a copper phosphate solid. Although the actual form of such a solid in distribution systems has not been established in the literature, the thermodynamic model for $Cu_3(PO_4)_2 \cdot 2H_2O$ gave reasonable predictions for dissolved copper release for all phosphate inhibitors. No silica was detected on any of the four coupons. Zinc was detected on all four coupons, being found in the form of zinc oxide for all phases.

Table 5.94
Compounds identified in scale on copper coupons in the presence of ZOP inhibitor

Element	Compound	Phase I	Phase II	Phase III	Phase IV
C	CO_3^{2-}	X	X	X	X
Ca	$CaCO_3$	X			X
	$Ca_3(PO_4)_2$				
	CaO			X	
Cu	Cu_2O	X	X	X	X
	CuO	X	X	X	X
	$Cu(OH)_2$	X	X	X	X
	Cu(II)	X	X		X
O	O^{2-}	X	X	X	X
	CO_3^{2-}	X	X	X	X
	OH^-	X	X	X	X
	PO_4^{3-}	X	X		X
P	PO_4^{3-}	X	X		X
	Meta				
	Pyro				
Si					
Zn	ZnO	X	X	X	X
	Zn				

Silicon. Five copper coupons exposed to Si inhibitor were analyzed for the elements shown in Table 5.95. One copper coupon was tested for phases I, II, and IV. Two coupons were tested for phase III, with one coupon receiving a dose of 6 mg/L as SiO_2 inhibitor (PDS11) and the other receiving a dose of 12 mg/L as SiO_2 inhibitor (PDS12).

Carbon was detected on all five coupons, always appearing as carbonate. Calcium was detected on two of five coupons, appearing as $CaCO_3$ in phase IV and CaO in phase III for the higher dose of 12 mg/L as SiO_2. No calcium was detected on either the phase I or phase III coupons. Thus, in the presence of Si inhibitor, $CaCO_3$ films were only detected for phase IV water quality conditions. The absence of a compound in an XPS scan does not imply the absence of the compound from the entire surface. One must keep in mind the nonuniform dispersion of deposits across the coupon surface.

Copper was detected on all five coupons, appearing as a mixture of Cu_2O, CuO, $Cu(OH)_2$, and Cu(II) salts for all phases. Cu(II) salts may be associated with carbonates (detected for five of five coupons) and cupric silicate (detected on three of five coupons). It has been suggested that $CuSiO_3$ can form from malachite exposed to SiO_2 inhibitor (Schock et al. 2005).

Oxygen was detected on all five coupons, appearing as a mixture of O^{2-}, CO_3^{2-}, and OH^-. No PO_4^{3-} was detected on any of the five coupons for Si inhibitor, indicating that phosphate inhibitor addition is the primary source for phosphorus on the copper surfaces for the given water qualities studied.

Table 5.95
Compounds identified in scale on copper coupons in the presence of Si inhibitor

Element	Compound	Phase I	Phase II	Phase III (PDS12)	Phase IV
C	CO_3^{2-}	X	X	X(X)	X
Ca	$CaCO_3$				X
	$Ca_3(PO_4)_2$				
	CaO			(X)	
Cu	Cu_2O	X	X	X(X)	X
	CuO	X	X	X(X)	X
	$Cu(OH)_2$	X	X	X(X)	X
	Cu(II)	X	X	X(X)	X
O	O^{2-}	X	X	X(X)	X
	CO_3^{2-}	X	X	X(X)	X
	OH^-	X	X	X(X)	X
	PO_4^{3-}				
P	PO_4^{3-}				
	Meta				
	Pyro				
Si	SiO_x	X		(X)	X
Zn					

Phosphorus was not detected on any of the five coupons, consistent with the results from oxygen. Even though OP was not present, some Cu(II) salts were still present in the Si inhibitor lines, comprising about 8% of all copper corrosion products, as shown in Figure 5.87. Average relative percent areas of copper for Cu(II) were approximately 23%, 13%, and 10% for BOP, OP, and ZOP inhibitors, respectively. Similarly, the average relative percent area for the pH lines was approximately 15%. The results from the phosphate inhibitors suggest that increasing phosphorus levels may correspond with increased percent areas of Cu(II); however, Cu(II) levels are also elevated for the pH lines, suggesting that sulfate, carbonate, silicate, or some other anion may also account for increased levels of Cu(II) deposits.

Silica was detected on three of five coupons, including phase I, phase III (PDS12 only), and phase IV. In all cases, Si was found as SiO_x. No silica was found on the coupons for phase II and phase III (PDS11). Absence of silica from the XPS results for phase II and phase III (PDS11) does not imply that silica is not present on the copper surface. Rather, nonuniformity in silica surface deposits may explain this result. The presence of SiO_x may indicate a mixed silica film of SiO_2 monomers and polymers, and/or the presence of a cupric silicate, $CuSiO_3$. Zinc was not detected on any of the five coupons. As mentioned in the scale modeling section, silicate may also be present as a copper silicate solid. Although the actual form of such a solid in distribution systems has not been established in the literature, the thermodynamic model for cupric silicate hydride, $CuSiO_3{\cdot}H_2O$, gave reasonable predictions for dissolved copper release with silicate inhibitor.

Table 5.96
Compounds identified in scale on copper coupons in the presence of pH_s-treated waters

Element	Compound	Phase I	Phase II	Phase III	Phase IV
C	CO_3^{2-}	X	X	X	X
Ca	$CaCO_3$				
	$Ca_3(PO_4)_2$				
	CaO				
Cu	Cu_2O	X	X		X
	CuO	X	X	X	X
	$Cu(OH)_2$	X	X	X	X
	Cu(II)	X	X	X	X
O	O^{2-}	X	X	X	X
	CO_3^{2-}	X	X	X	X
	OH^-	X	X	X	X
	PO_4^{3-}				
P	PO_4^{3-}				
	Meta				
	Pyro				
Si					
Zn					

pH. Four copper coupons exposed to the pH_s treatment were analyzed for the elements shown in Table 5.96. One copper coupon was tested for each of the four phases. Carbon was detected on four out of four coupons, always appearing as carbonate. Calcium was detected on none of the four coupons, indicating that the addition of phosphate and/or silicate inhibitors may help promote the formation of $CaCO_3$ films.

Copper was detected on all four coupons, appearing as a mixture of Cu_2O, CuO, $Cu(OH)_2$, and Cu(II) salts for all phases. Cu(II) salts may be present as a cupric carbonate compound such as $CuCO_3$ or $Cu_2(OH)_2CO_3$, given that CO_3^{2-} is present and no significant calcium is detected. Oxygen was detected on all four coupons, appearing as a mixture of O^{2-}, CO_3^{2-}, and OH^-. No PO_4^{3-} was detected on any of the four coupons for the pH_s treatment, confirming that phosphate inhibitor addition is the primary source for phosphorus on the copper surfaces for the given water qualities studied.

Phosphorus was not detected on any of the four coupons, consistent with the results from oxygen. Silica was not detected on any of the four coupons, indicating that silicate inhibitor addition is the primary source for silicate compounds on the copper surfaces for the given water qualities studied. Zinc was not detected on any of the four coupons.

Four copper coupons exposed to the pH_s+0.3 treatment were analyzed for the elements shown in Table 5.97. One copper coupon was tested for each of the four phases. Carbon was detected on all four coupons, always appearing as carbonate. Calcium was detected on one of the four coupons, appearing in phase IV as $CaCO_3$. This result is counterintuitive, given the lower calcium and alkalinity levels of phase IV.

Table 5.97
Compounds identified in scale on copper coupons in the presence of pH_s+0.3-treated waters

Element	Compound	Phase I	Phase II	Phase III	Phase IV
C	CO_3^{2-}	X	X	X	X
Ca	$CaCO_3$				X
	$Ca_3(PO_4)_2$				
	CaO				
Cu	Cu_2O	X	X	X	X
	CuO	X	X	X	X
	$Cu(OH)_2$	X	X	X	X
	Cu(II)	X	X	X	X
O	O^{2-}	X	X	X	X
	CO_3^{2-}	X	X	X	X
	OH^-	X	X	X	X
	PO_4^{3-}				
P	PO_4^{3-}				
	Meta				
	Pyro				
Si					
Zn					

Copper was detected on all four coupons, appearing as a mixture of Cu_2O, CuO, $Cu(OH)_2$, and Cu(II) salts for all phases. Cu(II) salts may be present as a cupric carbonate compound such as $CuCO_3$ or $Cu_2(OH)_2CO_3$, given that CO_3^{2-} is present and no significant calcium is detected in phases I, II, and III. Oxygen was detected on all four coupons, appearing as a mixture of O^{2-}, CO_3^{2-}, and OH^-. No PO_4^{3-} was detected on any of the four coupons for the pH_s+0.3 treatment, establishing that phosphate inhibitor addition is the primary source for phosphorus on the copper surfaces for the given water qualities studied. Phosphorus was not detected on any of the four coupons, consistent with the results from oxygen.

No silica was detected on any of the four coupons, confirming that silicate inhibitor addition is the primary source for silicate compounds on the copper surfaces for the given water qualities studied. Zinc was not detected on any of the four coupons.

Summary for Surface Characterization of Copper Coupons

Thermodynamic models were developed describing copper release and assuming that different solids might be expected to form in copper pipes receiving no inhibitor, phosphate inhibitor, or silicate inhibitor. The thermodynamic models were developed using thermodynamic data for each solid and relating the predicted dissolved concentration of cupric ion, Cu^{2+}, to the cupric ion complexing ligands present in the water. The sum of the free dissolved cupric ion and the complex cupric ion gives the total dissolved cupric ion concentration. The primary ligands responsible for increases in total copper were all related to alkalinity, including $CuHCO_3^-$, $CuCO_3^0$, $Cu(CO_3)_2^{2-}$, $Cu(OH)CO_3^-$, and $Cu(OH)_2CO_3^{2-}$. Equilibrium calculations demonstrated the profound effect of alkalinity on copper release. When combined, these complexes accounted for approximately 90% of all soluble copper species.

For the pH_s+0.3 control, actual dissolved and total copper release data were compared to predicted dissolved copper release from tenorite (CuO) and cupric hydroxide ($Cu(OH)_2$) models. The predicted copper release from the $Cu(OH)_2$ model was typically 0.3 mg/L as Cu less than the actual copper release observed for pH_s+0.3. Similarly, the tenorite model underpredicted the dissolved copper concentrations by about 0.7 mg/L as Cu. A similar underprediction with the cupric hydroxide model was noted by Xiao (2004) using the same PDS. For the model presented in this report, the log equilibrium constant for cupric hydroxide dissolution, 8.89, had been chosen to match that reported in previous cupric hydroxide models (Shock, Lytle, and Clement 1995; Xiao 2004). Although copper samples were collected after a 6-hour stagnation period, wherein equilibration would not be expected (Shock, Lytle, and Clement 1995), copper samples at equilibration would have likely been greater than or equal to those observed at 6 hours. Hence, the discrepancy is not likely due to the sample collection method.

A log equilibrium constant of 9.36 ± 0.02 has been reported for freshly precipitated cupric hydroxide (Hidmi and Edwards 1999). This equilibrium constant would yield a higher predicted copper release that is more reasonable, given the actual copper release observed with the pH_s+0.3 control. A unit variation in log equilibrium constant would result in a log-order variation in predicted copper release. Other reported values for the $Cu(OH)_2$ log equilibrium constant include 8.3, 8.7, 9.2, and 9.4 (King 1959, Hahn and Welcher 1968, Hogness and Johnson 1957, Schenk 1996). The variability in reported log equilibrium constants across the literature may be due to the age of the $Cu(OH)_2$ solids (Hidmi and Edwards 1999).

BOP, OP, and ZOP thermodynamic models predicting dissolved copper were developed for the phosphate inhibitors, assuming $Cu(OH)_2$, CuO, and $Cu_3(PO_4)\cdot 2H_2O$ each as controlling solids. For all phosphate inhibitors, the $Cu(OH)_2$ and $Cu_3(PO_4)\cdot 2H_2O$ models provided the best prediction of copper release, with each model overpredicting by no more than about 0.3 mg/L Cu. The CuO model consistently underpredicted copper release by about 0.25 mg/L Cu. If $Cu(OH)_2$ is the controlling solid for copper release in the pH_s+0.3 control, and a higher reported value of the log equilibrium constant for cupric hydroxide were used instead of 8.89, the predicted copper concentration from the cupric hydroxide model would exceed the observed copper release when phosphate inhibitors were added. In contrast, the values predicted by a cupric phosphate dihydride, $Cu_3(PO_4)\cdot 2H_2O$, would be a closer match to the data. A review of the literature confirmed the lack of information identifying the form of copper OP solids in water distribution systems. However, the forms $Cu_3(PO_4)_2\cdot 2H_2O$ and $Cu_3(PO_4)_2$ were modeled because thermodynamic data was available for these copper forms (Shock, Lytle, and Clement 1995).

For the silicate inhibitor, thermodynamic models predicting dissolved copper were developed, assuming $Cu(OH)_2$, CuO, and $CuSiO_3\cdot H_2O$ each as controlling solids. For all silicate inhibitors, the $Cu(OH)_2$ and $CuSiO_3$ models provided the best prediction of copper release, with each model overpredicting by no more than about 0.3 mg/L Cu. As mentioned with phosphorus, adjustment of the log equilibrium constant for cupric hydroxide solubility to reported values consistent with copper release in pH_s+0.3 would result in $CuSiO_3$ providing the closest match to the data, only slightly overpredicting the dissolved copper release.

Survey and high-resolution XPS spectra were collected to identify the corrosion products of copper and determine other chemical compounds present. Because of nonuniformity in the surface composition of copper scales, failure to detect a compound by XPS in one scan does not imply that the compound was not in the scale.

Cu_2O, CuO, $Cu(OH)_2$, and Cu(II) salts were detected as the predominant corrosion compounds in the scale formed on copper coupons in the presence of inhibitors. This procedure is modeled after the method of Xiao (2004). $CuCl_2$, $CuSO_4$, $CuCO_3$, and $CuSiO_3$ are all cupric salts (Schock et al. 2005) and may be included under the Cu(II) curve given by Xiao, according to published XPS binding energies (Wagner et al. 2003). In addition $Cu_3(PO4)_2$, $Cu_2(OH)_2CO_3$ (malachite), and $Cu_4SO_4(OH)\cdot 6H_2O$ (cupric hydroxysulfate) are all feasible forms of Cu(II) (Schock, Lytle, and Clement 1995); however, no published XPS binding energies were identified for these three compounds. The addition of corrosion inhibitors did not affect the relative distribution of copper corrosion products, as measured by the four copper compound curves. ZOP inhibitor addition was the primary source of zinc on the copper surface.

Phosphate inhibitor addition was the primary source of phosphorus compounds on the copper surface. Phosphate was detected on many coupons that did not have any $Ca_3(PO_4)_2$ scales. Because it was not precipitated with calcium, phosphate may be present as $Cu_2(PO_4)_3$ or adsorbed to the scale surface as a liquid–solid film. The thermodynamic model for cupric phosphate dihydride, $Cu_3(PO_4)_2\cdot 2H_2O$, predicted dissolved copper release similar to the reduced levels of copper release observed in the phosphate lines.

For phase IV water quality, no phosphorus was detected on the coupon surface for BOP and OP inhibitors. In contrast, phosphorus was detected on the coupon surface for ZOP inhibitor. Thus, the presence of zinc may aid in the precipitation of phosphates in water quality conditions similar to that of phase IV.

Silicate inhibitor addition was the primary source of silicates on the copper surface. XPS analysis indicated the presence of a polymeric silica film. XPS results were also consistent with the presence of $CuSiO_3$ in the surface scale. It has been suggested that $CuSiO_3$ can form from malachite exposed to SiO_2 inhibitor (Schock et al. 2005). The thermodynamic model for cupric silicate hydride, $CuSiO_3\cdot H_2O$, predicted dissolved copper release similar to the reduced levels of copper release observed in the silicate lines.

Lead Surface Characterization

Thermodynamic Model Development for Lead Release

Following the identification of the lead complexes contributing to dissolved lead during the study, a thermodynamic model was developed that would predict dissolved lead at equilibrium. Predictions based on PDS water quality and dissolved lead concentrations provide insight into the controlling solid phase.

Various forms of lead corrosion scales were identified by XPS in an effort to determine the controlling solid phase(s). Possible controlling solid phases were adapted to the possible soluble lead species typical of water distribution systems through thermodynamic modeling. Assuming the presence of Pb(II) species, the dissolved lead ions and complexes shown in Equation 5.61 were incorporated into the thermodynamic modeling efforts. Equilibrium modeling considered the same dissolved lead species as that of TBW I (Taylor et al. 2005) with the exception of phosphate and silica. However, after a thorough review of lead coordination literature, there is no evidence to suggest complexation of lead and silica.

$$Pb_T = [Pb^{2+}] + 4[Pb_4(OH)_4^{4+}] + 2[Pb_2OH^{3+}] + [PbOH^+] + 3[Pb_3(OH)_4^{2+}] + 6[Pb_6(OH)_8^{4+}] + [Pb(OH)_2^0] + [Pb(OH)_3^-] + [Pb(OH)_4^{2-}] + 3[Pb_3CO_3^{4+}] + 2[Pb_2CO_3^{2+}] + [PbCO_3^0] + [Pb(CO_3)_2^{2-}] + [PbHCO_3^+] + [PbSO_4^0] + [Pb(SO_4)_2^{2-}] + [PbCl^+] + [PbCl_2^0] + [PbCl_3^-] + [PbCl_4^{2-}] + [PbH_2PO_4^+] + [PbHPO_4^0] \quad (5.61)$$

The complexes considered within Equation 5.61 demonstrate the various water quality dependencies governing lead release. The speciation of dissolved lead suggests that such water quality parameters as alkalinity, chloride, sulfate, and phosphate may promote further lead release. From the empirical model defined in the previous section of the report describing lead release, pH, chloride, sulfate, and temperature were established as the main effects attributed to water quality. Both an increase in sulfate and pH predicted a mitigation of lead release, whereas raising chlorides or the temperature increased lead release.

The undesirable effect of chloride is supported by the dependency of chloride ligands for the formation of $PbCl^+$, $PbCl_2^0$, $PbCl_3^-$, and $PbCl_4^{2-}$,and agrees well with the empirical model. As previously discussed, the empirical model for lead predicts a benefit of increasing pH and sulfate. Equation 5.61 implies that as pH increases, all of the lead complexes incorporating carbonate and hydroxide-dependent ligands should increase, thus speciously contradicting the benefit of elevated pH. However, the formation of these complexes is dependent on Pb^{2+} as well as carbonate and hydroxide ligands. Given that the activity of Pb^{2+} is governed by the controlling solid phase, the main effect of pH is likely related to the acid/base properties of the controlling solid rather than the manipulation of pH-dependent ligands.

The presence of the soluble sulfate complexes, $PbSO_4^0$ and $Pb(SO_4)_2^{2-}$, appears to be inconsistent with the empirical model that suggests sulfate retards lead release. Consideration of these lead complexes suggests that sulfate promotes lead release, while during the study it was determined that increasing sulfate had in fact hindered lead release. The same trend with sulfate and lead release was observed during the previous blend study (Taylor et al. 2005). The role of sulfate was not apparent from the chemical composition of lead/tin coupons analyzed by XPS. Of the 24 lead/tin coupons analyzed, there was no detection of any sulfur-containing scales on the coupons.

Empirical modeling efforts did not define alkalinity as a significant factor for regulating lead release. During TBW I, alkalinity was found to promote lead release. The lead complexes shown in Equation 5.61 support the adverse impact of alkalinity on lead release. Lead complexes dependent of the carbonate system included $Pb_3CO_3^{4+}$, $Pb_2CO_3^{2+}$, $Pb(CO_3)_2^{2-}$, $PbCO_3^0$, and $PbHCO_3^+$ for the thermodynamic model.

Lead phosphate complexes were considered for describing the expected impact of applying BOP, OP, and ZOP, assuming a common controlling solid. Both $PbH_2PO_4^+$ and $PbHPO_4^0$ were not considered during TBW I because of the relatively low OP concentrations of the untreated (exclusion of phosphate inhibitor) blend. However, with the addition of BOP, OP, and ZOP, the presence of these complexes may represent a significant portion of the dissolved lead species. Assuming the presence of the same controlling solid phase for all of the PDSs, the presence of active phosphate ligands actually contradicts the benefit of adding a phosphate-based inhibitor. Any improvement in dissolved lead release from the control would imply either the formation of a different controlling solid phase or a chemical interaction with the solid originally present.

Table 5.98
Comparison of observed lead release from inhibitor PDSs to pH_s+0.3

Phase	Inhibitor	Average release (mg/L)		% Ratio to pH_s+0.3	
		Dissolved Pb	Total Pb	Dissolved Pb	Total Pb
I	pH_s+0.3	0.001	0.003		
	BOP	0.000	0.001	29	45
	OP	0.000	0.001	31	18
	ZOP	0.000	0.000	46	15
	Si	0.001	0.002	105	69
II	pH_s+0.3	0.004	0.008		
	BOP	0.000	0.001	6	8
	OP	0.000	0.000	8	3
	ZOP	0.000	0.000	11	5
	Si	0.000	0.000	10	5
III	pH_s+0.3	0.005	0.010		
	BOP	0.000	0.001	6	7
	OP	0.000	0.000	8	4
	ZOP	0.000	0.001	6	6
	Si	0.000	0.001	9	12
IV	pH_s+0.3	0.001	0.003		
	BOP	0.000	0.001	30	19
	OP	0.000	0.000	24	12
	ZOP	0.000	0.001	28	18
	Si	0.001	0.001	43	29

Inhibitor Control of Lead Release

The average lead release observed during the study is shown in Table 5.98 by phase for each inhibitor and relative to pH_s+0.3 (the control).

As noted previously, inhibitors reduced lead release relative to the control. There is one observation in Table 5.98 where the addition of an inhibitor did not reduce lead release relative to the control, which is dissolved lead for phase I Si. Total lead release, however, was reduced relative to the control for all phosphate- and silica-based inhibitors. This may have been due to the high alkalinity and lack of equilibrium relative to inhibition in the initial phase. Relatively high lead release also occurred in phase III, which had practically equal water quality to phase I.

Expected Lead Release From Thermodynamic Models

pHs+0.3. As demonstrated in Table 5.98, lead release observed during the addition of an inhibitor was compared with that of the initial source water. This analysis was applied with the consideration of lead release governed by a lead-controlling solid phase. However, this comparison required the prediction of lead release from pH_s+0.3 while applying the same method for

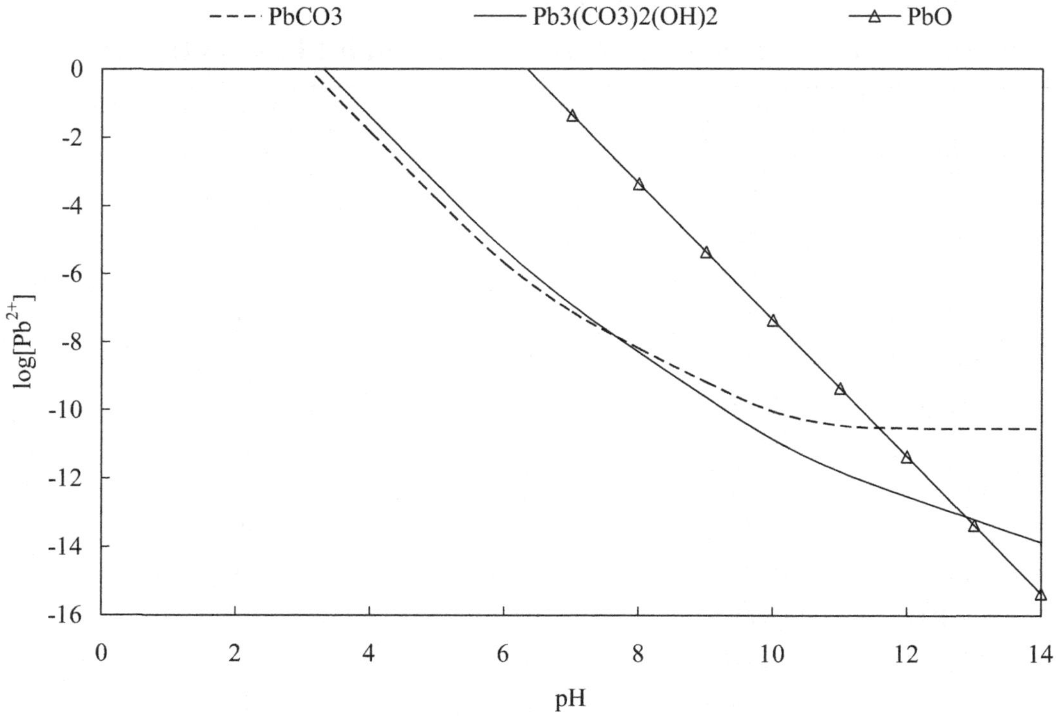

Figure 5.89 Solubility limit of lead for pH_s and pH_s+0.3

predicting lead release for inhibitor-treated PDSs. The prediction of lead release was determined from the previously defined thermodynamic model for lead release.

From the XPS analysis, it was determined that both cerussite ($PbCO_3$) and hydrocerussite ($Pb_3(OH)_2(CO_3)_2$) were the predominant lead corrosion scales present on the lead/tin coupons. However, a significant portion of lead oxide (PbO) was detected as well. Assessment of the solubility diagram, without complexation, shown in Figure 5.89 suggests that hydrocerussite limits the solubility of lead within the pH range of the study. From Figure 5.89, hydrocerussite is predominant between a pH of approximately 7.6 and 12.9 at an alkalinity of 100 mg/L as $CaCO_3$. Furthermore, Figure 5.89 illustrates the relatively high solubility of lead when governed by the dissolution of PbO.

Both cerussite and hydrocerussite were considered as the controlling solid phases during thermodynamic modeling. It was apparent that PbO would not exist as the controlling solid phase because of the relatively high solubility of lead permitted by the solid. As shown in Table 5.99, the thermodynamic model predicted lead releases much higher than any observed during the study for both cerussite and hydrocerussite. The results imply that the system was not at equilibrium during the collection of the sample. Tang (2003) developed similar fundamental equilibrium models during TBW I and established the same conclusion (Tang 2003, Taylor et al. 2005). Tang noted that the lead release in the project, and typically in the water community, was diffusion limited because of the lack of lead surface area in distribution systems. The dissolved lead concentration could not reach equilibrium between the lead coupon and the volume of water contained in the copper corrosion loops during the 6-hour stagnation time. Hence, diffusion theory was used to reduce the lead equilibrium concentrations by 0.001.

Table 5.99
Lead thermodynamic modeling of pH_s+0.3

Inhibitor	Phase	Actual lead release (mg/L)		Modeled lead release (mg/L)	
		Dissolved Pb	Total Pb	Hydrocerussite	Cerussite
pH_s+0.3	I	0.001	0.003	0.183	0.200
	II	0.004	0.008	0.170	0.202
	III	0.005	0.010	0.170	0.200
	IV	0.001	0.003	0.178	0.201
	All	0.003	0.006	0.177	0.200

Table 5.100
Lead thermodynamic modeling of BOP

Inhibitor	Phase	Actual lead release (mg/L)		Modeled lead release (mg/L)		
		Dissolved Pb	Total Pb	Hydroxypyromorphite	Hydrocerussite	Cerussite
BOP	I	0.000	0.001	0.064	0.181	0.200
	II	0.000	0.001	0.041	0.176	0.202
	III	0.000	0.001	0.064	0.181	0.200
	IV	0.000	0.001	0.049	0.187	0.201
	All	0.000	0.001	0.056	0.182	0.200

As shown in Table 5.99, soluble lead predicted at equilibrium for all phases was relatively constant and varied from 0.200 and 0.202 mg/L when cerussite was assumed to be the controlling solid phase.

Some variation in predicted lead release can be seen for hydrocerussite, ranging from 0.170 and 0.183 mg/L. Correcting for diffusion decreases the predicted lead release to approximately 0.2 mg/L for both hydrocerussite and cerussite. Nearly 40% of the actual dissolved lead data was measured below the detection limit (1 mg/L). Either cerussite or hydrocerussite could be the controlling solid phase. Both provided relatively similar predictions and have a predominance boundary near the operating pH.

Blended Ortho-polyphosphate. The XPS analysis for the four lead/tin coupons exposed to BOP during four phases of operation determined that both cerussite and hydrocerussite were the more prevalent lead corrosion scales. Equilibrium modeling was pursued for the medium dose (lead/tin coupons were exposed to medium-dose PDSs) of BOP for four phases. Cerussite and hydrocerussite were considered separately as the predominant lead corrosion scales or controlling solid phase.

The results from the thermodynamic modeling of lead/tin coupons exposed to BOP are shown in Table 5.100. These predicted equilibrium lead concentrations assume the same speciation of lead complexes applied for pH_s+0.3. Water quality was relatively similar between pH_s+0.3 and BOP, disregarding the addition of orthophosphate and polyphosphate for BOP, within a common phase. Hence, predictions of dissolved lead release would be expected to resemble what was calculated for cerussite and hydrocerussite in pH_s+0.3. With the incorporation of lead phosphate

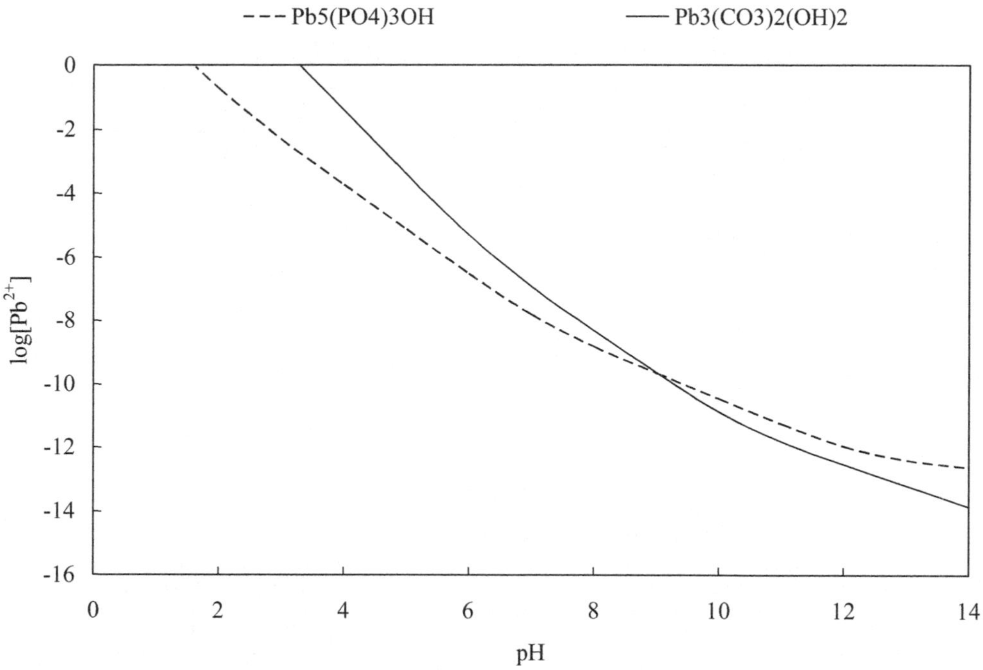

Figure 5.90 Solubility limit of lead for medium-dose phosphate-based inhibitors

complexes, the thermodynamic model may actually predict a slightly higher dissolved lead concentration. The equilibrium dissolved lead concentrations are nearly identical to the predicted lead concentrations displayed in Table 5.99 for pH_s+0.3.

Because the equilibrium model predicts dissolved lead concentrations similar to pH_s+0.3, the actual dissolved lead release observed for BOP should be similar to pH_s+0.3 as well. This assumption is valid if either cerussite or hydrocerussite do indeed represent the controlling solid phase for BOP. However, comparing actual lead release data for BOP and pH_s+0.3 from Table 5.99, it is apparent that BOP demonstrated a significant improvement in the control of lead. This observation would imply either the existence of a differing controlling solid phase or the formation of a chemical barrier promoted by the presence of BOP.

Further investigation regarding the possibility of a controlling solid phase that was undetected during the XPS analysis was pursued. The XPS analysis of the four lead/tin coupons exposed to BOP clearly indicated the presence of small amounts of phosphate on all coupons. However, the analysis was inadequate in establishing the identification of a phosphate compound. The presence of a less soluble lead OP compound would explain the decrease in lead release with the application of BOP. Postulating the presence of a lead orthophosphate, the equilibrium model was applied assuming hydroxypyromorphite ($Pb_5(PO_4)_3OH$) as the controlling solid phase. Conditions within a water distribution system likely favor hydroxypyromorphite as the stable lead OP form (Schock 1989).

The solubility limit of lead, without complexation, when considering all forms of lead found in the XPS analysis and hydroxypyromorphite is shown in Figure 5.90 at an alkalinity of 100 mg/L as $CaCO_3$ and phosphate of 1.0 mg/L as P. The solubility diagram would suggest that under equilibrium conditions, hydroxypyromorphite was the solubility-limiting solid during the study. Dissolved lead was significantly decreased when applying the thermodynamic model, as

Table 5.101
Lead thermodynamic modeling of OP

Inhibitor	Phase	Actual lead release (mg/L)		Modeled lead release (mg/L)		
		Dissolved Pb	Total Pb	Hydroxypyromorphite	Hydrocerussite	Cerussite
OP	I	0.000	0.001	0.064	0.179	0.200
	II	0.000	0.000	0.042	0.178	0.202
	III	0.000	0.000	0.061	0.176	0.200
	IV	0.000	0.000	0.048	0.190	0.201
	All	0.000	0.000	0.056	0.184	0.200

seen in Table 5.100. The decrease in dissolved lead indicates that the lead OP solid could explain the improvement observed with PDSs treated with BOP.

Orthophosphate. Lead corrosion products identified by the XPS analysis for the four lead/tin coupons exposed to OP found cerussite and hydrocerussite as the predominant lead scales. Phosphate was found on three of the four lead/tin coupons. Of the three lead/tin coupons, only one of the three identified a corresponding cation for the phosphate detected, as calcium phosphate. On the remaining two lead/tin coupons in which phosphate was found, calcium was not detected. As explained in the previous section describing thermodynamic modeling of lead/tin coupons exposed to BOP, the formation of hydroxypyromorphite is thermodynamically favorable.

Equilibrium calculations are provided in Table 5.101 considering hydroxypyromorphite, hydrocerussite, and cerussite as the possible controlling solid phases. From Table 5.99 there is a clear improvement with the control of lead release with the addition of OP when compared to pH_s+0.3. When considering the same possible controlling solids for OP with those proposed for pH_s+0.3 (cerussite and hydrocerussite), the equilibrium calculations suggest an insignificant difference. However, with the consideration of the lead OP solid, the mitigation of lead release is apparent.

Zinc Orthophosphate. As with the lead/tin coupons treated by BOP and OP, the XPS analysis for the four lead/tin coupons exposed to ZOP found cerussite and hydrocerussite as the predominant lead scales. Phosphate was found on three of the four lead/tin coupons exposed to ZOP. For the three lead/tin coupons with which phosphate was detected, the corresponding cation was not identified.

Equilibrium modeling was completed while considering hydroxypyromorphite, hydrocerussite, and cerussite as the controlling solid phases. The results shown in Table 5.102 demonstrate similarities to the other phosphate-based inhibitor. The assumption of including either cerussite or hydrocerussite as the controlling solid phase was insufficient in describing the differences in ZOP treatment to pH_s+0.3. Also, equilibrium modeling of hydroxypyromorphite demonstrated a significant improvement in controlling the release of dissolved lead.

Contrasting observed lead release between ZOP and OP, as shown in Table 5.101, suggests no significant differences in effectiveness. Any difference could be assumed to be completely attributable to the addition of zinc. From the XPS analysis, zinc was detected on all lead/tin coupons, predominantly as zinc oxide (ZnO). However, it was demonstrated that with exposure to water contaminated by galvanized steel, the formation of ZnO products was even possible for lead/tin coupons exposed to OP.

Table 5.102
Lead thermodynamic modeling of ZOP

		Actual lead release (mg/L)		Modeled lead release (mg/L)		
Inhibitor	Phase	Dissolved Pb	Total Pb	Hydroxypyromorphite	Hydrocerussite	Cerussite
	I	0.000	0.000	0.062	0.188	0.200
	II	0.000	0.000	0.040	0.196	0.204
ZOP	III	0.000	0.001	0.058	0.187	0.200
	IV	0.000	0.001	0.046	0.199	0.202
	All	0.000	0.001	0.054	0.196	0.201

Table 5.103
Lead thermodynamic modeling of Si

		Actual lead release (mg/L)		Modeled lead release (mg/L)		
Inhibitor	Phase	Dissolved Pb	Total Pb	Hydrocerussite	Cerussite	$PbSiO_4$
	I	0.001	0.002	0.147	0.203	979
	II	0.000	0.000	0.137	0.202	797
Si	III	0.000	0.001	0.147	0.203	971
	IV	0.001	0.001	0.144	0.202	920
	All	0.001	0.001	0.146	0.202	964

Silicon. Although not as effective as the phosphate-based inhibitors for the control of lead release, the empirical model describing lead release suggests a beneficial effect of increasing Si. Surface composition of the four lead/tin coupons exposed to Si identified cerussite and hydrocerussite as the predominant lead corrosion products. Of the four lead/tin coupons analyzed by XPS, two of the coupons contained forms of amorphous silica.

Following precipitation of silica, metal ions may be incorporated into the lattice of the amorphous structure. This property of silica may enable bonding between the lead surface of the lead/tin coupon. Lead silicate ($PbSiO_4$) was modeled to represent a possible bond between lead and silica. As shown in Table 5.103, lead concentrations were nearly 6 orders of magnitude greater than the actual lead release. Thermodynamic modeling of hydrocerussite demonstrated the effect of an increasing pH with the application of Si. However, the improvement seen from adjusting the pH with hydrocerussite as the controlling solid phase does not correspond with the actual improvement shown in Table 5.99.

Lead Scale Elemental Analysis

A typical survey scan for lead/tin coupons is shown in Figure 5.91. Peaks were dependent on the inhibitor to which the coupon had been exposed and were typically pronounced for the lead, tin, oxygen, carbon, calcium, silicon, phosphorus, and zinc.

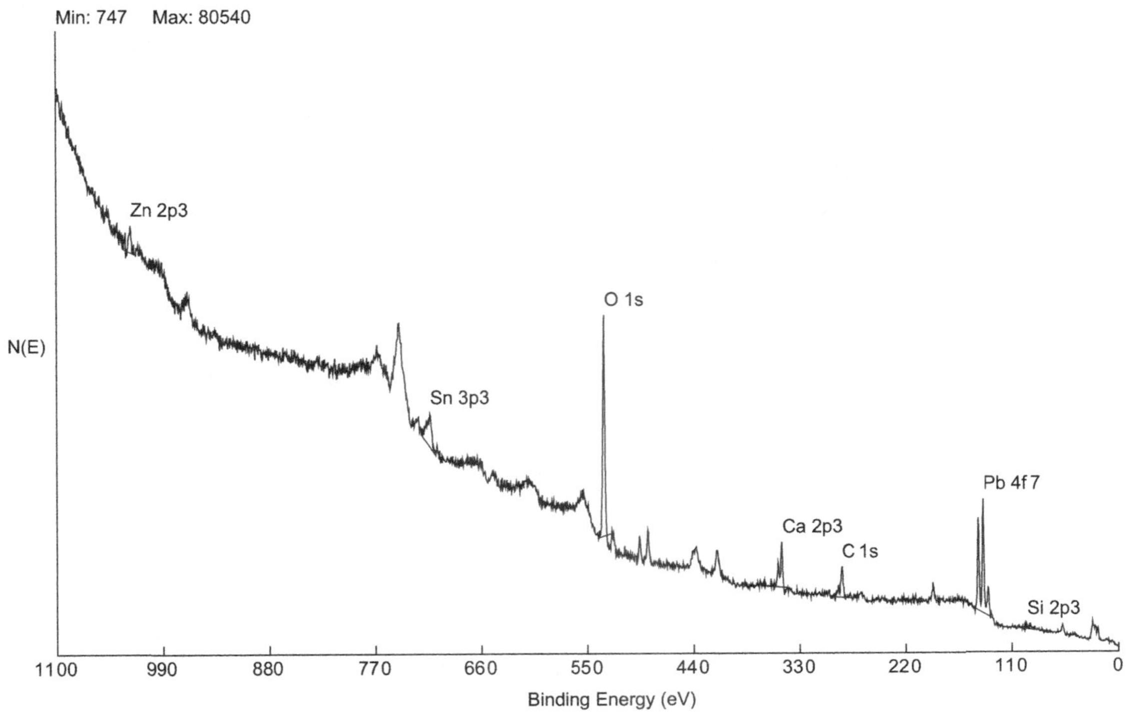

Figure 5.91 XPS survey scan for lead/tin coupon

Identification of Lead Corrosion Scale Products

Deconvolution of the lead high-resolution scans were conducted to determine the composition of scales on lead coupons exposed to the various inhibitors. Lead corrosion products $Pb_3(OH)_2(CO_3)_2$ (hydrocerussite), $PbCO_3$ (cerussite), PbO, and lead dioxide (PbO_2) are possible in an aqueous environment (Pankow 1991), and were identified by XRD and verified by XPS in scales on lead coupons (Taylor et al. 2005). Binding energies for each lead compound were consistent with those used during TBW I and were 137.4, 137.7, 138.1, and 138.7 eV for PbO_2, PbO, $Pb_3(OH)_2(CO_3)_2$, and $PbCO_3$, respectively. As shown in Figure 5.92, these binding energies were used to identify PbO_2, PbO, $Pb_3(OH)_2(CO_3)_2$, and $PbCO_3$ in the scale on the lead coupon exposed to BOP from phase IV.

Consideration of the lead compounds PbO_2, PbO, $Pb_3(OH)_2(CO_3)_2$, and $PbCO_3$ were sufficient to describe the high-resolution peaks for all lead/tin coupons. This would imply that introducing the inhibitors used during the study into the system did not significantly change the composition of the lead corrosion products. As shown in Figure 5.92, cerussite and hydrocerussite are the dominant solid forms of lead in the surface scale on the BOP lead coupon. However, PbO was found in all scales on lead coupons regardless of inhibitor, which suggests that PbO is involved with lead release with or without the presence of inhibitors for these water quality conditions. During the study, the percentage of PbO_2 was the least of all lead compounds identified.

Average percent areas for lead corresponding to PbO_2 were below 4% for all inhibitors, excluding OP. From Figure 5.93, OP displays the widest variation in PbO_2, ranging from 2.7% to 31.5% and averaging 13.9% through all phases. The relative small fraction of PbO_2 in the lead scale indicates that relatively significant amounts of lead are not released directly from PbO_2

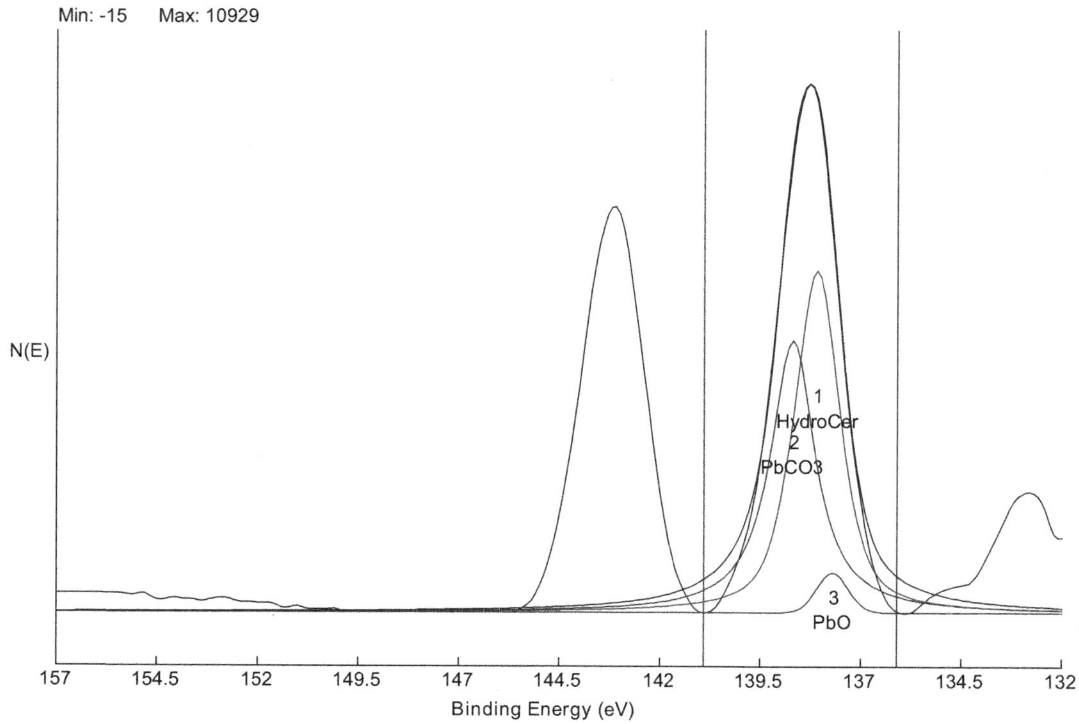

Figure 5.92 Deconvolution of lead for coupon exposed to BOP during phase IV

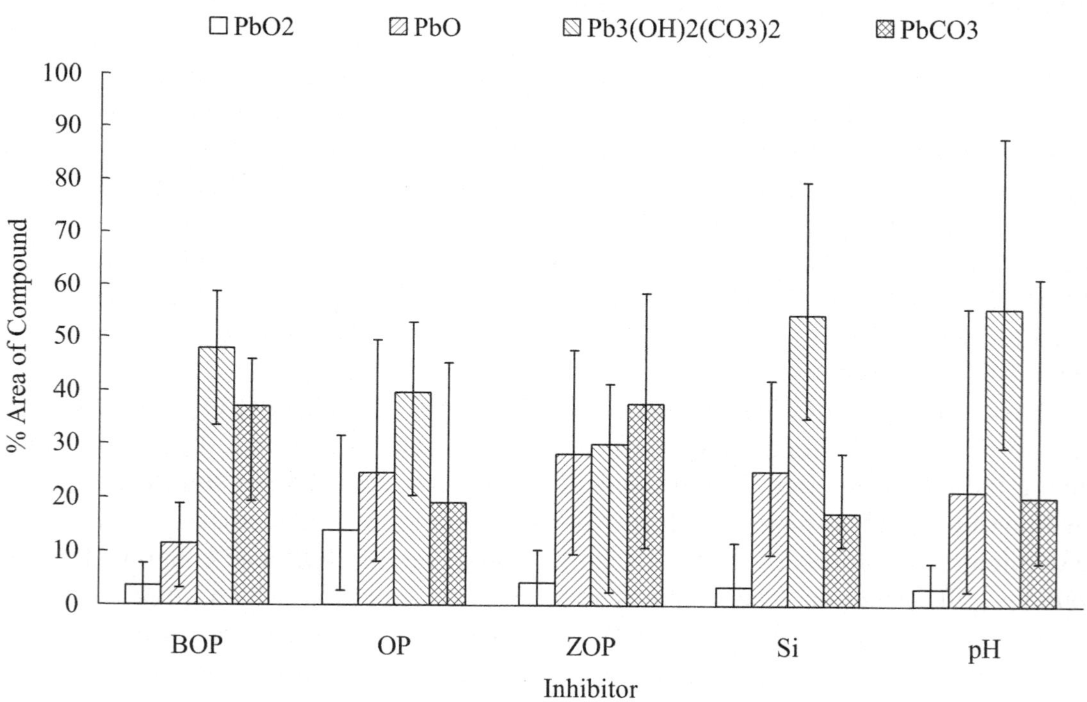

Figure 5.93 Distribution of lead compounds for all phases

surfaces and that PbO_2 was not the direct controlling phase for lead in this work. PbO accounted for 11.5% and 28.2% of lead scales on average. The maximum observation of PbO in the surface scale of lead coupons was greater than 40%. Excluding BOP, there was at least one observation for each treatment in which the percentage of PbO accounted for more than 30% of the lead in the scale on the lead coupons, which demonstrates the formation of PbO scale in lead coupons for all conditions for this work. The fraction of hydrocerussite in the scale for all lead/tin coupons was greater than 20%, which shows hydrocerussite as the major solid form in the scale of lead/tin coupons for this work. The fraction of cerussite was less than the fraction of hydrocerussite in the scale on all lead/tin coupons except for the scale on the lead/tin coupons in the presence of ZOP. However, there were observations in which cerussite ($PbCO_3$) accounted for more than 45% of the lead contained within a coupon. The relative distribution of solid lead forms indicates that hydrocerussite, cerussite, or PbO was the controlling solid phase for lead release in this work.

Effect of Phase on Lead Corrosion Scale Products

A plot of the relative variation in scale compositions for hydrocerussite and cerussite are shown in Figures 5.94 and 5.95. Similar plots were constructed for PbO_2 and PbO. Since the percent area for PbO_2 was relatively low, variations in the percentage were less distinguishable between phase and random error associated with the analysis. The only trend observed with respect to PbO was that the fraction of PbO in the surface scale on the lead coupons was consistently greater for phase I than for phase IV for all observations. However, there is no apparent relationship between the distribution of hydrocerussite and cerussite by water quality change. This result indicates that none of the inhibitors used in this work influenced the relative formation of hydrocerussite and cerussite for the water quality conditions (84 to 175 mg/L as $CaCO_3$ alkalinity, 35.4 to 122.9 mg/L chlorides, and 136 to 252 mg/L as $CaCO_3$ total hardness) in this work.

Effect of Inhibitor on Lead Corrosion Scale Products

Blended Ortho-polyphosphate. Four lead/tin coupons exposed to BOP during the four phases of the study were analyzed for the elements shown in Table 5.104. Analysis of carbon was conclusive in identifying carbonates on all four samples. Calcium was detected for all phases, excluding phase I. When calcium was significant within the survey scan, it could be deconvoluted as solely $CaCO_3$, implying that calcium is predominantly in the form of calcium carbonate when BOP is utilized within the PDS. Calcium carbonate was not determined from the analysis for phase I, even though phases I and III exhibited similar water quality. The deviation from expected calcium carbonate deposition for phase I could be due to temperature variations between phase I and phase III or the lack of uniformity in surface corrosion. Given that the solubility of calcium carbonate increases as temperature decreases, the colder temperatures of phase I would deter precipitation. Deconvolution of oxygen further supports the presence of carbonate compounds. The results also demonstrate the presence of O^{2-}, OH^-, and, for phase II, PO_4^{3-}. All the forms of oxygen correspond to various compounds of calcium, lead, tin, and zinc.

Deconvolution of phosphorus suggests that a majority of phosphorus was present as phosphate. Phosphate was found for all phases, whereas polyphosphate forms (meta- and pyrophosphates) were detected during phases II, III, and IV. Though phosphate was only found for phase II within the oxygen deconvolution, the high-resolution scan of phosphorus ensures the presence of phosphates on the scale. Although solubility relationships suggest that phosphates would precipitate

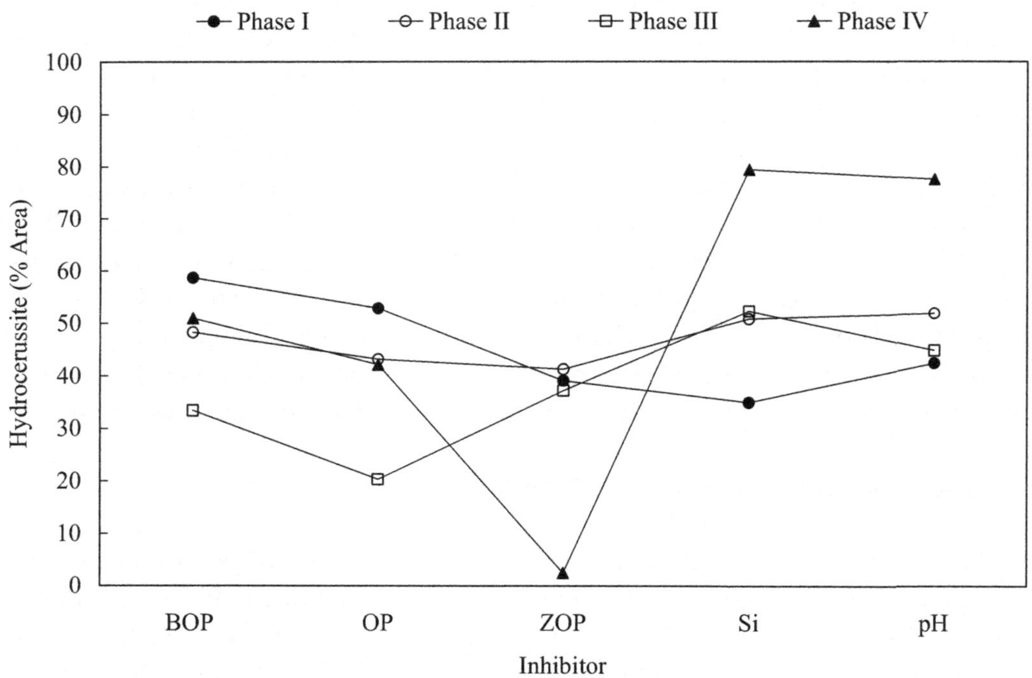

Figure 5.94 Percent area of hydrocerussite versus inhibitors for all phases

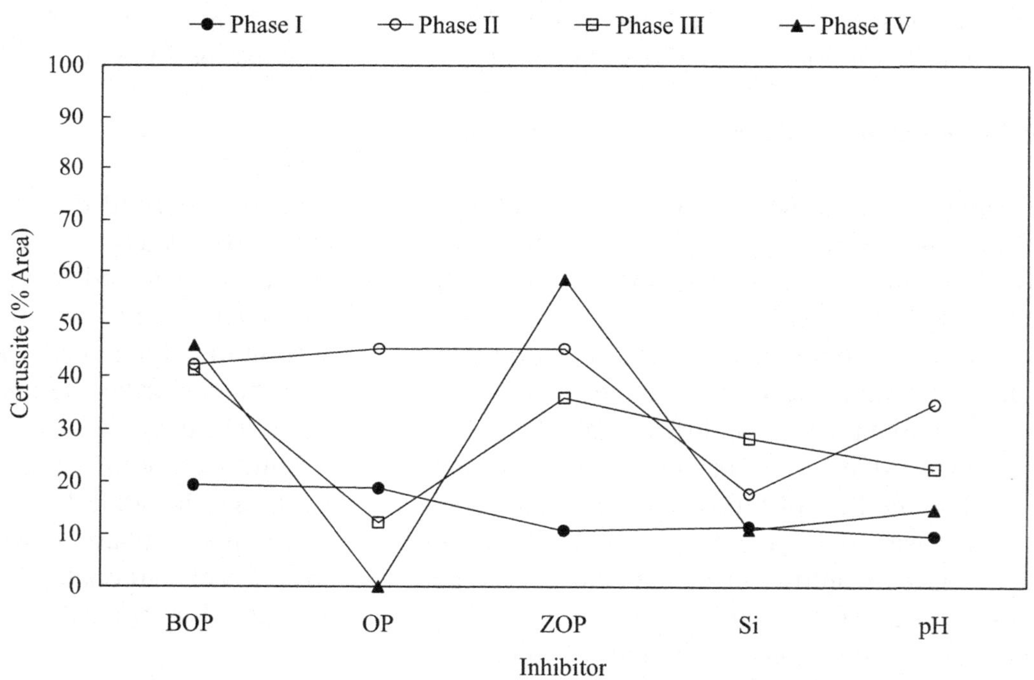

Figure 5.95 Percent area of cerussite versus inhibitors for all phases

Table 5.104
Compounds identified for BOP coupons

Element	Compound	Phase I	Phase II	Phase III	Phase IV
C	CO_3^{2-}	X	X	X	X
Ca	$CaCO_3$		X	X	X
O	O^{2-}	X	X	X	X
	CO_3^{2-}	X	X	X	X
	OH^-	X	X	X	X
	PO_4^{3-}		X		
P	PO_4^{3-}	X	X	X	X
	P_4O_{10}				X
	Meta		X		
	Pyro			X	X
Pb	PbO_2	X		X	
	PbO	X	X	X	X
	$Pb_3(OH)_2(CO_3)_2$	X	X	X	X
	$PbCO_3$	X	X	X	X
Si					
Sn	SnO	X	X	X	X
	Sn	X			X
Zn	ZnO		X	X	X
	Zn				X

as calcium phosphate, scanning of the lead/tin coupons from phase I did not detect calcium despite the presence of phosphate. The absence of calcium phosphate in the scale could be due to nonuniform scales or the interaction of phosphate with species other than calcium that hindered precipitation. Deconvolution of calcium in surface scales from the other phases did not establish the presence of calcium phosphate. It appears that the phosphorus constituents detected are not directly incorporated into the surface scale but rather form a surface film that contributes to inhibition.

Lead corrosion products were identified as PbO_2, PbO, hydrocerussite, and cerussite. Although PbO_2 was determined for phases I and III, both samples contributed less than 8% of total lead within the scale. There was no evidence to suggest a lead compound that incorporated phosphorus, even though phosphorus was observed for all four phases.

Tin was found to corrode as SnO (stannous oxide). Elemental tin remained on the samples from phases I and IV, in which Sn accounted for 8% and 5%, respectively. Zinc deposits were also present within the scale for phases II, III, and IV as ZnO and, for phase IV, elemental zinc. The zinc in the surface scale on the lead/tin coupons came from the GS pipe used in the PDSs and was not precipitated from trace zinc in the water prior to inhibitor addition.

Orthophosphate. A summary of the compounds identified in the scale on the lead/tin coupons exposed to OP during four phases of study are shown in Table 5.105. Analysis of carbon revealed the presence of carbonate for all phases, excluding phase II. The absence of carbonate represents an aberration in the analysis that behaves contrary to the presence of hydrocerussite

Table 5.105
Compounds identified for OP coupons

Element	Compound	Phase I	Phase II	Phase III	Phase IV
C	CO_3^{2-}	X		X	X
Ca	$CaCO_3$			X	
	CaO				X
	$Ca_3(PO_4)_2$				X
O	O^{2-}	X	X	X	X
	CO_3^{2-}	X	X	X	X
	OH^-	X	X	X	X
	PO_4^{3-}				X
P	PO_4^{3-}	X	X		X
	$Ca_3(PO_4)_2$				X
Pb	PbO_2	X	X	X	X
	PbO	X	X	X	X
	$Pb_3(OH)_2(CO_3)_2$	X	X	X	X
	$PbCO_3$	X	X	X	
	Pb				X
Si	SiO_x			X	
Sn	SnO	X	X	X	X
	Sn	X	X		X
	SnO_2				X
Zn	ZnO		X	X	
	ZnCl			X	

and cerussite in phase II. Calcium was found as $CaCO_3$ for phase III and both calcium phosphate ($Ca_3(PO_4)_2$) and calcium oxide (CaO) for phase IV. However, the high solubility of CaO would not permit the persistence of this solid form within an aqueous environment. Formation of the compound likely occurred while preparing the sample for analysis, during which constituents initially in soluble form remained on the coupon surface after drying. From the analysis, it may seem as though calcium carbonate is more likely to form while exposed to BOP than OP. Similarly, OP appears to stabilize calcium phosphate. However, these observations do not necessarily suggest a distinct difference in calcium scales in lead coupons in the presence of BOP and OP, but rather the lack of uniformity between the distribution of solid scales.

Oxygen was found to exhibit similar products to those of BOP. The forms of O^{2-}, CO_3^{2-}, and OH^- were determined for every phase, whereas PO_4^{3-} was found only for scale on the lead/tin coupons exposed to OP inhibition from phase IV. Generally, the detection of phosphate is less sensitive to the deconvolution of oxygen, given the presence of more abundant oxide and carbonate products; however, this observation likely coincides with the presence of a significant deposit of calcium phosphate. Phosphorus was determined to be present in the surface scale on lead/tin coupons for phases I, II, and IV, as opposed to BOP in which phosphorus was observed in the scale of lead/tin coupons for all phases. This difference in frequency should not be implied as a relevant distinction between BOP and OP. Rather the distinction between inhibitors pertains to the differences of phosphorus forms. As would be expected from application of OP in the system, the phosphorus present was as phosphates, whereas forms of polyphosphate and phosphate were found for coupons exposed to BOP.

Lead corrosion products were similar to those of BOP, with the exception of elemental lead that was detected during phase IV. Although PbO_2 was detected during phases I and III for BOP, it was relatively low. For the scale on lead/tin coupons exposed to OP, PbO_2 was found to contribute more as a lead corrosion product than BOP. During phases III and IV, PbO_2 accounted for 18% and 31% of lead. PbO, hydrocerussite, and cerussite were identified in the scale on lead/tin coupons exposed to OP for phases I, II, and III. Phase IV represented the only observation in which cerussite was not identified. Of the 23 lead/tin coupons that detected lead during the study, only the coupon exposed to OP during phase IV lacked the presence of cerussite. This same lead/tin coupon also represented the only sample during the study to detect elemental lead. The presence of elemental lead coincided with the identification of calcium phosphate. This observation may be relevant in demonstrating the ability of calcium phosphate to suppress the corrosion of lead. Of the 12 lead/tin coupons in which calcium was identified, calcium phosphate was only found for the lead/tin coupon exposed to OP during phase IV.

Silica was found present as SiO_x for phase III, which came from precipitation of silica in the raw water. Tin was identified as SnO and, for phase IV, stannic oxide (SnO_2). Elemental tin remained on the samples from phases I, II, and IV. Zinc deposits were also present within the scale for phases II, III, and IV as ZnO and, for phase IV, elemental zinc. The source of zinc was from the GS pipe used to build the PDSs.

Zinc Orthophosphate. A summary of the compounds identified in the scale on the lead/tin coupons exposed to ZOP during four phases of study are shown in Table 5.106. Carbon was analyzed and detected for all four phases for lead/tin coupons exposed to ZOP. The deconvolution of carbon was similar to OP in that it identified carbonates for all phases except for phase II. However, the peak associated with carbonate for the carbon deconvolution was generally less pronounced than the main peak. The presence of carbonates during phase II was validated by the identification of cerrusite, $PbCO_3$. Carbonates associated with calcium were identified for phases III and IV. The occurrence of calcium carbonate is similar to both BOP and OP. Calcium phosphate was not detected in the surface scale for lead/tin coupons exposed to ZOP and was found in only 1 of 12 scales for all lead/tin coupons exposed to phosphate-based inhibitors. Utilization of phosphate-based inhibitors or inhibition of lead release via phosphate-based inhibitors did not depend on the formation of a calcium phosphate scale in this work.

Analysis of oxygen found O^{2-}, CO_3^{2-}, and OH^- for all four phases, with the identification of PO_4^{3-} for phase IV. The detection of these forms of oxygen are similar to BOP and OP regarding the presence of O^{2-}, CO_3^{2-}, and OH^-. This observation is expected because the composition of the scales on the lead/tin coupons consist primarily of both lead and tin corrosion products. The corrosion products of lead account for a fraction of O^{2-}, CO_3^{2-}, and OH^-, whereas tin was associated with O^{2-}.

XPS scans of the lead/tin coupons exposed to ZOP identified phosphorus for phases I, III, and IV. Approximately 25% of the phosphorus was present as phosphate when phosphorus was detected on the lead/tin coupons. The remaining phosphorus was in metaphosphate, for phase IV, or pyrophosphate forms. Meta- and pyrophosphate solid forms were found in surface scales on lead/tin coupons in BOP and ZOP environments. As with both BOP and OP, the lead corrosion products determined for ZOP did not suggest the incorporation of phosphate. PbO, hydrocerussite,

Table 5.106
Compounds identified for ZOP coupons

Element	Compound	Phase I	Phase II	Phase III	Phase IV
C	CO_3^{2-}	X		X	X
Ca	$CaCO_3$			X	X
	CaO				X
O	O^{2-}	X	X	X	X
	CO_3^{2-}	X	X	X	X
	OH^-	X	X	X	X
	PO_4^{3-}				X
P	PO_4^{3-}	X		X	X
	Meta				X
	Pyro	X		X	X
Pb	PbO_2	X	X		X
	PbO	X	X	X	X
	$Pb_3(OH)_2(CO_3)_2$	X	X	X	X
	$PbCO_3$	X	X	X	X
Si	SiO			X	
Sn	SnO	X	X	X	X
	Sn	X		X	X
	SnO_2				X
Zn	ZnO	X	X	X	X
	ZnCl		X		
	$Zn(OH)_2$				X
	Zn				X

and cerussite were found present on each coupon exposed to ZOP for all phases. PbO_2 was determined for all phases, with the exception of phase III. However, even when PbO_2 was identified, it was less than 11% of the lead scale.

Silica was found only for phase III as SiO. Silica was not found for any lead/tin coupon exposed to BOP, but was determined during the same phase in which silica was detected for OP. The observation of silica was independent of ZOP addition and was rather influenced by silica present within the blended source water. Tin was found as SnO for all phases, while SnO_2 was found only for phase IV. Elemental tin remained on the samples from phases I, III, and IV but was relatively insignificant, accounting for less than 10% of tin.

Zinc was detected for all of the ZOP samples, predominantly as ZnO. The presence of zinc hydroxide ($Zn(OH)_2$) represented the only observation of the compound. $Zn(OH)_2$ was only observed in the surface scale for ZOP lead/tin coupons in phase IV. During phase IV, elemental zinc was also identified. The elemental form only accounted for 8% of the zinc analyzed for phase IV. Other than the sole identification of $Zn(OH)_2$, there were no differences among the zinc solids found on lead/tin coupons for all other phosphate inhibitors.

Table 5.107
Compounds identified for Si coupons

Element	Compound	Phase I	Phase II	Phase III	Phase IV
C	CO_3^{2-}	X		X	X
Ca	$CaCO_3$			X	X
	CaO				X
	$CaSiO_3$				X
O	O^{2-}	X	X	X	X
	CO_3^{2-}	X	X	X	
	OH^-	X	X	X	
	$CaSiO_3$				X
P					
Pb	PbO_2	X	X		
	PbO	X	X	X	X
	$Pb_3(OH)_2(CO_3)_2$	X	X	X	X
	$PbCO_3$	X	X	X	X
Si	SiO				X
	SiO_2				X
	SiO_x			X	X
	$CaSiO_3$				X
Sn	SnO	X	X	X	X
	Sn	X		X	X
Zn	ZnO	X		X	

Silicon. A summary of the compounds identified in the scale on the lead/tin coupons exposed to OP during four phases of study are shown in Table 5.107. The carbon scan identified carbonate during phases I, III, and IV for lead/tin coupons exposed to Si during four phases. The absence of carbonate for phase II is similar to the results obtained for BOP, OP, and ZOP. In comparison to BOP, OP, and ZOP, the detection of calcium carbonate was more probable for the lead/tin coupons exposed to Si because of the slightly higher pH. Addition of Si rendered an increasingly alkaline solution that would facilitate calcium carbonate precipitation. There was, however, no significant difference between Si and the phosphate-based inhibitors with respect to the occurrence and/or amount of calcium carbonate. Calcium carbonate was found for phases III and IV. $CaSiO_3$ was found in the scale on the lead/tin coupon from phase IV and accounted for 12% of the calcium analyzed. This observation would appear specific to the use of Si, given that the compound was not found within any other treatment.

From the deconvolution of oxygen, O^{2-}, CO_3^{2-}, and OH– were detected for all phases. As with BOP, OP, and ZOP, lead corrosion products were detected as PbO_2, PbO, hydrocerussite, and cerussite. PbO_2 was not detected for phases III and IV. For phases in which the compound was detected, it was found to account for less than 12% of the lead scale.

Detection of silica for phases I and II was indistinguishable from the background noise inherent in the lead/tin coupon survey scans. Despite failing to detect silica in the survey scan, a high-resolution scan was generated for phase II in which background noise was still prevalent. The forms of silica that were detected for phases III and IV suggest that silica may directly bond to the metal surface in addition to depositing SiO_2 and $CaSiO_3$ found during phase IV. Although silica was detected during phase III for OP, ZOP, and Si, the silica scale present on the lead/tin coupon exposed to Si was different from the phosphate inhibitor. A quantitative comparison of the silica scans suggests that the lead/tin coupons exposed to Si developed a denser scale. Providing that silica is a constituent of the source water, the study suggests that silica can exist within the scale, regardless of Si addition. However, application of Si would ensure the maintenance of a more predominant silica scale that seems to account for the differences in lead release observed between Si and pH control lines.

The relative fraction of ZnO in the surface scales of lead/tin coupons exposed to Si for phases I and III were similar to the relative fraction of ZnO in the surface scales of the lead/tin coupons exposed to ZOP. The detection of zinc was related to water quality by phase for surface scales produced on lead/tin coupons in the presence of BOP, OP, or Si.

pH. Tables 5.108 and 5.109 present the results from the XPS scans for lead/tin coupons that were not exposed to a corrosion inhibitor. These lead/tin coupons were exposed to source waters from PDS13 (pH_s) and PDS14 (pH_s+0.3) during four phases of operation. Deconvolution of carbon for pH_s and pH_s+0.3 denoted an effect due to the presence of carbonate for all phases, excluding phase II. This pattern was identical to OP, ZOP, and Si inhibitors. Given the identification of $PbCO_3$ for phase II, this would appear to represent a fallacy within the carbon scan. Calcium was detected as calcium carbonate during phases III and IV for both pH_s and pH_s+0.3. Similar observations were identified for BOP, OP, ZOP, and Si; however, there was no evidence of a calcium form other than calcium carbonate within pH_s or pH_s+0.3. The detection frequency between pH_s and pH_s+0.3 may seem contrary to the positive LSI of pH_s+0.3. Although calcium carbonate was observed during the same phases for both pH_s and pH_s+0.3, a quantitative assessment of the samples suggests that more deposition occurred within pH_s+0.3.

Compounds associated with oxygen were determined as O^{2-}, CO_3^{2-}, and OH^- for all phases, excluding phase III for pH_s and phase IV for pH_s+0.3. The coupon analyzed for pH_s during phase III only detected CO_3^{2-} within the oxygen scan. The failure to detect any form of lead, tin, or zinc (elements commonly identified with O^{2-}) suggests a predominance of calcium carbonate on the scale. As shown in Tables 5.108 and 5.109, phosphate-based compounds were not detected by XPS for pH_s and pH_s+0.3. The same observation was made regarding lead/tin coupons taken from PDSs receiving Si-based inhibitors. The absence of phosphorus from lead/tin coupons taken from the pH and Si PDSs indicates that phosphorus scale came from the phosphate-based inhibitors.

Lead corrosion products for pH_s and pH_s+0.3 were found to be similar regarding the detection of PbO_2, PbO, hydrocerussite, and cerussite. Differences between the inhibitors could be attributed to the absence of PbO_2 during phases I and IV for pH_s+0.3. However, the detection of PbO_2 is consistently low in all lead scales for all pH_s and pH_s+0.3 samples. Relative fractions of PbO_2, PbO, hydrocerussite, and cerussite for pH_s and pH_s+0.3 lead/tin coupon scales and those formed in the presence of other inhibitors are similar. Given the similarities in lead corrosion products between lead/tin coupons exposed to inhibitors and pH control lines, it would appear that the stabilization of a particular lead corrosion product does not represent the mechanism of any of the inhibitors used during the study.

Table 5.108
Compounds identified for pH_s coupons

Element	Compound	Phase I	Phase II	Phase III	Phase IV
C	CO_3^{2-}	X		X	X
Ca	$CaCO_3$			X	X
O	O^{2-}	X	X		X
	CO_3^{2-}	X	X	X	X
	OH^-	X	X		X
P					
Pb	PbO_2	X	X		
	PbO	X	X		X
	$Pb_3(OH)_2(CO_3)_2$	X	X		X
	$PbCO_3$	X	X		X
Si	SiO			X	
Sn	SnO	X	X		X
	Sn	X	X		
	SnO_2				X
Zn	ZnO				X

Table 5.109
Compounds identified for pH_s+0.3 coupons

Element	Compound	Phase I	Phase II	Phase III	Phase IV
C	CO_3^{2-}	X		X	X
Ca	$CaCO_3$				X
O	O^{2-}	X	X	X	X
	CO_3^{2-}	X	X	X	X
	OH^-	X	X	X	
P					
Pb	PbO_2		X	X	
	PbO	X	X	X	X
	$Pb_3(OH)_2(CO_3)_2$	X	X	X	X
	$PbCO_3$	X	X	X	X
Si	SiO			X	
Sn	SnO	X	X	X	X
	Sn	X			X
Zn	ZnO			X	

Silica was detected for both pH_S and pH_S+0.3 during phase III as SiO, which may have been due to the relatively high average silica concentration (11 mg/L as SiO_2) in the blended source water for phase III. The silica was also found in the lead surface scales formed in the presence of OP and ZOP during the same phase.

Corrosion products of tin were found in the form of oxides for both pH_S and pH_S+0.3. Results were consistent with other inhibitors in that a majority of the corrosion products were found as SnO. However, SnO_2 was detected during phase IV for pH_S. Elemental tin was identified during phases I and II for pH_S, and during phases I and IV for pH_S+0.3. Variations of elemental tin composition in the scale for the pH-only inhibitors were relatively similar between samples and accounted for approximately 9% (average of elemental tin when identified) of the tin formed in the scale.

Zinc was identified during phase IV for pH_S and during phase III for pH_S+0.3. Quantitatively, the amount of zinc was comparable to that of ZOP. The forms of zinc could be described by ZnO. It appears that the distinction between ZOP and other inhibitors pertains to surface uniformity found in the surface composition of the lead/tin coupons exposed to ZOP.

Summary for Lead Scale Surface Characterization

All lead corrosion products found in the scale on 50%/50% lead/tin coupons consisted of lead dioxide (PbO_2), lead oxide (PbO), cerussite ($PbCO_3$), or hydrocerussite ($Pb_3(CO_3)_2(OH)_2$). Average peak areas obtained from deconvolution of XPS-generated scans found that the predominant forms of lead corrosion products were lead oxide, cerussite, and hydrocerussite. Solubility properties regarding lead oxide would imply that either cerussite or hydrocerussite exists as the controlling solid phase for both pH_S and pH_S+0.3. Evidence of the high solubility of lead oxide is provided from the solubility diagram shown in Figure 5.89. Addition of a corrosion inhibitor did not affect the distribution of lead corrosion products in the surface scale of lead coupons as detected by XPS. However, solubility and equilibrium relationships suggest the possibility of a lead OP solid that would describe the effectiveness of phosphate-based inhibitors. The effect of Si was not well described from thermodynamic modeling. It is likely that the effectiveness of Si originates from a solid barrier that impedes an electrochemical connection.

Phosphorus was detected for all but two samples that were exposed to phosphate-based inhibitors. Coupons exposed to Si and pH inhibitors did not contain any form of phosphorus. Hence, a chemical reaction of phosphorus with lead occurred. The identification of silica on all surface scales formed on lead coupons exposed to all inhibitors shows that silica in the raw water is adequate to chemically react with lead and form a scale product on the lead coupons.

SURFACE ROUGHNESS

The following results are separated into sections that address galvanized steel, lined cast iron, polyvinyl chloride, unlined cast iron, copper, and lead/tin individually. Surface roughness of iron coupons was measured before exposure to inhibitors; however, after exposure, the newly formed scales on the iron surfaces were prone to cracking and weakly adhered to the coupon. Therefore, surface roughness of iron coupons was not measured after exposure to inhibitors.

To assess the effect of inhibitors on surface roughness, the roughness of coupons exposed to varying inhibitors is compared to the surface roughness of the coupon exposed to the control, pH_S+0.3. Similarly, to assess the effect of pH on surface roughness, the surface roughness of the

Table 5.110
Water quality by phase

Water quality	Phase I	Phase II	Phase III	Phase IV
GW (%)	62	27	62	40
SW (%)	27	62	27	40
RO (%)	11	11	11	20
Alkalinity (mg/L as $CaCO_3$)	160	103	150	123
Chlorides (mg/L Cl^-)	45	67	68	59
Sulfates (mg/L SO_4^{2-})	62	103	67	76
Temperature (°C)	21.3	26.2	25.7	21.2

coupons exposed to the pH_s and pH_s+0.3 environments are compared. To assess the effect of water quality on surface roughness, changes in surface roughness are compared by phase, considering major differences in water quality by phase, as shown in Table 5.110.

Phases I and III had the same blend of mostly groundwater (62%) with surface water (27%) and desalinated water by RO (11%). Phases I and III had elevated alkalinities (160 mg/L as $CaCO_3$ and 150 mg/L as $CaCO_3$, respectively) and relatively low sulfates (62 mg/L and 67 mg/L, respectively). In contrast, chlorides (45 mg/L and 68 mg/L, respectively) and temperature (21.3°C and 25.7°C, respectively) were different for each phase. Because phases I and III have the same blend of source waters, comparison of roughness from these two phases shows the effect of temperature and chlorides on surface roughness.

Phase II had a blend of mostly surface water (62%) with groundwater (27%), and desalinated water by RO (11%). Because it was predominately surface water, phase II had lower alkalinity (103 mg/L as $CaCO_3$) and higher sulfate (103 mg/L). Phase IV had a blend of desalinated water by RO (40%) with groundwater (20%) and surface water (20%). Phase IV had moderate alkalinity (123 mg/L as $CaCO_3$) and relatively low sulfate (76 mg/L).

Surface Roughness of Galvanized Steel

All GS coupons used in this study were cut from sections of pipe that had been excavated from within the distribution system and held in indoor storage for a few years, exposed to the air. Consequently, there is variation in the initial surface roughness of GS coupons. Over all four phases, 71 GS coupons were measured with an average initial roughness of 24.7 μm and a minimum and maximum of 11.9 μm and 40.4 μm, respectively. Figure 5.96 shows examples of a relatively smooth and rough GS coupon at actual size. The average initial roughness, R_i, for each coupon shown is 23.8 μm and 40.4 μm, respectively.

A summary of the GS surface roughness data is shown in Table 5.111. Replicates coupons for phases III and IV are shown in Table 5.112. A statistical comparison of the initial and final roughness averages was completed using a two-tailed pooled-data t-test. The statistical analysis is summarized in Table 5.111. The *p*-values are provided for the hypothesis tests for equality of initial and final roughness. A reported *p*-value of 0.05 or less indicates there is sufficient evidence to conclude that the final roughness is different from the initial roughness (level of significance of 95%). If the *p*-value of the t-test comparing the means of the surface roughness before and after

Figure 5.96 Photographs before exposure to inhibitors of a typical smooth GS coupon (left) and a rougher GS coupon (right)

exposure to the inhibitor is greater than 0.05, then there is no statistically significant difference in coupon surface roughness due to exposure to the inhibitor.

Effect of Inhibitors

Phase I had a predominately groundwater blend with elevated alkalinity and low temperature. The average water temperature for phase I was 21.2°C. In phase I, there was no evidence that inhibitors changed GS surface roughness relative to no inhibitor addition. This is shown in Table 5.111, as the surface roughness of only 4 of 14 coupons changed significantly following exposure in the coupon cradles. Surface roughness of GS did not change significantly for the pH_S+0.3 control or for any of the inhibitors. In contrast, roughness increased significantly for the pH_S coupon, from 15.5 to 28.3 μm. There was considerable variation in GS surface roughness prior to exposure, which was expected for the aged GS coupons. Even at a 0.05 *p*-value, there is a 5% probability the surface roughness of the GS coupon exposed to the pH_S environment did not actually change. The phase I roughness data indicates that inhibition had no practical effect on surface roughness.

Phase II had a predominately surface water blend with relatively low alkalinity and elevated sulfate. The average water temperature for phase II was 26.2°C. The surface roughness of 4 of the 14 GS coupons exposed to inhibitors changed significantly in phase II. Surface roughness of the GS coupon decreased from 26.6 to 18.7 μm in the presence of 1.0 mg/L as P-BOP, from 40.4 to 25.9 μm in the presence of 1.0 mg/L as P-OP, and from 24.1 to 16.7 μm in the presence of 2.0 mg/L as P-ZOP. Similarly, surface roughness of the GS coupon for the pH_S+0.3 control decreased significantly from 27.9 to 17.3 μm. The surface roughness for all remaining eight coupons—which included two from each phosphate-based inhibitor, three coupons from the silicate inhibitor, and one from the pH_S inhibitor—did not change significantly. The phase I and II data indicate that the inhibitors do not practically affect surface roughness, when compared to the no-inhibitor control, pH_S+0.3.

Phase III water quality was similar to phase I water quality and was dominated by groundwater (62%). However, the average temperature of phase III water was 25.8°C. Surface roughness for nearly all of the GS coupons changed significantly in phase III. Ten of the 14 GS coupon

Table 5.111
Statistical comparison of galvanized steel initial and final surface roughness

		PDS01	PDS02	PDS03	PDS04	PDS05	PDS06	PDS07	PDS08	PDS09	PDS10	PDS11	PDS12	PDS13	PDS14	Average
Phase	Parameter	BOP 0.5	BOP 1.0	BOP 2.0	OP 0.5	OP 1.0	OP 2.0	ZOP 0.5	ZOP 1.0	ZOP 2.0	Si 3.0	Si 6.0	Si 12.0	pH_s	pH_s+0.3	
I	R_i	19.6	11.9	18.7	26.9	21.9	23.2	21.5	16.3	19.7	19.3	17.2	26.5	15.5	23	20.1
	N	3	3	3	3	3	5	3	3	3	3	3	3	3	3	-
	R_f	18.9	18.4	16.7	22.9	20.7	22.6	17.9	15.5	23.9	15.8	17.3	23.7	28.3	27.1	20.7
	N	3	3	5	5	3	5	5	3	5	3	5	5	5	6	-
	ΔR	–0.7	6.5	–2.0	–4.1	–1.3	–0.7	–3.6	–0.8	4.2	–3.4	0.1	–2.8	12.8	4.1	0.6
	ΔR *p*-value	0.55	0.19	0.40	0.26	0.69	0.78	0.41	0.79	0.10	0.46	0.98	0.41	**0.05**	0.57	-
	ΔR/R_i	–0.04	0.55	–0.11	–0.15	–0.06	–0.03	–0.17	–0.05	0.21	–0.18	0.01	–0.11	0.83	0.18	0.06
II	R_i	25	26.6	22.9	17.1	40.4	22.7	28.9	33.3	24.1	16.7	26.3	17	14.7	27.9	24.5
	N	5	5	5	5	5	5	5	5	5	5	5	5	5	5	-
	R_f	21.6	18.7	27.7	20.3	25.9	-	21.1	27.4	16.7	21.4	22.3	16.5	15.1	17.3	20.9
	N	6	5	5	5	6	-	5	6	5	6	5	6	5	5	-
	ΔR	–3.4	–7.8	4.8	3.2	–14.5	-	–7.8	–5.9	–7.4	4.7	–4.0	–0.5	0.4	–10.5	–3.7
	ΔR *p*-value	0.11	**0.03**	0.60	0.06	**0.02**	-	0.16	0.35	**0.03**	0.24	0.34	0.84	0.80	**0.03**	-
	ΔR/R_i	–0.14	–0.29	0.21	0.19	–0.36		–0.27	–0.18	–0.31	0.28	–0.15	–0.03	0.03	–0.38	–0.11
III	R_i	22.6	34.1	19.9	21.7	20.2	36	22.1	21.4	26.5	20.2	23.8	20.4	26.1	24.4	24.2
	N	8	8	8	8	8	8	8	8	8	8	8	8	8	8	-
	R_f	19.3	27.1	24.8	30	26.7	28.7	26.2	27.9	22.1	29.1	29.9	30.4	25.9	28.5	26.9
	N	8	8	8	8	8	8	9	8	8	8	9	8	8	8	-
	ΔR	–3.3	–7.0	4.9	8.3	6.5	–7.2	4.2	6.5	–4.3	8.9	6.1	10.0	–0.2	4.1	2.7
	ΔR *p*-value	0.28	0.05	**<0.01**	**<0.01**	**<0.01**	**0.01**	**0.03**	**0.04**	**0.02**	**<0.01**	0.09	**<0.01**	0.93	0.14	-
	ΔR/R_i	–0.15	–0.21	0.25	0.38	0.32	–0.20	0.19	0.30	–0.16	0.44	0.26	0.49	–0.01	0.17	0.15
IV	R_i	23.7	25.9	25.9	33.8	31.8	31.9	30.2	26.1	25.4	37.3	26.1	23.8	23.8	28.9	28.2
	N	8	8	8	8	7	8	8	8	8	8	8	8	8	8	-
	R_f	27.0	24.2	20.7	34.0	27.6	37.2	25.1	30.4	26.6	34.8	27.0	26.3	29.4	25.5	28.3
	N	8	8	8	8	8	8	8	8	8	8	8	8	8	8	-
	ΔR	3.3	–1.7	–5.3	0.2	–4.2	5.2	–5.1	4.3	1.3	–2.5	1	2.6	5.6	–3.4	0.1
	ΔR *p*-value	0.07	0.30	**<0.01**	0.92	0.10	0.18	0.10	0.25	0.66	0.27	0.70	0.20	**0.04**	0.42	-
	ΔR/R_i	0.14	–0.07	–0.20	0.01	–0.13	0.16	–0.17	0.16	0.05	–0.07	0.04	0.11	0.24	–0.12	0.01

NOTES: R_i = initial roughness; R_f = final roughness; ΔR = delta roughness. Boldface indicates *p*-values < 0.05.

Table 5.112
Replicate roughness data for GS coupons

		PDS01	PDS02	PDS03	PDS04	PDS05	PDS06	PDS07	PDS08	PDS09	PDS10	PDS11	PDS12	PDS13	PDS14
Phase	Parameter	BOP 0.5	BOP 1.0	BOP 2.0	OP 0.5	OP 1.0	OP 2.0	ZOP 0.5	ZOP 1.0	ZOP 2.0	Si 0.5	Si 01.0	Si 2.0	pH_s	pH_s+0.3
III	R_i	-	21.4	-	-	32.6	-	-	18.8	-	-	17.4	-	17.6	24.7
	N	-	8	-	-	8	-	-	8	-	-	8	-	8	7
	R_f	-	33.8	-	-	32.7	-	-	23.4	-	-	29.5	-	20.5	37.8
	N	-	8	-	-	8	-	-	9	-	-	8	-	8	8
	ΔR	-	12.3	-	-	0.1	-	-	4.6	-	-	12.2	-	2.9	13.1
	ΔR *p*-values	-	**0.01**	-	-	0.94	-	-	0.17	-	-	**0.04**	-	0.09	**0.02**
IV	R_i	-	31.2	26.2	-	26.8	-	-	32.8	33.4	-	27.1	28.4	34.3	25.3
	N	-	8	8	-	8	-	-	8	8	-	8	8	8	8
	R_f	-	26.7	33.5	-	27.9	-	-	26.5	30.5	-	32.5	33.1	24.1	30.0
	N	-	8	8	-	8	-	-	8	8	-	8	8	8	8
	ΔR	-	–4.5	7.4	-	1.1	-	-	–6.2	–2.9	-	5.4	4.7	–10.2	4.7
	ΔR *p*-values	-	**0.02**	0.16	-	0.65	-	-	0.08	0.43	-	0.12	0.07	**<0.01**	0.06

NOTES: R_i = initial roughness; R_f = final roughness; ΔR = delta roughness. Boldface indicates *p*-values < 0.05.

surface roughness reported in Table 5.111 changed significantly; whereas surface roughness of 4 of the 14 GS coupons did not change. There was no relationship between GS changing surface roughness and inhibitor or inhibitor dose. For BOP inhibitor, GS surface roughness increased at 1.0 mg/L as P and 2.0 mg/L as P. For both OP and ZOP inhibitors, GS surface roughness increased at 0.5 and 1.0 mg/L as P, while decreasing at 2.0 mg/L as P. For silicate inhibitor, GS surface roughness increased significantly at every dose. The average surface roughness of the 10 GS coupons that significantly changed was 24.2 μm and 27.3 μm before and after inhibitor exposure, respectively; whereas the average surface roughness of the 4 GS coupons that did not change was 24.2 μm and 25.9 μm before and after incubation, respectively. GS surface roughness decreased in five and increased in nine observations. For the pH_s+0.3 control, one coupon had a significant increase in GS surface roughness, whereas the other did not. In contrast, surface roughness of GS did not change significantly for either of the pH_s coupons. The major difference between phase III and phases II and IV was water quality, and the major difference between phase I and phase III was temperature. Hence, these results suggest that the change in GS surface roughness was correlated to relatively high alkalinity and temperature and not to the inhibitor or inhibitor dose.

In phase IV, only 2 of 14 GS coupons had a significant change in surface roughness during exposure to inhibitors. For BOP inhibitor, the GS surface roughness changed significantly during exposure, decreasing from 25.9 to 20.7 μm. However, the matching coupon for the BOP replicate did not have any significant change in GS surface roughness at 1.0 mg/L as P. None of the other GS coupons, including replicates, exposed to P or Si inhibitor environments had a significant change in surface roughness. Similarly, for the pH_s+0.3 control, surface roughness did not change significantly for either coupon. In contrast, GS surface roughness did change significantly for both coupons exposed to the pH_s environment; however, it increased for one coupon while decreasing for the other coupon. The phase IV GS surface roughness data indicates that inhibitor type or dose had no effect on surface roughness.

The change in ΔR and $\Delta R/R_i$ was analyzed with respect to initial roughness, R_i, as shown in Figures 5.97 and 5.98. As noted in chapter 3, the measurement of roughness was enhanced in the beginning of the third phase. The data was normalized with respect to initial roughness using the ratio $\Delta R/R_i$ to remove any bias that may have been caused by measurement technique.

Even with this normalization, Figures 5.97 and 5.98 show the same trend for both analyses. Coupons with lower initial roughness tended to have positive changes in surface roughness, and coupons with higher initial roughness tended to have negative changes in surface roughness. The trend lines shown in Figures 5.97 and 5.98 are for all data, although the data is separated by phase in these figures. The average GS roughness was approximately 25 μm, which deviated by as much as 15 μm or from –40% to 80% of the initial roughness following inhibitor exposure. The higher change in roughness is clearly shown in Figures 5.97 and 5.98 for the phase III GS coupons. The initial roughness of the GS coupons for phases I, II, III, and IV were 20.7 μm, 24.5 μm, 24.2 μm, and 28.2 μm, respectively, as shown in Table 5.111. Hence, the greater change of the phase III GS coupons cannot be attributed to greater initial roughness, but rather to the effects of higher temperature and chlorides.

A model was developed to describe the final roughness of the GS coupons as function of the mean, or average roughness, as shown in Equation 5.62. This simple model allows the predicted change of roughness to vary indirectly with initial roughness, which is what was observed for the actual data.

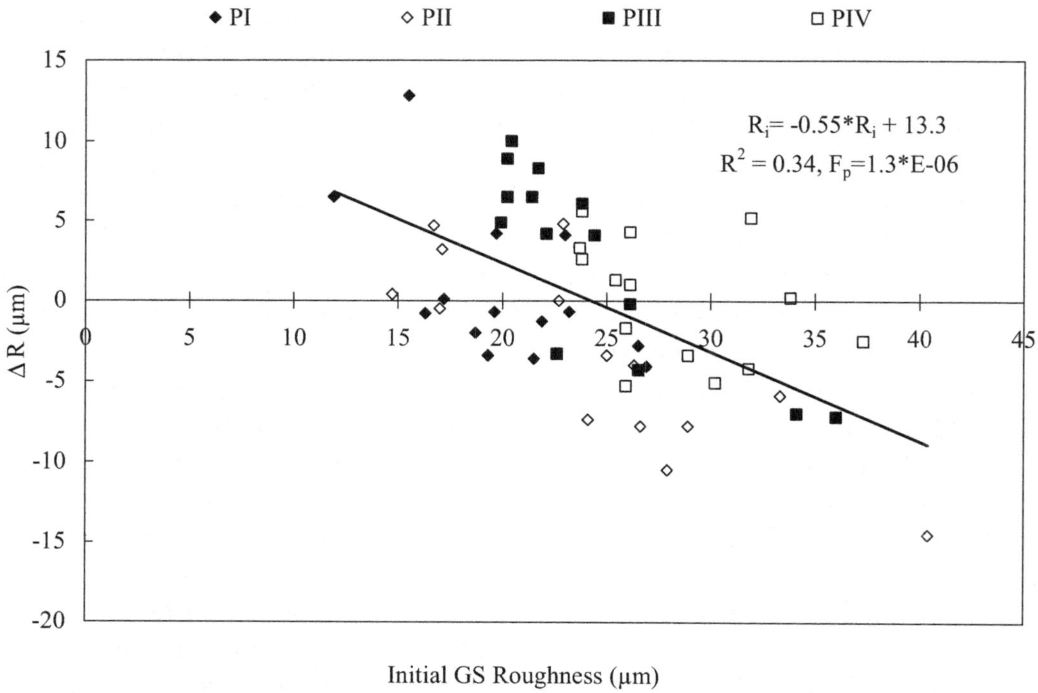

Figure 5.97 Change of initial roughness, ΔR, versus initial roughness, R_i, for GS coupons by phase

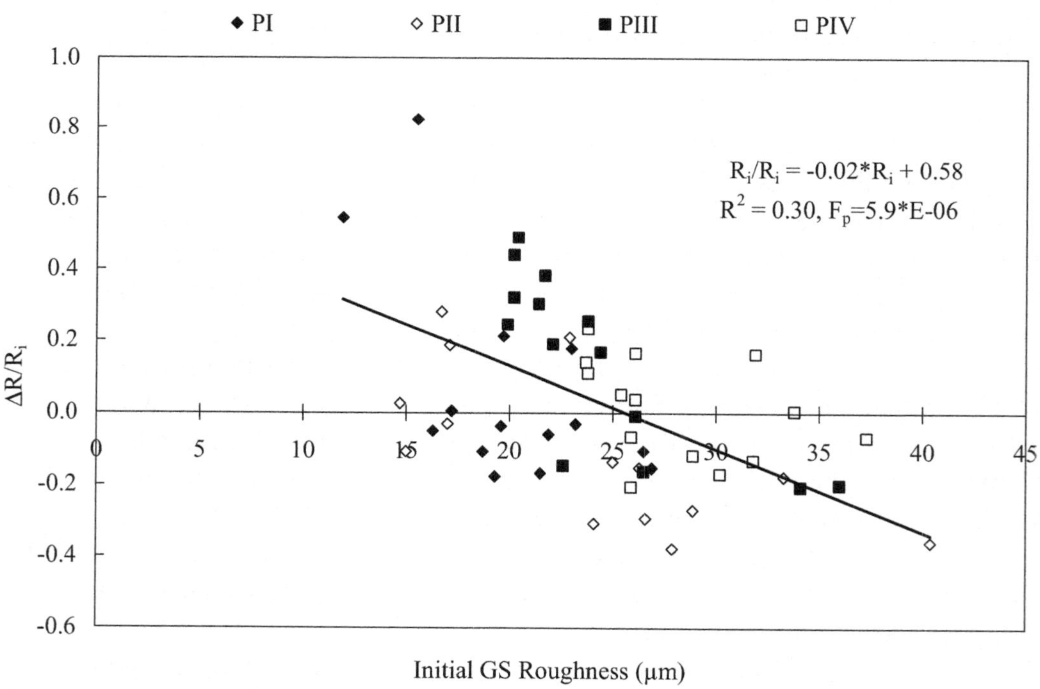

Figure 5.98 Change of the ratio of roughness to initial roughness, $\Delta R/R_i$, versus initial roughness, R_i, for GS coupons by phase

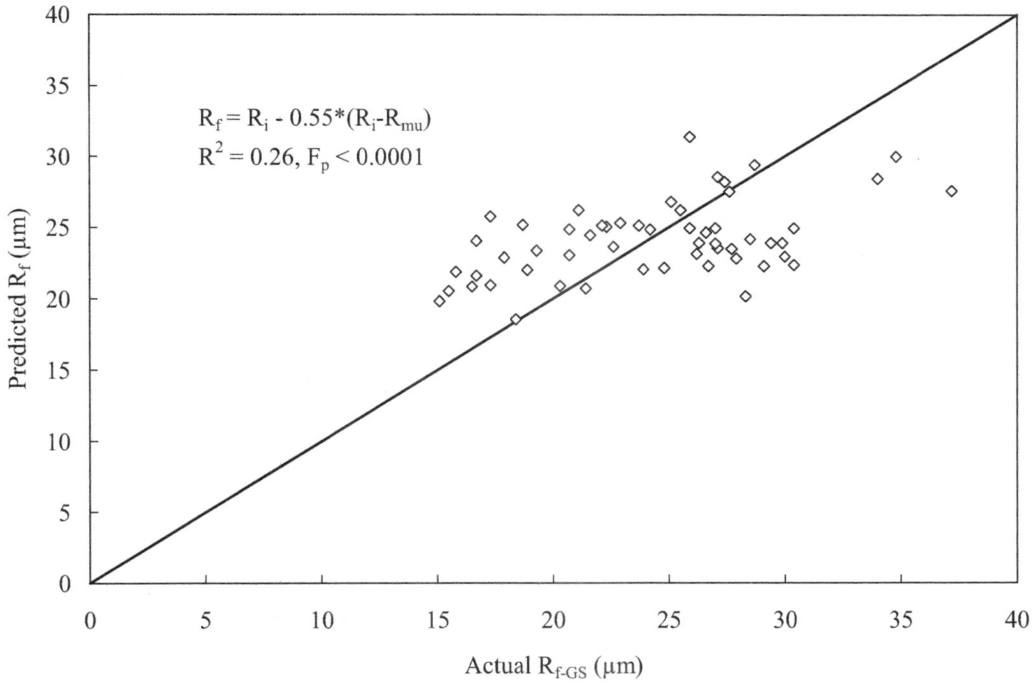

Figure 5.99 Predicted versus final GS surface roughness

$$R_f = R_i - m(R_i - R_{mu}) \quad (5.62)$$

where R_f = final roughness, µm
R_{mu} = mean roughness, µm
m = gain term defined by regression (0.55 for GS)

The R^2 for the model was 0.26, which is low, but F probability was less than 0.0001. The magnitude of these statistical parameters means that the model did not fit the data well but confirms that the data had a trend. The data shown in Figures 5.97 and 5.98 and by the model F_p values (probability based on F statistic) show that the change in roughness of the GS coupons always moved towards a mean or average value, which generated positive and negative changes of surface roughness regardless of inhibitor or inhibitor dose. The predicted and actual final surface roughness in shown in Figure 5.99.

Replicates

A series of replicate GS coupons were analyzed in phases III and IV to gain insight into variability of the roughness experiments. The results of those analyses are shown in Tables 5.113 and 5.114. A visual comparison of the surface roughness of GS coupon versus the surface roughness of the replicated GS coupon is shown in Figure 5.100.

Six replicates of GS surface roughness are shown in Table 5.113. The change in surface roughness in both the replicate and the original GS coupon was statistically significant in only one of six pairs for phase III replicates. However, both GS coupons in four of the replicate pairs

Table 5.113
Paired replicate surface roughness data for GS coupons from phase III

	PDS02	PDS02-R	PDS05	PDS05-R	PDS08	PDS08-R	PDS11	PDS11-R	PDS13	PDS13-R	PDS14	PDS14-R
Parameter	BOP 1.0	BOP 1.0	OP 1.0	OP 1.0	ZOP 1.0	ZOP 1.0	Si 1.0	Si 01.0	pH_s	pH_s	pH_s+0.3	pH_s+0.3
R_i	34.1	21.4	20.2	32.6	18.8	21.4	23.8	17.4	14.7	24.7	27.9	24.7
N	8	8	8	8	8	8	8	8	5	7	5	7
R_f	27.1	33.8	26.7	32.7	23.4	27.9	29.9	29.5	15.1	37.8	17.3	37.8
N	8	8	8	8	9	8	9	8	5	8	5	8
ΔR	–7	12.3	6.5	0.1	4.6	6.5	6.1	12.2	0.4	13.1	–10.5	13.1
ΔR *p*-value	0.05	**0.01**	**<0.01**	0.94	0.17	**0.04**	0.09	**0.04**	0.80	**0.02**	**0.03**	**0.02**

NOTES: R_i = initial roughness; R_f = final roughness; ΔR = delta roughness. Boldface indicates p-values < 0.05.

Table 5.114
Paired replicate surface roughness data for GS coupons from phase IV

	PDS 02	PDS 02-R	PDS 05	PDS 05-R	PDS 08	PDS 08-R	PDS 11	PDS 11-R	PDS 12	PDS 12-R	PDS 13	PDS 13-R	PDS 14	PDS 14-R
Parameter	BOP 1.0	BOP 1.0	OP 1.0	OP 1.0	ZOP 1.0	ZOP 1.0	Si 1.0	Si 1.0	Si 2.0	Si 2.0	pH_s	pH_s	pH_s +0.3	pH_s +0.3
R_i	23.7	90.8	31.8	99	26.1	112.3	26.1	122.8	23.8	79.9	23.8	92.7	28.9	95.2
N	8	5	7	8	8	8	8	5	8	5	8	5	8	5
R_f	27	81.6	27.6	71.6	30.4	76.8	27.0	91.3	26.3	98.2	29.4	100.4	25.5	90.6
N	8	6	8	5	8	5	8	5	8	5	8	6	8	8
ΔR	3.3	–9.2	–4.2	–27.4	4.3	–35.5	1.0	–31.4	2.6	18.2	5.6	7.6	–3.4	–4.6
ΔR *p*-value	0.07	0.46	0.1	0.06	0.25	**<0.01**	0.70	**<0.01**	0.20	0.16	**0.04**	0.63	0.42	0.44

NOTES: R_i = initial roughness; R_f = final roughness; ΔR = delta roughness. Boldface indicates p-values < 0.05.

experienced an increase in GS surface roughness during exposure, whereas in the remaining two of the six pairs, surface roughness increased for one coupon and decreased for the paired coupon. The replicate data for phase IV is shown in Table 5.114. Four of the seven pairs of GS replicate coupons experienced similar changes, with surface roughness of both coupons increasing or decreasing together.

ΔR of the replicate coupon pairs are plotted in Figure 5.100 and are significantly variable, which shows that replication of surface analysis was only achievable within ±30 to 40 μm for the GS coupons. Differences in R_i of the replicate coupon pairs make direct comparison of the ΔR ineffective to establish systems-level precision. Traditional calculation of precision for ΔR and $ΔR/R_i$ yielded RPDs, or errors averaging 300% and 100%, respectively.

The RPD is a measure of precision and is calculated by taking the absolute difference of two values, dividing it by the average of the same two values, and expressing the result as a percentage. Such large RPDs generated a useful comparison of differences in ΔR and $ΔR/R_i$ of the original and replicate, which were assessed, as shown in Table 5.115 for the replicate GS coupon

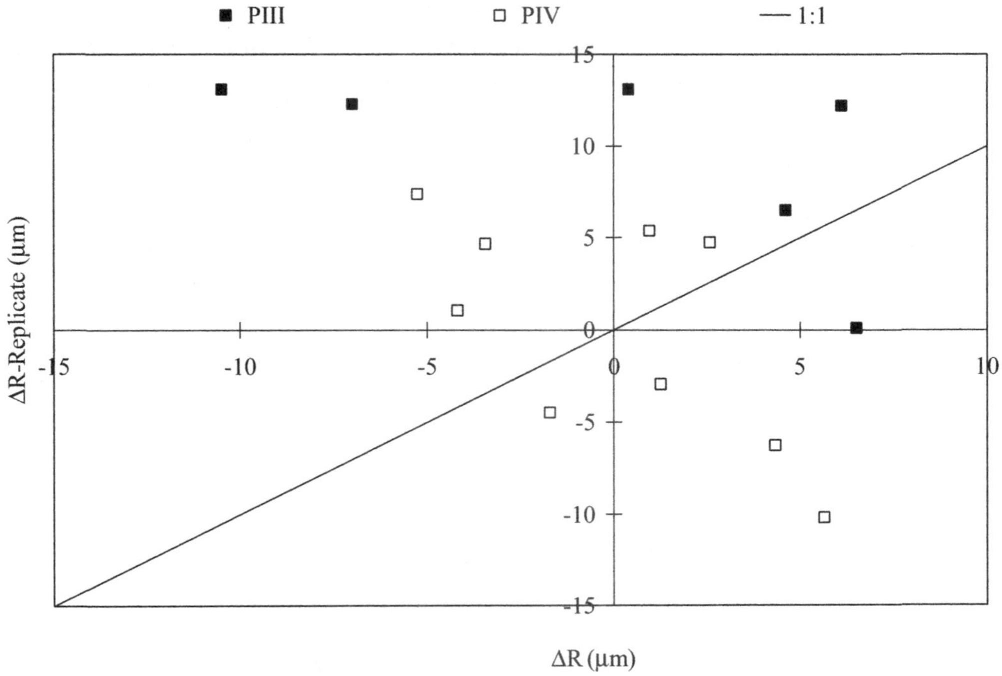

Figure 5.100 Comparison of the change in surface roughness among replicate GS coupons during phases III and IV

in PDS1 (BOP-1.0). The RPD for ΔR and $\Delta R/R_i$, as shown in Table 5.115, is –728% and –422%, which is based on a change in surface roughness of –7 μm for the original GS coupon and +12.3 μm for the replicate coupon. Because the average change in roughness is small, the errors calculated using traditional RPD methods are large. This suggests that traditional RPD methods are not well suited for making practical assessments of changes in surface roughness of GS coupons. However, ΔR was replicated within ±30 to 40 μm for the GS coupons. As noted previously, the GS coupon with higher initial surface roughness experienced a decrease in surface roughness, and the coupon with the lower initial surface roughness experienced an increase in surface roughness. In this case, the initial surface roughness of the original coupon was greater than the mean, and the initial surface roughness of the replicate was less than the mean. The surface roughness of both coupons tended to change towards the mean. That trend was shown in Figure 5.97 and confirmed mathematically by statistical verification of the model shown in Equation 5.62. Those analyses demonstrated that GS surface roughness approached an equilibrium value, which is represented by R_{mu}. Nonetheless, the changes in surface roughness of galvanized steel were not correlated to inhibitor or inhibitor dose in this work and are probably influenced by other factors that were not measured.

Figures 5.101–5.104 present an alternative measure of experimental precision at the systems level. The first pair of figures shows ΔR versus R_i for GS coupons (Figure 5.101) and GS replicate coupons (Figure 5.102). The second pair of figures shows $\Delta R/R_i$ versus R_i for GS

Table 5.115
Example RPD calculation assessing variation in difference of roughness, GS

		R_i	R_f	ΔR	$\Delta R/R_i$	RPD-ΔR	RPD-$\Delta R/R_i$
PDS02	BOP-1.0	34.1	27.1	–7	–0.21	–728%	–422%
PDS02-R	BOP-1.0	21.4	33.8	12.3	0.57	—	—

NOTES: R_i = initial roughness; R_f = final roughness; ΔR = delta roughness.

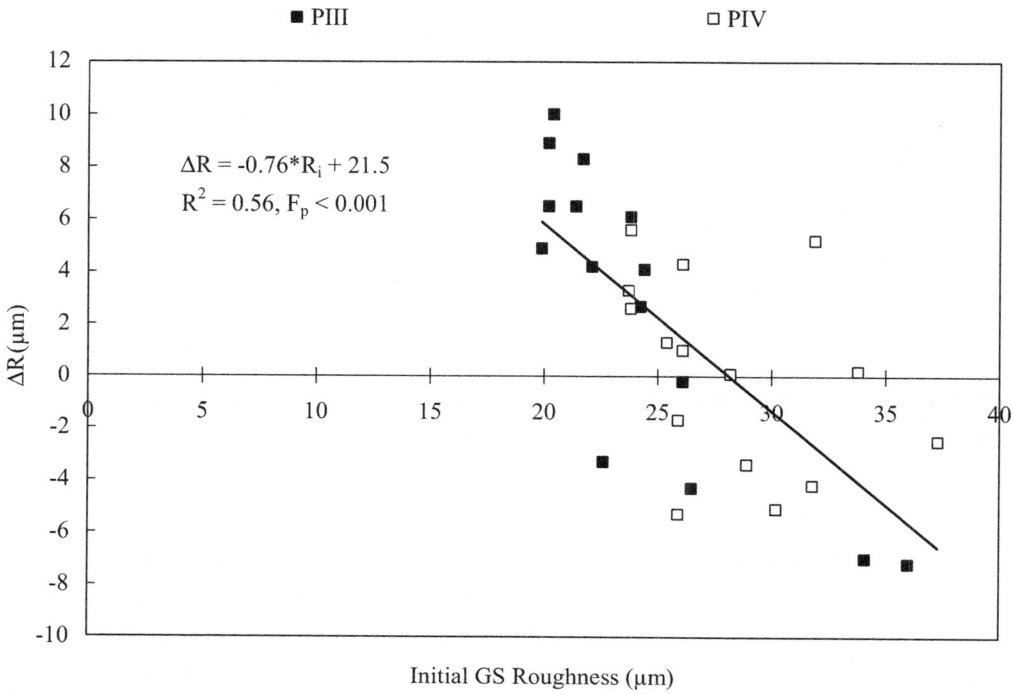

Figure 5.101 Change of initial roughness, ΔR, versus initial roughness, R_i, for GS coupons for phases III and IV

coupons (Figure 5.103) and GS replicate coupons (Figure 5.104). All figures show the same trend, with ΔR being positive when R_i is less than the mean GS surface roughness, and ΔR being negative when R_i is greater than the mean GS surface roughness. Experimental precision can be estimated by comparing the estimated mean GS surface roughness from the data shown in Figure 5.101 (mean roughness for GS coupons estimated at 28 μm) and Figure 5.102 (mean roughness of replicate GS coupons estimated at 30 μm), which gives an RPD of 7%.

Effect of Water Quality

Although inhibitors, including pH elevation, had no affect on surface roughness of GS coupons, there was considerable variation in GS surface roughness when water quality was varied by phase. Hence, the effect of water quality on GS surface roughness was evaluated.

ΔR versus the average alkalinity by phase is shown in Figure 5.105, and it increases directly with alkalinity. The trend line and associated statistical parameters are also shown in

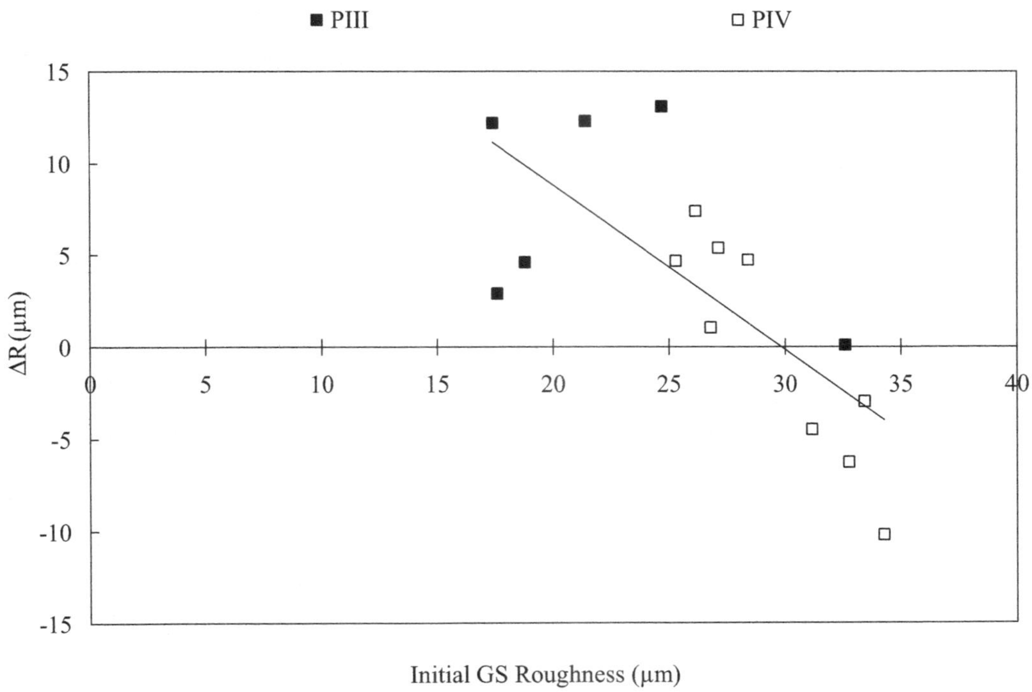

Figure 5.102 Change of initial roughness, ΔR, versus initial roughness, R_i, for replicate GS coupons for phases III and IV

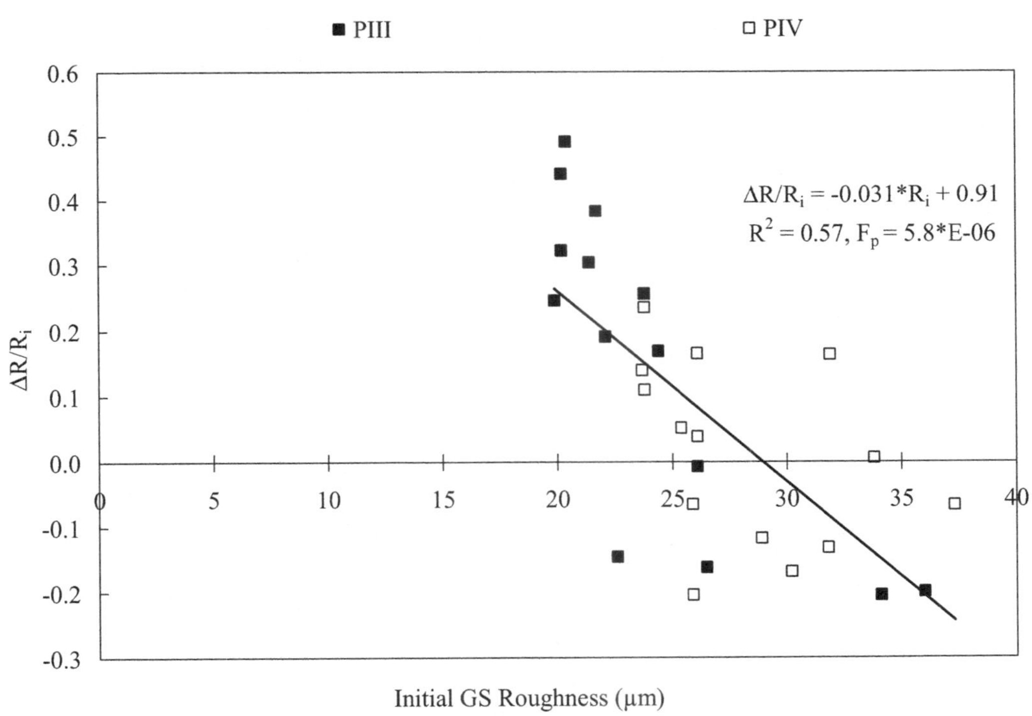

Figure 5.103 Change of the ratio of delta roughness to initial roughness, $\Delta R/R_i$, versus initial roughness, R_i, for GS coupons for phases III and IV

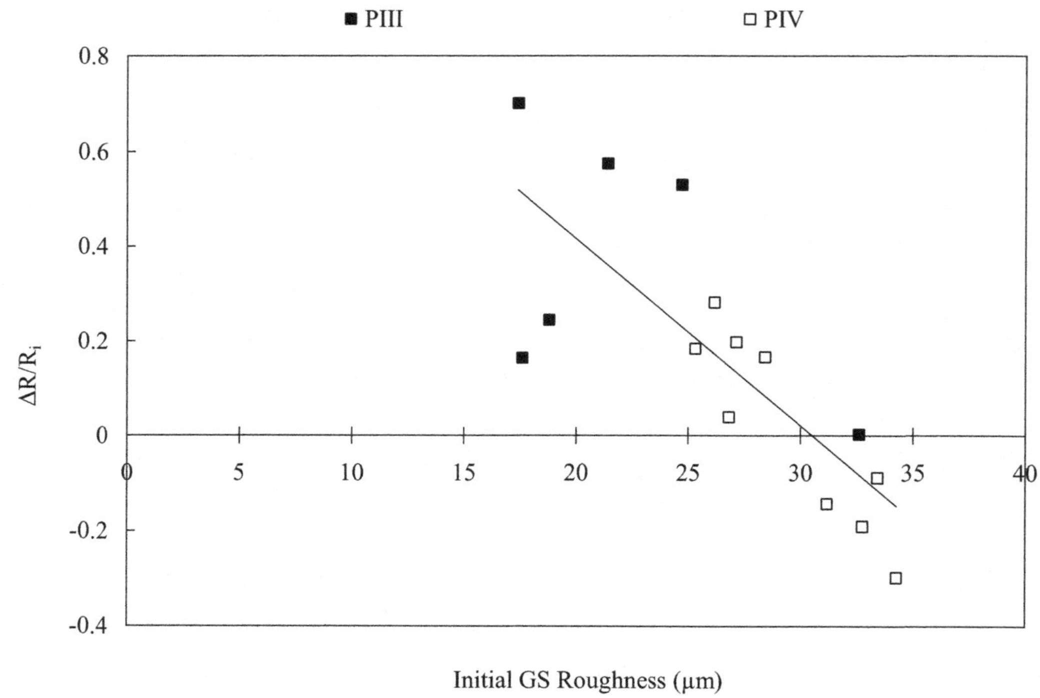

Figure 5.104 Change of initial roughness, ΔR, versus initial roughness, R_i, for replicate GS coupons for phases III and IV

Figure 5.105. Within each phase, there is much variability in the change in roughness; however, the trend with alkalinity and roughness across all phases is significant at the 95% level, with an F_p of 0.03. However, Figure 5.105 only considers the relationship with alkalinity, which may be confounded when the effects of the temperature, sulfates, and chlorides are considered.

Linear and nonlinear models, as shown in Equations 5.63 and 5.64, were evaluated for the effects of alkalinity, temperature, sulfates, and chlorides on the change in GS surface roughness during exposure to inhibition. R_i was moved to the left side of Equation 5.62 and the water quality parameters were incorporated in these models to develop Equations 5.63 and 5.64.

$$\Delta R = m \times (R_i - R_v) + X_1 \times alk + X_2 \times T + X_3 \times SO_4 + X_4 \times Cl \qquad (5.63)$$

$$\Delta R = m \times (R_i - R_v) \times alk^a \times T^b \times SO_4{}^c + Cl^d \qquad (5.64)$$

where X_i = statistically determined regression constants
a,b,c,d = statistically determined regression coefficients
alk = PDS alkalinity, mg/L as $CaCO_3$
T = PDS average phase temperature, °C
SO_4 = PDS average phase sulfates, mg/L
Cl = PDS average phase chlorides, mg/L

The result from the regression of Equation 5.63 is shown in Equation 5.65. None of the water quality parameters were significant in the linear model, which is nearly identical to the

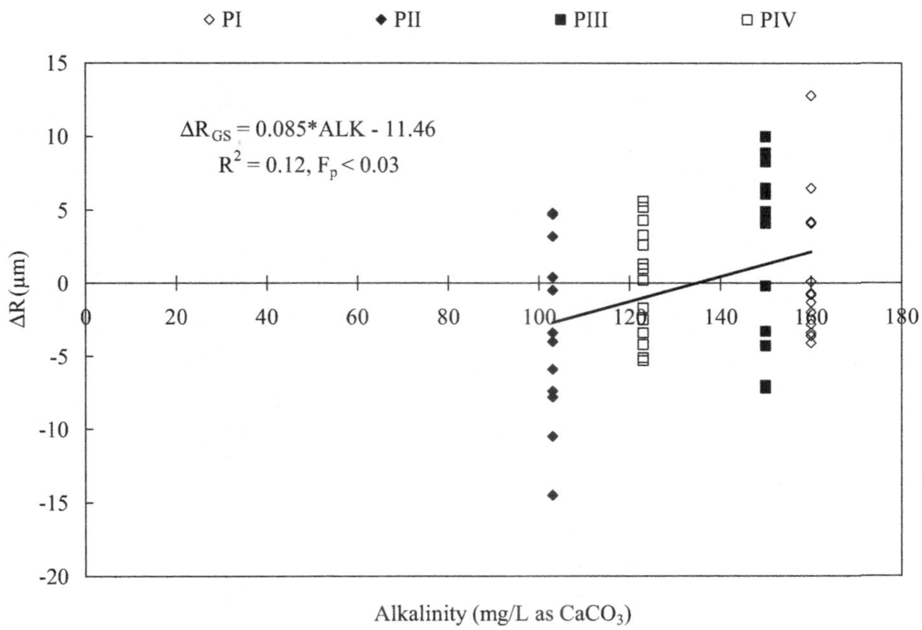

Figure 5.105 Change of GS surface roughness, ΔR, versus alkalinity

model presented in the "Effect of Inhibitors" section. However, the effects of chemical parameters are often better described by a power model because of its similarity to equilibrium quotients.

$$\Delta R = 0.42 \times (R_i - 24.3) \tag{5.65}$$

where $R^2 = 0.14$, $p < 0.009$, $F_p < 0.003$

The results of the regression of Equation 5.64 are shown in Equation 5.66. Temperature was the only water quality parameter that was found to be significant, with the m term dropping out of the final model. For this study, the analyses described in this section indicate that chemical water quality, including temperature, does clearly influence the change in roughness of galvanized steel.

$$\Delta R_{GS} = (R_i - 24.3) \times T^{-0.27} \tag{5.66}$$

where $R^2 = 0.14$, $p < 0.009$, $F_p < 0.003$

Summary for Surface Roughness of Galvanized Steel

The surface roughness of GS coupons was measured for phases I, II, III, and IV. GS coupons were cut from GS pipe that had been in service underground for more than two decades. Before cutting, the pipe had been stored in an open atmosphere for several years. The initial surface roughness of GS coupons varied from 11.9 to 37.3 μm and averaged 24.7 μm.

Figure 5.106 Photographs before exposure to inhibitors of a typical smooth LCI coupon (left) and a rougher LCI coupon (right)

The number of statistically significant changes in GS surface roughness by phase was 1 of 14 for phase I, 4 of 13 for phase II, 10 of 14 for phase III, and 2 of 14 for phase IV. The change in surface roughness of GS coupons varied from +12.8 to –14.5 µm and averaged –0.4 µm. There was no relationship to inhibitor or inhibitor dose.

The change in GS surface roughness, ΔR, and the ratio of the change in GS surface roughness to the initial GS surface roughness, $\Delta R/R_i$, decreased as surface roughness increased. ΔR was described by a linear model that depended on the initial roughness, R_i, and approached a mean, R_{mu}, after exposure to inhibition and was independent of inhibitors and water quality.

Linear and nonlinear (power) models relating water quality and R_i to ΔR and R_f were evaluated for all phases. Some models were presented, including alkalinity and temperature effects on water quality; however, the effect of these water quality variables is not clear for this data set.

Surface Roughness of Lined Cast Iron

All LCI coupons used in this study were cut from sections of pipe that had been excavated from within the distribution system and held in indoor storage for a few years, exposed to the air. Over all four phases, 54 LCI coupons were measured with average initial roughness of 25.2 µm, with a minimum and maximum of 13.3 µm and 68.5 µm, respectively. However, 48 of 54 coupons (89%) had a surface roughness less than 32 µm. The six coupons with the largest surface roughness had visible differences in surface morphology when compared to the rest of the LCI coupons, as shown in Figure 5.106. The average R_i for each coupon shown is 18.6 µm and 68.5 µm, respectively. The surface in the typical LCI coupon is noticeably smoother than the rough LCI coupon that has a degraded lining, exposing small bits of gravel. Hence, there is moderate variation in R_i of LCI coupons, with exceptionally high R_i being due to degradation of the lining. A summary of the LCI surface roughness data is shown in Table 5.116.

A statistical comparison of the initial and final roughness averages was completed using a two-tailed pooled-data t-test. The statistical analysis is summarized in Table 5.116. Data for LCI replicate coupons is shown in Table 5.117. *P*-values are provided for the hypothesis tests for

Table 5.116
Statistical comparison of LCI initial and final surface roughness

Phase	Parameter	PDS01	PDS02	PDS03	PDS04	PDS05	PDS06	PDS07	PDS08	PDS09	PDS10	PDS11	PDS12	PDS13	PDS14	Average
		BOP 0.5	BOP 1.0	BOP 2.0	OP 0.5	OP 1.0	OP 2.0	ZOP 0.5	ZOP 1.0	ZOP 2.0	Si 3.0	Si 6.0	Si 12.0	pH_s	pH_s +0.3	
I	R_i	-	13.3	-	-	17.5	-	-	24.9	-	-	16.1	-	18.7	21.7	18.7
	N	-	5	-	-	3	-	-	3	-	-	3	-	3	3	-
	R_f	-	17.3	-	-	21.3	-	-	23.8	-	-	16.0	-	21.9	24.9	20.9
	N	-	5	-	-	5	-	-	5	-	-	5	-	5	5	-
	ΔR	-	4.0	-	-	3.9	-	-	–1.1	-	-	0.0	-	3.3	3.2	2.2
	ΔR *p*-values	-	**0.02**	-	-	0.16	-	-	0.76	-	-	0.99	-	0.32	0.53	-
	$\Delta R/R_i$	-	0.30	-	-	0.22	-	-	–0.04	-	-	0.00	-	0.17	0.15	0.13
II	R_i	22.5	34.7	29.7	22.1	18.9	16.1	25.8	28.1	23.1	20.3	19.0	26.8	23.1	16.9	23.4
	N	5	5	5	5	5	5	5	5	5	5	5	5	5	5	-
	R_f	21.5	32.2	29.9	18.2	15.1	16.7	30.0	24.8	21.8	24.4	18.1	27.0	26.6	18.9	23.2
	N	5	5	5	5	5	5	5	5	5	5	5	5	5	5	-
	ΔR	–1.0	–2.5	0.2	–3.9	–3.8	0.6	4.2	–3.2	–1.3	4.1	–0.9	0.2	3.6	2.0	–0.1
	ΔR *p*-values	0.74	0.54	0.96	0.25	0.22	0.68	0.26	0.45	0.67	0.08	0.56	0.96	0.11	0.22	-
	$\Delta R/R_i$	–0.04	–0.07	0.01	–0.18	–0.20	0.04	0.16	–0.12	–0.06	0.20	–0.05	0.01	0.15	0.12	0.00
III	R_i	45.3	27.7	18.6	24.7	42.7	19.4	31.2	19.0	29.1	40.1	68.5	23.4	25.7	19.4	31.1
	N	5	5	5	5	5	5	5	5	5	5	5	5	5	5	-
	R_f	48.6	33.3	24.0	26.6	43.6	23.9	25.8	20.7	22.2	24.5	43.4	26.6	32.4	28.1	30.3
	N	5	5	5	5	5	7	5	5	5	5	5	5	5	6	-
	ΔR	3.3	5.6	5.3	1.9	0.9	4.5	–5.4	1.7	–6.9	–15.5	–25.1	3.2	6.7	8.7	–0.8
	ΔR *p*-values	0.28	0.17	**0.01**	0.36	0.89	0.18	0.14	0.33	0.22	0.24	**0.03**	0.18	**<0.01**	**<0.01**	-
	$\Delta R/R_i$	0.07	0.20	0.29	0.08	0.02	0.23	–0.17	0.09	–0.24	–0.39	–0.37	0.13	0.26	0.45	0.05
IV	R_i	22.9	19.9	29.1	19.1	28.4	22.0	23.8	19.9	24.3	20.7	26.8	23.6	27.7	19.6	23.4
	N	5	5	5	5	5	5	5	5	5	5	5	5	5	5	-
	R_f	31.8	24.1	31.9	21.7	32.8	30.1	31.6	23.1	27.3	22.0	36.0	23.3	30.3	20.4	27.6
	N	5	5	5	5	5	5	5	5	5	5	5	5	5	5	-
	ΔR	8.9	4.2	2.7	2.5	4.4	8.0	7.8	3.1	3.0	1.4	9.2	–0.4	2.6	0.8	4.2
	ΔR *p*-values	**0.01**	0.14	0.48	0.15	**0.02**	0.17	0.06	0.09	0.33	0.45	0.07	0.89	0.44	0.53	-
	$\Delta R/R_i$	0.39	0.21	0.09	0.13	0.15	0.37	0.33	0.16	0.12	0.07	0.35	–0.02	0.09	0.04	0.18

NOTES: R_i = initial roughness; R_f = final roughness; ΔR = delta roughness. Boldface indicates *p*-values < 0.05.

Table 5.117
Replicate roughness data for LCI coupons

Phase	Parameter	PDS 01	PDS 02	PDS 03	PDS 04	PDS 05	PDS 06	PDS 07	PDS 08	PDS 09	PDS 10	PDS 11	PDS 12	PDS 13	PDS 14
		BOP 0.5	BOP 1.0	BOP 2.0	OP 0.5	OP 1.0	OP 2.0	ZOP 0.5	ZOP 1.0	ZOP 2.0	Si 3.0	Si 6.0	Si 12.0	pH_s	pH_s +0.3
III	R_i	-	-	-	-	-	-	-	-	-	-	-	-	-	-
	N	-	-	-	-	-	-	-	-	-	-	-	-	-	-
	R_f	-	-	-	-	-	-	-	-	-	-	-	-	-	-
	N	-	-	-	-	-	-	-	-	-	-	-	-	-	-
	ΔR	-	-	-	-	-	-	-	-	-	-	-	-	-	-
	ΔR *p*-values	-	-	-	-	-	-	-	-	-	-	-	-	-	-
IV	R_i	-	35.2	-	-	24.7	-	-	21.2	-	-	28.0	-	23.1	27.8
	N	-	5	-	-	5	-	-	5	-	-	5	-	5	5
	R_f	-	43.7	-	-	25.7	-	-	20.5	-	-	44.7	-	22.0	26.9
	N	-	4	-	-	5	-	-	5	-	-	6	-	5	5
	ΔR	-	8.4	-	-	1.0	-	-	-0.7	-	-	16.7	-	-1.1	-0.9
	ΔR *p*-values	-	**0.04**	-	-	0.82	-	-	0.50	-	-	**<0.01**	-	0.47	0.80

NOTES: R_i = initial roughness; R_f = final roughness; ΔR = delta roughness. Boldface indicates *p*-values < 0.05.

equality of initial and final roughness. A reported p-value of 0.05 or less indicates there is sufficient evidence to conclude that the final roughness is different from the initial roughness (level of significance of 95%). If the p-value of the t-test comparing the means of the surface roughness before and after exposure to the inhibitor is greater than 0.05, then there is no statistically significant difference in coupon surface roughness due to exposure to the inhibitor.

Effect of Inhibitors

Phase I had a predominately groundwater blend with elevated alkalinity and low temperature. The average water temperature for phase I was 21.2°C. In phase I, six coupons were each placed in the middle-dose PDSs for all four inhibitors and in the pH_s and pH_s+0.3 PDSs. Table 5.116 shows that the surface roughness of LCI changed significantly for only one of six coupons.

Phase II had a predominately surface water blend with relatively low alkalinity and elevated sulfate. The average water temperature for phase II was 26.2°C. None of the 14 LCI coupons changed significantly in phase II. Phase III water quality was similar to phase I water quality and was dominated by groundwater (62%). However, the average temperature of phase III water was 25.8°C. Surface roughness for 4 of 14 LCI coupons changed significantly in phase III. Surface roughness changed significantly for the 6.0-mg/L-as-SiO_2 dose of Si inhibitor, decreasing from 68.5 to 43.4 μm. This coupon is the rough coupon shown on the right in Figure 5.106. This data indicates that the roughest LCI coupons may have had decreases in surface roughness due to loss of rough surface elements. Phase IV water quality had the highest proportion of desalinated water by RO (20%) among all the phases. The average temperature of phase IV water was 21.2°C. In phase IV, only 2 of 14 LCI coupons had a significant change in surface roughness during exposure to inhibitors. Considering data from all phases, surface roughness of LCI never changed significantly for any of the 11 coupons receiving ZOP inhibitor. All other inhibitors had at least one coupon with a significant change in surface roughness. Also, the surface roughness of LCI increased more frequently with BOP inhibitor than for any other inhibitor, with 4 of the 11 coupons having significant increases in surface roughness. Apart from these observations, the addition of inhibitors did not appear to have any influence on LCI surface roughness. The change in ΔR and $\Delta R/R_i$ was analyzed with respect to initial roughness, R_i, as shown in Figures 5.107 and 5.108. Even though the method used to measure surface roughness of LCI was similar across all phases, the data was still normalized with respect to initial roughness using the ratio $\Delta R/R_i$ to demonstrate the effect of this transformation.

A model was developed to describe the final roughness of the LCI coupons as a function of the mean or average roughness, as shown in Equation 5.67. The R^2 for the model was 0.26, which is low, but the F probability was less than 0.0001. This model did not demonstrate any visual improvement in predictive power versus a simple plot, as shown in Figure 5.109, where the R_i is used as the predictor of R_f. This corresponds to the case where m = 0.

$$R_f = R_i - m(R_i - R_{mu}) \qquad (5.67)$$

where R_f = final roughness, μm
R_{mu} = mean roughness, μm
m = gain term defined by regression (0.41 for LCI)

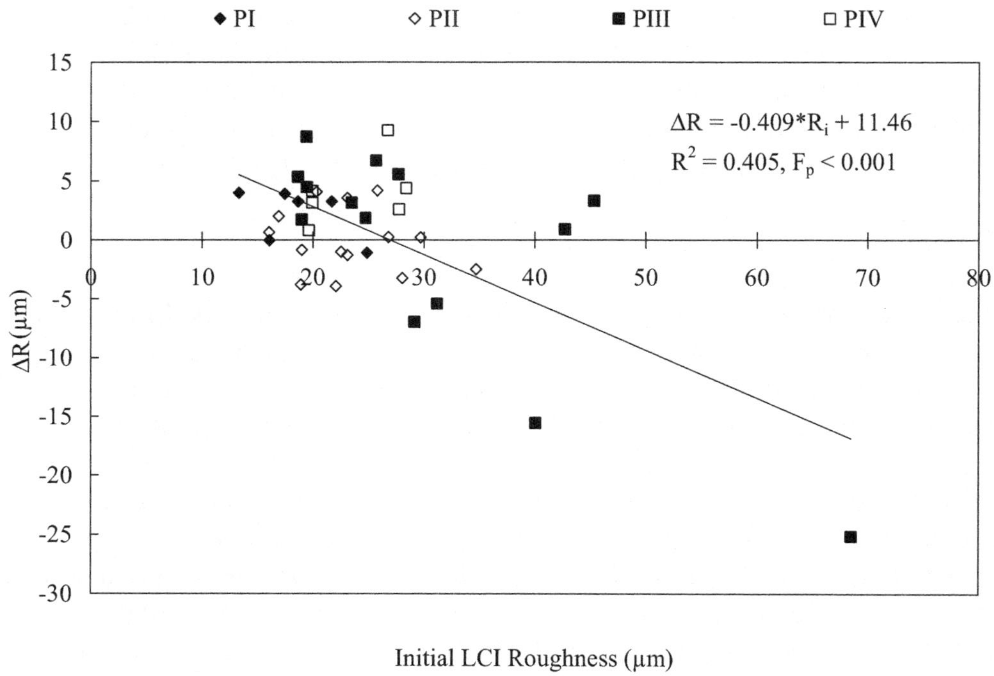

Figure 5.107 Change of initial roughness, ΔR, versus initial roughness, R_i, for LCI coupons by phase

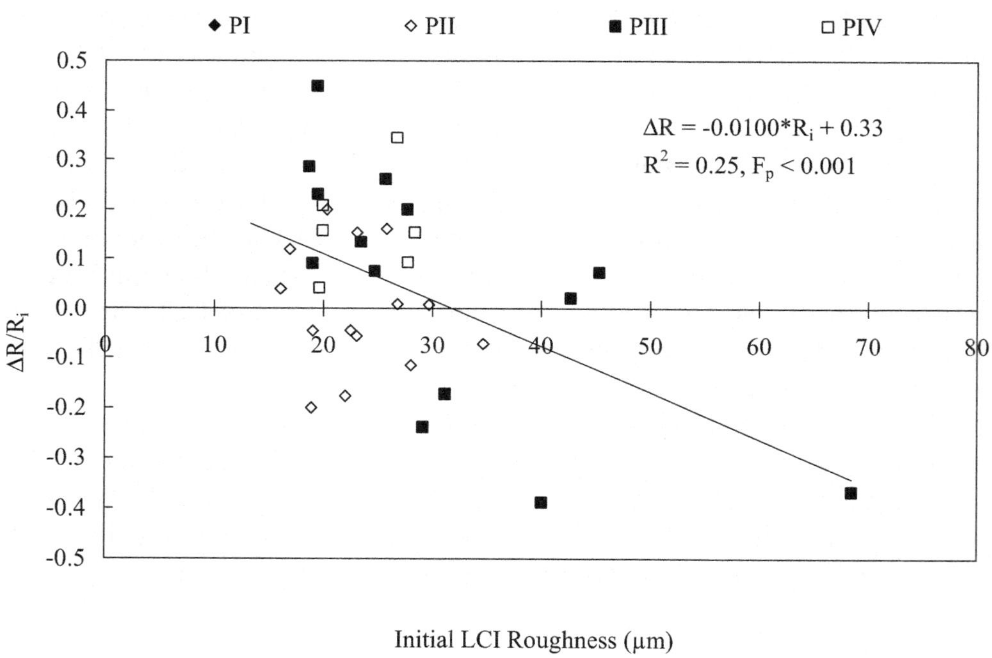

Figure 5.108 Change of the ratio of delta roughness to initial roughness, $\Delta R/R_i$, versus initial roughness, R_i, for LCI coupons by phase

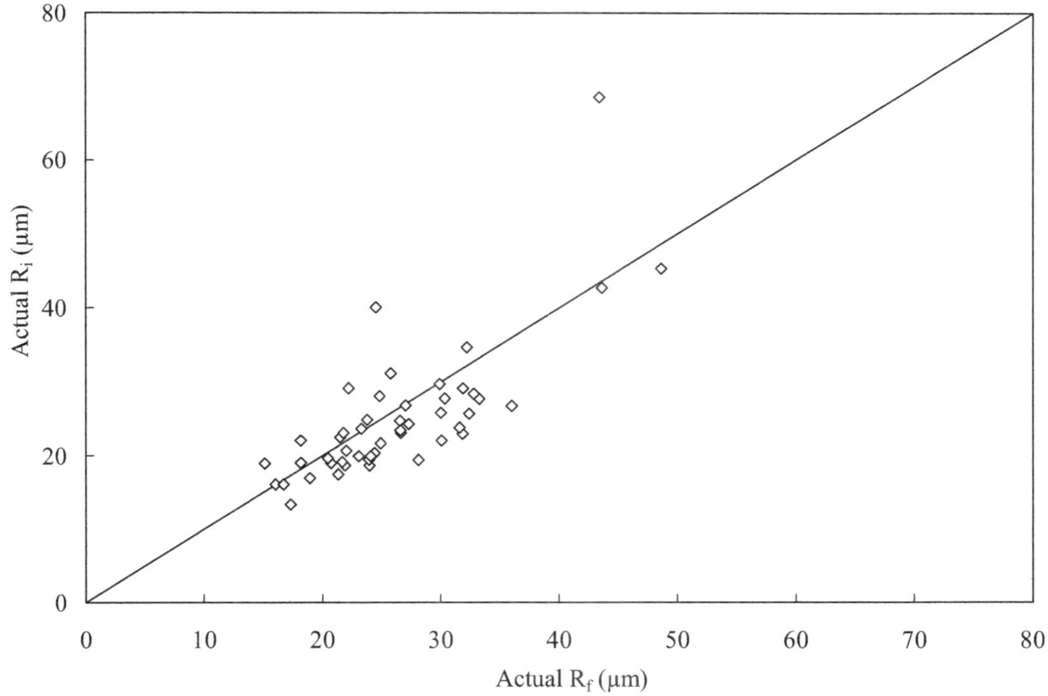

Figure 5.109 Initial versus final LCI surface roughness

Replicates

Six replicates of LCI surface roughness are shown in Table 5.118. The change in surface roughness in both the replicate and the original LCI coupon was not statistically significant for any of the six pairs for phase IV replicates. However, for three of the six pairs, both coupons did not have any significant change in LCI surface roughness.

ΔR of the replicate coupon pairs is plotted in Figure 5.110, which shows that replication of surface analysis was achievable within ±8 μm for the LCI coupons. Traditional calculation of precision for ΔR and $\Delta R/R_i$ yielded RPDs, or errors, averaging –861% and 426%, respectively.

The RPD is a measure of precision and is calculated by taking the absolute difference of two values, dividing by the average of the same two values, and expressing the result as a percentage. Table 5.119 shows the quantities used to assess RPD of ΔR and $\Delta R/R_i$ for the replicate LCI coupon pair in PDS08 (ZOP-1.0). The RPD for ΔR and $\Delta R/R_i$ as shown in Table 5.119 is –315% and 306%, which is based on a change in surface roughness of +3.1 μm for the original LCI coupon and –0.7 μm for the replicate coupon. Although the actual changes are relatively small, the errors calculated using traditional RPD methods are large, which infers that traditional RPD methods are not well suited for making practical assessments of changes in surface roughness of LCI coupons.

An alternative measure of experimental precision at the systems level is illustrated using Figures 5.111–5.114. These figures show both ΔR and $\Delta R/R_i$ versus R_i for LCI coupons and LCI replicate coupons; however, they do not exhibit the same trend observed with similar plots for GS, where the change in roughness was dependent on the initial roughness, with a mean roughness at which roughness tended to neither increase. In these figures for LCI, the change in roughness

Table 5.118
Paired replicate surface roughness data for LCI coupons from phase IV

Parameter	PDS 02 BOP 1.0	PDS 02-R BOP 1.0	PDS 05 OP 1.0	PDS 05-R OP 1.0	PDS 08 ZOP 1.0	PDS 08-R ZOP 1.0	PDS 11 Si 1.0	PDS 11-R Si 1.0	PDS 13 pH_s	PDS 13-R pH_s	PDS 14 pH_s +0.3	PDS 14-R pH_s +0.3
R_i	19.9	35.2	28.4	24.7	19.9	21.2	26.8	28.0	27.7	23.1	19.6	27.8
N	5	5	5	5	5	5	5	5	5	5	5	5
R_f	24.1	43.7	32.8	25.7	23.1	20.5	36.0	44.7	30.3	22.0	20.4	26.9
N	5	4	5	5	5	5	5	6	5	5	5	5
ΔR	4.2	8.4	4.4	1.0	3.1	–0.7	9.2	16.7	2.6	–1.1	0.8	–0.9
ΔR *p*-values	0.14	**0.04**	**0.02**	0.82	0.09	0.50	0.07	**<0.01**	0.44	0.47	0.53	0.80

NOTES: R_i = initial roughness; R_f = final roughness; ΔR = delta roughness. Boldface indicates *p*-values < 0.05.

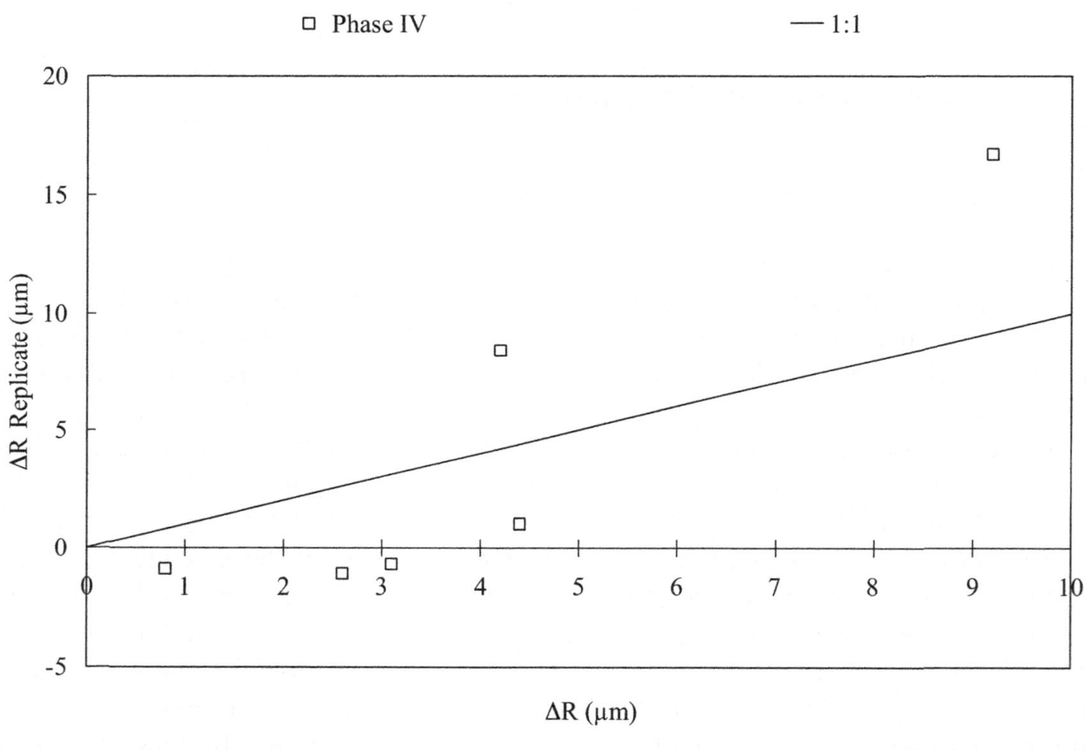

Figure 5.110 Comparison of the change in surface roughness among replicate LCI coupons during phase IV

Table 5.119
Example RPD calculation assessing variation in difference of roughness, LCI

		R_i	R_f	ΔR	$ΔR/R_i$	RPD-ΔR	$RPD\text{-}ΔR/R_i$
PDS02	BOP-1.0	34.1	27.1	–7	–0.21	–315%	–306%
PDS02-R	BOP-1.0	21.4	33.8	12.3	0.57	—	—

NOTES: R_i = initial roughness; R_f = final roughness; ΔR = delta roughness.

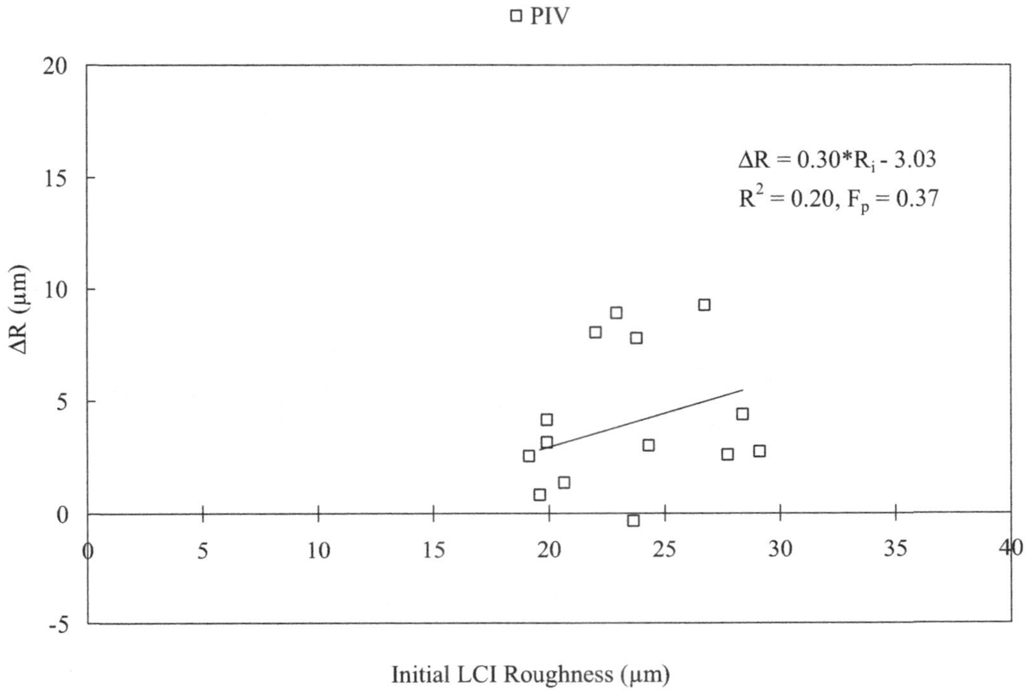

Figure 5.111 Change of initial roughness, ΔR, versus initial roughness, R_i, for LCI coupons for phase IV only

seems to be independent of the initial roughness. Hence, the approach used to assess GS precision is not appropriate for LCI.

Systems precision in measurement of LCI surface roughness may be assessed by comparing the average roughness between the six original coupons and six replicate coupons. The average initial roughness of the original LCI coupons and the replicate LCI coupons was 23.7 μm and 26.7 μm, respectively, with an RPD of 12%. Similarly, the average final roughness of the original LCI coupons and the replicate LCI coupons was 27.8 μm and 30.6 μm, respectively, with an RPD of 10%. Finally, the average Δ roughness of the original LCI coupons and the replicate LCI coupons was 4.1 μm and 3.9 μm, respectively, with an RPD of 4.1%. Thus, there is reasonable precision in measurement of LCI surface roughness on a systems level.

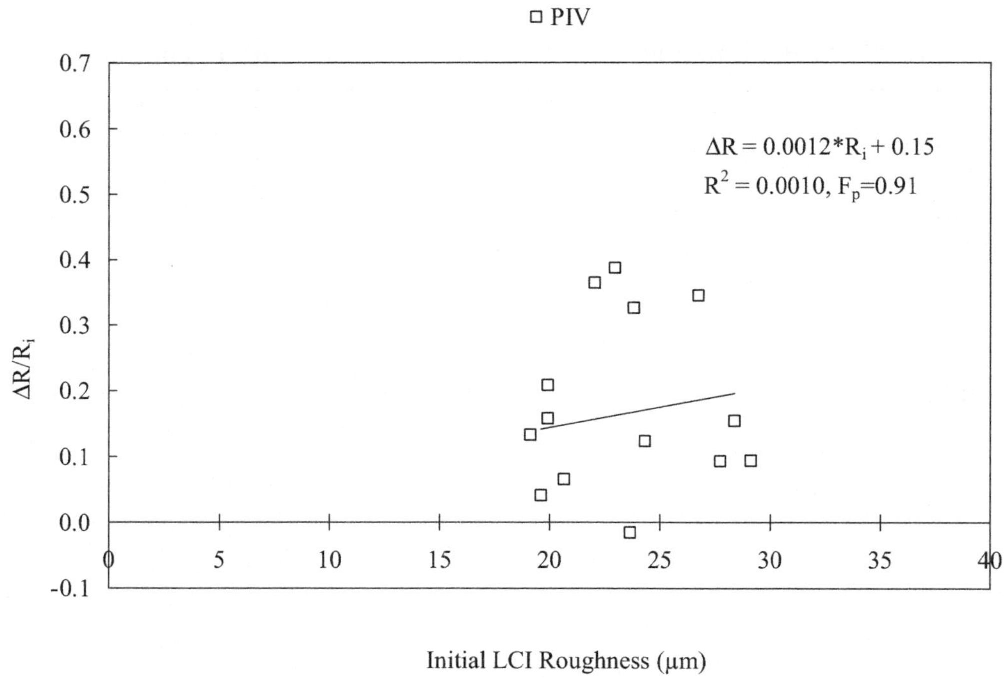

Figure 5.112 Change of initial roughness, ΔR, normalized to R_i, versus initial roughness, R_i, for LCI coupons for phase IV only

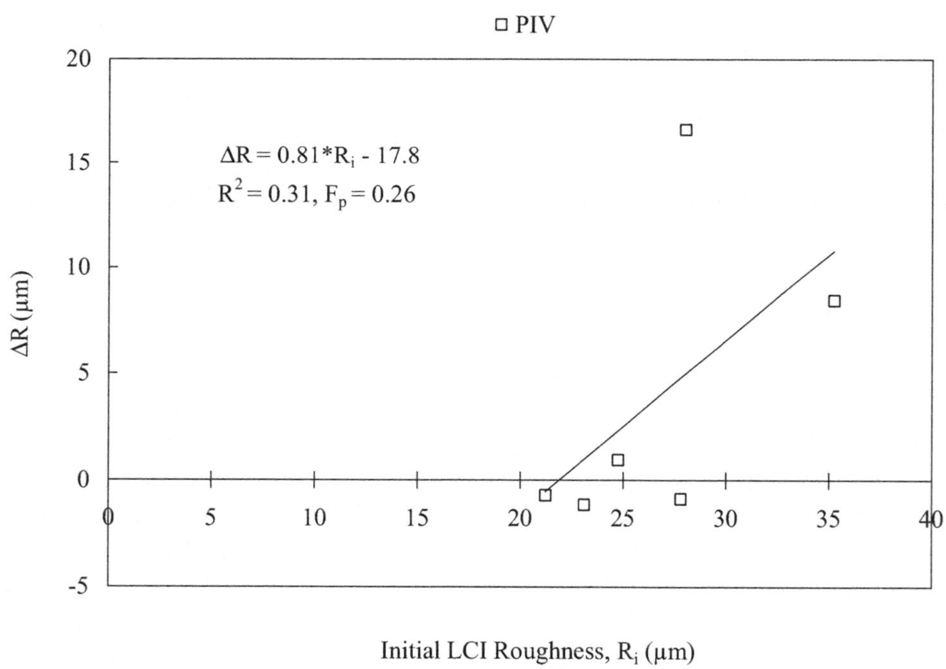

Figure 5.113 Change of initial roughness, ΔR, versus initial roughness, R_i, for replicate LCI coupons for phase IV only

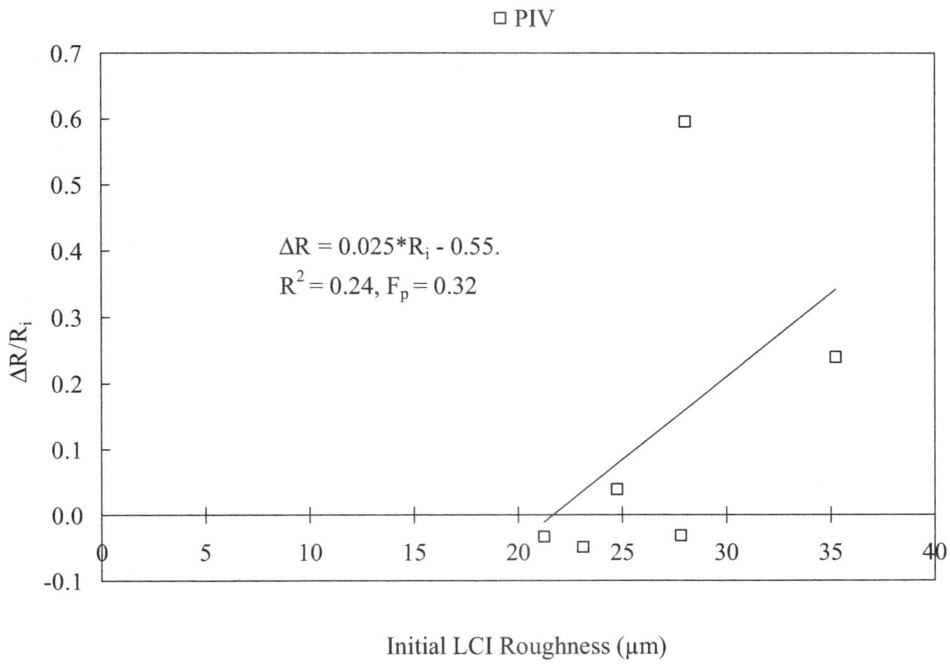

Figure 5.114 Change of initial roughness, ΔR, normalized to R_i, versus initial roughness, R_i, for replicate LCI coupons for phase IV only

Effect of Water Quality

Linear and nonlinear (power) models for correlating water quality and R_i to ΔR and R_f of LCI were evaluated for all phases. The variation of R_f for LCI was generally described by a linear model that depended on the initial LCI roughness, R_i, and a mean roughness term, R_{mu}, and was independent of inhibitors or water quality. The form of this model was shown previously in Equation 5.67. This model suggests that LCI is relatively inert, which was confirmed by experimental data in Tables 5.116 and 5.117, which demonstrated that only 17% of LCI coupons had a significant change in surface roughness during incubation.

Summary for Surface Roughness of Lined Cast Iron

Surface roughness of LCI coupons was measured in all four phases. LCI coupons were cut from LCI pipe that had been in the ground for more than two decades, then was extracted and stored in an open atmosphere. The initial surface roughness of LCI coupons varied from 13.3 to 68.5 μm and averaged 25.2 μm. The change of surface roughness of LCI coupons varied from +16.7 to –25.1 μm and averaged +1.52 μm.

Including replicates, there were a total of 54 observations of LCI surface roughness. The number of statistically significant changes of LCI surface roughness by phase was 1 of 6 for phase I, 0 of 14 for phase II, 4 of 14 for phase III, and 4 of 20 for phase IV. The change was both positive and negative and varied from +16.7 to –25.1 μm. Surface roughness of LCI tended to increase more frequently for BOP inhibitor than for any other inhibitor. Surface roughness of LCI

Figure 5.115 PVC coupon before (left) and after (right) incubation, demonstrating significant metal deposition

never changed significantly for ZOP inhibitor. There was no clear relationship between inhibitor dose and surface roughness. The average surface roughness of all LCI coupons was 25.2 μm before exposure and 26.7 μm after exposure.

Linear and nonlinear (power) models for correlating water quality and R_i to ΔR and R_f of LCI were evaluated for all phases. The variation of R_f for LCI was generally described by a linear model that depended on the initial LCI roughness, R_i, and a mean roughness term, R_{mu}, and was independent of inhibitors or water quality.

Surface Roughness of Polyvinyl Chloride

All PVC coupons used in this study were cut from sections of pipe that had been excavated from within the distribution system and held in indoor storage for a few years, exposed to the air. Over all four phases, 66 PVC coupons were measured with average initial roughness of 1.52 μm, with a minimum and maximum of 0.64 μm and 2.98 μm, respectively. Hence, there was significant variation of surface roughness among the PVC coupons. After incubation, the surface of most PVC coupons was covered with a scatter of orange-brown precipitates, indicating deposition of a metal present in the source water or iron released from biofilm coupons upstream of the roughness coupons. Figure 5.115 illustrates this deposition as it occurred for the PVC coupon receiving the 6.0-mg/L-as-SiO_2 dose of Si inhibitor. On the left, the coupon is shown as it appeared before being placed in the cradles, whereas on the right, the coupon is shown as it appeared after removal from the cradles. Metal precipitation is presumed to explain significant increases in surface roughness of PVC, which is a relatively inert material.

A summary of the PVC surface roughness data is shown in Table 5.120. A statistical comparison of the initial and final roughness averages was completed using a two-tailed, pooled-data t-test. The statistical analysis is summarized in Table 5.120. Similar information for the PVC replicate coupons is shown in Table 5.121. The *p*-values are provided for the hypothesis tests for

Table 5.120
Statistical comparison of PVC initial and final surface roughness

Phase	Parameter	PDS01 BOP 0.5	PDS02 BOP 1.0	PDS03 BOP 2.0	PDS04 OP 0.5	PDS05 OP 1.0	PDS06 OP 2.0	PDS07 ZOP 0.5	PDS08 ZOP 1.0	PDS09 ZOP 2.0	PDS10 Si 3.0	PDS11 Si 6.0	PDS12 Si 12.0	PDS13 pH_s	PDS14 pH_s +0.3	Average
I	R_i	-	1.62	-	-	1.64	-	-	1.59	-	-	1.35	-	-	1.51	1.54
	N	-	3	-	-	3	-	-	3	-	-	3	-	-	3	-
	R_f	-	1.44	-	-	1.60	-	-	1.76	-	-	1.54	-	-	1.53	1.57
	N	-	5	-	-	5	-	-	5	-	-	5	-	-	5	-
	ΔR	-	–0.18	-	-	–0.04	-	-	0.17	-	-	0.19	-	-	0.03	0.03
	ΔR *p*-values	-	0.06	-	-	0.70	-	-	0.59	-	-	0.29	-	-	0.90	-
	$\Delta R/R_i$	-	–0.11	-	-	–0.03	-	-	0.10	-	-	0.14	-	-	0.02	0.02
II	R_i	1.43	1.62	1.61	1.71	1.36	1.35	1.42	1.63	1.28	1.38	1.35	1.33	1.45	1.58	1.47
	N	5	5	5	5	5	5	5	5	5	5	5	5	5	5	-
	R_f	1.48	1.47	1.61	1.58	1.44	1.35	1.74	1.48	1.35	1.39	1.90	1.36	1.59	1.49	1.52
	N	5	5	5	5	5	5	5	5	5	5	5	5	5	5	-
	ΔR	0.04	–0.15	0.00	–0.13	0.08	0.00	0.32	–0.15	0.07	0.01	0.55	0.03	0.14	–0.09	0.05
	ΔR *p*-values	0.75	0.48	0.98	0.44	0.58	0.98	0.18	0.31	0.50	0.95	**0.01**	0.76	0.53	0.38	-
	$\Delta R/R_i$	0.03	–0.09	0.00	–0.08	0.06	0.00	0.22	–0.09	0.06	0.01	0.40	0.02	0.09	–0.06	0.04
III	R_i	1.61	2.31	1.71	1.96	1.87	1.97	2.15	2.63	2.19	1.86	1.83	2.98	2.08	2.38	2.11
	N	5	5	5	5	5	5	5	5	5	5	5	5	5	5	-
	R_f	2.00	2.21	2.02	2.03	1.83	2.04	2.53	2.74	2.28	1.64	1.90	2.74	2.13	2.54	2.19
	N	5	5	5	5	5	5	5	5	5	5	5	5	5	5	-
	ΔR	0.39	–0.09	0.31	0.07	–0.05	0.07	0.38	0.11	0.09	–0.22	0.07	–0.24	0.05	0.16	0.08
	ΔR *p*-values	**0.02**	0.68	**0.04**	0.65	0.82	0.80	0.17	0.75	0.74	**<0.01**	0.76	0.63	0.69	0.72	-
	$\Delta R/R_i$	0.24	–0.04	0.18	0.03	–0.03	0.04	0.18	0.04	0.04	–0.12	0.04	–0.08	0.02	0.07	0.04
IV	R_i	0.73	0.76	0.86	1.17	0.64	1.15	1.84	0.82	1.69	0.90	0.87	0.87	0.97	1.74	1.07
	N	5	5	5	5	5	5	5	5	5	5	5	5	5	5	-
	R_f	0.91	1.01	1.01	1.36	1.04	1.22	2.07	1.05	2.01	1.02	1.26	1.20	1.12	1.91	1.30
	N	5	5	5	5	5	5	5	5	5	5	5	5	5	5	-
	ΔR	0.18	0.25	0.15	0.19	0.40	0.07	0.24	0.23	0.32	0.12	0.39	0.34	0.15	0.17	0.23
	ΔR *p*-values	**<0.01**	**<0.01**	**<0.01**	**<0.01**	**<0.01**	0.50	**0.01**	**<0.01**	**0.01**	**0.01**	**<0.01**	0.27	0.27	0.17	-
	$\Delta R/R_i$	0.24	0.32	0.17	0.16	0.62	0.06	0.13	0.28	0.19	0.14	0.45	0.39	0.16	0.10	0.24

NOTES: R_i = initial roughness; R_f = final roughness; ΔR = delta roughness. Boldface indicates *p*-values < 0.05.

Table 5.121
Replicate roughness data for PVC coupons

Phase	Parameter	PDS01	PDS02	PDS03	PDS04	PDS05	PDS06	PDS07	PDS08	PDS09	PDS10	PDS11	PDS12	PDS13	PDS14
		BOP 0.5	BOP 1.0	BOP 2.0	OP 0.5	OP 1.0	OP 2.0	ZOP 0.5	ZOP 1.0	ZOP 2.0	Si 3.0	Si 6.0	Si 12.0	pH_s	pH_s +0.3
III	R_i	-	1.75	-	-	2.02	-	-	2.04	-	-	1.76	-	1.59	1.74
	N	-	5	-	-	5	-	-	5	-	-	5	-	5	5
	R_f	-	1.90	-	-	2.06	-	-	2.10	-	-	2.30	-	2.01	2.04
	N	-	5	-	-	5	-	-	5	-	-	5	-	5	5
	ΔR	-	0.15	-	-	0.04	-	-	0.06	-	-	0.55	-	0.43	0.31
	ΔR *p*-values	-	0.32	-	-	0.68	-	-	0.82	-	-	0.06	-	**0.03**	**0.03**
IV	R_i	-	2.00	1.24	1.42	0.71	1.49	0.93	1.30	1.51	1.44	1.62	1.08	1.18	0.84
	N	-	5	5	5	5	5	5	5	5	5	5	5	5	5
	R_f	-	1.69	1.47	1.47	1.14	1.99	1.21	1.56	2.08	1.78	2.81	1.09	1.52	1.17
	N	-	5	5	5	5	5	5	5	5	5	5	5	5	5
	ΔR	-	–0.31	0.23	0.05	0.43	0.50	0.28	0.26	0.57	0.34	1.19	0.02	0.34	0.33
	ΔR *p*-values	-	0.16	0.26	0.58	**<0.01**	**<0.01**	**0.02**	**<0.01**	0.08	0.11	**0.03**	0.94	**0.01**	**<0.01**

NOTES: R_i = initial roughness; R_f = final roughness; ΔR = delta roughness. Boldface indicates *p*-values < 0.05.

equality of initial and final roughness. A reported p-value of 0.05 or less indicates there is sufficient evidence to conclude that the final roughness is different from the initial roughness (level of significance of 95%). If the p-value of the t-test comparing the means of the surface roughness before and after exposure to the inhibitor is greater than 0.05, then there is no statistically significant difference in coupon surface roughness due to exposure to the inhibitor.

Effect of Inhibitors

Phase I had a predominately groundwater blend with elevated alkalinity and low temperature. The average water temperature for phase I was 21.2°C. In phase I, there was no evidence that inhibitors changed PVC surface roughness relative to no inhibitor addition. This is shown in Table 5.120, as the surface roughness in all six coupons did not change significantly following exposure in the coupon cradles.

Phase II had a predominately surface water blend with relatively low alkalinity and elevated sulfate. The average water temperature for phase II was 26.2°C. The surface roughness of 1 of the 14 PVC coupons exposed to inhibitors changed significantly in phase II. Surface roughness increased from 1.35 to 1.90 μm for the PVC coupon in the presence of 6.0 mg/L as SiO_2 of the Si inhibitor. The coupon is shown in Figure 5.115, both before and after incubation. The increase in surface roughness is presumed to be due to the precipitation of a metal on the surface. Phase III water quality was similar to phase I water quality and was dominated by groundwater (62%). However, the average temperature of phase III water was 25.8°C. Surface roughness for 3 of 14 PVC coupons changed significantly in phase III. Phase IV water quality had the highest proportion of desalinated water by RO (20%) among all the phases. The average temperature of phase IV water was 21.2°C. In phase IV, 10 of 14 PVC coupons had a significant change in surface roughness during exposure to inhibitors. Roughness increased significantly for all coupons receiving phosphate inhibitors, except for the coupon receiving 2.0 mg/L as P-OP. Surface roughness increased for the 3.0- and 6.0-mg/L-as-SiO_2 dose of silicate inhibitor, while not changing significantly for the 12.0-mg/L-as-SiO_2 dose. Surface roughness did not increase significantly for the PVC coupon in the pH_s+0.3 control or the pH_s environment. The phase IV PVC surface roughness data indicates that inhibitor type or dose had no effect on surface roughness. However, because surface roughness of PVC increased most frequently in phase IV, this suggests that higher proportions of desalinated water may correspond with increased surface deposits on PVC pipe.

The change in ΔR and $\Delta R/R_i$ was analyzed with respect to initial roughness, R_i, and is shown in Figures 5.116 and 5.117. Even though the method used to measure surface roughness of PVC was similar across all phases, the data was still normalized with respect to initial roughness, using the ratio $\Delta R/R_i$ to demonstrate the effect of this transformation.

Figures 5.116 and 5.117 show the same trend for both analyses. As the initial roughness of PVC increases, the change in surface roughness of PVC coupons tends to decrease. The trend lines shown in Figures 5.116 and 5.117 are for all data, although the data is separated by phase in these figures. The average PVC roughness was approximately 1.5 μm, which after inhibitor exposure changed by as much as 1.5 μm, or from –15% to 65% of the initial roughness. The higher change in roughness is clearly shown in both Figures 5.116 and 5.117 for the phase IV PVC coupons. The initial roughness of the PVC coupons for phases I, II, III, and IV were 1.54 μm, 1.47 μm, 2.11 μm, and 1.07 μm, as shown in Table 5.120.

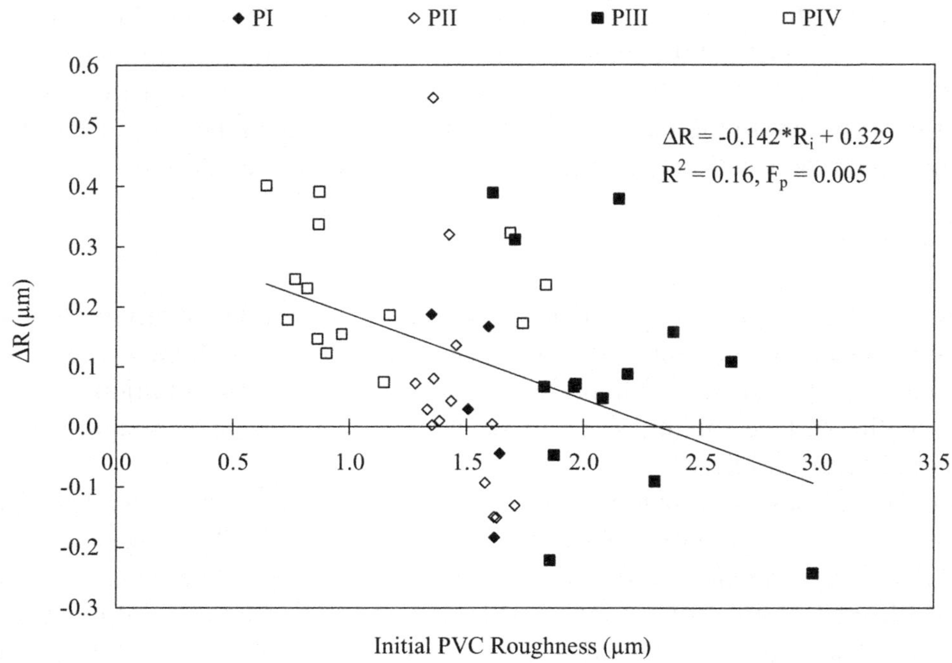

Figure 5.116 Change of initial roughness, ΔR, versus initial roughness, R_i, for PVC coupons by phase

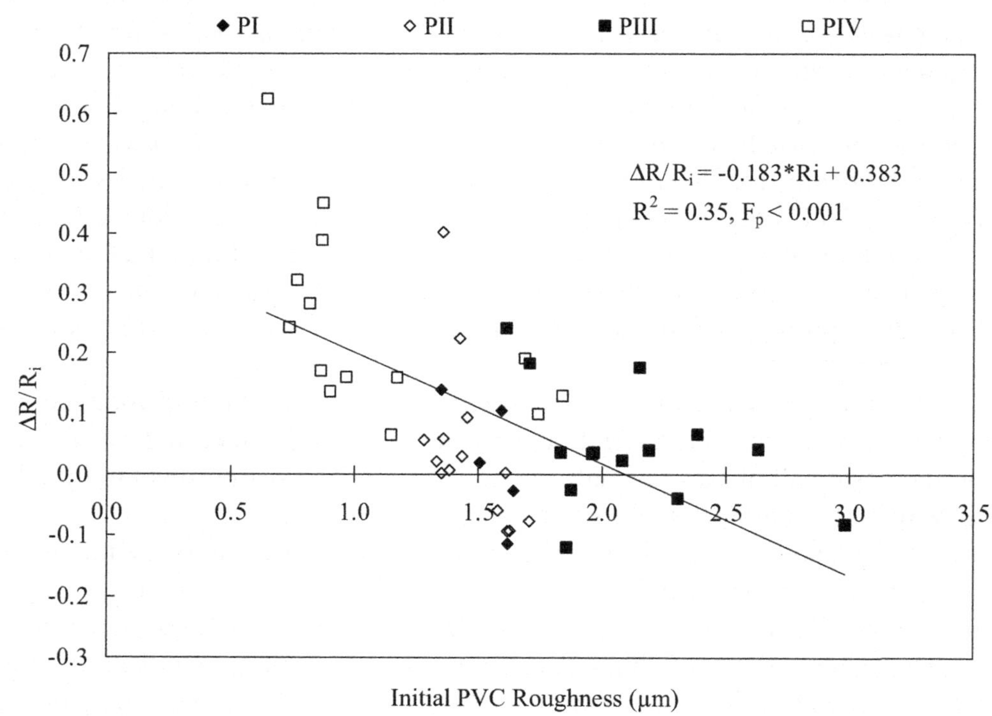

Figure 5.117 Change of the ratio of delta roughness to initial roughness, $\Delta R/R_i$, versus initial roughness, R_i, for PVC coupons by phase

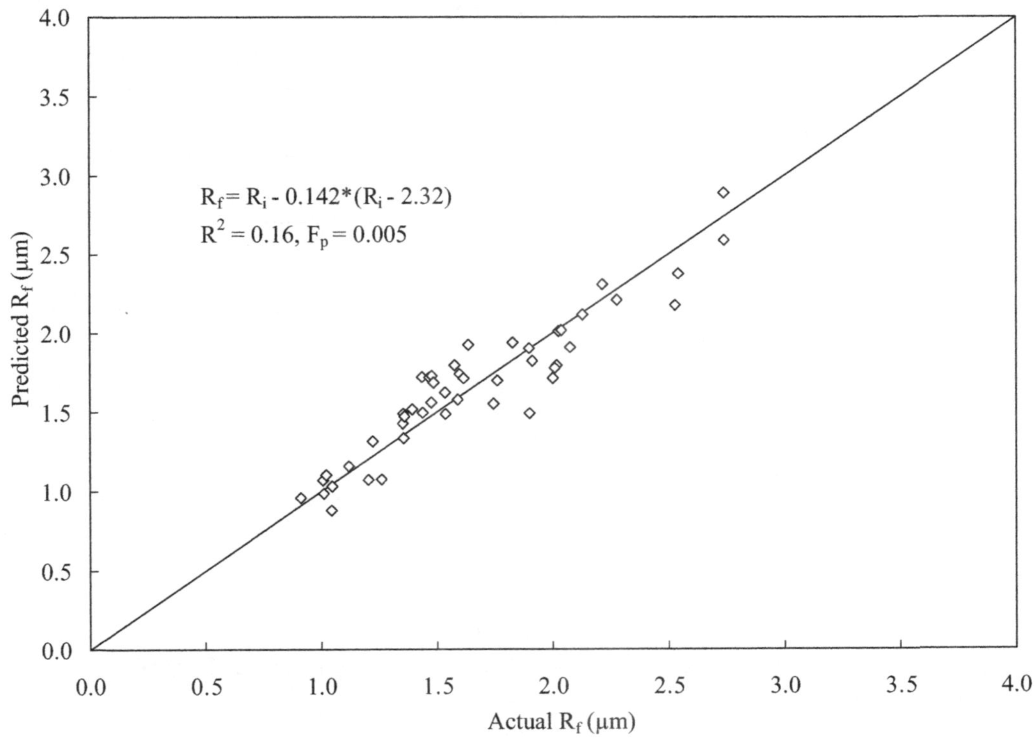

Figure 5.118 Predicted versus final PVC surface roughness

A model was developed to describe the final roughness of the PVC coupons as a function of the mean, or average roughness, as shown in Equation 5.68.

$$R_f = R_i - m(R_i - R_{mu}) \tag{5.68}$$

where R_f = final roughness, μm
R_{mu} = mean roughness, μm
m = gain term defined by regression (0.14 for PVC)

This simple model allows the predicted change of roughness to vary indirectly with initial roughness, which is what was observed for the actual data. The mean roughness for PVC, at which roughness tended not to change, was 2.32 μm. The R^2 for the model was 0.16, which is low, but both m and the R_{mu} term were significant in the model at *p*-value of 0.01 or less. Figure 5.118 shows that the model predicted the R_f of PVC within ±0.45 μm of the actual roughness without considering inhibitors or water quality.

Replicates

A series of replicate PVC coupons were analyzed in phases III and IV to gain insight into variability of the roughness experiments. The results of those analyses are shown in Tables 5.122, 5.123, and 5.124. A visual comparison of the surface roughness of the original PVC coupon versus the surface roughness of the replicated PVC coupon is shown in Figure 5.119.

Table 5.122
Paired replicate surface roughness data for PVC coupons from phase III

	PDS 02	PDS 02-R	PDS 05	PDS 05-R	PDS 08	PDS 08-R	PDS 11	PDS 11-R	PDS 13	PDS 13-R	PDS 14	PDS 14-R
Parameter	BOP 1.0	BOP 1.0	OP 1.0	OP 1.0	ZOP 1.0	ZOP 1.0	Si 6.0	Si 6.0	pH_s	pH_s	pH_s +0.3	pH_s +0.3
R_i	2.31	1.75	1.87	2.02	2.63	2.04	1.83	1.76	2.08	1.59	2.38	1.74
N	5	5	5	5	5	5	5	5	5	5	5	5
R_f	2.21	1.90	1.83	2.06	2.74	2.10	1.90	2.30	2.13	2.01	2.54	2.04
N	5	5	5	5	5	5	5	5	5	5	5	5
ΔR	–0.09	0.15	–0.05	0.04	0.11	0.06	0.07	0.55	0.05	0.43	0.16	0.31
ΔR *p*-values	0.68	0.32	0.82	0.68	0.75	0.82	0.76	0.06	0.69	**0.03**	0.72	**0.03**

NOTES: R_i = initial roughness; R_f = final roughness; ΔR = delta roughness. Boldface indicates *p*-values < 0.05.

Table 5.123
Paired replicate surface roughness data for PVC coupons PDS02–PDS08 from phase IV

	PDS 02	PDS 02-R	PDS 03	PDS 03-R	PDS 04	PDS 04-R	PDS 05	PDS 05-R	PDS 06	PDS 06-R	PDS 07	PDS 07-R	PDS 08	PDS 08-R
Parameter	BOP 1.0	BOP 1.0	BOP 2.0	BOP 2.0	OP 0.5	OP 0.5	OP 1.0	OP 1.0	OP 2.0	OP 2.0	ZOP 0.5	ZOP 0.5	ZOP 1.0	ZOP 1.0
R_i	0.76	2.00	0.86	1.24	1.17	1.42	0.64	0.71	1.15	1.49	1.84	0.93	0.82	1.30
N	5	5	5	5	5	5	5	5	5	5	5	5	5	5
R_f	1.01	1.69	1.01	1.47	1.36	1.47	1.04	1.14	1.22	1.99	2.07	1.21	1.05	1.56
N	5	5	5	5	5	5	5	5	5	5	5	5	5	5
ΔR	0.25	–0.31	0.15	0.23	0.19	0.05	0.40	0.43	0.07	0.50	0.24	0.28	0.23	0.26
ΔR *p*-values	**<0.01**	0.16	**<0.01**	0.26	**<0.01**	0.58	**<0.01**	**<0.01**	0.50	**<0.01**	**0.01**	**0.02**	**<0.01**	**<0.01**

NOTES: R_i = initial roughness; R_f = final roughness; ΔR = delta roughness. Boldface indicates *p*-values < 0.05.

Table 5.124
Paired replicate surface roughness data for PVC coupons PDS09–PDS14 from phase IV

	PDS 09	PDS 09-R	PDS 10	PDS 10-R	PDS 11	PDS 11-R	PDS 12	PDS 12-R	PDS 13	PDS 13-R	PDS 14	PDS 14-R
Parameter	ZOP 2.0	ZOP 2.0	Si 3.0	Si 3.0	Si 6.0	Si 6.0	Si 12.0	Si 12.0	pH_s	pH_s	pH_s +0.3	pH_s +0.3
R_i	1.69	1.51	0.90	1.44	0.87	1.62	0.87	1.08	0.97	1.18	1.74	0.84
N	5	5	5	5	5	5	5	5	5	5	5	5
R_f	2.01	2.08	1.02	1.78	1.26	2.81	1.20	1.09	1.12	1.52	1.91	1.17
N	5	5	5	5	5	5	5	5	5	5	5	5
ΔR	0.32	0.57	0.12	0.34	0.39	1.19	0.34	0.02	0.15	0.34	0.17	0.33
ΔR *p*-values	**0.01**	0.08	**0.01**	0.11	**<0.01**	**0.03**	0.27	0.94	0.27	**0.01**	0.17	**<0.01**

NOTES: R_i = initial roughness; R_f = final roughness; ΔR = delta roughness. Boldface indicates *p*-values < 0.05.

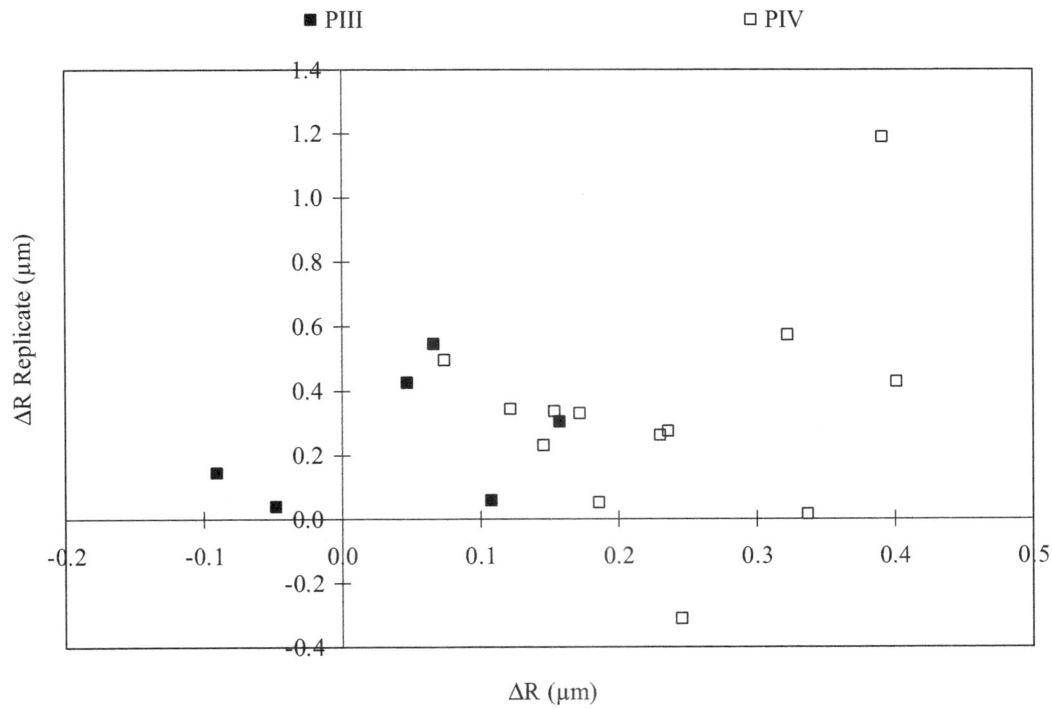

Figure 5.119 Comparison of the change in surface roughness among replicate PVC coupons during phases III and IV

Six replicates of PVC surface roughness are shown in Table 5.122. None of the six pairs of PVC coupons had a significant change in surface roughness for both the original and replicate coupons. However, roughness increased for both coupons in four of the six PVC coupon pairs, whereas in the remaining two of the six pairs, surface roughness increased for one coupon and decreased for its paired coupon. The replicate data for phase IV is shown in Tables 5.123 and 5.124. The change in surface roughness in both the replicate and the original PVC coupon was statistically significant in 4 of 13 pairs for phase IV replicates. Twelve of the 13 pairs of PVC phase IV replicate coupons experienced similar changes, with surface roughness of both coupons increasing or decreasing together.

ΔR of the replicate coupon pairs is plotted in Figure 5.119, which shows that replication of surface analysis was achievable within ±0.8 μm for the PVC coupons. Differences in R_i of the replicate coupon pairs make direct comparison of the ΔR ineffective to establish systems-level precision. Traditional calculation of precision for ΔR and $\Delta R/R_i$ yielded RPDs, or errors, averaging 325% and 220%, respectively.

The RPD is a measure of precision and is calculated by taking the absolute difference of two values, dividing it by the average of the same two values, and expressing the result as a percentage. Table 5.125 shows the quantities used to assess RPD of ΔR and $\Delta R/R_i$ for the replicate PVC coupon pair in PDS02 (BOP-1.0). The RPD for ΔR and $\Delta R/R_i$, as shown in Table 5.125, is 860% and 560%, respectively, which is based on a change in surface roughness of –0.09 μm for the original PVC coupon and +0.15 μm for the replicate coupon. Although the actual changes are relatively small, the errors calculated using traditional RPD methods are large, which infers that

Table 5.125
Example RPD calculation assessing variation in difference of roughness, PVC

		R_i	R_f	ΔR	$\Delta R/R_i$	RPD-ΔR	RPD-$\Delta R/R_i$
PDS02	BOP-1.0	2.31	2.21	–0.09	–0.04	860%	560%
PDS02-R	BOP-1.0	1.75	1.90	0.15	0.08		

NOTES: R_i = initial roughness; R_f = final roughness; ΔR = delta roughness.

traditional RPD methods are not well suited for making practical assessments of changes in surface roughness of PVC coupons.

As noted in the "Effect of Inhibitors" section, surface roughness of the PVC coupon with the higher initial surface roughness was more likely to decrease while the coupon with the lower initial surface roughness was more likely to increase. This trend was shown in Figure 5.116 and confirmed statistically with the model in Equation 5.68. The changes in PVC surface roughness were not correlated to inhibitor or inhibitor dose in this work, and are probably influenced by factors that promoted metal deposits.

Another method of assessing precision of the change in PVC surface roughness is to determine the mean roughness of the original PVC coupons and compare it to the mean roughness of the replicate PVC coupons. These values may be calculated using plots like Figure 5.116 and 5.117, shown previously. Similar plots were constructed for phase III and phase IV original data and replicate data, and R_{mu} was calculated by linear regression. For the ΔR plots, the R_{mu} parameters for the original and replicate data were 2.70 µm and 3.67 µm, respectively. Similarly, for the $\Delta R/R_i$ plots, the ΔR_{mu} parameters for the original and replicate data were 2.36 µm and 2.37 µm, respectively. Thus, for the ΔR and the $\Delta R/R_i$ plots, the RPDs of R_{mu} were 30.5% and 0.38%, respectively, indicating reasonable precision in the measurement of PVC surface roughness on a systems level.

Effect of Water Quality

Linear and nonlinear (power) models relating water quality and R_i to ΔR and R_f were evaluated for all phases. The variation of R_f for PVC was generally described by a linear model that depended on the initial PVC roughness, R_i, and a mean roughness term, R_{mu}, and was independent of inhibitors or water quality. The form of this model was shown previously in Equation 5.68.

Summary for Surface Roughness of Polyvinyl Chloride

PVC coupons were cut from PVC pipe that had been in the ground for more than two decades, and then extracted and stored in an open atmosphere. The initial surface roughness of PVC coupons varied from 0.64 to 2.98 µm and averaged 1.52 µm. The change of surface roughness of PVC coupons varied from +1.19 to –0.31 µm and averaged 0.16 µm.

There were a total of 66 observations of PVC surface roughness. The number of statistically significant changes of PVC surface roughness by phase was 0 of 6 for phase I, 1 of 14 for phase II, 5 of 20 for phase III, and 17 of 24 for phase IV. Overall, 36% of the PVC coupons experienced a significant change in surface roughness. The change was both positive and negative and varied from +1.19 to –0.32 µm. There was no relationship to inhibitor or inhibitor dose.

Figure 5.120 Photographs before exposure to inhibitors of a typical smooth UCI coupon (left) and a rougher UCI coupon (right)

The change in PVC surface roughness and the ratio of the initial PVC surface roughness to the change in PVC surface roughness decreased as surface roughness increased. The variation of R_f for PVC was generally described by a linear model that depended on the initial PVC roughness, R_i, and a mean roughness term, R_{mu}, and was independent of inhibitors, including pH, or inhibitor dose or water quality.

Photographs of the PVC coupon surfaces showed a scattered red-brown precipitate on several PVC coupons. After incubation, the surface of most PVC coupons was covered with a scatter of orange-brown precipitates, indicating deposition of a metal present in the source water or iron released from biofilm coupons upstream of the roughness coupons. Metal precipitation is presumed to explain significant increases in surface roughness of PVC, which is a relatively inert material.

Surface Roughness of Unlined Cast Iron

All UCI coupons used in this study were cut from sections of pipe that had been excavated from within the distribution system and held in indoor storage for a few years, exposed to the air. Consequently, there is variation in the initial surface roughness of UCI coupons across all four phases.

Two examples of UCI coupons are shown in Figure 5.120. The coupon on the left had a smooth orange-brown scale that was soft. Underlying this scale were red, dark brown, and black scales that were hard. The average surface roughness of the left coupon was 52.1 μm. The coupon on the right had a nonuniform rough scale. In the center were dark brown scales surrounded by yellow scales. Near the top of the coupon there was a porous red-brown scale. The average surface roughness of the coupon on the right was 87.5 μm. Both coupons were scanned in phase III before exposure to inhibitors. The coupons are shown at actual size.

When the optical profiler scans a surface, it begins at a starting elevation and descends a specified distance, creating a map of the surface. Portions of the surface that are not between the start elevation and the finish elevation do not scan and appear as noise. In phases I and II, surface

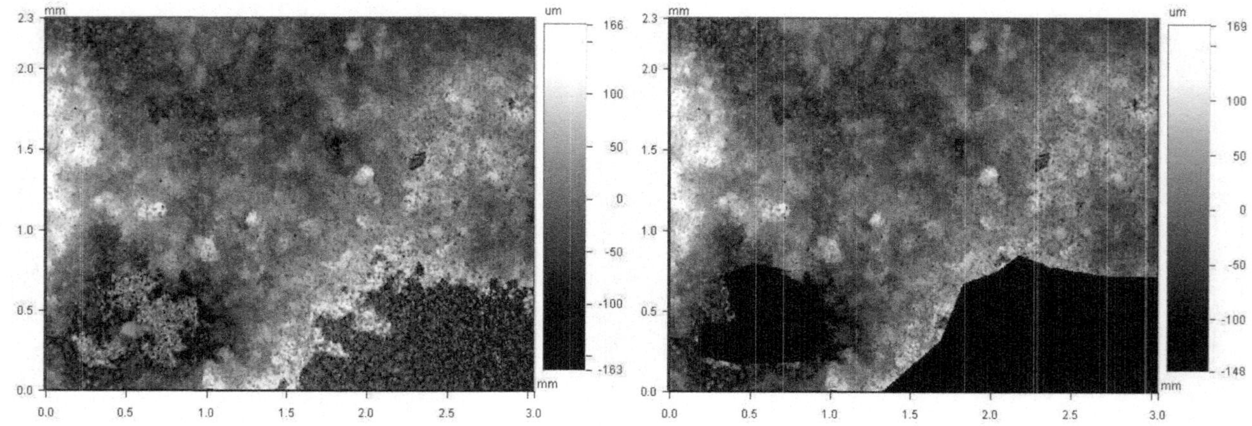

Figure 5.121 Effect of short scan length on a phase I UCI coupon roughness scan

roughness of UCI was consistently less than surface roughness of UCI in phases III and IV. The average initial surface roughness for phases I and II was 43.5 μm, whereas for phases III and IV it was 85.5 μm. The surface roughness in phases III and IV was greater because the scan lengths for UCI only were longer than they had been in phases I and II.

Figure 5.121 shows an example scan of a UCI coupon from phase I before incubation in the coupon cradle. The scan was taken over a 2.3 × 3.0 mm rectangular area. An elevation bar is shown on the right of the scan, with higher elevations in white, middle elevations in gray, and lower elevations in black. The roughness of this coupon was 28.7 μm.

The left image in Figure 5.121 shows the original scan with noise included. The lower left corner of the scan shows a depression, indicated by black, about 1 μm square. Within this depression, some locations were below the finish elevation and hence appear as noise. Similarly, there is noise in the lower right corner of the scan. This region is surrounded by white, indicating that the noise is generated due to that area being above the start elevation. During analysis, regions outside the scan length were removed by the use of an analysis mask, as shown in the right image in Figure 5.121. All UCI coupons were analyzed in this way; if any portion of the surface extended outside the scan length, it was masked to exclude it from the roughness calculation.

Isoelevation lines were generated during a scan that crossed the scan area at a constant elevation. These lines corresponded with the interference fringes observed at each surface elevation. These lines moved across the scanned area by height. No isoelevation lines would appear for any point outside of the predetermined range. Therefore, the adequacy of the predetermined range was proved by the absence or presence of isoelevation lines across the scanned area. The previously discussed analyses masks, as shown in Figure 5.121, demonstrate the inadequate scan length used for the phase I and phase II UCI coupons. Hence, the scan length was increased only for the UCI coupons in phases III and IV.

The scan length in phases I and II was frequently shorter than the distance from the highest to the lowest elevation on the UCI coupon. Therefore, when roughness was calculated over the full 2.3 mm × 3.0 mm scan area, there was a bias towards a lower surface roughness because some of the highest and lowest points on the surface were excluded from the scan. This was undesirable because it did not give a full representation of the surface roughness over the scan area. Beginning with phase III, the scan length was increased to get a more complete picture of surface roughness over the scan area.

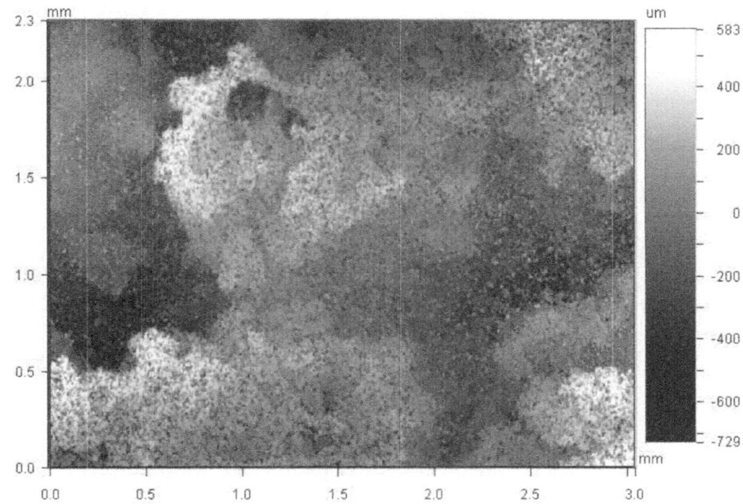

Figure 5.122 Complete scan of a phase III UCI coupon with a very rough surface

Figure 5.122 shows an example scan of a UCI coupon from phase III after removal from the coupon cradle. This coupon is different from that shown in Figure 5.121. The scan was taken over a 2.3 × 3.0 mm rectangular area. Although the surface was very rough, the entire scan area was captured. The surface roughness for this scan was 140.2 μm. If the scan length had not been increased, the only way to measure surface roughness over a full scan area would have been to reduce the scan area. Figure 5.122 demonstrates how for phases III and IV the full surface was captured at all elevations, indicating that the surface roughness calculated in these phases provides a more accurate quantitative measure.

The disadvantage of this approach is that the UCI surface roughness values for phases III and IV for the 2.3 mm × 3.0 mm scan area are greater than the UCI surface roughness values for phases I and II. Hence, UCI surface roughness data from phases I and II cannot be compared quantitatively with data from phases III and IV, and can only be used to assess the relative impact of inhibitors on surface roughness within each phase.

A summary of the UCI surface roughness data is shown in Table 5.126. A statistical comparison of the initial and final roughness averages was completed using a two-tailed pooled-data t-test. The statistical analysis is summarized in Table 5.126. The *p*-values are provided for the hypothesis tests for equality of initial and final roughness. A reported *p*-value of 0.05 or less indicates there is sufficient evidence to conclude that the final roughness is different from the initial roughness (level of significance of 95%). Similar information is shown for replicate UCI coupons in Table 5.127.

A plot of initial roughness versus final roughness segregated by phase is shown in Figure 5.123. The initial roughness values of phase III were smoother than those of phase IV, introducing a bias into the results.

Figure 5.123 shows that the R_f values for phases III and IV are not significantly different and do not depend on R_i. Table 5.126 shows that the average values of R_f for phase III and phase IV were 94.2 μm and 93.7 μm, respectively. A statistical comparison of the phase III and phase IV final roughness averages was completed using a two-tailed pooled-data t-test. The *p*-value of the test was 0.10, indicating that there is insufficient evidence to conclude that final UCI roughness

Table 5.126
Statistical comparison of UCI initial and final surface roughness

Phase	Parameter*	PDS01 BOP 0.5	PDS02 BOP 1.0	PDS03 BOP 2.0	PDS04 OP 0.5	PDS05 OP 1.0	PDS06 OP 2.0	PDS07 ZOP 0.5	PDS08 ZOP 1.0	PDS09 ZOP 2.0	PDS10 Si 3.0	PDS11 Si 6.0	PDS12 Si 12.0	PDS13 pH_s	PDS14 pH_s +0.3	Average
	R_i	29.9	32.1	28.5	37.3	34.8	43.2	34.6	38.0	33.5	24.6	23.9	42.2	39.3	41.4	34.5
	N	5	3	3	3	3	3	3	4	4	3	3	3	3	3	-
	R_f	29.2	37.0	29.3	31.0	36.4	38.7	31.7	39.9	34.0	29.7	22.6	34.1	38.1	37.5	33.5
I†	N	6	5	5	5	5	5	5	5	5	5	5	5	5	5	-
	ΔR	–0.7	4.8	0.8	–6.3	1.5	–4.5	–2.9	1.9	0.5	5.0	–1.3	–8.1	–1.1	–3.9	–1.0
	ΔR *p*-values	0.85	0.20	0.89	0.36	0.80	0.46	0.61	0.71	0.91	0.31	0.78	0.18	0.81	0.52	-
	$\Delta R/R_i$	–0.02	0.15	0.03	–0.17	0.04	–0.10	–0.08	0.05	0.02	0.20	–0.05	–0.19	–0.03	–0.10	–0.02
	R_i	53.5	64.0	40.7	43.7	60.8	53.0	59.8	40.6	44.2	58.9	41.0	52.7	58.5	62.1	52.4
	N	9	5	5	5	5	5	5	5	5	5	5	5	5	5	-
	R_f	35.7	44.4	48.6	31.0	34.4	41.4	38.0	22.9	39.4	44.3	42.7	45.0	43.2	62.4	41.0
II†	N	5	5	5	6	8	5	6	7	7	5	8	6	5	5	-
	ΔR	–17.8	–19.6	8.0	–12.7	–26.4	–11.6	–21.8	–17.7	–4.8	–14.6	1.7	–7.6	–15.3	0.3	–11.4
	ΔR *p*-values	**<0.01**	**0.03**	0.19	**0.05**	**<0.01**	0.23	**0.01**	**0.01**	0.69	0.20	0.81	0.25	0.17	0.97	-
	$\Delta R/R_i$	–0.33	–0.31	0.20	–0.29	–0.43	–0.22	–0.36	–0.44	–0.11	–0.25	0.04	–0.14	–0.26	0.00	–0.21
	R_i	46.6	72.8	68.3	74.6	66.2	68.5	82.3	78.0	47.6	86.4	87.6	75.5	71.6	74.5	71.5
	N	5	5	5	5	5	5	5	5	5	5	6	5	5	5	-
	R_f	90.3	102.0	91.2	95.7	116.7	98.8	90.8	91.6	76.3	96.4	88.7	92.0	87.7	101.6	94.2
III	N	5	6	5	5	5	5	5	5	5	5	6	5	5	5	-
	ΔR	43.6	29.1	22.9	21.1	50.4	30.2	8.5	13.6	28.7	10.0	1.1	16.5	16.0	27.1	22.8
	ΔR *p*-values	**0.04**	**0.01**	0.14	0.14	**<0.01**	**0.03**	0.43	0.23	0.11	0.31	0.90	0.14	0.12	**<0.01**	-
	$\Delta R/R_i$	0.94	0.40	0.34	0.28	0.76	0.44	0.10	0.17	0.60	0.12	0.01	0.22	0.22	0.36	0.36
	R_i	93.3	91.2	115.7	99.4	96.4	93.4	98.4	86.7	89.4	100.3	108.5	99.5	103.5	95.8	98.0
	N	5	5	5	5	5	5	5	5	5	5	5	5	5	5	-
	R_f	82.7	101.1	90.2	102.8	89.3	97.7	108.7	73.4	97.5	72.8	100.0	101.2	115.4	78.9	93.7
IV	N	8	5	5	5	5	5	5	5	6	5	6	5	5	5	-
	ΔR	–10.6	9.9	–25.5	3.3	–7.1	4.4	10.3	–13.3	8.1	–27.5	–8.5	1.7	11.9	–16.9	–4.3
	ΔR *p*-values	**0.05**	0.27	**<0.01**	0.48	0.23	0.58	0.08	0.18	0.26	**0.03**	0.28	0.83	0.35	0.06	-
	$\Delta R/R_i$	–0.11	0.11	–0.22	0.03	–0.07	0.05	0.10	–0.15	0.09	–0.27	–0.08	0.02	0.12	–0.18	–0.04

*R_i = initial roughness; R_f = final roughness; ΔR = delta roughness. Boldface indicates *p*-values < 0.05.
†Surface roughness of UCI in phases I and II was measured using a small scan length, hence, underestimating the actual surface roughness.

Table 5.127
Replicate roughness data for UCI coupons

Phase	Parameter	PDS 01 BOP 0.5	PDS 02 BOP 1.0	PDS 03 BOP 2.0	PDS 04 OP 0.5	PDS 05 OP 1.0	PDS 06 OP 2.0	PDS 07 ZOP 0.5	PDS 08 ZOP 1.0	PDS 09 ZOP 2.0	PDS 10 Si 3.0	PDS 11 Si 6.0	PDS 12 Si 12.0	PDS 13 pH_s	PDS 14 pH_s +0.3
III	R_i	-	77.8	-	-	52.1	-	-	81.0	-	-	65.3	-	87.8	76.8
	N	-	5	-	-	5	-	-	5	-	-	5	-	5	5
	R_f	-	102.7	-	-	75.1	-	-	94.0	-	-	111.8	-	110	95.4
	N	-	5	-	-	5	-	-	5	-	-	6	-	5	5
	ΔR	-	24.9	-	-	23.1	-	-	13.1	-	-	46.5	-	22.1	18.6
	ΔR *p*-values	-	**0.03**	-	-	**0.02**	-	-	0.20	-	-	**<0.01**	-	**0.04**	0.06
IV	R_i	-	90.8	-	-	99.0	-	-	112.3	-	-	122.8	79.9	92.7	95.2
	N	-	5	-	-	8	-	-	8	-	-	5	5	5	5
	R_f	-	81.6	-	-	71.6	-	-	76.8	-	-	91.3	98.2	100	90.6
	N	-	6	-	-	5	-	-	5	-	-	5	5	6	8
	ΔR	-	–9.2	-	-	–27.4	-	-	–35.5	-	-	–31.4	18.2	7.6	–4.6
	ΔR *p*-values	-	0.46	-	-	0.06	-	-	**<0.01**	-	-	**<0.01**	0.16	0.63	0.44

NOTES: R_i = initial roughness; R_f = final roughness; ΔR = delta roughness. Boldface indicates *p*-values < 0.05.

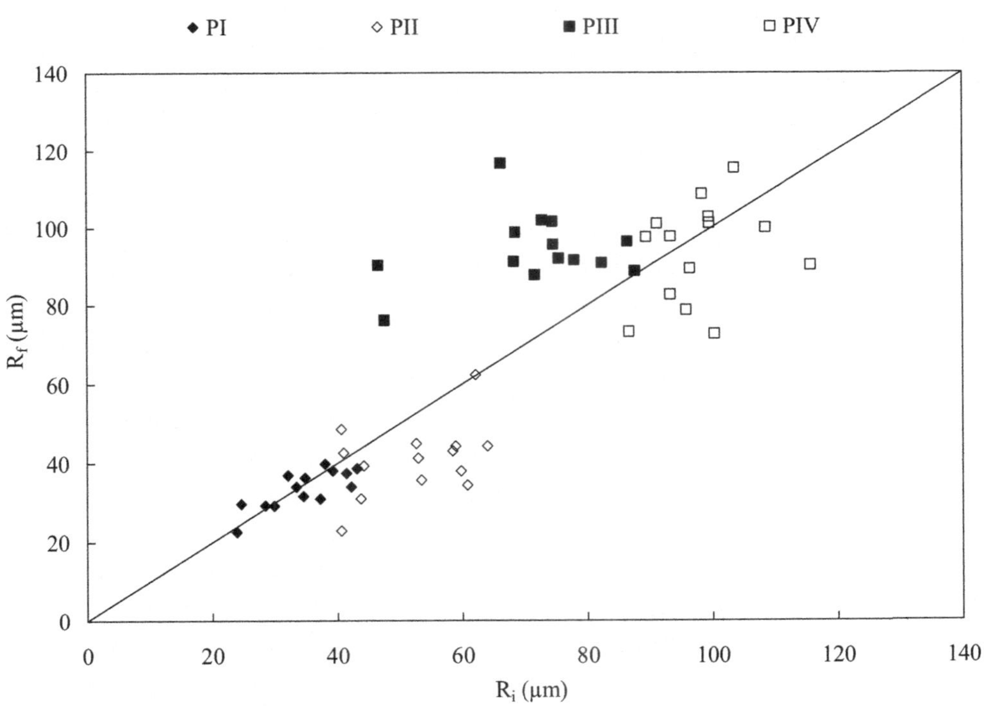

Figure 5.123 Plot of initial roughness versus final roughness by phase for UCI coupons

values were different in phases III and IV (level of significance of 95%). Any comparison of the initial roughness by phase must be premised by difference in the initial roughness of the coupons between phases. Hence, surface roughness of UCI coupons is compared among and within all phases to assess the effect of inhibition and water quality.

Effect of Inhibitors

Phase I had a predominately groundwater blend with elevated alkalinity and low temperature. The average water temperature for phase I was 21.2°C. In phase I, there was no evidence that inhibitors changed UCI surface roughness relative to no inhibitor addition. This is shown in Table 5.126, as the surface roughness of none of 14 coupons changed significantly following exposure in the coupon cradles. The two-tailed t-tests may not have been as sensitive in phase I because usually roughness was only measured at three locations per coupon before incubation in the coupon cradles. In contrast, for all other phases, roughness was measured at five locations per coupon, making the t-tests more sensitive to differences in roughness. The plot shows no significant trends for UCI roughness by inhibitor or dose during phase I.

Phase II had a predominately surface water blend with relatively low alkalinity and elevated sulfate. The average water temperature for phase II was 26.2°C. The surface roughness of 6 of the 14 UCI coupons exposed to inhibitors changed significantly in phase II. These six coupons all had significant decreases in UCI surface roughness. Surface roughness decreased significantly for the 0.5- and 1.0-mg/L-as-P doses of BOP, OP, and ZOP inhibitors. Surface roughness did not change significantly for coupons receiving silicate inhibitors or no inhibitors. The average initial roughness and average final roughness of the nine UCI coupons receiving phosphate inhibitor were 51.1 μm and 37.3 μm, respectively. In contrast, the average initial roughness and average final roughness of the five UCI coupons not receiving phosphate inhibitor were 54.6 μm and 47.5 μm, respectively. Hence, for phase II, these data indicate that phosphate inhibitors have a practical effect on UCI surface roughness.

Figure 5.124 shows the surface detail of a UCI coupon from phase II before incubation in PDS01 at 0.5 mg/L as P-BOP. Each image corresponds to the same 25-mm^2 area. The image on the left shows the coupon surface before exposure to BOP inhibitor. The surface is covered with a mixture of orange-brown, black, and red scales. The image on the right shows the same location after exposure to BOP inhibitor. The surface appears smoother and has been covered with a more uniform scale, mostly yellow and light brown. According to Table 5.126, the initial surface roughness was 53.5 μm and decreased significantly to 35.7 μm. These photographs provide additional evidence that surface roughness decreased in phase II for UCI coupons.

Phase III water quality was similar to phase I water quality and was dominated by groundwater (62%). However, the average temperature of phase III water was 25.8°C. Including replicates, the surface roughness of 9 of the 20 UCI coupons exposed to inhibitors changed significantly in phase III. Six of eight BOP and OP coupons had significant increases in surface roughness. The average initial roughness and final roughness of the eight UCI coupons receiving BOP or OP inhibitors were 65.9 μm and 96.5 μm, respectively. In contrast, none of the four UCI coupons receiving ZOP inhibitor had significant changes in surface roughness.

For the UCI coupons in the pH_s and pH_s+0.3 environments, surface roughness increased significantly for one of two coupons, while not increasing significantly for the matching coupon. Hence, for phase III, these data indicate that BOP and OP inhibitors correspond with significant

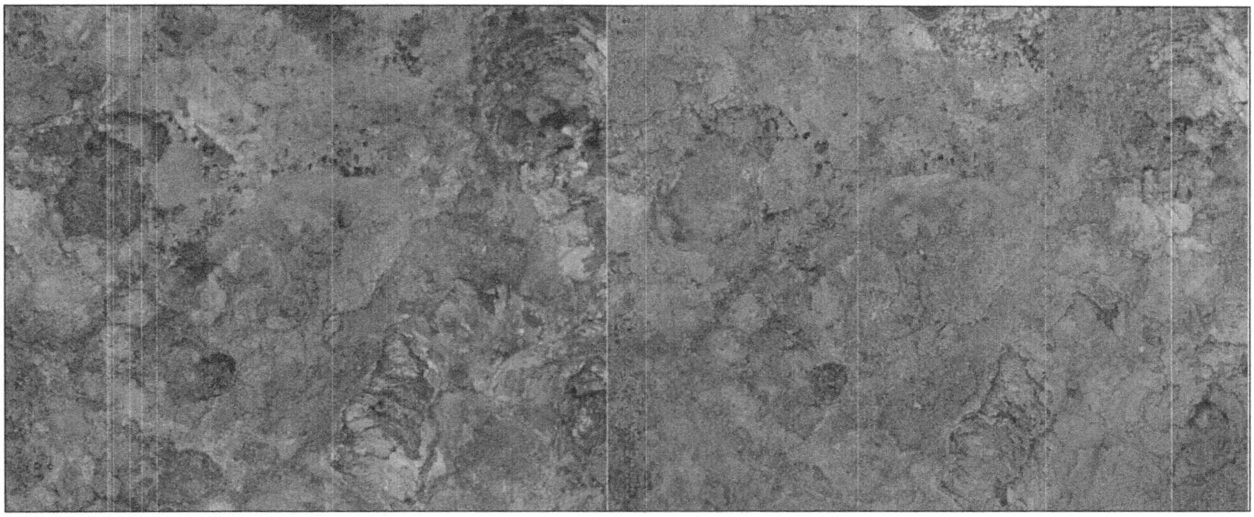

Figure 5.124 Phase II UCI coupon, detail before (left) and after (right) incubation in coupon cradle

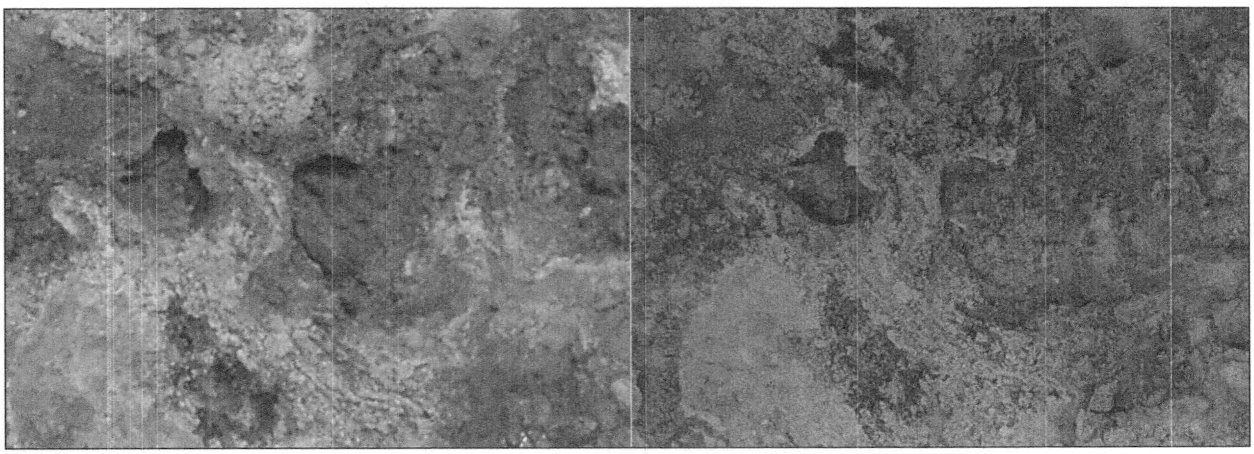

Figure 5.125 Phase III UCI coupon detail before (left) and after (right) incubation

increases in UCI surface roughness. Also, ZOP was the only inhibitor with no significant change in UCI surface roughness.

Figure 5.125 shows the surface detail of a UCI coupon from phase III before incubation in PDS05 at 1.0 mg/L as P-OP. Each image corresponds to the same 25-mm^2 area. The image on the left shows the coupon surface before exposure to the OP inhibitor in phase III. The surface is covered with a mixture of orange-brown, dark brown, and red scales. The image on the right shows the same location after exposure to OP inhibitor. The surface appears rougher and less uniform, having been covered with fewer yellow and red scales, but more black scale. The roughening effect appears evenly distributed over most scales. According to Table 5.126, the initial surface roughness was 66.2 μm and increased significantly to 116.7 μm. These photographs

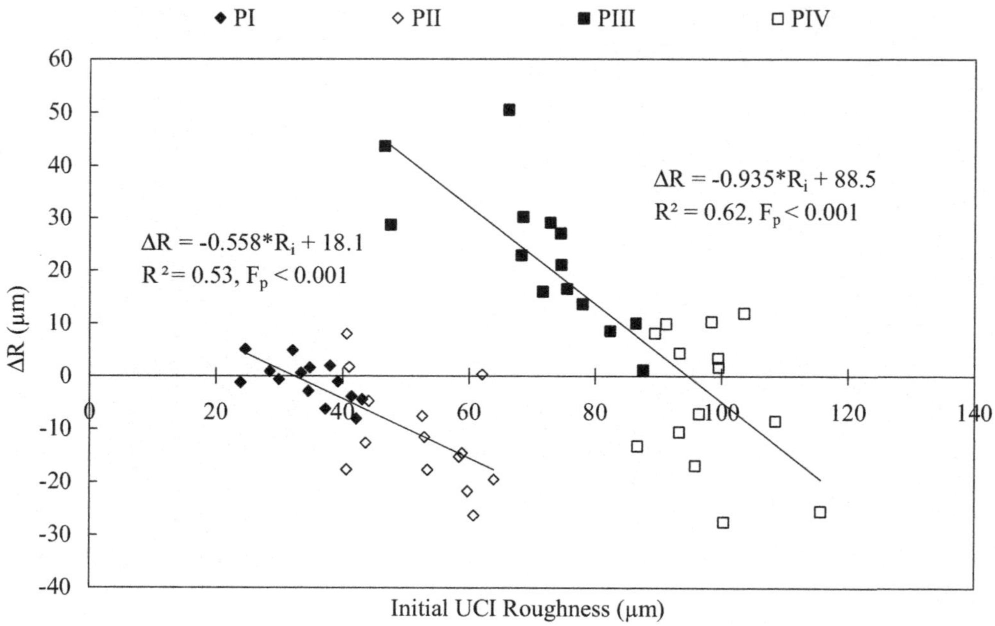

Figure 5.126 Change of initial roughness, ΔR, versus initial roughness, R_i, for UCI coupons by phase

provide additional evidence that surface roughness increased in phase III for UCI coupons receiving phosphate inhibitors.

Phase IV water quality had the highest proportion of desalinated water by RO (20%) among all the phases. The average temperature of phase IV water was 21.2°C. Including replicates, 5 of 21 UCI coupons had a significant change in surface roughness during exposure to inhibitors. Surface roughness decreased for at least one coupon for all inhibitors except OP. Surface roughness decreased for two BOP coupons. Surface roughness did not change significantly for the coupons in the pH_S and pH_S+0.3 environments. Hence, the phase IV UCI surface roughness data indicate that inhibitor had an effect on surface roughness.

Examination of Tables 5.126 and 5.127 for significant changes in surface roughness by inhibitor overall phases confirms the following basic patterns. Surface roughness changes least frequently for the pH_S and pH_S+0.3 environments, each time changing for one of six coupons. For the silicate inhibitors, roughness changes for 3 of 15 coupons. Roughness tended to change more frequently for the phosphate inhibitors, changing for 15 of 42 coupons. This change is most frequent for the BOP inhibitor, changing for 7 of 14 coupons. UCI surface roughness for OP and ZOP inhibitors changes for 5 of 14 and 3 of 14 coupons, respectively.

These data imply that the mechanism of phosphate inhibitor action on the surface of UCI pipe involves a phenomenon that is dependent on water quality. There was not sufficient evidence to indicate that the effect of phosphate inhibitor on surface roughness varied with dose.

The change in ΔR and ΔR/R_i was analyzed with respect to initial roughness, R_i, as shown in Figures 5.126 and 5.127. As discussed earlier in this section, the scan length for UCI was increased starting in phase III to improve the accuracy of the roughness scan.

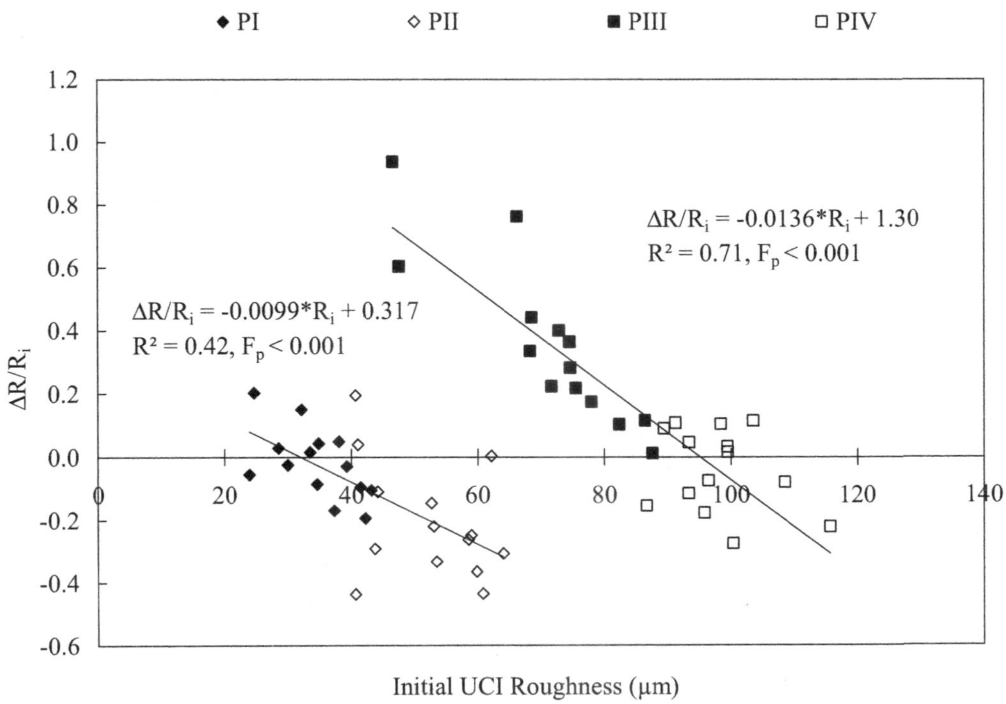

Figure 5.127 Change of the ratio of roughness to initial roughness, $\Delta R/R_i$, versus initial roughness, R_i, for UCI coupons by phase

The data was normalized with respect to initial roughness using the ratio $\Delta R/R_i$ to assess the relative change in surface roughness for both scanning techniques. Figure 5.127 shows good agreement between phases I, II, and IV, with the relative change in roughness varying from –45% to 20%. Thus, normalization of UCI surface roughness to R_i is effective in accounting for differences in scan length.

As shown in Figure 5.125, UCI roughness increased visibly in phase III. Phase III had more frequent and significant roughness changes than phase IV, because of temperature and alkalinity effects in phase III. After examining the final roughness, where instrumental noise was not a factor, the final surface roughness for phase III, 95.4 μm, and phase IV, 91.5 μm, were similar.

Even with this normalization, Figures 5.126 and 5.127 show the same trend for both analyses. Coupons with lower initial roughness tended to have positive increases of surface roughness, and coupons with higher initial roughness tended to have negative increases in surface roughness. The trend lines shown in Figures 5.126 and 5.127 are for phases I and II and phases III and IV. For phases I and II, the average UCI roughness was approximately 43.5 μm, with deviations as much as –30 to +10 μm or from –45% to 20% of the initial roughness following inhibitor exposure. For phases II and IV, the average UCI roughness was approximately 85.5 μm, with deviations as much as –30 to +50 μm or from –30% to 95% of the initial roughness following inhibitor exposure.

A pseudo-equilibrium model was developed to describe the final roughness of the UCI coupons as function of the initial roughness and the mean roughness, as shown in Equation 5.69. Separate coefficients were calculated for phases I and II and phases III and IV. This simple model allows the predicted change of roughness to vary indirectly with initial roughness, which is what was observed for the actual data.

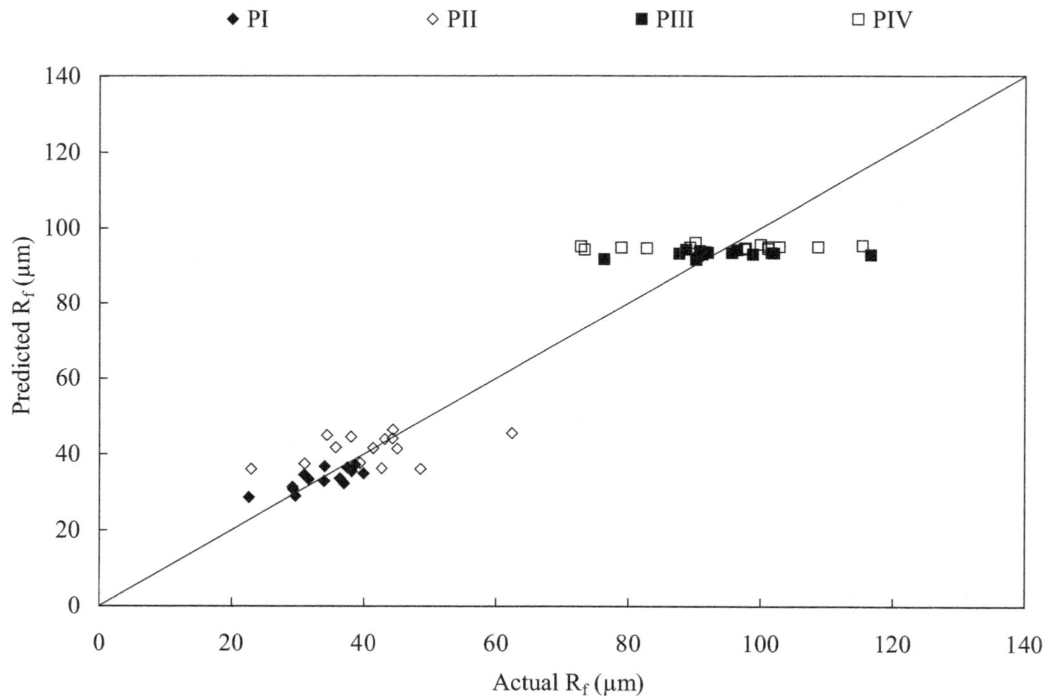

Figure 5.128 Predicted versus actual final UCI surface roughness

$$R_f = R_i - m(R_i - R_{mu}) \quad (5.69)$$

where R_f = final roughness, µm
R_{mu} = mean roughness, µm (32.4 for phases I and II; 94.7 for phases III and IV)
m = gain term defined by regression (0.558 for phases I and II; 0.935 for phases III and IV)

Predicted versus actual roughness for UCI using separate models, as defined in Equation 5.69 for phases I and II and phases III and IV, are shown in Figure 5.128. The model is somewhat descriptive for phases I and II and not descriptive for phases III and IV. For phases I and II, the R^2 for the model was 0.41, with the *p*-value for both parameters less than 0.001. For phases III and IV, the R^2 for the model was 0.01, with the *p*-value for both parameters less than 0.001, indicating statistical significance despite lack of practical fit.

The data for all phases previously shown in Figures 5.126 and 5.127 show a tendency for the change in roughness of UCI to move towards a mean, or average value, generating positive and negative changes in surface roughness. R_f and R_i are positively correlated for phases I and II and independent for phase III and IV, which explains why the model tended to predict similar R_f values regardless of R_i for phases III and IV, resulting in lack of fit shown in Figure 5.128.

Replicates

A series of replicate UCI coupons were analyzed in phases III and IV to gain insight into variability of the roughness experiments. The results of those analyses are shown in Tables 5.128

Table 5.128
Paired replicate surface roughness data for UCI coupons from phase III

	PDS 02	PDS 02-R	PDS 05	PDS 05-R	PDS 08	PDS 08-R	PDS 11	PDS 11-R	PDS 13	PDS 13-R	PDS 14	PDS 14-R
Parameter	BOP 1.0	BOP 1.0	OP 1.0	OP 1.0	ZOP 1.0	ZOP 1.0	Si 6.0	Si 6.0	pH_s	pH_s	pH_s +0.3	pH_s +0.3
R_i	72.8	77.8	66.2	52.1	78.0	81.0	87.6	65.3	71.6	87.8	74.5	76.8
N	5	5	5	5	5	5	6	5	5	5	5	5
R_f	102.0	102.7	116.7	75.1	91.6	94.0	88.7	111.8	87.7	109.9	101.6	95.4
N	6	5	5	5	5	5	6	6	5	5	5	5
ΔR	29.1	24.9	50.4	23.1	13.6	13.1	1.1	46.5	16.0	22.1	27.1	18.6
ΔR *p*-values	**0.01**	**0.03**	**<0.01**	**0.02**	0.23	0.20	0.90	**<0.01**	0.12	**0.04**	**<0.01**	0.06

NOTES: R_i = initial roughness; R_f = final roughness; ΔR = delta roughness. Boldface indicates *p*-values < 0.05.

Table 5.129
Paired replicate surface roughness data for UCI coupons from phase IV

	PDS 02	PDS 02-R	PDS 05	PDS 05-R	PDS 08	PDS 08-R	PDS 11	PDS 11-R	PDS 12	PDS 12-R	PDS 13	PDS 13-R	PDS 14	PDS 14-R
Parameter	BOP 1.0	BOP 1.0	OP 01.0	OP 1.0	ZOP 1.0	ZOP 1.0	Si 1.0	Si 01.0	Si 12.0	Si 12.0	pH_s	pH_s	pH_s +0.3	pH_s +0.3
R_i	91.2	90.8	96.4	99.0	86.7	112.3	108.5	122.8	99.5	79.9	103.5	92.7	95.8	95.2
N	5	5	5	8	5	8	5	5	5	5	5	5	5	5
R_f	101.1	81.6	89.3	71.6	73.4	76.8	100.0	91.3	101.2	98.2	115.4	100.4	78.9	90.6
N	5	6	5	5	5	5	6	5	5	5	5	6	5	8
ΔR	9.9	−9.2	−7.1	−27.4	−13.3	−35.5	−8.5	−31.4	1.7	18.2	11.9	7.6	−16.9	−4.6
ΔR *p*-values	0.27	0.46	0.23	0.06	0.18	**<0.01**	0.28	**<0.01**	0.83	0.16	0.35	0.63	0.06	0.44

NOTES: R_i = initial roughness; R_f = final roughness; ΔR = delta roughness. Boldface indicates *p*-values < 0.05.

and 5.129. A visual comparison of the surface roughness of UCI coupon versus the surface roughness of the replicated UCI coupon is shown in Figure 5.129.

Six replicate pairs of UCI surface roughness are shown in Table 5.128. The change in surface roughness in both the replicate and the original UCI coupon was statistically significant in two of six pairs for phase III replicates. Only one of six pairs in the phase III replicates had no significant change for both coupons. Even though the increase was not always statistically significant, all six replicate pairs experienced an increase in UCI surface roughness for both coupons during exposure. This is illustrated by the plot in Figure 5.129. The replicate data for phase IV is shown in Table 5.129. None of the seven pairs in the phase IV replicates had a significant change in UCI surface roughness for both the original and replicate coupons. Six of the seven pairs of UCI replicate coupons experienced similar changes, with surface roughness of both coupons increasing or decreasing together.

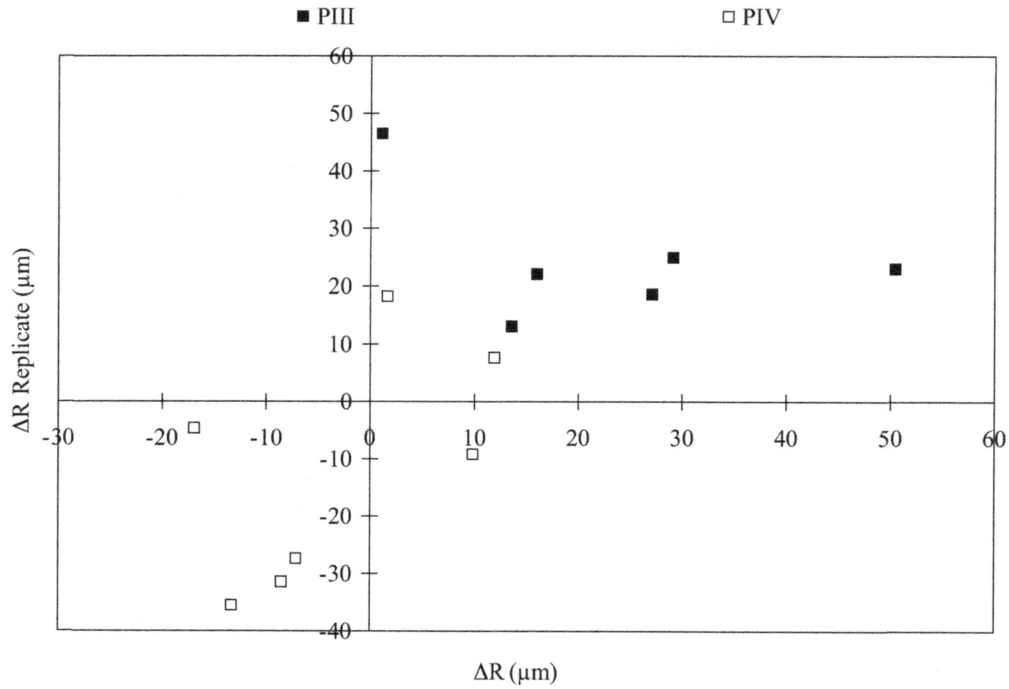

Figure 5.129 Comparison of the change in surface roughness among replicate UCI coupons during phases III and IV

ΔR values of the replicate coupon pairs are plotted in Figure 5.129 and are significantly variable, which shows that replication of surface analysis was only achievable within ±45 μm for the UCI coupon. Differences in R_i of the replicate coupon pairs make direct comparison of the ΔR ineffective to establish systems-level precision. Traditional calculation of precision for ΔR and $\Delta R/R_i$ yielded RPDs, or errors, averaging 520% and 540%, respectively.

The RPD is a measure of precision and is calculated by taking the absolute difference of two values, dividing it by the average of the same two values, and expressing the result as a percentage. Such large RPDs are generated because of differences in ΔR and $\Delta R/R_i$ of the original and replicate and were assessed as shown in Table 5.130 for the replicate UCI coupon in PDS02 (BOP-1.0). The RPD for ΔR and $\Delta R/R_i$, as shown in Table 5.130, is 5,710% and 6,070%, which is based on a change in surface roughness of +9.9 μm for the original UCI coupon and –9.2 μm for the replicate coupon. Because the actual changes are relatively small, the errors calculated using traditional RPD methods are large, which infers that traditional RPD methods are not well suited for making practical assessments of changes in surface roughness of UCI coupons. However, ΔR generally was replicated within ±45 μm for the UCI.

A better measure of system experimental precision relative to roughness change of UCI coupons during exposure to inhibition can be illustrated using Figure 5.126, mentioned previously, and Figure 5.130. These figures show ΔR versus R_i for the original UCI coupons, Figure 5.126, and the UCI replicate coupons, Figure 5.130. The same trend of increasing ΔR for R_i less than the mean UCI surface roughness and decreasing ΔR for R_i more than the mean UCI surface roughness is shown in both figures. Experimental precision can be estimated by comparing the estimated means for initial roughness from the data shown in Figure 5.126 (mean

Table 5.130
Example RPD calculation assessing variation in difference of roughness, UCI

		R_i	R_f	ΔR	$ΔR/R_i$	RPD-ΔR	RPD-$ΔR/R_i$
PDS02	BOP-1.0	91.2	101.1	9.9	0.11	5,710%	6,070%
PDS02-R	BOP-1.0	90.8	81.6	–9.2	–0.10		

NOTES: R_i = initial roughness; R_f = final roughness; ΔR = delta roughness.

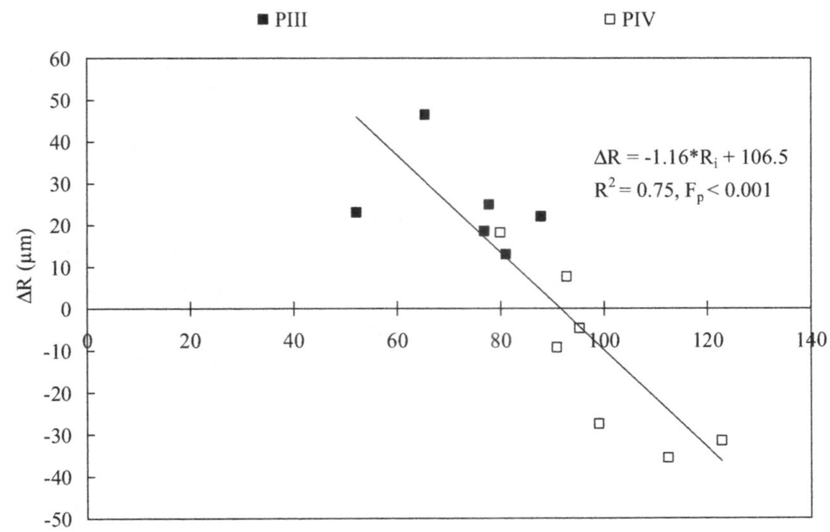

Figure 5.130 Change of initial roughness, ΔR, versus initial roughness, R_i, for UCI replicate coupons for phases III and IV

roughness for original UCI coupons estimated at 94.7 μm) and Figure 5.130 (mean roughness for replicate UCI coupons estimated at 91.5 μm), which gives a relative RPD of 3.4%.

Effect of Water Quality

Linear and nonlinear (power) models relating water quality and R_i to ΔR and R_f were evaluated for all phases and for phases I–II and III–IV. These models indicated that alkalinity, chlorides, or temperature were equally significantly correlated to ΔR and R_f and did not provide any practical interpretation of relationships among water quality and R_i to ΔR and R_f. The lack of significance may be due to the reduced database that resulted from to the need to change the scan length.

Summary for Surface Roughness of Unlined Cast Iron

Surface roughness of UCI coupons was measured for phases I, II, III, and IV. UCI coupons were cut from sections of pipe that had been excavated from within the distribution system and held in indoor storage for a few years, exposed to air. Therefore, the initial roughness of UCI coupons varied significantly between coupons. Also, the initial roughness varied between phases I and II and between phases III and IV because the scan length in phases I and II was often not great enough to capture the entire highest and lowest portions in the scan area. Scan length was increased starting in phase III, providing a more accurate representation of UCI surface roughness. In phases III and IV, the initial surface roughness of UCI was, on average, 85.5 μm, with a minimum and maximum of 46.6 μm and 122.8 μm, respectively. In phase I, surface roughness of none of 14 UCI coupons changed significantly. In phase II, surface roughness of 6 of 14 UCI coupons decreased significantly. In phase III, surface roughness of 9 of 20 UCI coupons increased significantly. In phase IV, surface roughness of 5 of 21 UCI coupons decreased significantly.

Surface roughness of UCI coupons receiving phosphate inhibitors had significant changes in surface roughness more often than UCI coupons receiving other inhibitors. Photographs of UCI coupons were presented, demonstrating decreases and increases in surface roughness for phases II and III.

Linear and nonlinear (power) models relating water quality and R_i to ΔR and R_f were evaluated for all phases and for phases I–II and III–IV. These models indicated that alkalinity, chlorides, or temperature were equally significantly correlated to ΔR and R_f, and did not provide any practical interpretation of relationships among water quality and R_i to ΔR and R_f. The lack of significance may be due to the reduced database that resulted from to the need to change the scan length.

Surface Roughness of Copper

All copper coupons used in this study were new coupons, which had not been previously exposed to water like GS, LCI, PVC, and UCI coupons. Therefore, there is little variation in the initial surface roughness, R_i, of the copper coupons. All copper coupons were of a standard size, 3 in. × 1/2 in. × 1/16 in. (7.6 cm × 1.27 cm × 0.16 cm). The surface roughness of copper was examined in phases III and IV only. A total of 35 copper coupons were measured with average initial roughness of 3.30 μm, with a minimum and maximum of 2.73 μm and 3.80 μm, respectively. The average initial roughness of copper for phases III and IV were 3.22 μm and 3.35 μm, respectively. A two-tailed paired-data t-test for initial roughness of phases III and IV had a p-value of 0.16, indicating that there is not a significant difference in initial surface roughness of copper. This result demonstrates the uniformity among the newly manufactured copper coupons. This is illustrated in Figure 5.131, which shows two copper coupons before exposure to inhibitors. The coupons are shown at actual size.

A summary of the copper surface roughness data is shown in Table 5.131. A statistical comparison of the initial and final roughness averages was completed using a two-tailed pooled-data t-test. The p-values are provided for the hypothesis tests for equality of initial and final roughness. A reported p-value of 0.05 or less indicates there is sufficient evidence to conclude that the final roughness is different from the initial roughness (level of significance of 95%). If the p-value of the t-test is greater than 0.05, then there is no statistically significant difference in coupon surface roughness due to exposure to the inhibitor. The same type of information is shown for replicate coupons in Table 5.132.

Figure 5.131 Photographs of typical copper coupons before exposure to inhibitors

Effect of Inhibitors

Phase III water quality had a predominately groundwater blend with high alkalinity and high temperature. The average temperature of phase III water was 25.8°C. Surface roughness for 11 of 14 copper coupons changed significantly in phase III. Surface roughness did not change significantly for the coupons receiving 1.0 mg/L as P-ZOP, 3.0 mg/L as SiO_2, and the pH_s environment. The data in Table 5.131 demonstrate that surface roughness of copper tended to increase in phase III, with average initial and final roughness of 3.22 μm and 3.95 μm, respectively.

Phase IV water quality had the highest proportion of desalinated water by RO (20%) among all the phases. The average temperature of phase IV water was 21.2°C. Nine of 14 copper coupons had a significant change in surface roughness during exposure to inhibitors. Surface roughness tended to increase in phase IV, with average initial and final roughness of 3.35 μm and 3.86 μm, respectively.

Figure 5.132 shows ΔR by inhibitor and phase. The average change in roughness was calculated over all coupons receiving each inhibitor. Figure 5.132 shows that coupons receiving phosphate inhibitors tended to have greater increases in roughness than coupons receiving other inhibitors. Also, the roughness increase with phosphate inhibitor was consistently greater in phase III than phase IV, suggesting an interaction between water quality, inhibitor, and copper surface roughness. The change in roughness for Si, pH_s, and pH_s+0.3 inhibitors tended to be less than 0.5 μm and did not appear to vary between phases.

In phase III, photographs of the copper coupons showed significant blue-green scales on BOP and OP inhibitors. Similar scales were only scattered across coupons receiving ZOP inhibitor. In contrast, no blue-green scales were observed on the coupons receiving Si, pH_s, and pH_s+0.3 inhibitor. The top coupon in Figure 5.133 received 0.5 mg/L as P-OP inhibitor. The bottom coupon in Figure 5.133 was in the pH_s+0.3 environment. In phase IV, photographs of the copper coupons showed a speckled blue-green scale on coupons receiving BOP, OP, ZOP, and Si inhibitor. No blue-green scales were observed for the pH_s and pH_s+0.3 inhibitors. The top coupon in Figure 5.134 received 0.5 mg/L as P-BOP inhibitor. The bottom coupon in Figure 5.134 was in the pH_s environment in phase IV. Therefore, it appears that surface roughness of copper coupons varies with inhibitor and water quality.

The change in ΔR and $\Delta R/R_i$ was analyzed with respect to initial roughness, R_i, as shown in Figures 5.135 and 5.136. The same method was used in both phases III and IV to measure scan length. Figure 5.135 shows that the change in surface roughness of copper coupons ranged from –0.1 to 2.1 μm. As shown in Figure 5.136, this corresponds to a –3% to 61% change in roughness of copper coupons during the incubation period. Both ΔR and $\Delta R/R_i$ were independent of the initial roughness.

Table 5.131
Statistical comparison of copper initial and final surface roughness

Phase	Parameter	PDS01	PDS02	PDS03	PDS04	PDS05	PDS06	PDS07	PDS08	PDS09	PDS10	PDS11	PDS12	PDS13	PDS14	Average
		BOP 0.5	BOP 1.0	BOP 2.0	OP 0.5	OP 1.0	OP 2.0	ZOP 0.5	ZOP 1.0	ZOP 2.0	Si 3.0	Si 6.0	Si 12.0	pH_s	pH_s +0.3	
III	R_i	2.73	3.36	3.27	3.40	3.16	3.80	3.33	3.44	3.45	3.29	2.80	2.84	3.54	2.76	3.23
	N	8	8	8	8	8	8	8	8	8	8	8	8	8	8	-
	R_f	2.90	4.19	5.22	5.45	3.62	5.59	4.53	3.76	3.98	3.27	3.15	3.45	3.74	3.18	4.00
	N	8	8	8	8	8	8	8	8	8	8	8	8	8	8	-
	ΔR	0.17	0.82	1.95	2.06	0.46	1.78	1.20	0.32	0.53	–0.03	0.34	0.61	0.20	0.43	0.77
	ΔR *p*-values	**0.01**	**0.04**	**<0.01**	**<0.01**	**0.05**	**<0.01**	**<0.01**	0.05	**<0.01**	0.56	**0.02**	**<0.01**	0.17	**<0.01**	-
	$ΔR/R_i$	0.06	0.24	0.60	0.61	0.14	0.47	0.36	0.09	0.15	–0.01	0.12	0.22	0.06	0.15	0.23
IV	R_i	3.37	3.50	3.38	3.38	3.40	3.16	3.67	3.38	3.50	3.56	3.09	3.43	3.24	3.12	3.37
	N	7	8	8	8	8	8	8	8	8	8	8	8	8	8	-
	R_f	3.68	4.35	4.65	4.41	3.89	4.72	3.95	3.37	3.89	3.46	3.22	3.57	3.37	3.17	3.83
	N	8	8	8	8	8	8	8	8	8	8	8	8	8	8	-
	ΔR	0.30	0.84	1.27	1.03	0.49	1.56	0.28	–0.01	0.40	–0.10	0.13	0.14	0.13	0.05	0.47
	ΔR *p*-values	**0.02**	**<0.01**	**<0.01**	**<0.01**	**0.01**	**<0.01**	**<0.01**	0.87	**<0.01**	**0.04**	0.06	0.37	0.07	0.53	-
	$ΔR/R_i$	0.09	0.24	0.37	0.30	0.14	0.49	0.08	0.00	0.11	–0.03	0.04	0.04	0.04	0.02	0.14

NOTES: The surface roughness of copper was not examined in phases I and II. R_i = initial roughness; R_f = final roughness; ΔR = delta roughness. Boldface indicates *p*-values < 0.05.

Table 5.132
Replicate roughness data for copper coupons

Phase	Parameter	PDS01	PDS02	PDS03	PDS04	PDS05	PDS06	PDS07	PDS08	PDS09	PDS10	PDS11	PDS12	PDS13	PDS14
		BOP 0.5	BOP 1.0	BOP 2.0	OP 0.5	OP 1.0	OP 2.0	ZOP 0.5	ZOP 1.0	ZOP 2.0	Si 3.0	Si 6.0	Si 12.0	pH_s	pH_s +0.3
III	R_i	-	-	-	-	-	-	-	-	-	-	-	-	3.16	-
	N	-	-	-	-	-	-	-	-	-	-	-	-	8	-
	R_f	-	-	-	-	-	-	-	-	-	-	-	-	3.23	-
	N	-	-	-	-	-	-	-	-	-	-	-	-	8	-
	ΔR	-	-	-	-	-	-	-	-	-	-	-	-	0.07	-
	ΔR *p*-values	-	-	-	-	-	-	-	-	-	-	-	-	0.37	-
IV	R_i	-	3.34	-	-	3.30	-	-	3.42	-	-	3.27	-	3.36	3.22
	N	-	8	-	-	8	-	-	8	-	-	8	-	8	8
	R_f	-	4.10	-	-	4.28	-	-	3.80	-	-	3.96	-	3.97	3.46
	N	-	8	-	-	8	-	-	8	-	-	8	-	8	8
	ΔR	-	0.76	-	-	0.98	-	-	0.39	-	-	0.70	-	0.61	0.24
	ΔR *p*-values	-	**<0.01**	-	-	**<0.01**	-	-	**0.03**	-	-	**<0.01**	-	**<0.01**	**0.01**

NOTES: R_i = initial roughness; R_f = final roughness; ΔR = delta roughness. Boldface indicates *p*-values < 0.05.

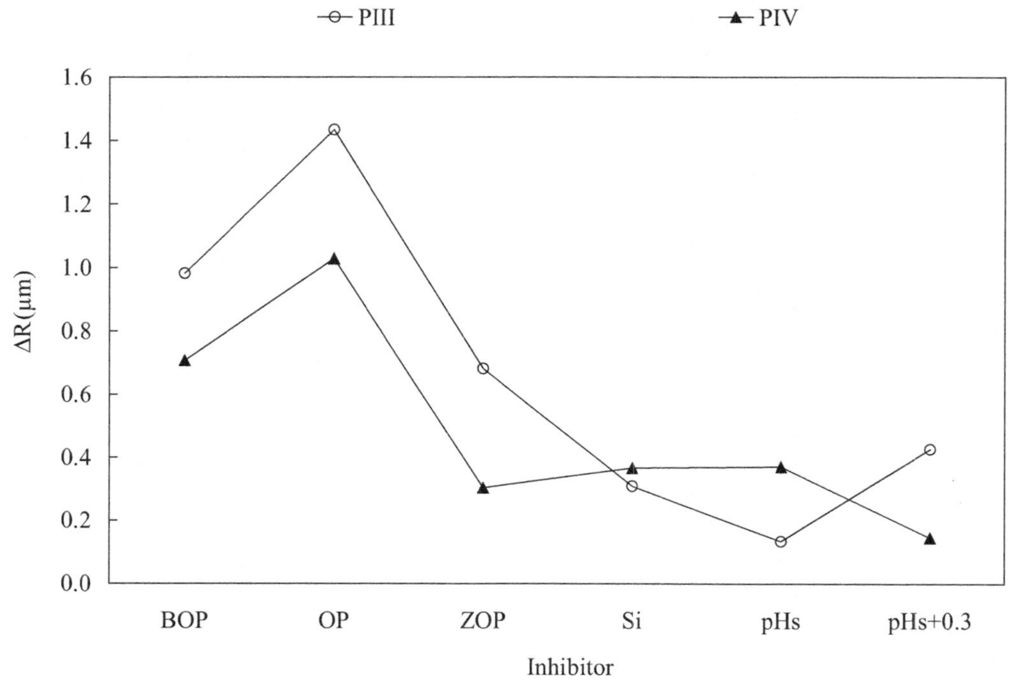

Figure 5.132 Change of initial roughness, ΔR, of copper coupons by inhibitor and phase

Figure 5.133 Visual comparison of surface scales on phase III copper coupons

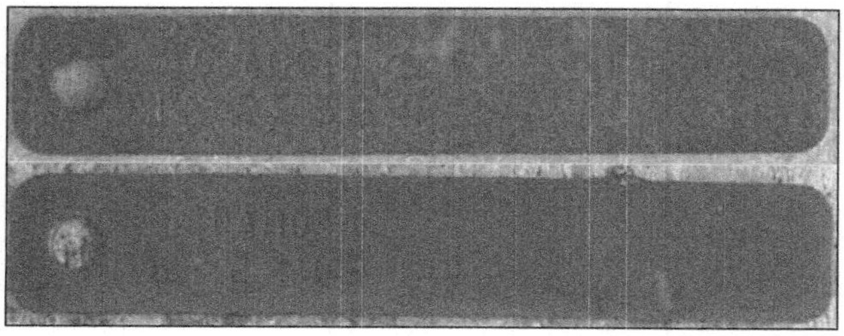

Figure 5.134 Visual comparison of surface scales on phase IV copper coupons

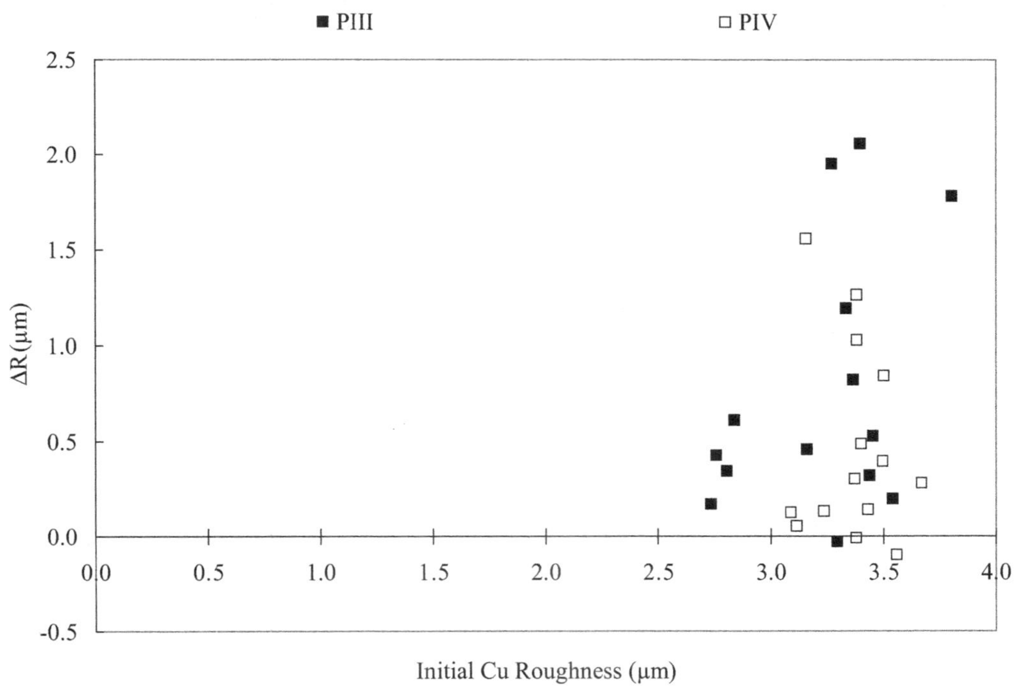

Figure 5.135 Change of initial roughness, ΔR, versus initial roughness, R_i, for copper coupons by phase

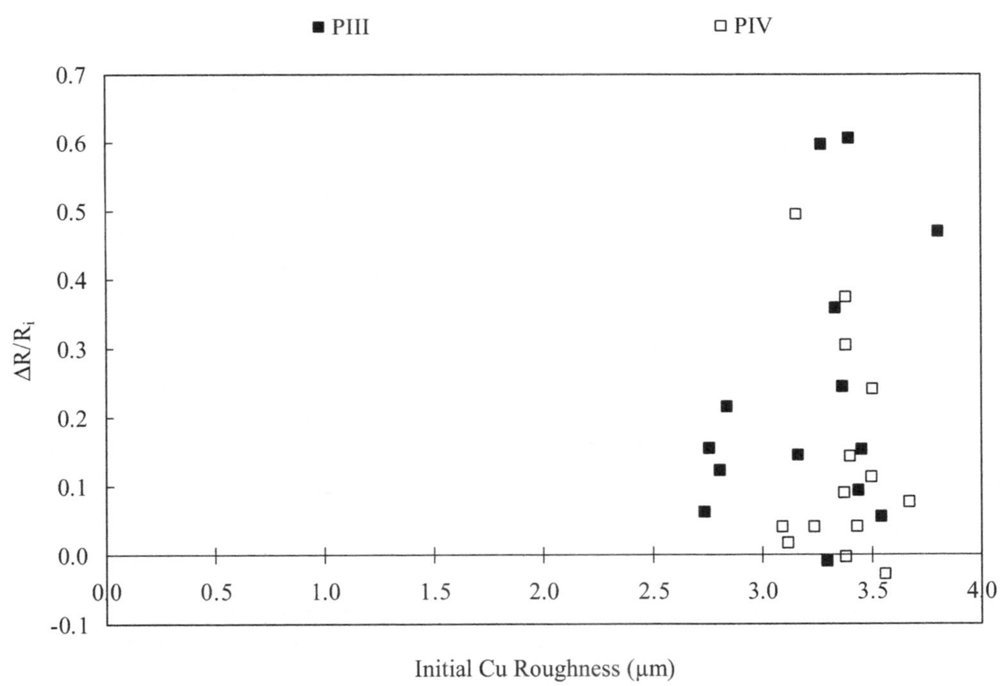

Figure 5.136 Change of the ratio of roughness to initial roughness, $\Delta R/R_i$, versus initial roughness, R_i, for copper coupons by phase

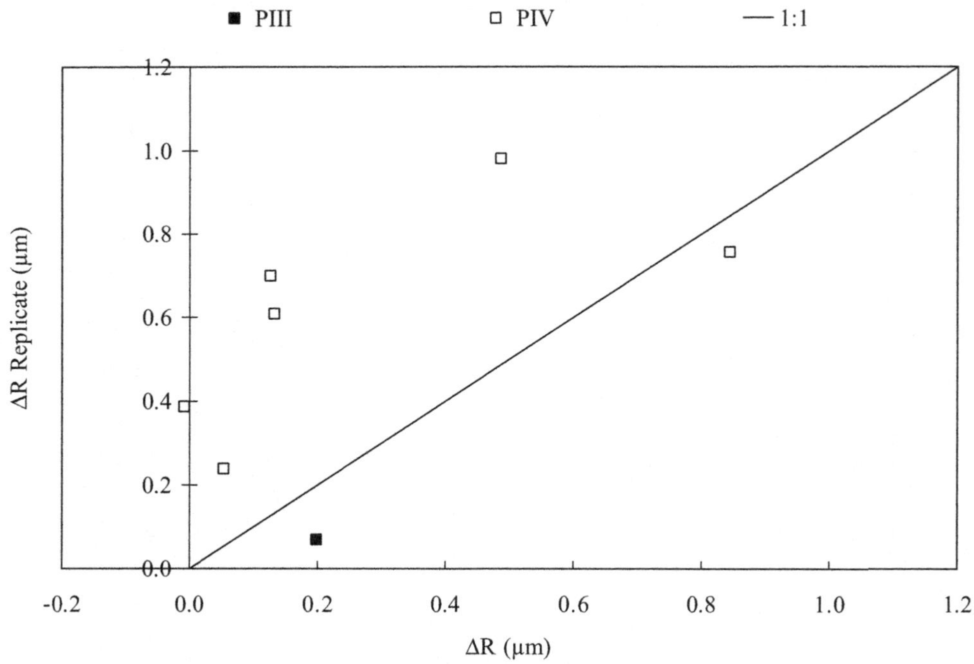

Figure 5.137 Comparison of the change in surface roughness among replicate copper coupons during phases III and IV

Replicates

A series of replicate copper coupons were analyzed in phases III and IV to gain insight into variability of the roughness experiments. The results of those analyses are shown in Table 5.132.

Surface roughness of the original copper coupon and the replicate copper coupon are plotted in Figure 5.137. The graph demonstrates that roughness tended to increase in both the original and replicate copper coupons. The change in surface roughness was measured consistently within 0.6 µm between replicates.

The surface roughness of original and replicate copper coupons is shown in Tables 5.133 and 5.134. For phase III one replicate coupon was placed in the pH_s environment. The surface roughness of the pair of copper coupons in the pH_s environment did not change significantly for either coupon. For phase IV, replicate coupons were placed in the middle-inhibitor dose, pH_s, and pH_s+0.3 PDSs. Surface roughness of both copper coupons increased significantly for the BOP and OP pairs. The other coupon pairs had only one coupon with a significant increase in copper surface roughness.

The RPD is a measure of precision and is calculated by taking the absolute difference of two values, dividing it by the average of the same two values, and expressing the result as a percentage. An example calculation of RPD of ΔR and $\Delta R/R_i$ for the pH_s environment is shown in Table 5.135. The RPD for ΔR and $\Delta R/R_i$, as shown in Table 5.135, is 96% and 87%, which is based on a change in surface roughness of 0.20 µm for the original copper coupon and 0.07 µm for the replicate coupon. Although the actual changes are relatively small, the errors calculated using traditional RPD methods are large, which indicates the natural variability in change in

Table 5.133
Paired replicate surface roughness data for copper coupons from phase III

Parameter	PDS13 pH_s	PDS13-R pH_s
R_i	3.54	3.16
N	8	8
R_f	3.74	3.23
N	8	8
ΔR	0.20	0.07
ΔR *p*-values	0.17	0.37

NOTES: R_i = initial roughness; R_f = final roughness; ΔR = delta roughness.

surface roughness of copper coupons. The RPD of ΔR ranged from 11% to 210%, whereas the RPD of $\Delta R/R_i$ ranged from 6% to 210%. In contrast, the RPD of R_i of the copper coupons ranged from 1% to 11%, whereas the RPD of R_f of the copper coupons ranged from 6% to 21%. The low RPDs for R_i show that R_i was uniform for the copper coupons before exposure to inhibitors. Because of this uniformity among the copper coupons, RPD of R_f is an appropriate measure of systems precision for measurement of copper surface roughness.

Effect of Water Quality

Linear and nonlinear (power) models for correlating water quality and R_i to ΔR and R_f of copper were evaluated in phases III and IV. In all cases, the water quality parameters were statistically insignificant. The lack of significance may be due to the limited data set.

Summary for Surface Roughness of Copper

Surface roughness of copper coupons was measured for phases III and IV. Unlike other materials, copper surface roughness was measured using new corrosion coupons. Hence, the initial surface roughness of copper coupons was relatively uniform, averaging 3.30 µm, with a minimum and maximum of 2.73 µm and 3.80 µm, respectively. Surface roughness of copper coupons tended to increase in both phases, with significant increases observed for 11 of 15 copper coupons in phase III and 14 of 20 copper coupons in phase IV.

The increase in copper surface roughness depended on inhibitor type and phase but was independent of the initial roughness. The greatest increases in copper surface roughness were observed for the phosphate inhibitors. In phase III, blue-green scales formed on the copper coupons that received phosphate inhibitors. Similarly, in phase IV, speckled blue-green scales formed on the copper coupons that received phosphate and silicate inhibitors. For coupons receiving pH_s and pH_s+0.3 inhibitors, no blue-green scales were observed in either phase. Although inhibitor and water quality influence the surface roughness of copper coupons, this relationship was not found to be significant in linear and nonlinear (power) models. This lack of significance may be due to the limited data set.

Table 5.134
Paired replicate surface roughness data for copper coupons from phase IV

Parameter	PDS02 BOP 1.0	PDS02-R BOP 1.0	PDS05 OP 1.0	PDS05-R OP 1.0	PDS08 ZOP 1.0	PDS08-R ZOP 1.0	PDS11 Si 1.0	PDS11-R Si 01.0	PDS13 pH_s	PDS13-R pH_s	PDS14 pH_s +0.3	PDS14-R pH_s +0.3
R_i	3.50	3.34	3.40	3.30	3.38	3.42	3.09	3.27	3.24	3.36	3.12	3.22
N	8	8	8	8	8	8	8	8	8	8	8	8
R_f	4.35	4.10	3.89	4.28	3.37	3.80	3.22	3.96	3.37	3.97	3.17	3.46
N	8	8	8	8	8	8	8	8	8	8	8	8
ΔR	0.84	0.76	0.49	0.98	–0.01	0.39	0.13	0.70	0.13	0.61	0.05	0.24
ΔR *p*-values	**<0.01**	**<0.01**	**0.01**	**<0.01**	0.87	**0.03**	0.06	**<0.01**	0.07	**<0.01**	0.53	**0.01**

NOTES: R_i = initial roughness; R_f = final roughness; ΔR = delta roughness. Boldface indicates p-values < 0.05.

Table 5.135
Example RPD calculation assessing variation in difference of roughness, copper

		R_i	R_f	ΔR	$\Delta R/R_i$	RPD-ΔR_i	RPD ΔR/R	RPD R_i	RPD R_f
PDS13	pH_s	3.54	3.74	0.20	0.06	96%	87%	11%	15%
PDS13-R	pH_s	3.15	3.23	0.07	0.02				

NOTES: R_i = initial roughness; R_f = final roughness; ΔR = delta roughness.

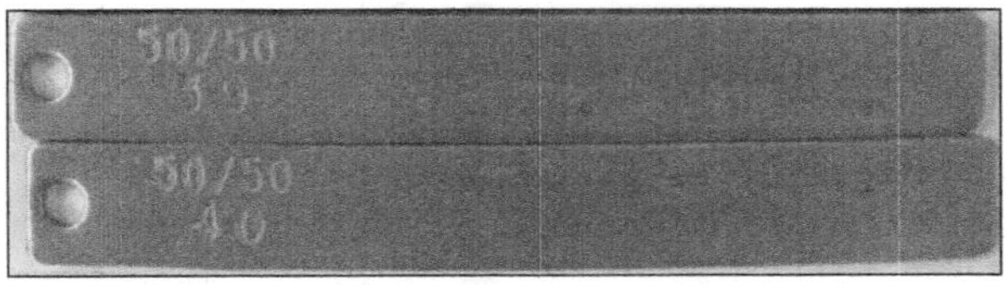

Figure 5.138 Photographs of typical lead/tin coupons before exposure to inhibitors

Surface Roughness of Lead/Tin

All lead/tin (Pb/Sn) coupons used in this study were new coupons (Figure 5.138) that had not been previously exposed to water like the GS, LCI, PVC, and UCI coupons. Therefore, there was little variation in the initial surface roughness, R_i, of the lead/tin coupons. All lead/tin coupons were of a standard size, 3 in. × 3/8 in. × 1/16 in. (7.62 cm × 1.27 cm × 0.16 cm). The surface roughness of lead/tin was examined in phases III and IV only. A total of 31 lead/tin coupons were measured with average initial roughness of 1.72 μm, 1.74 μm for phase III, and 1.71 μm for phase IV, with a minimum and maximum of 0.98 μm and 2.62 μm, respectively.

A two-tailed paired-data t-test for initial roughness of phases III and IV had a *p*-value of 0.73, indicating that there is not a significant difference in initial surface roughness of lead/tin. This result demonstrates the uniformity among the newly manufactured lead/tin coupons. This is illustrated in Figure 5.138, which shows two lead/tin coupons before exposure to inhibitors. The coupons are shown at actual size.

A summary of the lead/tin surface roughness data is shown in Table 5.136. A statistical comparison of the initial and final roughness averages was completed using a two-tailed pooled-data t-test. The statistical analysis is summarized in Table 5.136. The *p*-values are provided for the hypothesis tests for equality of initial and final roughness. A reported *p*-value of 0.05 or less indicates there is sufficient evidence to conclude that the final roughness is different from the initial roughness (level of significance of 95%). If the *p*-value of the t-test comparing the means of the surface roughness before and after exposure to the inhibitor is greater than 0.05, then there is no statistically significant difference in coupon surface roughness due to exposure to the inhibitor.

Effect of Inhibitors

Phase III water quality had a predominately groundwater blend with high alkalinity and high temperature. The average temperature of phase III water was 25.8°C. Surface roughness for 3 of 14 lead/tin coupons changed significantly in phase III. The data in Table 5.136 demonstrate

Table 5.136
Statistical comparison of lead/tin initial and final surface roughness

Phase	Parameter	PDS01	PDS02	PDS03	PDS04	PDS05	PDS06	PDS07	PDS08	PDS09	PDS10	PDS11	PDS12	PDS13	PDS14	Average
		BOP 0.5	BOP 1.0	BOP 2.0	OP 0.5	OP 1.0	OP 2.0	ZOP 0.5	ZOP 1.0	ZOP 2.0	Si 3.0	Si 6.0	Si 12.0	pH_s	pH_s +0.3	
III	R_i	1.52	2.07	1.61	1.84	2.02	1.72	1.64	2.00	1.78	1.60	1.48	1.50	1.85	1.70	1.74
	N	8	8	8	8	8	8	8	8	8	8	8	8	8	8	
	R_f	1.29	5.18	1.60	1.98	3.27	1.82	1.78	1.65	1.85	2.50	2.14	2.63	2.30	2.94	2.35
	N	8	8	8	7	8	8	8	8	8	8	8	8	8	8	
	ΔR	–0.23	3.11	–0.02	0.14	1.25	0.10	0.14	–0.35	0.07	0.90	0.66	1.13	0.45	1.24	0.61
	R *p*-values	0.52	**<0.01**	0.96	0.69	0.09	0.64	0.57	0.43	0.73	**0.01**	0.09	**0.01**	0.32	0.06	
	$\Delta R/R_i$	–0.15	1.51	–0.01	0.07	0.62	0.06	0.09	–0.17	0.04	0.56	0.44	0.75	0.24	0.73	
IV	R_i	1.66	2.62	2.02	1.80	1.30	0.98	2.19	1.16	2.29	1.85	2.34	1.58	1.17	1.10	1.72
	N	8	8	8	8	8	8	8	8	8	8	8	8	8	8	
	R_f	1.37	1.69	1.22	1.30	1.97	2.25	2.38	1.73	2.72	2.08	2.12	1.87	1.19	1.61	1.82
	N	8	8	8	8	8	8	8	8	8	8	8	8	8	8	
	ΔR	–0.29	–0.93	–0.80	–0.50	0.68	1.27	0.19	0.57	0.43	0.23	–0.22	0.29	0.02	0.50	0.10
	ΔR *p*-values	0.11	0.14	0.19	0.08	**0.01**	**<0.01**	0.75	0.08	0.38	0.42	0.57	0.33	**0.04**	**0.01**	
	$\Delta R/R_i$	–0.17	–0.35	–0.40	–0.28	0.52	1.30	0.08	0.49	0.19	0.12	–0.09	0.18	–0.76	0.45	

NOTES: Surface roughness was not measured in phases I and II. R_i = initial roughness; R_f = final roughness; ΔR = delta roughness. Boldface indicates *p*-values < 0.05.

that surface roughness of lead/tin tended to increase in phase III, with average initial and final roughness of 1.74 μm and 2.35 μm, respectively. Replicate data is shown in Table 5.137 and was only collected during phase IV.

Phase IV water quality had the highest proportion of desalinated water by RO (20%) among all the phases. The average temperature of phase IV water was 21.2°C. Surface roughness for 4 of 14 lead/tin coupons had a significant change in surface roughness during exposure to inhibitors. Surface roughness tended to increase slightly in phase IV, with average initial and final roughness of 1.72 μm and 1.82 μm, respectively.

Figure 5.139 shows ΔR by inhibitor and phase. The average change in roughness was calculated over all coupons receiving each inhibitor. From Figure 5.139 it appears that the change in roughness of lead/tin coupons is independent of inhibitor type and water quality.

In phases III and IV, photographs of the lead/tin coupons showed no significant difference in appearance before and after receiving inhibitors. This was consistent for all inhibitor types and pH environments.

The change in ΔR and $\Delta R/R_i$ was analyzed with respect to initial roughness, R_i, as shown in Figures 5.140 and 5.141. The same method was used in both phases III and IV to measure scan length. Figure 5.140 shows that the change in surface roughness of lead/tin coupons ranged from –0.9 to 3.1 μm.As shown in Figure 5.141, this corresponds to a –40% to 151% change in roughness of lead/tin coupons during the incubation period. The change in surface roughness, ΔR, was independent of R_i at a 95% level of significance with an F_p-value of 0.23. In contrast, $\Delta R/R_i$ was dependent on R_i at the 95% level of significance with an F_p value of 0.05.

Replicates

A series of replicate lead/tin coupons were analyzed in phases III and IV to gain insight into variability of the roughness experiments. The results of those analyses are shown in Table 5.138. Surface roughness of the original lead/tin coupon and the replicate lead/tin coupon are plotted in Figure 5.142. The graph demonstrates that the change in roughness varied between the replicate coupons. The change in surface roughness was measured consistently within 0.8 μm between replicates.

The surface roughness of original and replicate copper coupons is shown in Table 5.138. For phase IV, replicate coupons were placed in PDSs with the 1.0 mg/L as P-BOP inhibitor, 6.0 mg/L as SiO_2 silicate inhibitor, and pH_s+0.3 inhibition. Only one coupon increased significantly in the pH_s+0.3 lead/tin coupon pair. For the BOP and silicate coupon pairs, no coupons had a significant change in surface roughness.

The RPD is a measure of precision and is calculated by taking the absolute difference of two values, dividing it by the average of the same two values, and expressing the result as a percentage. An example calculation of RPD of ΔR and $\Delta R/R_i$ for the pH_s+0.3 environment is shown in Table 5.139. The RPD for ΔR and $\Delta R/R_i$, as shown in Table 5.139, is 712% and 371%, respectively, which is based on a change in surface roughness of 0.20 μm for the original lead/tin coupon and 0.07 μm for the replicate coupon. Although the actual changes are relatively small, the errors calculated using traditional RPD methods are large, which indicates the natural variability in change in surface roughness of lead/tin coupons. The RPD of ΔR ranged from 141% to 712%, whereas the RPD of $\Delta R/R_i$ ranged from 100% to 371%. In contrast, the RPD of R_i of the lead/tin coupons ranged from 42% to 62%, whereas the RPD of R_f of the lead/tin coupons ranged from 7% to 35%. The RPDs for R_i were relatively high compared to those of copper, which did not

Table 5.137
Replicate roughness data for lead/tin coupons

		PDS01	PDS02	PDS03	PDS04	PDS05	PDS06	PDS07	PDS08	PDS09	PDS10	PDS11	PDS12	PDS13	PDS14
Phase	Parameter	BOP 0.5	BOP 1.0	BOP 2.0	OP 0.5	OP 1.0	OP 2.0	ZOP 0.5	ZOP 1.0	ZOP 2.0	Si 3.0	Si 6.0	Si 12.0	pH_s	pH_s +0.3
IV	R_i	-	1.35	-	-	-	-	-	-	-	-	1.53	-	-	2.07
	N	-	8	-	-	-	-	-	-	-	-	8	-	-	8
	R_f	-	1.19	-	-	-	-	-	-	-	-	1.97	-	-	1.79
	N	-	8	-	-	-	-	-	-	-	-	8	-	-	8
	ΔR	-	–0.16	-	-	-	-	-	-	-	-	0.44	-	-	–0.28
	ΔR *p*-values	-	0.16	-	-	-	-	-	-	-	-	0.30	-	-	0.10

NOTES: R_i = initial roughness; R_f = final roughness; ΔR = delta roughness.

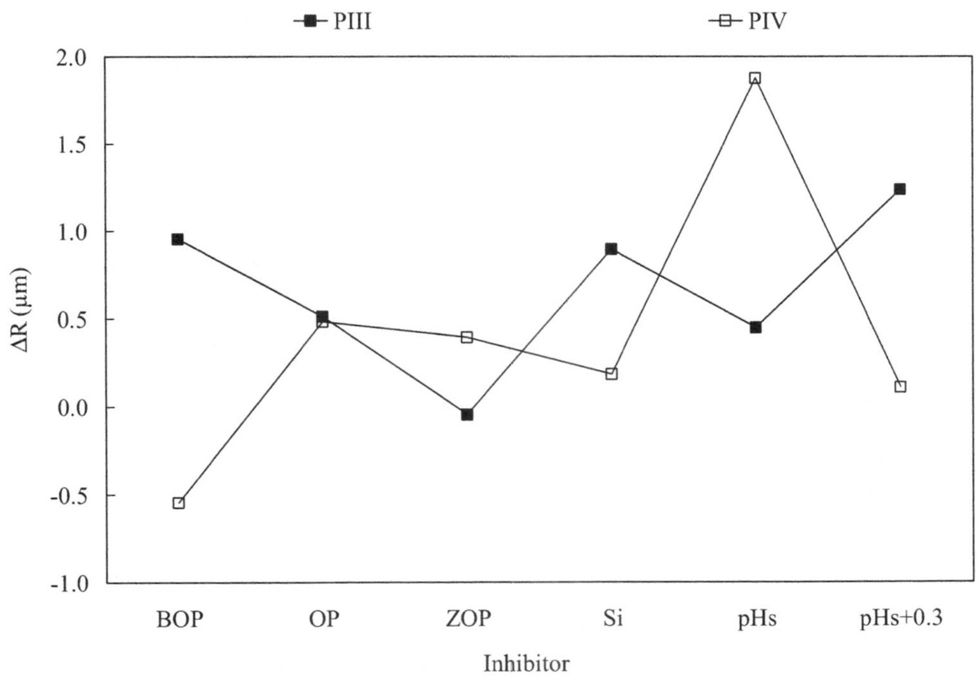

Figure 5.139 Change of initial roughness, ΔR, of lead/tin coupons by inhibitor and phase

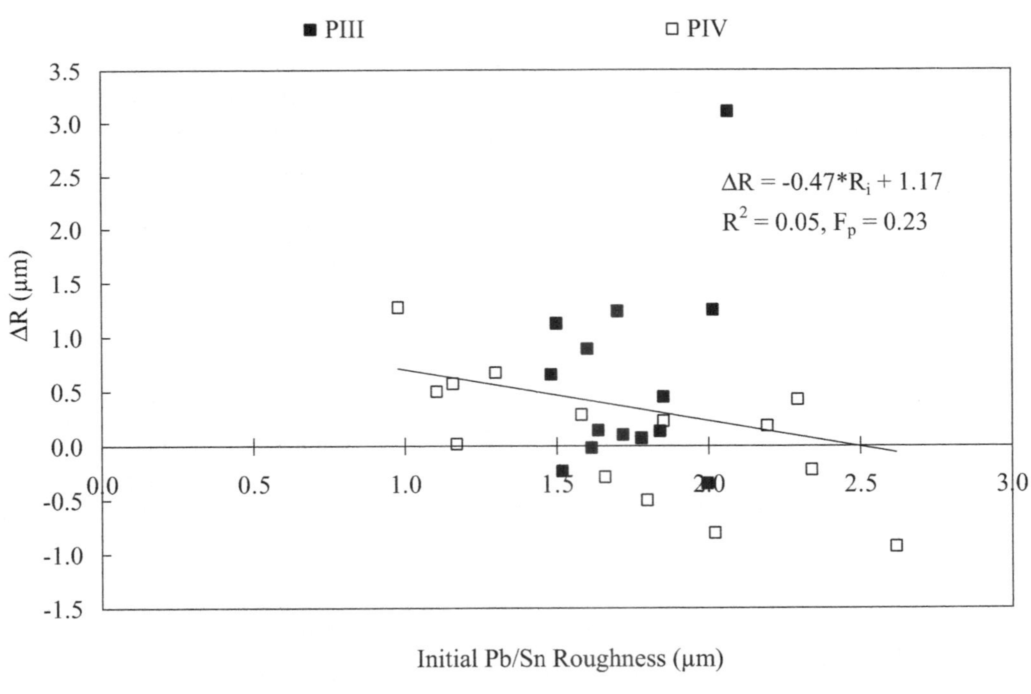

Figure 5.140 Change of initial roughness, ΔR, versus initial roughness, R_i, for lead/tin coupons by phase

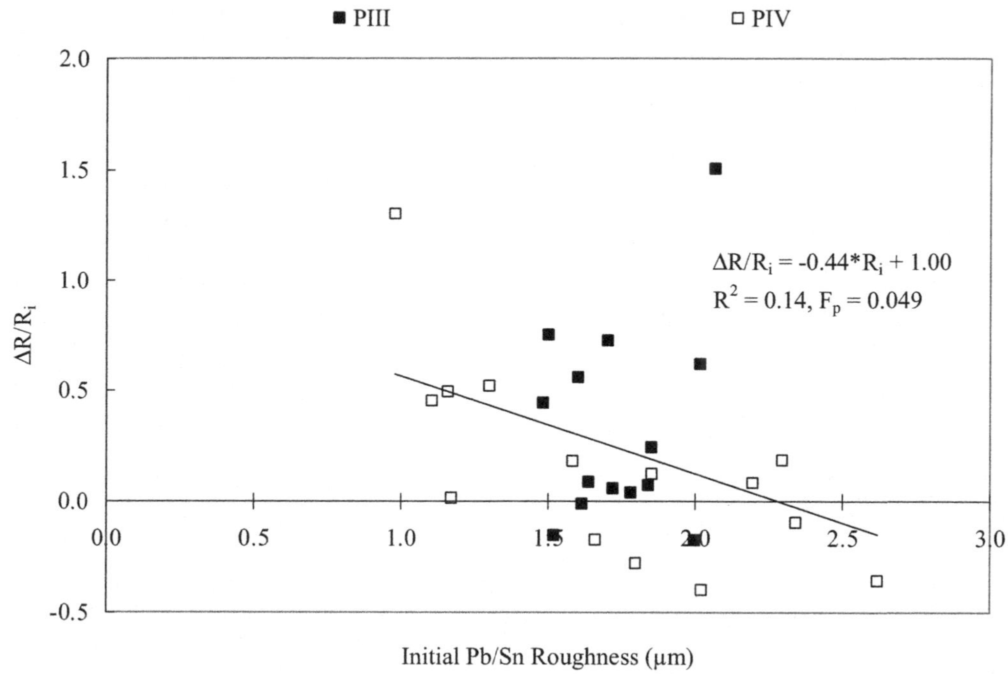

Figure 5.141 Change of the ratio of roughness to initial roughness, $\Delta R/R_i$, versus initial roughness, R_i, for lead/tin coupons by phase

Table 5.138
Paired replicate surface roughness data for copper coupons from phase IV

Parameter	PDS02 BOP 1.0	PDS02-R BO 1.0	PDS11 Si 3.0	PDS11-R Si 3.0	PDS14 pH_s+0.3	PDS14-R pH_s+0.3
R_i	2.62	1.35	2.34	1.53	1.10	2.07
N	8	8	8	8	8	8
R_f	1.69	1.19	2.12	1.97	1.61	1.79
N	8	8	8	8	8	8
ΔR	–0.93	–0.16	–0.22	0.44	0.50	–0.28
ΔR *p*-values	0.14	0.16	0.57	0.30	**0.01**	0.10

NOTES: R_i = initial roughness; R_f = final roughness; ΔR = delta roughness. Boldface indicates *p*-values < 0.05.

exceed 11%. Hence, measurement of R_i for lead/tin coupons indicates that there were differences in the initial roughness of replicate lead/tin coupons. The RPD of R_f of lead/tin coupons did not exceed 35%, indicating reasonable agreement between replicate coupons. However, these precision data must be taken with caution as they are based on three replicate pairs.

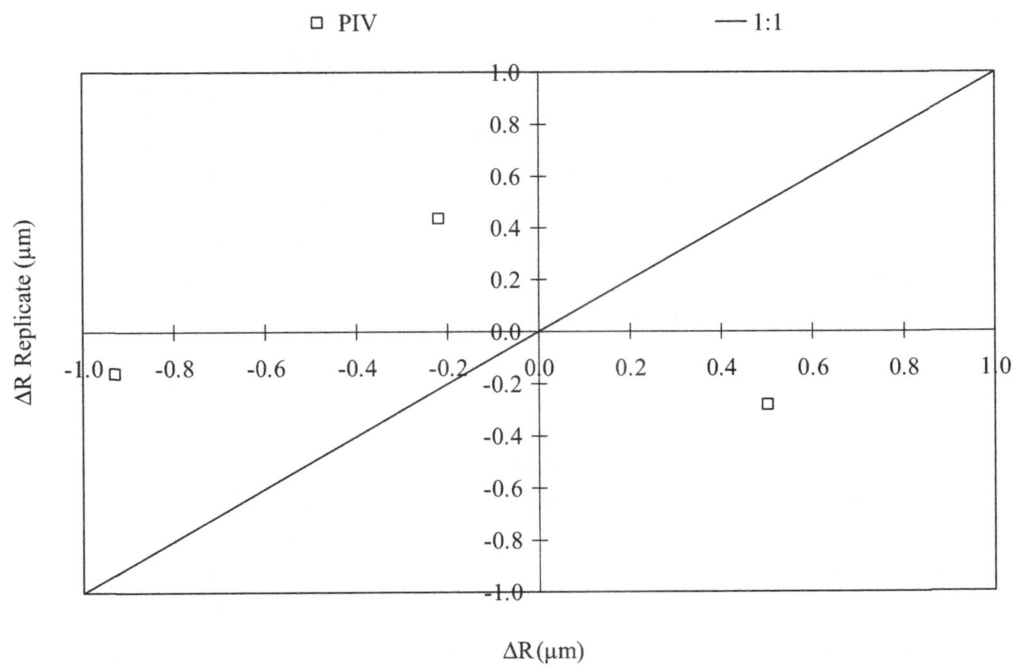

Figure 5.142 Comparison of the change in surface roughness among replicate lead/tin coupons during phase IV

Table 5.139
Example RPD calculation assessing variation in difference of roughness, lead/tin

		R_i	R_f	ΔR	$ΔR/R_i$	RPD ΔR	RPD $ΔR/R_i$	RPD R_i	RPD R_f
PDS14	pH_s+0.3	1.10	1.61	0.50	0.45	712%	371%	61%	11%
PDS14-R	pH_s+0.3	2.07	1.79	–0.28	–0.14	-	-	-	-

NOTES: R_i = initial roughness; R_f = final roughness; ΔR = delta roughness.

Effect of Water Quality

Linear and nonlinear (power) models for correlating water quality and R_i to ΔR and R_f of lead/tin were evaluated in phases III and IV. In all cases, the water quality parameters were statistically insignificant. The lack of significance may be due to the limited data set.

Summary for Surface Roughness of Lead/Tin

Surface roughness of lead/tin was measured for phases III and IV. As with copper, lead/tin surface roughness was measured using new corrosion coupons. Hence, the initial surface roughness of lead/tin coupons was relatively uniform, averaging 1.72 µm, with a minimum and maximum of 1.0 µm and 2.6 µm, respectively, although not as uniform as that of copper. Surface roughness of lead/tin coupons tended to remain unchanged in both phases, with no significant change observed for 11 of 14 lead/tin coupons in phase III, and 13 of 17 lead/tin coupons in phase IV.

The change in lead/tin surface roughness, ΔR, did not depend on inhibitor type or phase and was independent of the initial roughness. In contrast, the change in roughness normalized to initial roughness, $\Delta R/R_i$, was significantly correlated with R_i. Photographs of lead/tin coupons in phases III and IV showed no visible differences in appearance of the lead/tin coupons, providing additional evidence that lead/tin surface roughness did not vary significantly by inhibitor type or by phase. Regression with linear and nonlinear (power) models confirmed the lack of significance of inhibitors and water quality on lead/tin surface roughness.

RESIDUAL

The loss of chloramine residual in drinking water distribution systems was investigated as influenced by dose, temperature, water quality, pipe material, pipe geometry, and turbulence within the distribution system. Initially, statistical analysis was conducted to prove there was no difference among the influent residuals to the PDSs. The influent chloramine concentration of the 14 PDSs was compared using a one-way analysis of variance (ANOVA) test for means. Influent and effluent residuals to the 14 PDSs were analyzed weekly during operation for 1 year. As described in the "Pilot Distribution System Operation" section in chapter 3, PDSs were fed from blend tanks that were continuously adjusted and monitored for residual maintenance. The ANOVA p-value was 0.3, which meant that there was not a statistically significant difference in the mean influent chloramine concentration among the PDSs, or more simply, an equal chloramine residual was maintained in all PDSs during operation. Table 5.140 shows the results of the ANOVA test, including the number of samples taken, the mean influent residual, and the standard deviation for the PDS. The minimum mean influent residual was 4.7 mg/L as Cl_2 for PDS06, PDS09, and PDS13. The maximum mean influent residual was 5.1 mg/L as Cl_2 for PDS01 and PDS14.

Residual Modeling

Background

The most accepted form to describe the decay of chlorine residual in the distribution systems is by reactions of chlorine within the bulk liquid phase and the surface, corrosion products, and biofilms on the pipe wall (Biswas, Lu and Clark 1993; Rossman, Clark, and Grayman 1994). Vasconcelos et al. (1997) postulated that chlorine decay models could be based on first-order kinetics for bulk reactions and either zero-order or first-order kinetics for pipe wall reactions. In 2000, Powell et al. compared six types of kinetic models and concluded that a first-order model was adequate to describe bulk and overall decay. The overall chlorine decay expression for water flowing through a distribution system is shown in Equation 5.70.

$$\left(\frac{dC}{dt}\right)_{Total} = \left(\frac{dC}{dt}\right)_{Bulk} + \left(\frac{dC}{dt}\right)_{Wall} \qquad (5.70)$$

Table 5.140
Results of ANOVA for mean of PDS influent residual concentration

PDS	Observations	Mean influent NH_2Cl (mg/L Cl_2)	Standard deviation
01	52	5.1	0.89
02	52	4.9	0.95
03	52	4.8	0.90
04	52	5.0	0.94
05	52	4.9	1.00
06	52	4.7	0.92
07	52	4.9	1.03
08	52	4.8	0.96
09	52	4.7	1.00
10	52	5.0	0.96
11	52	5.0	1.00
12	52	5.0	0.95
13	52	4.7	0.94
14	52	5.1	0.96

Previous studies have shown that temperature, pH, and organic content of the water affect the kinetic constants for reactions in the bulk liquid (Kiéné, Lu, and Lévi 1998; Hua et al. 1999; Powell et al. 2000; Duirk et al. 2002; Vikesland, Ozekin, and Valentine 2001). Kiéné, Lu, and Lévi (1998) have described bulk chlorine decay using a first-order model with respect to chlorine, as shown in Equation 5.71 where the bulk decay constant (K_B) was positively related with organics concentration and water temperature (Equation 5.72). The expression used to describe the temperature effect is similar to the Arrhenius equation. Several studies have shown that chloramine decay rate decreases when the pH of the water increases (Margerum et al. 1994; Duirk et al. 2002; Vikesland and Valentine 2002b).

$$\left(\frac{dC}{dt}\right)_{Bulk} = -K_B \times C \tag{5.71}$$

$$K_B = a \times [TOC] \times \exp^{\left(\frac{-b}{T}\right)} \tag{5.72}$$

where K_B = bulk decay constant
C = chlorine concentration in the bulk
$a = 1.8 \times 106$ L/mg-hr
$b = 6{,}050$ K
T = temperature, K

Synthetic materials such as PVC, medium- and high-density polyethylene, cement-lined iron, and polypropylene have been reported as having a very low chlorine demand (Kiéné, Lu, and Lévi 1998; Hallam et al. 2002). Vikesland and Valentine (2002a, b) showed that combined chlorine is depleted quickly in metallic pipes because of reactions with iron in the aqueous phase and at the pipe surface. Residual loss due to reactions at the pipe wall must take into account the limitations of mass transport of residual to the pipe wall, the rate of the reactions at the pipe wall, and the ratio of wetted surface area to volume (Frateur et al. 1999; Vikesland, Ozekin, and Valentine 2001; Vikesland and Valentine 2002a, b; Hallam et al. 2002). The most commonly accepted expression for the wall decay rate is shown in Equation 5.73 as a function of pipe diameter, D_p, concentration, C, and mass transfer constants, and is used in several computer prediction models. The mass transfer constant, K_F, and the wall reaction constant, K_{W1}, can be then combined into an overall wall decay constant, W.

$$\left(\frac{dC}{dt}\right)_{Wall} = -\frac{4}{D_p}\frac{K_{W1}K_F}{(K_{W1} - K_F)} \times C = -\frac{W}{D_p} \times C \qquad (5.73)$$

A model was developed during TBW I to predict chloramine dissipation within the PDSs and is shown in Equation 5.74. The model was developed using nonlinear regression with data obtained from the same PDSs used during TBW II. During the TBW I project, chloramine decay was found to best fit a first-order kinetics model with respect to chloramine concentration and showed that pipe material, total organic carbon (TOC), temperature, pipe geometry, and time significantly affected demand (Taylor et al. 2005). As is shown in the model, chloramine decay and the rate of chloramine decay increases with time, temperature, UV_{254}, bulk water demand, and wall demand. The pipe material was found to be the main factor affecting the rate of chloramine decay. Unlined metallic pipes were found to have a faster rate of decay than LCI and PVC pipes. The wall decay constant for UCI and GS pipes was 8 to 10 times greater than the same for the LCI and PVC pipes.

$$Cl_{2(t)} = Cl_{2(0)} * \exp\left(-\left(K_B * UV_{254} + \frac{W}{D}\right) * A^{(T-20)} * time\right) \qquad (5.74)$$

where K_B = bulk decay constant, cm/hr
W = wall decay constant, in./hr
$Cl_{2(t)}$ = chlorine concentration at time t, mg/L as Cl_2
$Cl_{2(0)}$ = initial chlorine concentration, mg/L as Cl_2
A = temperature correction coefficient
T = temperature, °C
time = time, expressed in hours
D = pipe diameter, inches

Effect of Temperature During TBW II Project

The rate of chloramine decay in the PDS increased as temperature increased. The natural logarithm of the ratio of initial and final chloramine concentration is plotted versus residence time in Figure 5.143 for selected PDSs at 15°C, 21°C, and 28°C. The order of the rates (slopes) of

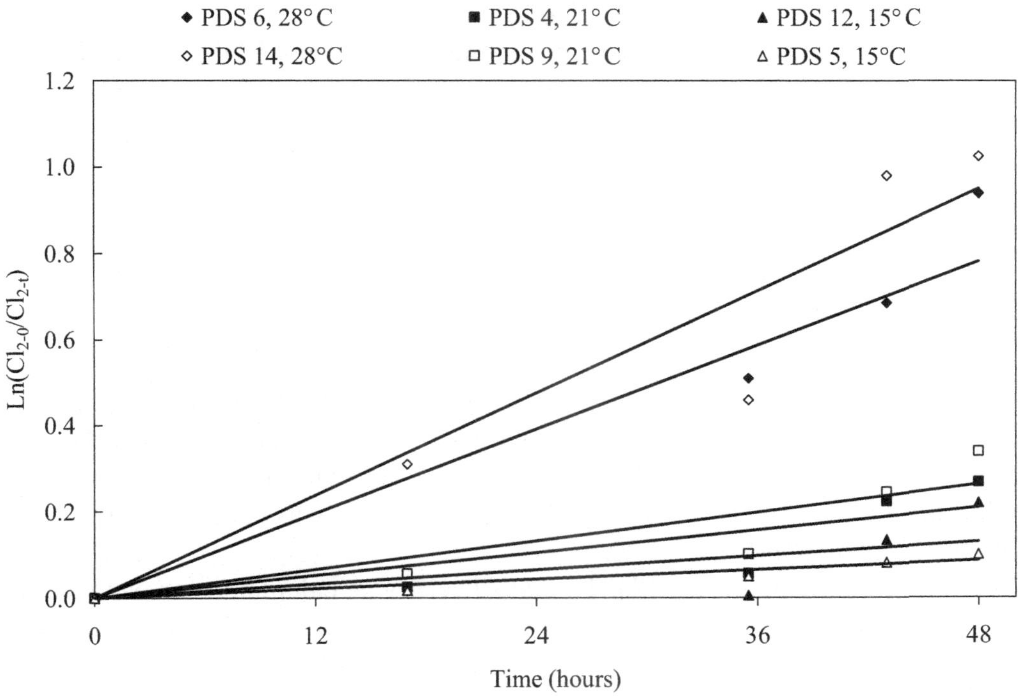

Figure 5.143 Effect of temperature on chloramine decay in the PDS

chloramine decay with respect to temperature was 28°C > 21°C > 15°C. Previous studies (Powell et al. 2000; Kiéne, Lu, and Lévi 1998; Jadas-Hécart et al. 1992) have shown that rate constants can be adjusted for temperature variations using a relation derived from the van't Hoff–Arrhenius equation shown in Equation 5.75. The Arrhenius equation can be manipulated to obtain Equation 5.76, which can be used to calculate the rate constant at operation temperature, K_{Temp}, from the rate constant defined at a base temperature (20°C). An Arrhenius-type rate constant for temperature correction was developed by nonlinear regression in combination with several other parameters and is presented later in this section.

$$K_T = A^*\exp^{\left(\frac{E_a}{RT}\right)} \tag{5.75}$$

$$K_{Temp} = K_{20}{}^*\theta^{(Temp - 20)} \tag{5.76}$$

where K_{Temp} = dissipation constant at temperature T
K_{20} = dissipation constant at 20°C
E_a = activation energy
R = ideal gas constant
T = absolute temperature, K or °R
θ = temperature correction factor

Table 5.141
Paired t-test comparing chloramine demand between PDS13 and PDS14

PDS	Observations	Mean NH_2Cl demand (mg/L Cl_2)	Standard deviation	p-value	Significant?
13	52	2.93	1.3	0.03	Yes
14	52	2.64	0.7		

Effect of pH

Two control lines were used during the TBW II project. PDS14 water quality was practically equivalent to the water quality of the groundwater (60%+) blend used in TBW I. By design, the pH of PDS14 influent was routinely 0.3 pH units higher than the saturation pH (pH_s). PDS13 water quality differed from PDS14 water quality because HCl was added to reduce the actual pH to pH_s, which was also intended.

The chloramine demand for PDS13 and PDS14 was calculated for each sampling event and a paired t-test was conducted to determine whether there was a difference in demand between the PDSs. The result of the paired t-test is shown in Table 5.141.

The average chloramine demand in PDS14 was 2.6 mg/L as Cl_2, whereas the demand for PDS13 averaged 2.9 mg/L as Cl_2. The difference was found to be significant with a p-value of 0.03. This result is consistent with literature that has found the chloramine decay rate to decrease with the increase of the water pH (Vikesland, Ozekin, and Valentine 2001; Duirk et al. 2002; Vikesland and Valentine 2002a, b; Duirk et al. 2005). As with temperature, a correction for varying pH was developed by nonlinear regression and is presented later in this section.

Effect of Pipe Material During TBW II Project

Samples were taken from the internal ports located between the different pipe materials of the PDS to assess the effect of each individual pipe material on chloramine decay. A total of 12 sampling events were spaced during a 6-month period (September 2006 to February 2007). Chlorine residual decayed at different rates in the PDS, depending on pipe material. Lower rates of chloramine decay were observed in PVC and LCI pipes while chloramine concentration declined rapidly in the UCI and GS pipes. Figures 5.144–5.147 show the average of the calculated overall first-order rate of chloramine decay (K) in PVC, LCI, UCI, and GS pipes by inhibitor type and dose. Error lines are also included for each PDS showing the maximum and minimum K observed.

Figure 5.144 shows that the chloramine decay rate varied from 0.007 hr^{-1} to 0.015 hr^{-1} in the PVC pipes. There is not much difference between the control lines, between the inhibitors and the control lines, or within the inhibitors by type or dose. The two control PDSs (PDS13 and PDS14) had K of 0.011 and 0.010 hr^{-1}, respectively. In the BOP PDS, K was 0.010 hr^{-1}, 0.012 hr^{-1}, and 0.010 hr^{-1} for PDS01 (low dose), PDS02 (medium dose), and PDS03 (high dose) with no correlation between inhibitor dose and K. For the OP PDS, K was 0.011 hr^{-1}, 0.008 hr^{-1}, and 0.010 hr^{-1} for PDS04 (low dose), PDS05 (medium dose), and PDS06 (high dose), respectively. No effect of ZOP inhibitor dose was observed in PDS07 (low dose), PDS08 (medium dose), and PDS09 (high dose) with K of 0.008 hr^{-1}, 0.010 hr^{-1}, and 0.010 hr^{-1}, respectively. The PDS receiving the SiO_2 inhibitor showed a K (0.015 hr^{-1}) greater than the control lines

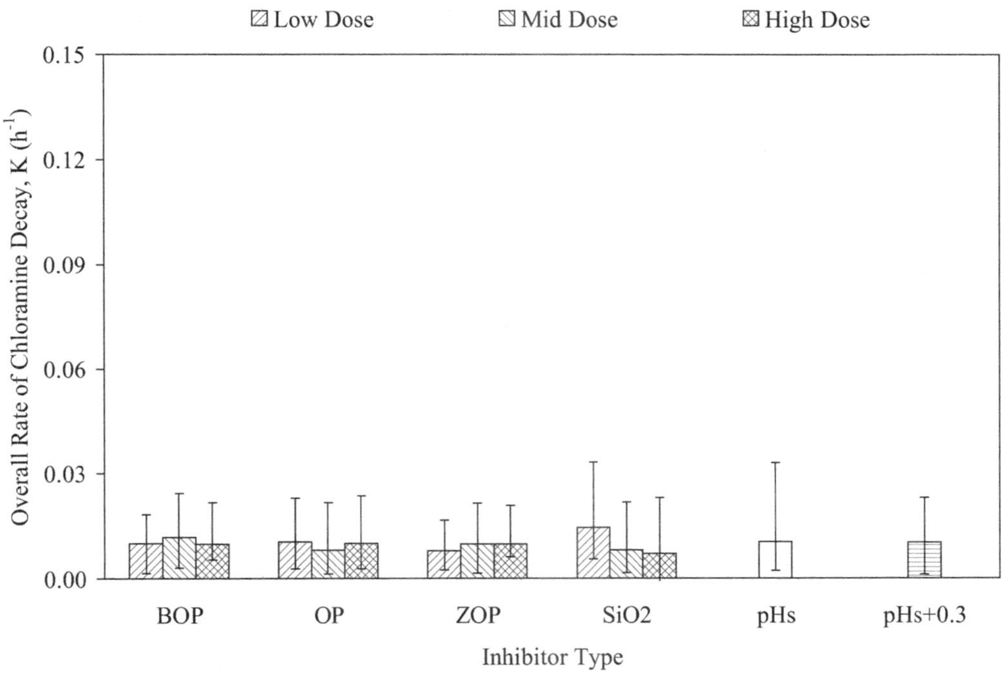

Figure 5.144 Average overall rate of chloramine decay in PVC pipe

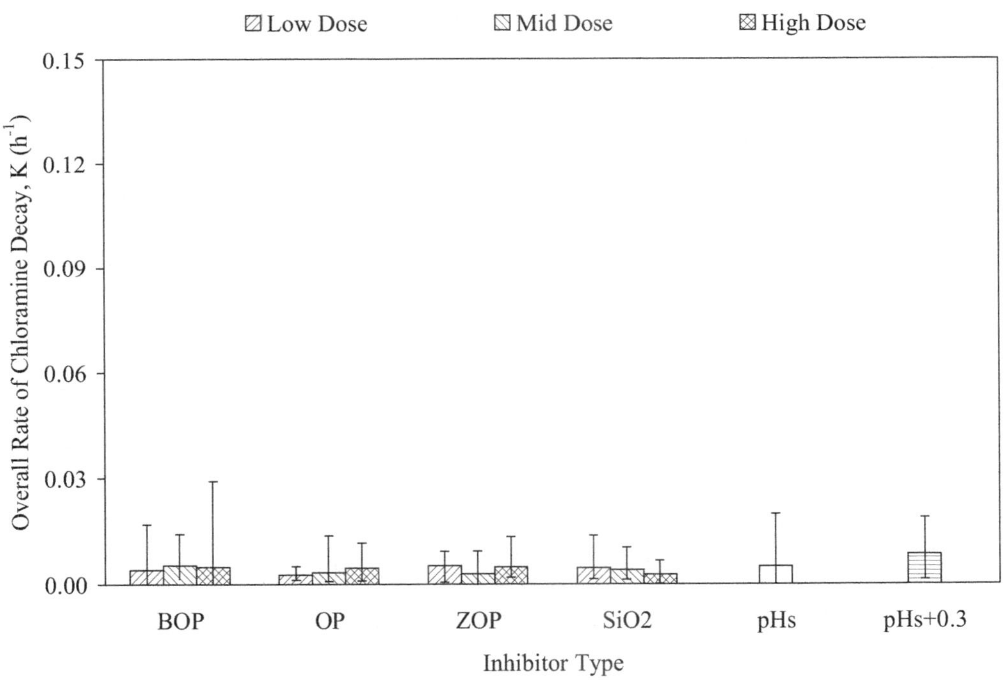

Figure 5.145 Average overall rate of chloramine decay in LCI pipe

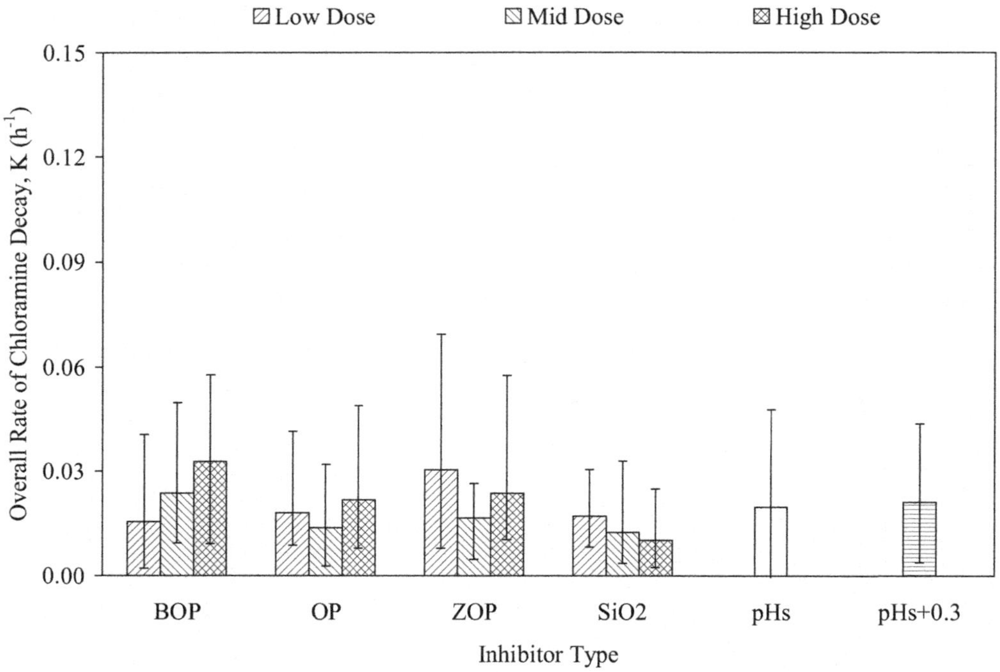

Figure 5.146 Average overall rate of chloramine decay in UCI pipe

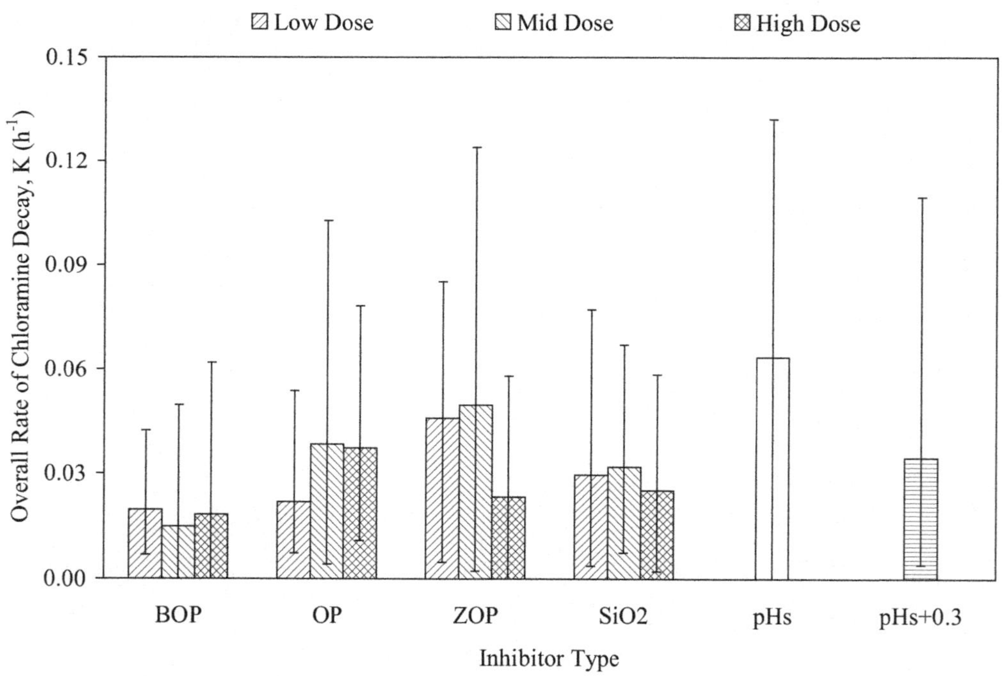

Figure 5.147 Average overall rate of chloramine decay in GS pipe

(0.010 hr^{-1}) for PDS10 (low dose), a K of 0.008 hr^{-1} for PDS11 (medium dose), and 0.007 hr^{-1} for PDS12 (high dose).

Figure 5.145 shows the rate of chloramine decay in the LCI pipes to be between 0.003 hr^{-1} and 0.008 hr^{-1}. Similar to the PVC pipes, there seems to be little variations between the PDSs by inhibitor type or dose. K for the control lines, PDS13 (pH_s) and PDS14 (pH_s+0.3), were 0.005 and 0.008 hr^{-1}, respectively. The BOP PDS had K of 0.004 hr^{-1}, 0.005 hr^{-1}, and 0.005 hr^{-1} for PDS01, PDS02, and PDS03, respectively. The OP PDS had K of 0.003 hr^{-1} for PDS04 and PDS05, and a K of 0.004 hr^{-1} for PDS 06. The low and high dose of OP, which were from PDS07 and PDS09, respectively, had K of 0.005 hr^{-1} while PDS08 had a K of 0.003 hr^{-1}. For the SiO_2 PDS, K was 0.005 hr^{-1} for PDS10, 0.004 hr^{-1} for PDS11, and 0.003 hr^{-1} for PDS12.

Figure 5.146 shows the rates of decay in the UCI pipes, with rates ranging from 0.010 hr^{-1} for PDS12 to 0.033 hr^{-1} for PDS03. However, most of the PDSs showed K that was around 0.020 hr^{-1}, which was also K for the control lines. The rate of decay seems to increase with the inhibitor dose for the BOP inhibitor with K of 0.015 hr^{-1} for PDS01, 0.024 hr^{-1} for PDS02, and 0.033 hr^{-1} for PDS03. K seems to decrease with dose for the silicone inhibitor with K of 0.017 hr^{-1}, 0.012 hr^{-1}, and 0.010 hr^{-1} for PDS10, PDS11, and PDS12, respectively; however, the error bars do not support those statements. No correlation was observed for OP and ZOP; for both inhibitors, K decreases going from the low-dose PDS to the medium-dose PDS before increasing again for the high-dose PDS. For OP, PDS04 has a K of 0.018 hr^{-1}, PDS05 has a K of 0.014 hr^{-1} and PDS06 has a K of 0.022 hr^{-1}. For ZOP PDSs, PDS07, PDS08, and PDS09 have K of 0.030 hr^{-1}, 0.017 hr^{-1}, and 0.024 hr^{-1}, respectively.

Figure 5.147 shows the rate of decay for the GS pipes. K ranges from 0.015 hr^{-1} for PDS02 to 0.064 hr^{-1} for PDS13, with 10 of the 14 K in the range of 0.020 hr^{-1} to 0.040 hr^{-1}. There are no defined trends except that BOP PDSs seem to have a slightly lower K than PDS14 (pH_s+0.3) while ZOP and pH_s seem to have a greater K than PDS14. BOP PDSs have K of 0.020 hr^{-1}, 0.015 hr^{-1}, and 0.018 hr^{-1} for PDS01, PDS02, and PDS03, respectively, compared with 0.034 hr^{-1} for PDS14. OP PDSs have K that are similar to PDS14, with K of 0.022 hr^{-1}, 0.038 hr^{-1}, and 0.037 hr^{-1} for PDS04 (low dose), PDS05 (medium dose), and PDS06 (high dose). The ZOP PDSs have K for PDS07 (low dose) at 0.046 hr^{-1} and PDS08 (medium dose) at 0.049 hr^{-1} while the high dose (PDS09) has a K of 0.023 hr^{-1}. The silicate PDSs have K of 0.030 hr^{-1}, 0.032 hr^{-1}, and 0.025 hr^{-1} for PDS10 (low dose), PDS11 (medium dose), and PDS12 (high dose), which are similar to K for PDS14 at 0.034 hr^{-1}. The highest K was observed in PDS13 with a K of 0.064 hr^{-1}.

Effect of Corrosion Inhibitors

The average chloramine demand in the PDSs for the entire study is shown in Figure 5.148. Chloramine concentrations were measured daily in the staging tanks, and weekly in the PDS influent and effluent over 1 year. As noted in the "Pilot Distribution System Operation" section in chapter 3, chloramine residuals were adequately controlled during the year of operation. This data has not been normalized for temperature but was developed from samples collected simultaneously from the PDSs so that the effects of temperature on chloramine dissipation are considered. PDS14 water quality was very similar to the water quality of the major groundwater blend water quality used in TBW I. Hence, results from PDS14 provide a basis for relating results from TBW I and TBW II. PDS13 water quality differed from PDS14 water quality because HCl was added to reduce the actual pH to pH_s.

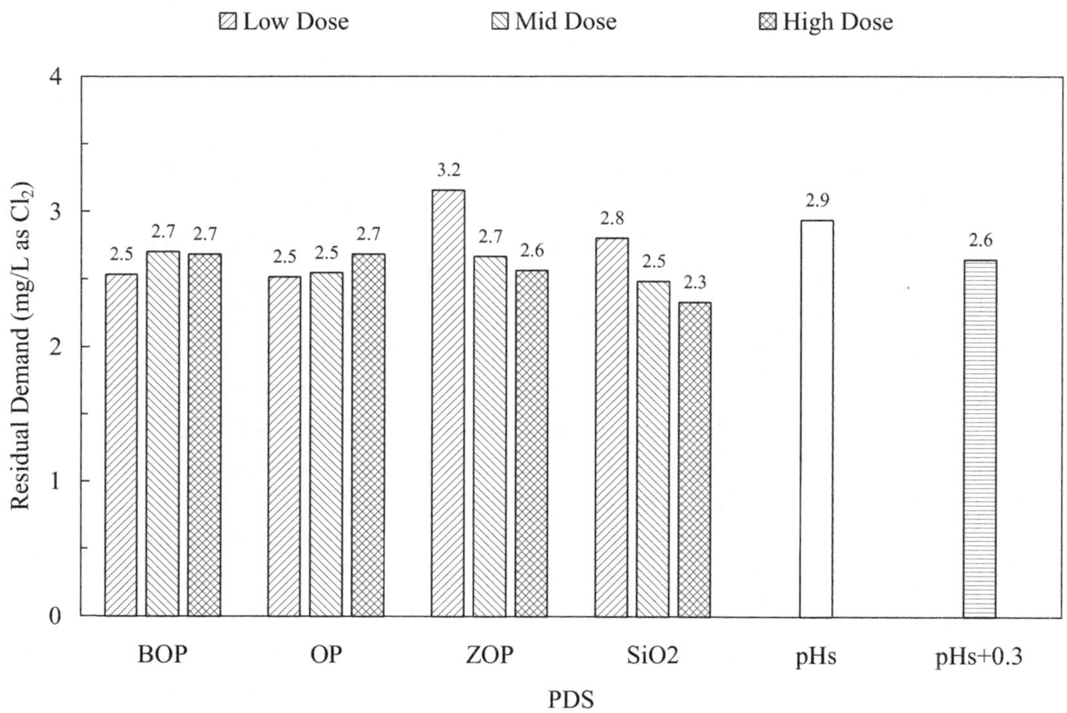

Figure 5.148 Average chloramine residual demand by PDS

Figure 5.148 shows that PDS14 (pH_s+0.3) had a lower chloramine demand (2.6 mg/L) than PDS13 (pH_s, 2.9 mg/L) and that there is no readily apparent difference in chloramine demand between PDS14 and the PDSs using inhibitors. The average chloramine demand for the 12 inhibitor PDSs varied from 2.3 to 3.2 mg/L as Cl_2. However, the chloramine demands for 9 of the 12 PDSs varied from 2.5 to 2.7 mg/L as Cl_2. The broader range of chlorine demand is due to the demand in PDS07 (3.2 mg/L as Cl_2), PDS10 (2.8 mg/L as Cl_2), and PDS12 (2.3 mg/L as Cl_2). The chloramine demand for the BOP and OP PDSs appears to be equal and possibly increases with inhibitor dose. Likewise, the chloramine demand for the ZOP and Si PDSs appears to be practically equal but decreases with inhibitor dose. Globally, the data shown in Figure 5.148 do not provide any obvious relationship between inhibitor use and chloramine demand in the PDSs.

The significance of the differences in chloramine residual demand was evaluated using paired t-test for means between PDS13 and PDS14 and the rest of the PDSs. The results of the t-tests are shown in Table 5.142. PDS with a *p*-value below 0.05, which corresponds to a 95% confidence level of certainty, were considered as being significantly different than the control PDS.

The comparison of PDS13 to the rest of the PDSs shows that the chloramine demand in PDS13 was significantly lower than the chloramine demand for PDS07 and significantly greater than all other PDSs.

The comparison of PDS14 to the rest of the PDSs shows that the chloramine demand was not significantly different for PDS14 and PDS01 through PDS06, PDS08 through PDS11, and PDS13. The chloramine demand in PDS14 was greater than PDS12 demand and lower than PDS07 demand. As did the global view, this indicates that there was no apparent relationship between inhibitor type or dose and effluent chlorine residual in any of the PDSs.

Table 5.142
Results of paired t-test for chloramine demand between PDS01–PDS12 and PDS13 and PDS14

PDS	Average NH_2Cl Demand ($mg/L\ Cl_2$)	Standard deviation	Comparison to PDS13		Comparison to PDS14	
			p-value	Significantly different?	*p*-value	Significantly different?
1	2.54	0.8	0.00	Yes	0.26	No
2	2.71	1.2	0.02	Yes	0.60	No
3	2.69	1.0	0.02	Yes	0.67	No
4	2.52	1.0	0.00	Yes	0.25	No
5	2.55	1.2	0.00	Yes	0.43	No
6	2.69	1.1	0.04	Yes	0.67	No
7	3.16	1.2	0.03	Yes	0.00	Yes
8	2.67	1.1	0.02	Yes	0.80	No
9	2.57	1.2	0.00	Yes	0.48	No
10	2.80	1.0	0.16	Yes	0.07	No
11	2.48	1.1	0.00	Yes	0.12	No
12	2.33	1.1	0.00	Yes	0.01	Yes
13	2.94	1.3	N/A	N/A	0.03	Yes
14	2.65	0.7	0.03	Yes	N/A	N/A

N/A = not applicable.

The pH of the water could explain the differences on chloramine demand between PDS12, PDS13, and PDS14. PDS12 showed the highest average pH of all the PDSs at 8.2, PDS13 had the lowest at 7.7, while PDS14 had a pH similar to all the other PDSs at 7.9. Literature indicates that chloramines are more stable at higher pH. PDS07 has constantly shown an abnormal behavior, such as elevated iron release, which could explain the elevated chloramine demand.

New Model Development

A new model to predict the decay of chloramines in the PDSs was developed using the data obtained during the TBW II project. The initial model form was based on the original model obtained from the TBW I project, but it was modified to include terms to account for effects of corrosion control strategies (pH control and corrosion inhibitor) and inhibitor dose. Dummy variables were used to separate the data by pipe material and corrosion inhibitor measures. The general form of the initial model is shown in Equation 5.77.

$$Cl_{2(t)} = Cl_{2(0)}*\exp\left[\left(\begin{array}{c} K_B*UV + K_H*H*DV_H + K_{TP}*TP*DV_{TP} \\ + K_{SiO_2}*SiO_2DV_{SiO_2} + \dfrac{W_{PVC}*DV_{PVC}}{D_{PVC}} \\ + \dfrac{W_{LCI}*DV_{LCI}}{D_{LCI}} + \dfrac{W_{UCI}*DV_{UCI}}{D_{UCI}} + \dfrac{W_{GS}*DV_{GS}}{D_{GS}} \end{array}\right)*\theta^{(T-20)}*time\right] \quad (5.77)$$

where $Cl_{2(t)}$ = chlorine concentration at time t, mg/L as Cl_2
$Cl_{2(0)}$ = initial chlorine concentration, mg/L as Cl_2
K_B = bulk decay constant, cm/hr
K_H = coefficient for H^+ term
K_{TP} = coefficient for total phosphorus parameter
K_{SiO2} = coefficient for silica parameter
W_{PVC} = wall decay constant for PVC pipe, in./hr
W_{LCI} = wall decay constant for LCI pipe, in./hr
W_{UCI} = wall decay constant for UCI pipe, in./hr
W_{GS} = wall decay constant for GS pipe, in./hr
UV = UV_{254}, cm^{-1}
H = molar concentration of H^+, mol/L
TP = total phosphorus, mg/L as P
SiO_2 = silica, mg/L as SiO_2
DV_i = dummy variable for parameter i (H^+, TP, SiO_2, PVC, LCI, UCI, and GS)
D_i = pipe diameter of pipe material, inches
θ = temperature correction coefficient
T = temperature, °C
time = time, expressed in hours

The dummy variables DV_{PVC}, DV_{LCI}, DV_{UCI}, and DV_{GS} are ones when the data corresponds to the respective pipe material and zeros when the data corresponds to other types of pipe material. DV_{TP}, DV_{SiO2}, and DV_H are the dummy variables for phosphate inhibitors (PDS01 to PDS09), silicate inhibitor (PDS10 to PDS12), and pH control (PDS13 and PDS14), respectively. The dummy variables are ones and zeros used to indicate the presence or absence of the respective corrosion control measure. This allows the model to be adjusted to achieve sensitivity to a specific pipe material and a specific corrosion inhibitor for a constant set of water quality parameters. The use of the variables TP, SiO_2, and H allows for the estimation of the dose effect of the inhibitors or pH change.

The wall decay rates were determined for all pipe materials; however, the bulk decay constant, K_B, was taken as an intrinsic property of the water and kept constant for all pipe materials. Any demand exerted by a contaminant released by the pipe wall is included in the pipe wall decay rate constant. This is a change from application of the residual model in TBW I but increases the significance of pipe material, which increases the practical inference and utilization of the model.

K_B was set at 0.02 cm/hr based on the K_B obtained for the PVC and LCI during the TBW I project. It was assumed that the water in the PVC and LCI pipes had not been contaminated by the pipe wall effect given that PVC and LCI are fairly inert materials and were located upstream of the unlined metallic pipes. W was calculated for each pipe material by nonlinear regression using

Table 5.143
Result of initial nonlinear regression

Parameter	Value	*p*-value
θ	1.11	<0.0001
W_{PVC}	–0.024	<0.0001
W_{LCI}	–0.013	0.0012
W_{UCI}	–0.103	<0.0001
W_{GS}	–0.051	<0.0001
K_H	–111898	0.04
K_{TP}	–2.3E–04	0.64
K_{SiO2}	1.54E–05	0.73

Table 5.144
Parameter values for chloramine decay model

Parameter	Value	*p*-value
θ	1.11	<0.0001
W_{PVC}	–0.025	<0.0001
W_{LCI}	–0.014	<0.0001
W_{UCI}	–0.103	<0.0001
W_{GS}	–0.051	<0.0001
K_H	–106403.7	0.022

dummy variables to separate the data. The effect of temperature was incorporated by using a temperature correction coefficient, θ, derived from the Arrhenius equation, which was also obtained using nonlinear regression.

Table 5.143 shows the result of the nonlinear least-squares regression to evaluate the constants in Equation 5.77. The decay constant associated with the silicate inhibitor is not significant, as shown by the *p*-value of 0.73 in Table 5.143. This result indicates that the use of the silicate inhibitor did not have an effect on effluent chloramine residual.

After elimination of the terms for the silicate inhibitor from the model, the regression was repeated. The decay constant for the phosphate inhibitor was found to be insignificant with a *p*-value of 0.4. The terms for the phosphate inhibitors were eliminated from the model, and new values for the decay constants were calculated and are shown in Table 5.144. All the terms are significant at the 95% confidence level, as indicated by the *p*-values being lower than 0.05. The R^2 obtained for the model was 0.75.

The resulting model indicates that the use of phosphate-based or silicate inhibitors had no appreciable effect on the rate of chloramine residual decay. Only pH control had an effect on the rate of chloramine decay. The general equation of the final model is shown in Equation 5.78.

$$Cl_{2(t)} = Cl_{2(0)}*\exp\left[\left(\begin{array}{c} K_B*UV + K_H*H*DV_H + \dfrac{W_{PVC}*DV_{PVC}}{D_{PVC}} \\ + \dfrac{W_{LCI}*DV_{LCI}}{D_{LCI}} + \dfrac{W_{UCI}*DV_{UCI}}{D_{UCI}} + \dfrac{W_{GS}*DV_{GS}}{D_{GS}} \end{array}\right)*\theta^{(T-20)}*time\right] \tag{5.78}$$

where $Cl_{2(t)}$ = chlorine concentration at time t, mg/L as Cl_2
$Cl_{2(0)}$ = initial chlorine concentration, mg/L as Cl_2
K_B = bulk decay constant, cm/hr
K_H = coefficient for H^+ term
W_{PVC} = wall decay constant for PVC pipe, in./hr
W_{LCI} = wall decay constant for LCI pipe, in./hr
W_{UCI} = wall decay constant for UCI pipe, in./hr
W_{GS} = wall decay constant for GS pipe, in./hr
UV = UV_{254}, cm^{-1}
H = molar concentration of H^+, mol/L
DV_i = dummy variable for parameter i (H^+, PVC, LCI, UCI, and GS)
D_i = pipe diameter of pipe material, inches
θ = temperature correction coefficient
T = temperature, °C
time = time, expressed in hours

Plots of predicted versus actual chloramine residual for the data used to develop the model are presented in Figures 5.149–5.152 for PVC, LCI, UCI, and GS, respectively. Each of these figures show an acceptable prediction of chloramine residual, as shown by the fit of the data to the "ideal" 45° trend line.

PVC and LCI pipes have the lowest demand on the pipe wall (W), which indicates that the wall surface of these materials has a small effect on chloramine residual. PVC has a W of 0.025 cm/hr while LCI has a W of 0.014 cm/hr. The inner surface of PVC and LCI pipes is smoother, less reactive, and less likely to promote biofilm growth than the UCI and GS pipe interior walls, which is the reason for the relatively lower wall decay rate. GS pipes have a W that is roughly 2 and 3 times greater than W for PVC and LCI. As shown in Table 5.144, the wall decay rate of the UCI pipe is greater than the wall decay rates for any pipe material used in this work. The UCI W is 0.103 in./hr and is 4 and 7 times greater than W for PVC and LCI pipe and twice the GS W. The higher wall demand of the UCI and GS pipe surfaces is due in part to the greater chemical reactivity of iron and zinc relative to LCI and PVC.

The effect of the pipe material can be observed in Figure 5.153. Figure 5.153 shows the predicted behavior with time of water with an organic content of 3 mg/L as C, a temperature of 25°C, a pH of 8.0, and an initial chloramine residual of 4 mg/L as Cl_2 running through 6-in. (0.15-m)-diameter PVC, LCI, UCI, and GS pipes. Under these conditions, it can be seen that the water would maintain a chloramine residual of about 1.0 to 1.5 mg/L as Cl_2 after 5 days in the PVC and LCI pipes, respectively. The surface of PVC and LCI pipes are considered inert, smooth, and with low propensity to develop biofilms, all of which help the pipe to have a low rate of chloramine decay at the pipe wall. UCI and GS pipes lose chloramine residual much faster than PVC and LCI pipes. Within the UCI pipe, the residual drops to 0.3 mg/L as Cl_2 after 80 hours and is nonexistent after 120 hours. The GS pipe maintains the chloramine residual a little longer with

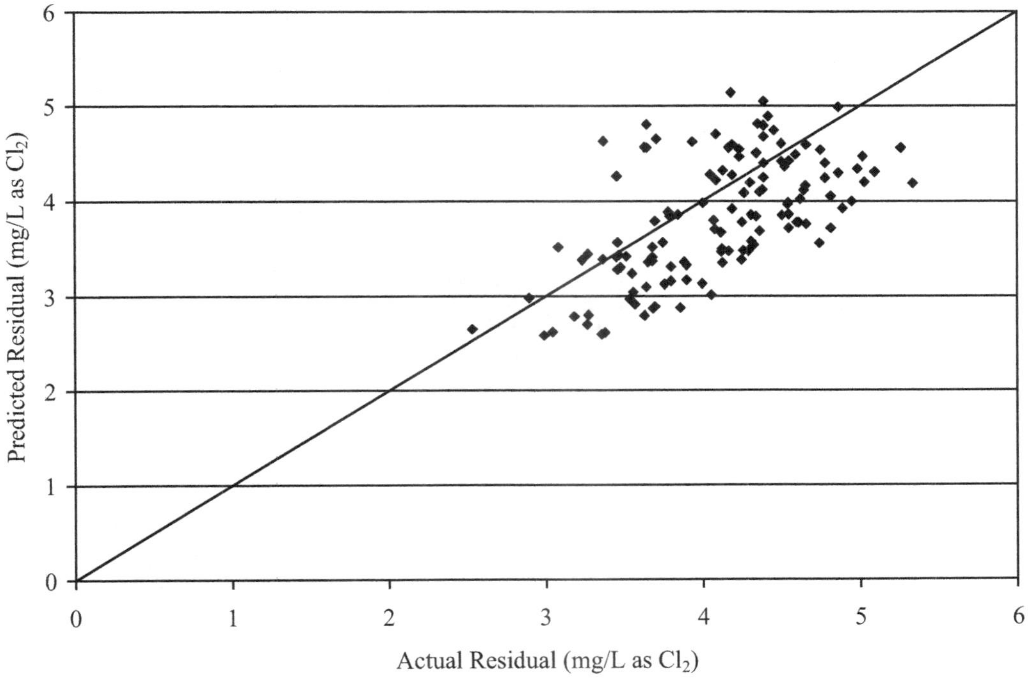

Figure 5.149 Predicted versus actual chloramine residual for PVC pipes

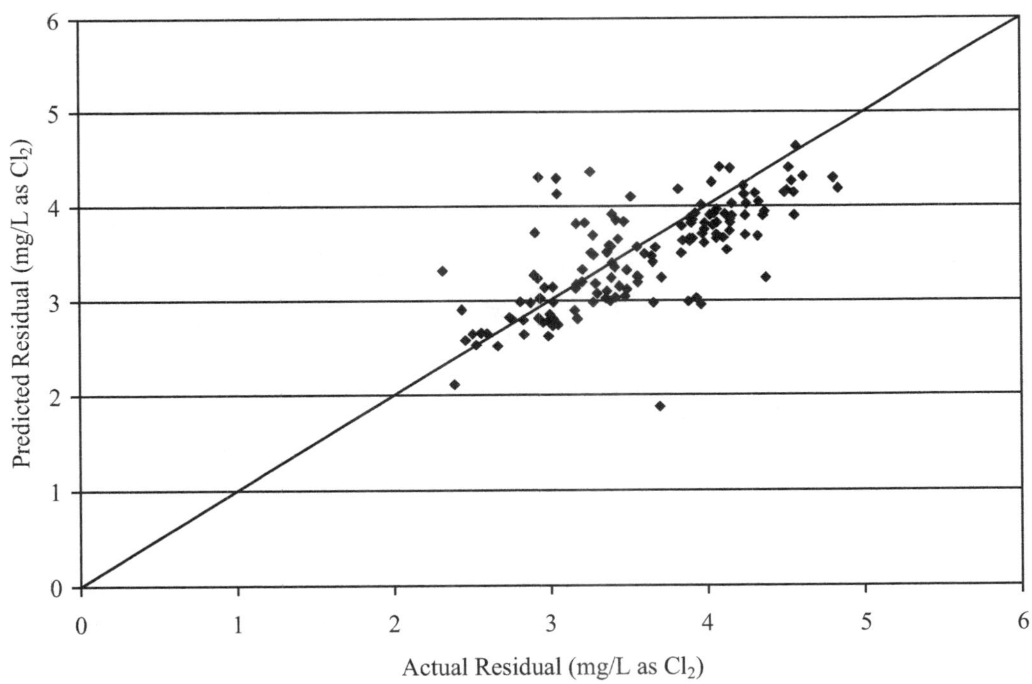

Figure 5.150 Predicted versus actual chloramine residual for LCI pipes

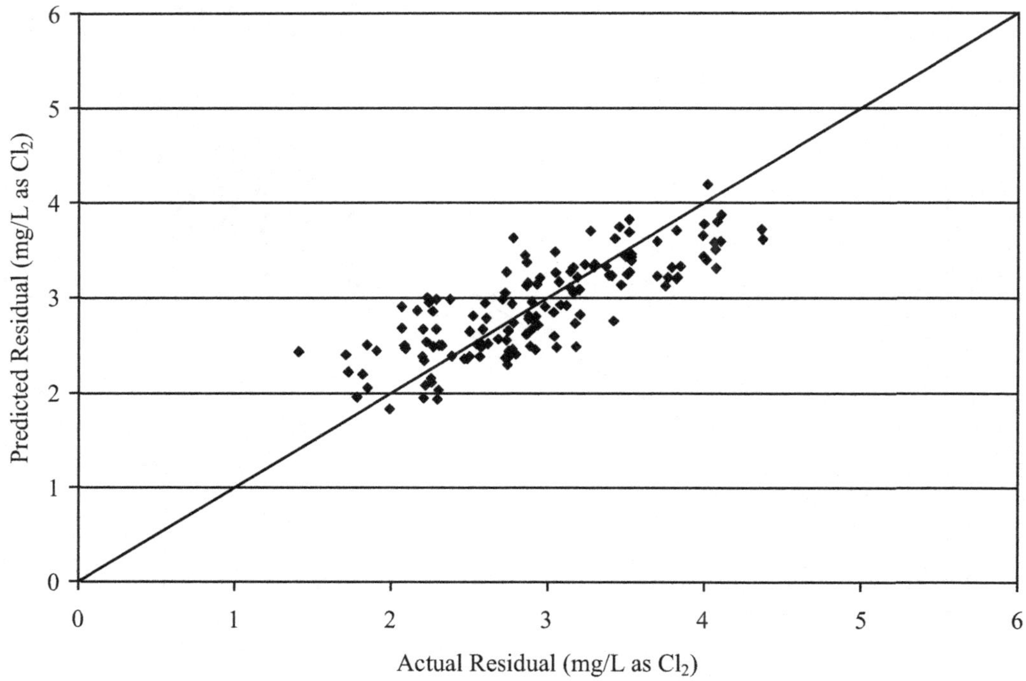

Figure 5.151 Predicted versus actual chloramine residual for UCI pipes

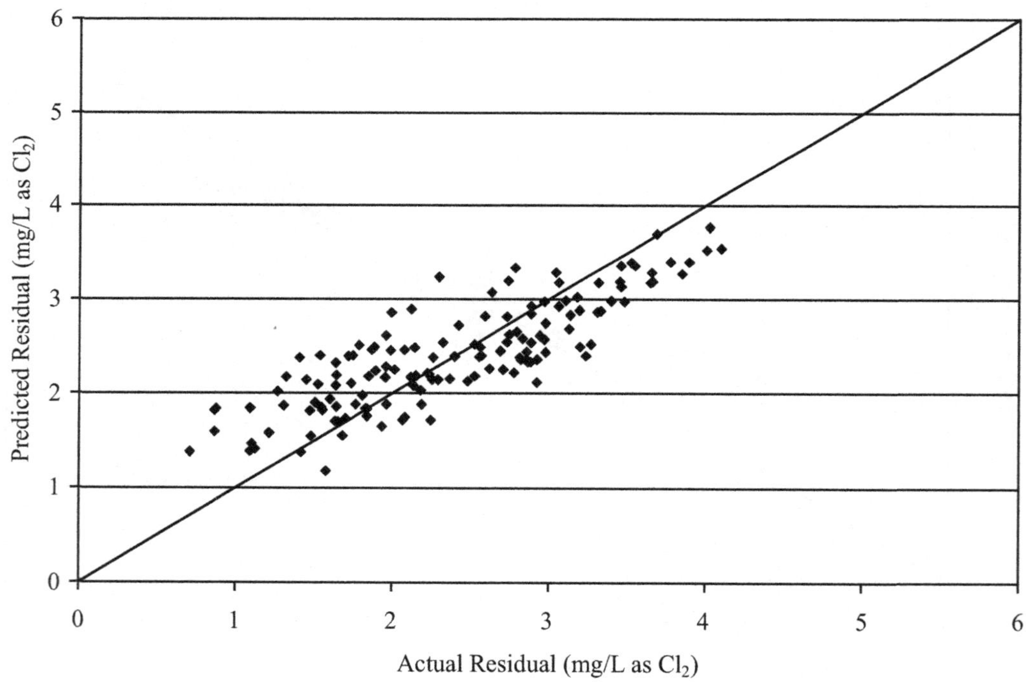

Figure 5.152 Predicted versus actual chloramine residual for GS pipes

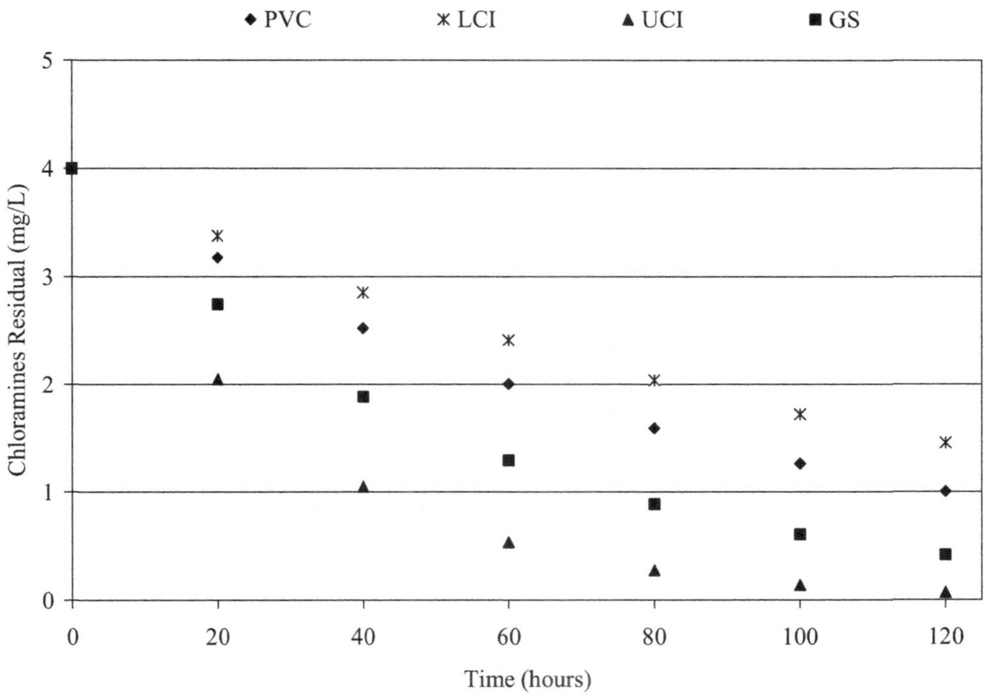

Figure 5.153 Effect of pipe material on predicted chloramine residual decay

a predicted residual of 0.4 mg/L as Cl_2 after 120 hours. As discussed in other sections of this report, UCI and GS pipes had higher surface roughness, more corrosion products, and higher biofilm densities than LCI and PVC pipes, which with the relatively higher chemical reactivity of UCI and GS pipe materials accounted for the difference in residuals between UCI or GS and PVC or LCI pipes.

The diameter of the pipe affects the rate of decay of chloramines inside the pipes given that the effect of the pipe wall is related to the surface-area-to-volume ratio. For a circular pipe running full, that ratio is equal to the diameter of the pipe. For larger-diameter pipes, the effect of the pipe wall is less than for small-diameter pipes, regardless of pipe material. In pipes with diameters less than 1 ft (0.3 m), the wall effect becomes more significant. Figure 5.154 shows the effect of the pipe diameter on chloramine decay for PVC and GS pipes with a diameter of 6 in., 24 in., and 48 in. (0.15 m, 0.61 m, and 1.22 m). The predictions were calculated for water at 25°C, an organic content of 3 mg/L, a pH of 8.0, and an initial chloramine residual of 4 mg/L as Cl_2.

Figure 5.154 shows a predicted chloramine residual of 1.0 mg/L as Cl_2 for a 6-in. (0.15-m) PVC pipe after 120 hours (5 days), and the residual increases to 1.9 mg/L and 2.1 mg/L as Cl_2 after the same time period when the diameters are 24 in. and 48 in. (0.61 m and 1.22 m), respectively. For the GS pipe, the predicted chloramine residual after 120 hours are 0.4 mg/L, 1.5 mg/L, and 1.9 mg/L as Cl_2 for pipes with diameters of 6 in., 24 in., and 48 in. (0.15 m, 0.61 m, and 1.22 m), respectively.

Even though the pipe diameter effect affects all the types of pipe, the pipe materials with the highest W show the largest improvements with the increase of pipe diameter. When the diameter of the pipes is big enough, the difference between the pipe materials is diminished and the effluent residual would be similar regardless of the type of pipe used. In Figure 5.154, the

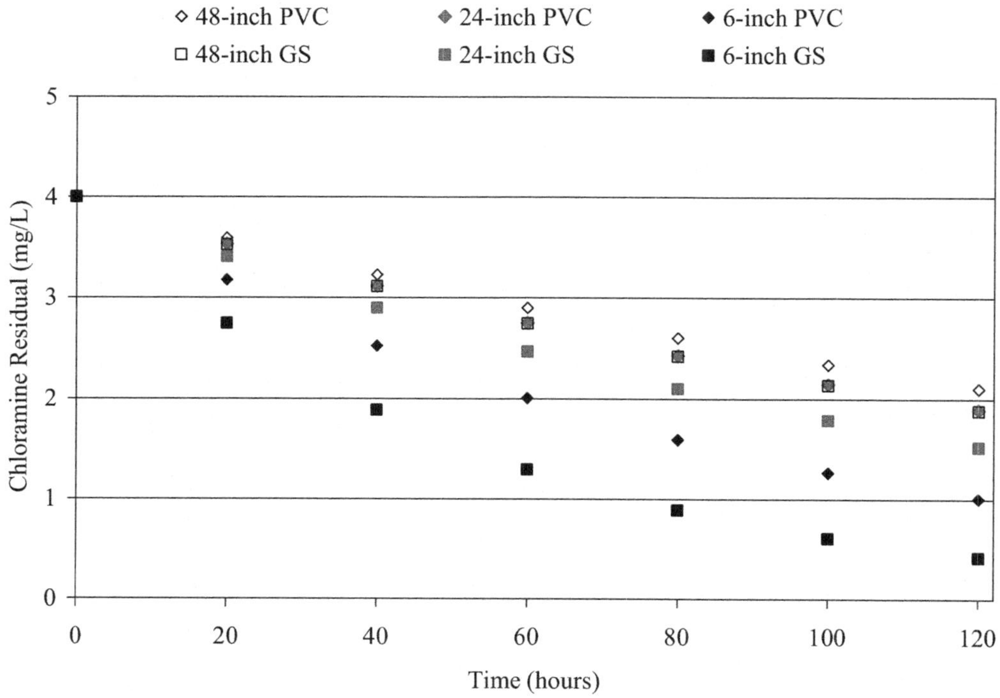

Figure 5.154 Effect of pipe diameter on predicted chloramine decay

predicted chloramine residuals are 2.1 mg/L and 1.9 mg/L as Cl_2 for 48-in. (1.22-m) PVC and GS pipes. As shown in Figure 5.154, the decrease in diameter from 48-in. (1.22-m) to 6-in. (0.15-m) PVC affects residual loss more than the change in material from PVC to GS over the same size range. The implications of the residual model are that hydraulic design, operation, and maintenance of distributions systems are essential regardless of pipe material.

The effect of temperature on chloramine decay was incorporated into the model using a temperature correction factor (θ) based on the Arrhenius equation. The value of θ was determined by regression and was 1.11, which means that the decay constants will double for a temperature increase of 7°C. Figure 5.155 shows the effect of temperature on water with an organic content of 3 mg/L, a pH of 8.0, and an initial chloramine residual of 4 mg/L as Cl_2 running through 6-in. (0.15-m) PVC and GS pipes.

In the PVC pipe, a temperature increase from 10°C to 25°C decreased the remaining chloramine residual from 3.0 mg/L as Cl_2 to 1.0 mg/L as Cl_2 after 120 hours. Similarly, in the GS pipe, the effluent chloramine residual was reduced from 2.5 mg/L as Cl_2 to 0.4 mg/L as Cl_2 after the same period of time.

The effect of pH on chloramine residual can be observed in Figure 5.156. Predictions of the chloramine residual were calculated at pH_s of 7.6, 7.9, and 8.2 for water with an influent chloramine residual of 5 mg/L as Cl_2, a temperature of 25°C, and of 0.08 cm^{-1} running through 6-in. (0.15-m) LCI and GS pipe.

Figure 5.156 shows that in LCI pipes, the chloramine residual would increase from 1.3 mg/L as Cl_2 to 1.7 mg/L as Cl_2 when the pH increases from 7.6 to 7.9, and it would be 2.0 mg/L as Cl_2 when the pH is 8.2. The chloramine residual would also increase with pH in GS pipes; however, the relative increase would be less than it was observed for the LCI. For an

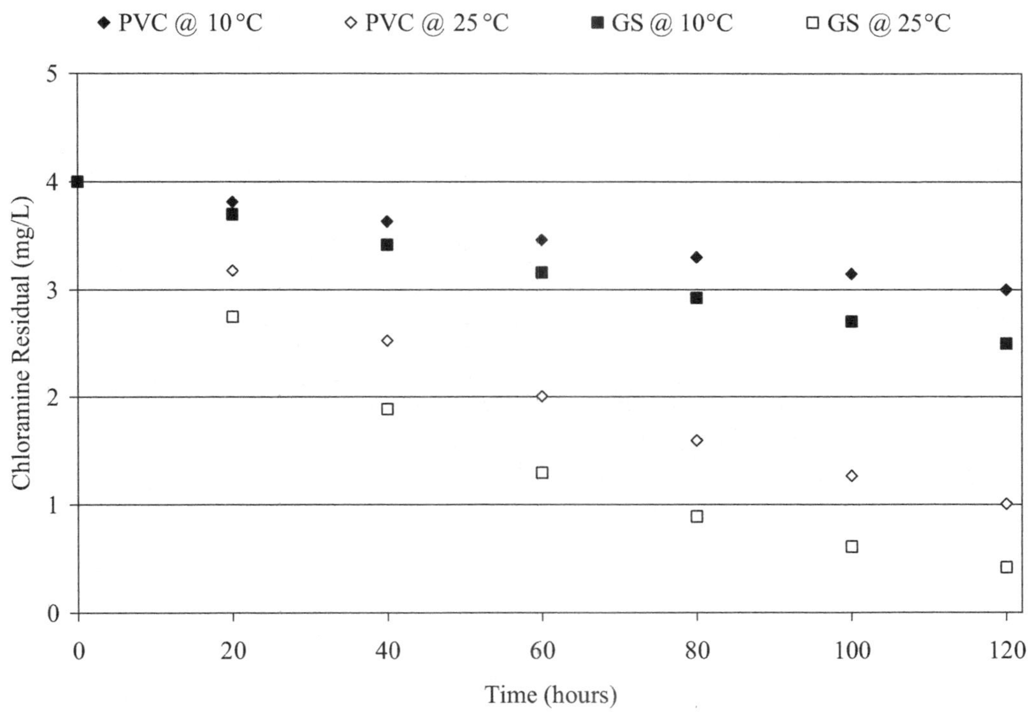

Figure 5.155 Effect of temperature on predicted chloramine residual decay

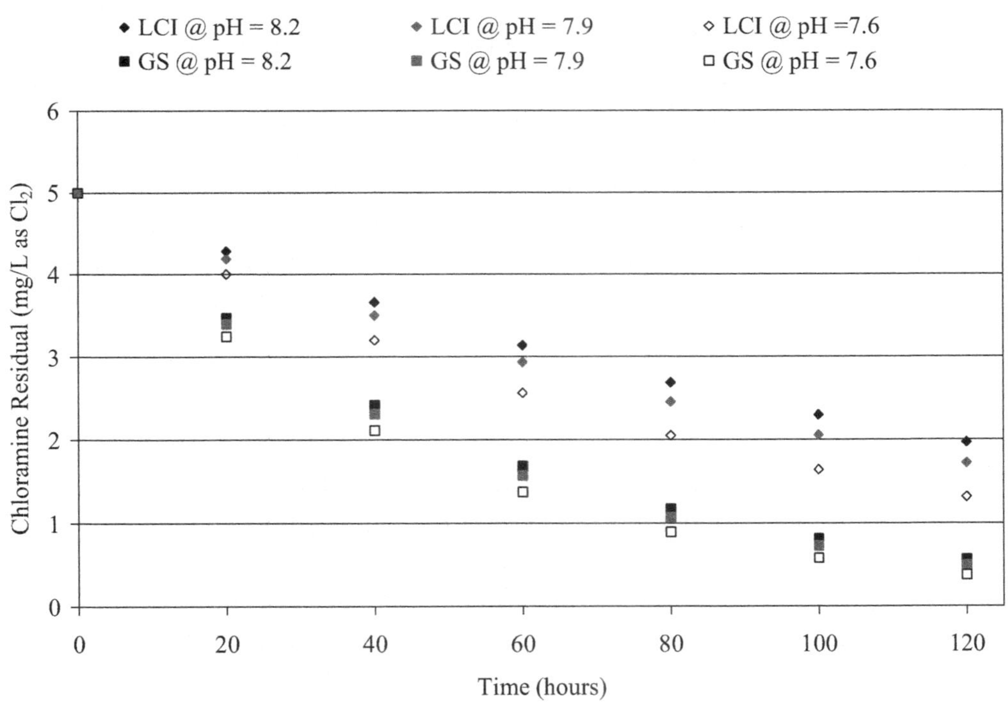

Figure 5.156 Predicted effect of pH on chloramine residual in LCI and GS pipes

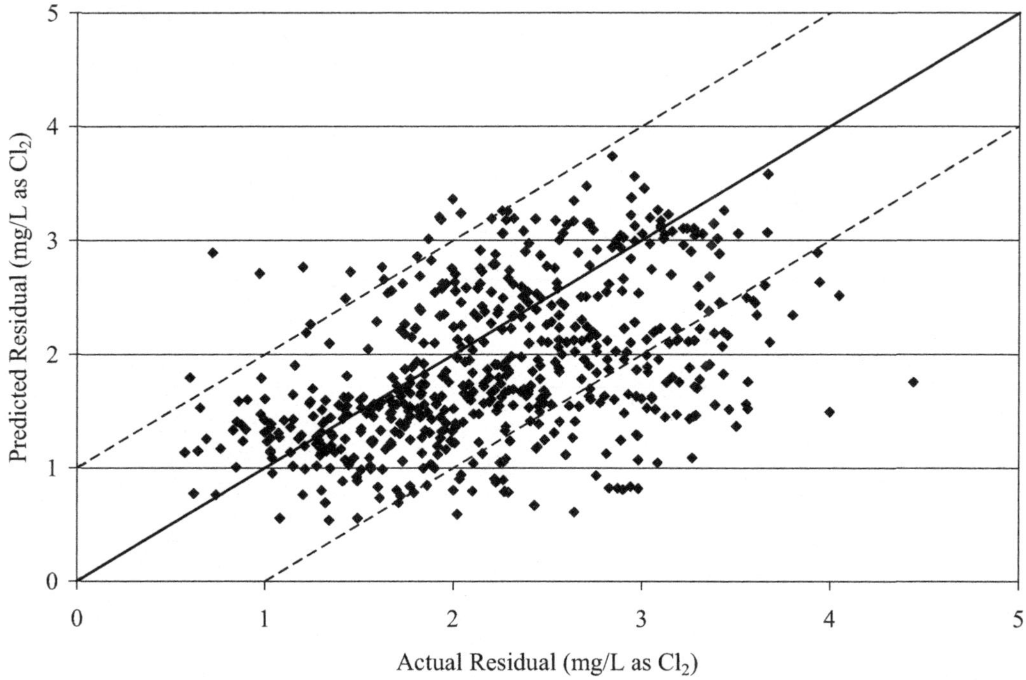

Figure 5.157 Verification of chloramine residual model using independent data

increase of pH from 7.6 to 8.2, the residual after 120 hours is predicted to only increase from 0.4 to 0.6 mg/L as Cl_2. This difference is a result of the greater W for GS pipes compared to the LCI, which makes the influence of pH change almost negligible. It is necessary to warn the reader not to use the model outside the pH range of 7.6 to 8.2.

No effect of inhibitor type or dose was found. Several variations of the model were tested that included the effect of inhibitor type and dose by using terms for total phosphorus and active silicone. In all the regressions, the terms used for inhibitor effect had to be discarded because the constants obtained were not significant. This result is not surprising based on the chloramine demand information supplied in the previous subsection and the overall rate of chloramine decay shown in Figures 5.144–5.147.

Model Verification

Figure 5.157 shows verification of the model using data obtained from the distribution system that was not used for model development. The data used was collected from the influent and the effluent of the PDS. Given that the PDS had sections of PVC, LCI, UCI, and GS pipes, an incremental approach was used to predict the effluent chloramine concentration. The chloramine concentration entering the PDS was used as the starting point, and the model for PVC was used to predict the residual of the water entering the LCI section of the PDS. The same was done for the LCI, UCI, and GS pipes using the residence time of the water in each section of pipe. The average of the absolute error between the actual and predicted effluent chloramine residual was 0.62 mg/L as Cl_2. The black solid line represents the line of perfect agreement between actual and predicted values, and the dotted lines are located plus and minus 1 mg/L as Cl_2 from that line.

Reynolds Number Effect on Chloramine Decay

Experiments were conducted as described in the chapter 3 to generate chloramine residuals paired with varying Reynolds numbers by varying velocity in the single-material PDSs to assess the effect of turbulence on chloramine decay. Several experiments were conducted at Reynolds numbers of 50, 100, 150, and 200 using the single-material lines of PVC, LCI, UCI, and GS. The experiments were conducted with and without the presence of 1.0 mg/L of BOP inhibitor. All data used to determine the effect of velocity and turbulence on loss of residual was normalized to remove the effect of temperature. Normalization was done by using the previously described temperature correction factor and adjusting the overall chloramine decay rate to 20°C from the actual temperature of the experiment. Rate constants were also developed from the Reynolds number experiments for UCI and GS pipes conducted during TBW I to broaden the database, which included Reynolds numbers from 50 to 8,000.

The overall decay rate constant K is commonly expressed as the combination of the rate of decay in the bulk liquid and the rate of decay at the pipe wall (Rossman, Clark, and Grayman 1994). The bulk reaction rate is intrinsic to the type of water and is dependent on water quality parameters such as temperature, pH, and organic content and not by hydraulic conditions. For these reasons, the bulk decay constant K_B can be assumed to be constant under laminar or turbulent flow conditions.

As shown in the previous discussion, the reaction rate K is the sum of a wall reaction rate and a bulk reaction rate as shown in Equation 5.70, which is summarized in Equation 5.79 and allows the wall reaction rate to be determined for varying TOC concentrations if the overall reaction rate, K, and the bulk reaction rate, K_B, are known. This is the method used to determine W in this work. The bulk decay constant, K_B, was subtracted from the overall decay constant, K, to obtain the overall wall decay constant, W, following Equation 5.79.

$$W = D^*(K - 0.02^*UV_{254} - 106644^*[H^+]) \qquad (5.79)$$

where
W = overall wall decay constant, in./hr
D = pipe diameter, inches
K_B = bulk decay constant, cm/hr
$[H^+]$ = molar concentration of H^+ ions
K = overall decay rate, hr^{-1}

Unlined Cast Iron

A total of 18 experiments were conducted using the UCI PDS. The Reynolds numbers used for the UCI pipe experiments range from 50 to 8,000, as shown in Table 5.144. Table 5.145 shows the Reynolds number for each experiment as well as the HRT, flow rate, flow velocity, overall decay rate constant (K), and overall wall decay constant (W). Table 5.145 also show which experiments were conducted using BOP inhibitor at a dose of 1.0 mg/L as P. Analysis of the overall and wall decay constant for the experiments at the same Reynolds number (Re) shows that the use of BOP inhibitor did not have an effect on the rate of chloramine decay in the UCI pipes. That is similar to what was observed when comparing PDS02 and PDS14 earlier in this report.

Table 5.145
General information about Reynolds number experiments in UCI pipes

Re	Flow rate (gpm)	Velocity (ft/min)	HRT (hr)	Inhibitor?	K (hr^{-1})	W (in./hr)
50	0.10	0.07	24	No	0.012	0.064
50	0.10	0.07	24	No	0.024	0.124
50	0.10	0.07	24	Yes	0.025	0.134
100	0.19	0.13	12	No	0.027	0.155
100	0.19	0.13	12	No	0.012	0.053
100	0.19	0.13	12	Yes	0.017	0.090
150	0.29	0.20	8	No	0.019	0.101
150	0.29	0.20	8	Yes	0.029	0.159
200	0.38	0.26	6	No	0.046	0.254
200	0.38	0.26	6	Yes	0.035	0.193
245	0.47	3.17	4.70	No	0.021	0.113
772	1.47	10.0	1.49	No	0.048	0.273
1,224	2.33	15.9	0.94	No	0.026	0.141
2,449	4.66	31.7	0.47	No	0.105	0.614
3,079	5.86	39.9	0.37	No	0.040	0.228
3,263	6.21	42.3	0.35	No	0.078	0.457
6,051	11.52	78.4	0.19	No	0.063	0.364
8,045	15.31	104.2	0.14	No	0.083	0.483

NOTES: 1 gpm = 0.063 L/sec; 1 ft/min = 0.005 m/sec; 1 in./hr = 0.025 m/hr.

W was plotted with relation to the Reynolds number, as shown in Figure 5.158. The overall wall decay constant, W, was found to be dependent on the Reynolds number. The data in Figure 5.158 is grouped into a low and high region, which is similar to the variation of Reynolds number with velocity in that laminar flow occurs at Reynolds numbers less than 200 and turbulent flow occurs at Reynolds numbers greater than 2,000. There may be a transition region for the effect of Reynolds number on W that cannot be seen in Figure 5.158 because of the lack of data. However, two regions can be observed depending on the flow regime, one for laminar flow and the second for turbulent flow.

During laminar flow, W increases rapidly with the increase of Reynolds number. The increase on W is more gradual during transition and fully turbulent flow. Linear regression was used to determine an equation to correlate Re and W. The results are shown in Equations 5.80 and 5.81. The slope of the lines fitted to the data shows that W increases 17 times faster during laminar flow than it does during turbulent flow for each increment in Reynolds number. The two trend lines intercept at a Reynolds number of around 500 and a W of 0.3 in./hr.

$$W_{Lam} = 0.00044*Re + 0.076 \qquad (5.80)$$

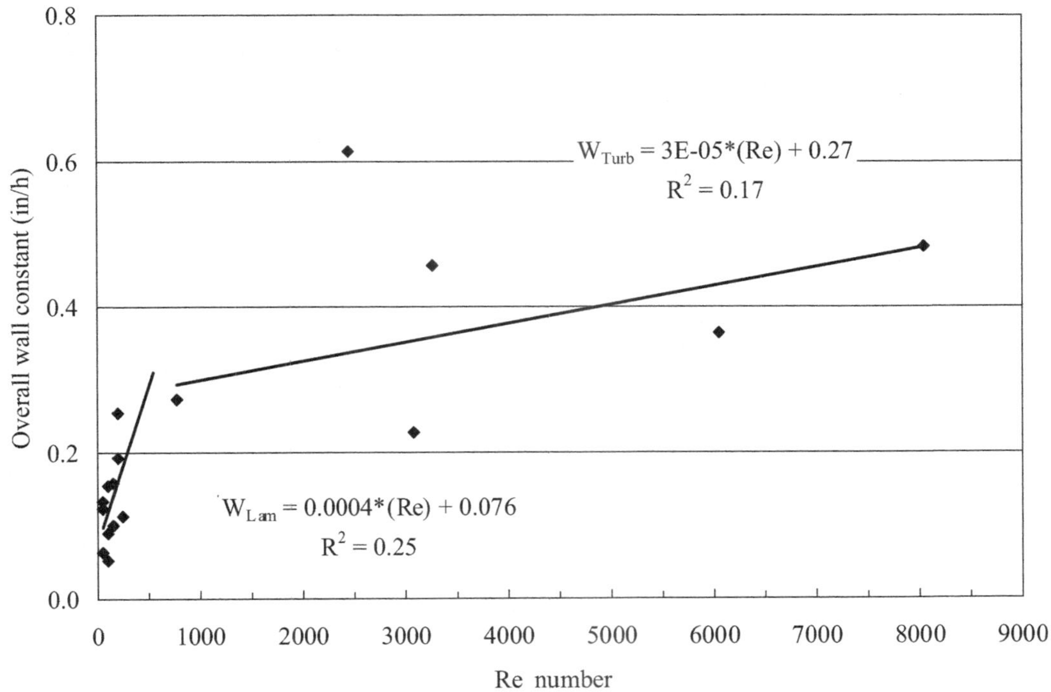

Figure 5.158 Effect of Reynolds number on overall wall decay rate, W, for UCI pipes

$$W_{Turb} = 0.000026*Re + 0.27 \tag{5.81}$$

where W_{Lam} = overall wall decay rate under laminar conditions, in./hr
W_{Turb} = overall wall decay rate under turbulent conditions, in./hr
Re = Reynolds number

Galvanized Steel Pipes

A total of 14 experiments were conducted using the GS PDS, and the general information is presented in Table 5.146. Table 5.146 shows the Reynolds number used, flow rate, flow velocity, HRT, and overall and wall rate of decay observed as well as whether inhibitor was used during the experiment. K increases with the increase of Reynolds number and ranges from 0.06 hr^{-1} to 0.13 hr^{-1} for Reynolds numbers of 50 and 391 up to 0.43 hr^{-1} for a Reynolds number of 5,025. The use of BOP inhibitor at a TP dose of 1 mg/L as P did not have an effect on the rate of chloramine decay in the GS pipes. At a Reynolds number of 150 and 200, the water with the BOP inhibitor had a wall rate greater than the water without the inhibitor; however, at a Reynolds number of 100, the opposite is true. The experiments run at a Reynolds number of 50 had identical overall and wall decay rates irrespective of the use of inhibitor.

The overall wall decay constant was calculated using Equation 5.79, and the results are shown in Table 5.146 and were plotted in relation to the Reynolds number used in Figure 5.159. In the same way as for UCI pipes, a stepwise correlation between W and Re was observed, depending on the flow regime. The linear regression of Re versus W for laminar flow and turbulent flow are presented in Equations 5.82 and 5.83, respectively.

Table 5.146
General information about Reynolds number experiments in GS pipes

Re	Flow rate (gpm)	Velocity (ft/min)	HRT (hr)	Inhibitor?	K (hr^{-1})	W (in./hr)
50	0.06	0.44	12	No	0.059	0.113
50	0.06	0.44	12	Yes	0.058	0.113
100	0.13	0.87	6	No	0.088	0.171
100	0.13	0.87	6	No	0.092	0.180
100	0.13	0.87	6	No	0.048	0.092
100	0.13	0.87	6	Yes	0.041	0.077
150	0.19	1.31	4	No	0.033	0.065
150	0.19	1.31	4	Yes	0.073	0.144
200	0.26	1.74	3	No	0.064	0.126
200	0.26	1.74	3	Yes	0.087	0.170
391	0.23	1.57	1.49	No	0.131	0.256
2,279	1.34	9.11	0.26	No	0.167	0.329
2,934	1.72	11.74	0.20	No	0.197	0.390
5,025	2.95	20.10	0.12	No	0.435	0.866

NOTES: 1 gpm = 0.063 L/sec; 1 ft/min = 0.005 m/sec; 1 in./hr = 0.025 m/hr.

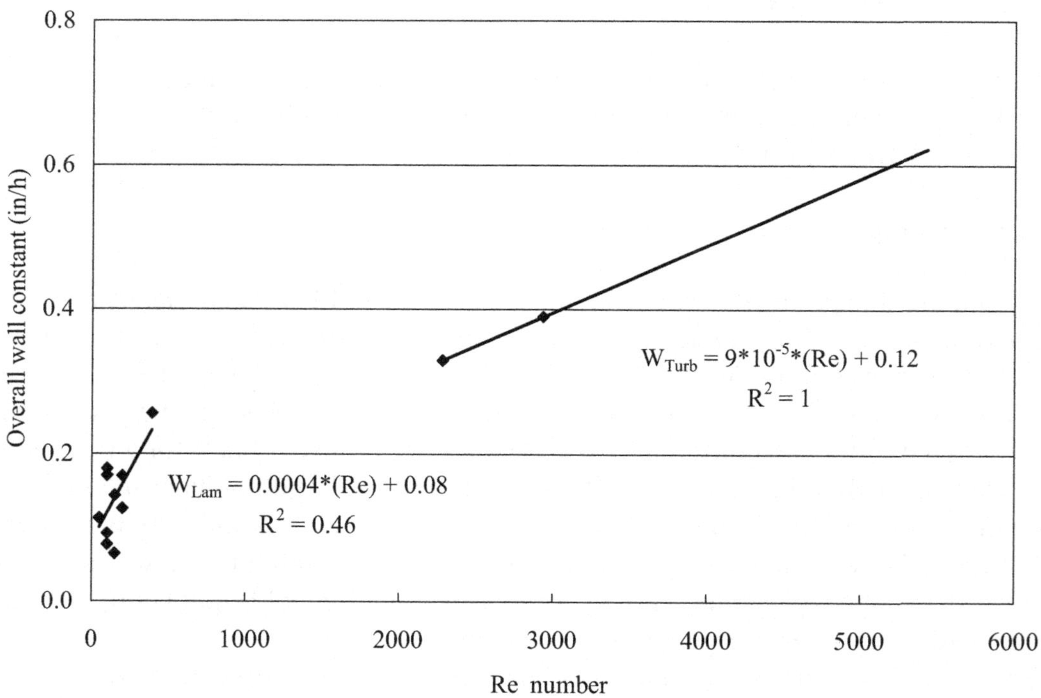

Figure 5.159 Effect of Reynolds number on overall wall decay rate, W, for GS pipes

Table 5.147
General information about Reynolds number experiments in PVC pipes

Re	Flow rate (gpm)	Velocity (ft/min)	HRT (hr)	Inhibitor?	K (hr^{-1})	W (in./hr)
50	0.10	0.07	24	No	0.006	0.025
50	0.10	0.07	24	Yes	0.000	0.000
100	0.19	0.13	12	No	0.018	0.092
100	0.19	0.13	12	Yes	0.001	0.000
150	0.29	0.20	8	Yes	0.001	0.000
200	0.38	0.26	6	No	0.018	0.094
200	0.38	0.26	6	Yes	0.000	0.000

NOTES: 1 gpm = 0.063 L/sec; 1 ft/min = 0.005 m/sec; 1 in./hr = 0.025 m/hr.

$$W_{Lam} = 0.0004*Re + 0.08 \quad (5.82)$$

$$W_{Turb} = 0.00009*Re + 0.12 \quad (5.83)$$

Analysis of the equation shows W increasing 40 times faster under laminar flow conditions than under fully turbulent flow. Under transitional flow conditions (500 < Re < 2,000), W appears to be around 0.3 in./hr.

Polyvinyl Chloride Pipes

A total of seven experiments were run for the PVC PDS. Table 5.147 shows the Reynolds number, flow rate, flow velocity, HRT, and calculated rate constants for each of the experiments. Half of the experiments used BOP at 1.0 mg/L as P and are identified in Table 5.147.

There seems to be no correlation between Reynolds number and the overall or wall rate of chloramine decay in the PVC pipes. This result is intuitive given that the chloramine demand at the wall of PVC pipes is very small and the bulk reaction rates are dominant. However, the results are not conclusive because for more than 50% of the experiments, there was no or little change on the residual between the influent and the effluent because some temperatures were too low (12° to 19°C) to observe a change in such a short retention time (6 to 24 hours). Figure 5.160 shows W in relation to Reynolds number for the PVC PDS.

Lined Cast Iron Pipes

Similar results to the PVC pipes were obtained for the LCI pipe. Table 5.148 shows the information for the experiments conducted, and Figure 5.161 shows W plotted versus the Re used for the LCI PDS. Both constants are very low and almost flat for the range of Reynolds number studied. As for the PVC pipes, the wall of the LCI pipes have a low demand, and the results are not surprising.

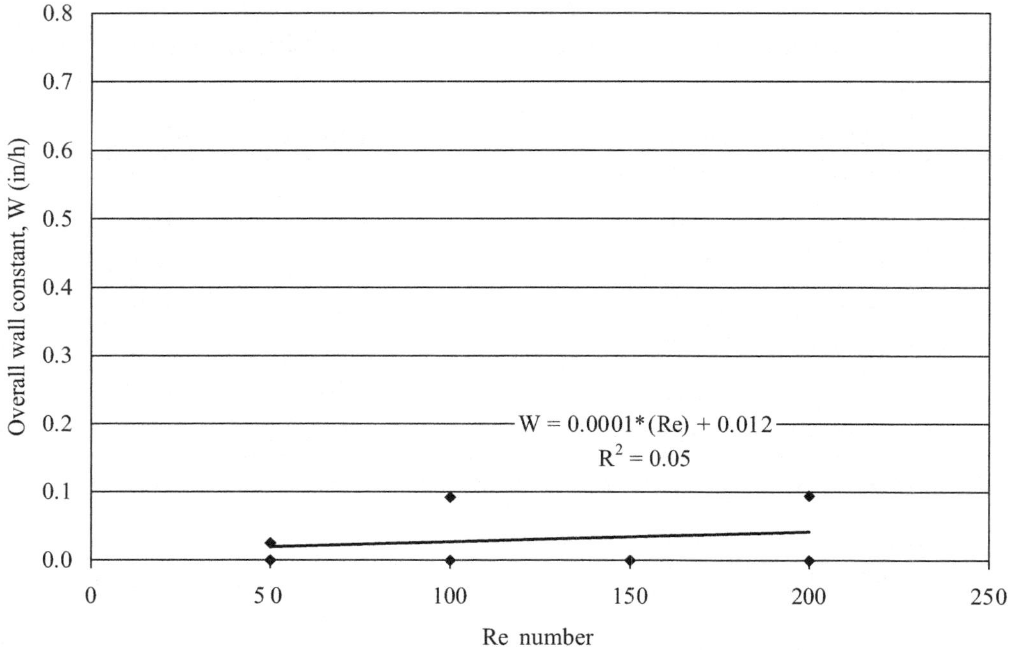

Figure 5.160 Effect of Reynolds number on overall wall decay rate, W, for PVC pipes

Table 5.148
General information about Reynolds number experiments in LCI pipes

Re	Flow rate (gpm)	Velocity (ft/min)	HRT (hr)	Inhibitor?	K (hr^{-1})	W (in./hr)
50	0.10	0.07	24	No	0.0045	0.008
50	0.10	0.07	24	No	0.0011	0.000
50	0.10	0.07	24	Yes	0.0000	0.000
100	0.19	0.13	12	No	0.0000	0.000
100	0.19	0.13	12	Yes	0.0037	0.007
150	0.29	0.20	8	No	0.0021	0.000
150	0.29	0.20	8	Yes	0.0000	0.000
200	0.38	0.26	6	No	0.0026	0.000
200	0.38	0.26	6	Yes	0.0000	0.000

NOTES: 1 gpm = 0.063 L/sec; 1 ft/min = 0.005 m/sec; 1 in./hr = 0.025 m/hr.

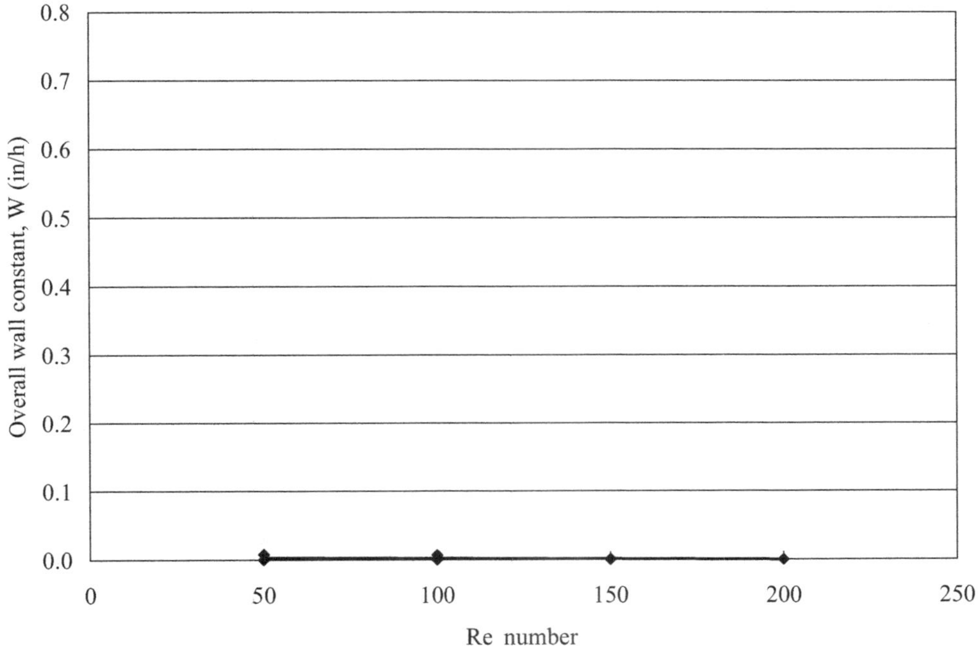

Figure 5.161 Effect of Reynolds number on overall wall decay rate, W, for LCI pipes

Predicted Effluent Residual Based on Reynolds Number

Figures 5.162 and 5.163 show the predicted residual for 6-in. (0.15-m)-diameter UCI and GS pipes, respectively, for Reynolds numbers ranging from 15 to 5,000, for water with an organic content of 3 mg/L, a water temperature of 20°C, a pH of 8.0, and an initial chloramine concentration of 4 mg/L as Cl_2. Figure 5.162 shows that for the UCI pipe, the chloramine residual is predicted to be below the regulatory minimum of 0.6 mg/L as Cl_2 in 40 hours under transitional (Re = 1,000) and fully turbulent flow conditions (Re = 5,000), and would be 1.2 mg/L as Cl_2 and 2.1 mg/L as Cl_2 for Reynolds numbers of 200 and 15, respectively, after the same 40 hours. Only when the Reynolds number is 15, the water is predicted to maintain the chloramine residual for a period of 120 hours or 5 days.

Figure 5.163 shows that for GS pipes with a diameter of 6 in. (0.15 m), the chloramine residual is predicted to be below the required minimum after 60 hours under transitional (Re = 1,000) and fully turbulent flow conditions (Re = 5,000), and would be 0.7 mg/L as Cl_2 and 1.4 mg/L as Cl_2 for Reynolds numbers of 200 and 15, respectively, after the same amount of time. Even under fully laminar conditions, the water is predicted to fail to maintain the required minimum residual of 0.6 mg/L as Cl_2 for 120 hours.

Chloramine Decay Model Incorporating Reynolds Number Effect

The effect of Reynolds number on chloramine decay could be incorporated into the model by combining the model shown in Equation 5.78 with Equations 5.80 to 5.83. A general form of the new model is presented in Equation 5.84.

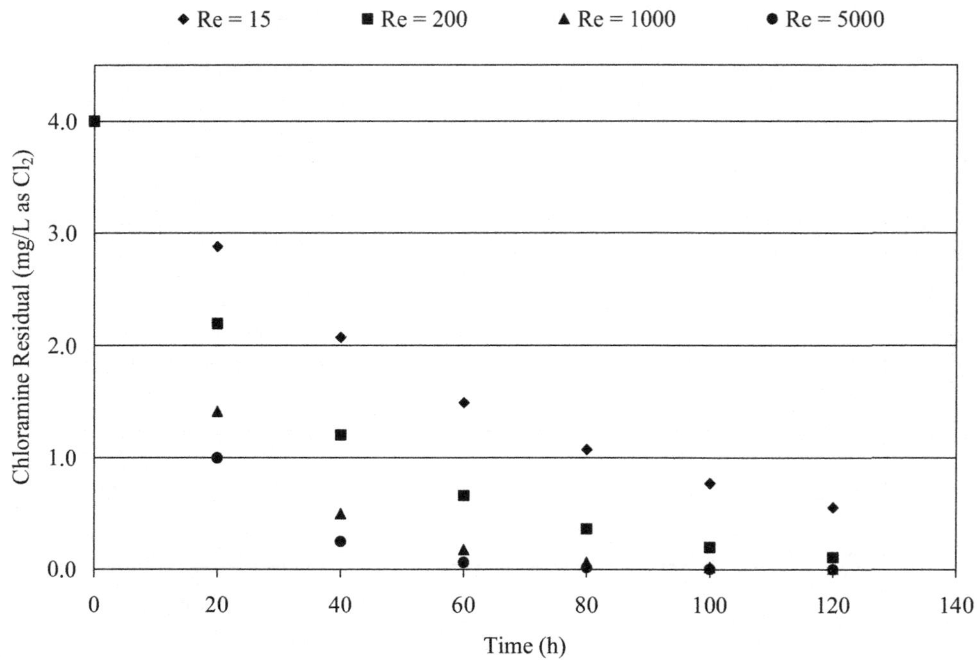

Figure 5.162 Predicted chloramine residual in UCI pipes at different Reynolds numbers

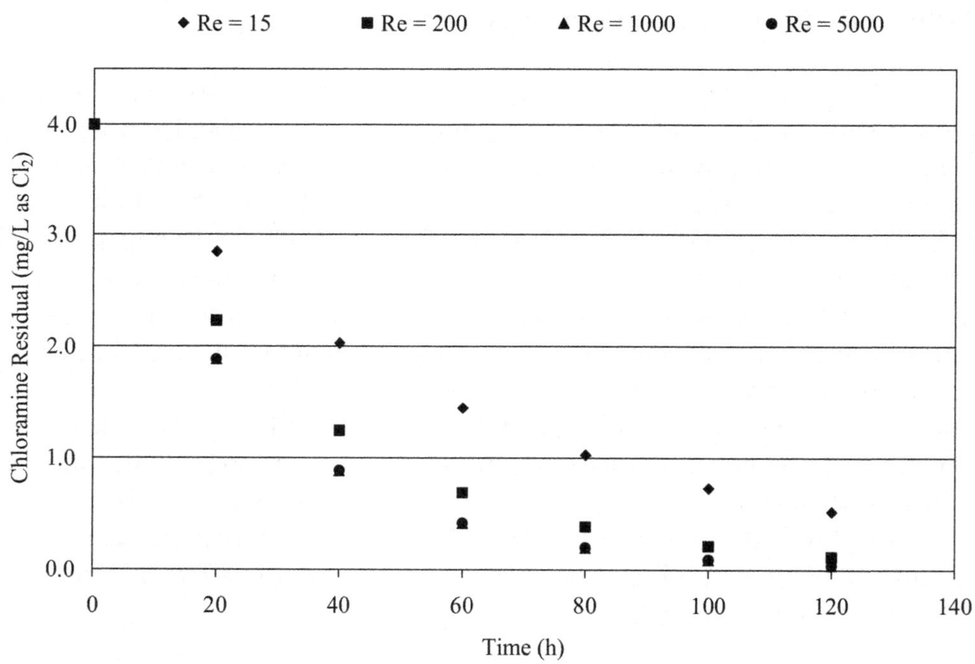

Figure 5.163 Predicted chloramine residual in GS pipes at different Reynolds numbers

Table 5.149
Model parameters for chloramine model modified using Reynolds number

Pipe material	K_B (cm/hr)	K_H (L/mol-hr)	θ	B (in./hr)	C (in./hr)
Polyvinyl chloride	0.02	106403.7	1.13	0	0.025
Lined cast iron	0.02	106403.7	1.13	0	0.014
Unlined cast iron					
Re < 500	0.02	106403.7	1.13	0.00044	0.076
Re > 500	0.02	106403.7	1.13	0.00003	0.27
Galvanized steel					
Re < 500	0.02	106403.7	1.13	0.0004	0.083
Re > 2,000	0.02	106403.7	1.13	0.00009	0.12

NOTES: 1 in. = 2.54 cm

$$Cl_{2(t)} = Cl_{2(0)}*\exp\left(-\left(K_B*UV_{254} + K_H*[H^+] + \frac{B*Re + C}{D}\right)*\theta^{(T-20)}*\text{time}\right) \quad (5.84)$$

where $Cl_{2(t)}$ = chlorine concentration at time t, mg/L as Cl_2
$Cl_{2(0)}$ = initial chlorine concentration, mg/L as Cl_2
K_B = bulk decay constant (cm/hr) = 0.02 cm/hr
K_H = coefficient for H^+ term (L/mol-hr) = 106,404 L/mol-hr
B = wall decay material coefficient for Reynolds number, in./hr
C = wall decay material constant, in./hr
UV = UV_{254}, cm^{-1}
$[H^+]$ = molar concentration of H^+, mol/L
D = pipe diameter, inches
θ = temperature correction coefficient
T = temperature, °C
time = time, expressed in hours

The values for the model parameters are included in Table 5.149. For PVC and LCI pipes, the parameter B takes a value of zero (0) because for those materials, K does not change with the Reynolds number.

Summary

The chloramine residual at the influent point of the PDS was not significantly different for any of the PDSs. The influent residual ranged from 4.8 to 5.1 mg/L as Cl_2. Chloramine demand did not change in the PDS with inhibitor type or dose during this study. PDS14 had the same chloramine demand as most of the PDSs using inhibitors with the exception of PDS12 (12 mg/L SiO_2). PDS13 had a greater chloramine demand than all the PDSs. Paired t-tests between PDS13 and the other PDSs showed that the differences in chloramine demand were significant at the 95%

confidence level. The same type of analysis for PDS14 concluded that the differences in chloramine demand between PDS01 to PDS11 and PDS14 were not significant.

Chloramine decayed at different rates depending on the material of the pipe. PVC and LCI showed the lowest rate of chloramine decay, and UCI and GS showed the highest. The order of the decay rate was LCI < PVC << UCI < GS. The rate of chloramine decay also increased with the increase in temperature and organic content of the water, and decreased with increased water pH. Chloramine was observed to decline faster during summer than in winter.

A first-order chloramine decay model was developed by nonlinear regression based on the model obtained from the TBW I project. Data obtained from the internal sampling ports of the hybrid lines were used to determine the rate of decay for each pipe material. The form of the model is shown in Equation 5.77. The model contains a bulk decay rate constant (K_B), a pH effect constant (K_H), a wall decay rate constant (W), and a temperature correction factor (θ). The model was adjusted by maintaining the bulk decay rate constant for all the pipe material to the K_B obtained for PVC and LCI during the TBW I project and calculating a common θ and K_H as well as an individual W for each pipe material.

The values for W are presented in Table 5.144. LCI and PVC have the lowest W at 0.014 in./hr and 0.025 in./hr, respectively, since those materials are fairly inert, do not promote bacteria, and, in general, have the lowest chloramine demand of the materials used in the study. UCI and GS have W that are 4 and 9 times greater than W for LCI. UCI has a W of 0.103 in./hr, and GS has a W of 0.051 in./hr. The higher wall demand of the LCI and GS pipe surfaces are due in part to the greater chemical reactivity of iron and zinc relative to LCI and PVC. Material wall demand also increases with roughness, biological film densities, and corrosion by-products. The diameter of the pipe affects the rate of chloramine decay because the effect of W is proportional to the ratio of the volume to wetted surface area. The relative effect of W decreases when the pipe diameter increases. This effect is more noticeable for unlined metallic pipes because their W is more elevated.

The rate of chloramine decay was found to increase with the increase of temperature. The temperature correction factor, θ, was obtained to be 1.11, which means that the rate of decay would double for each 7°C increment of water temperature. The pH of the water was the only corrosion control measure to have an effect on the rate of chloramine decay in the PDSs. The decay of chloramine residual was found to be slower when the water pH increased. The effect of pH is more noticeable in LCI and PVC pipes than in UCI and GS pipes.

The chloramine decay model was verified using data that was not used during model development. The available data included only the influent and effluent residual for the PDS. For that reason, the model was used in increments with the calculated effluent for the PVC section of the PDS being used as the influent for the LCI section. The same approach was used for the UCI and GS sections. The verification showed a good agreement between the predicted and actual chloramine residual with an average absolute error of 0.6 mg/L as Cl_2.

The chloramine decay model was subsequently modified to include the effect of the Reynolds number (flow velocity). Flow velocity experiments in the single-material lines showed that UCI and GS wall decay rates increase with Reynolds number. PVC and LCI did not show variation in W with the Reynolds number. UCI and GS showed a region of rapid increase of W under laminar conditions, followed by a more gradual increase under transitional and fully turbulent flow. The modified model is presented in Equation 5.84, and the parameter values are shown in Table 5.149.

Table 5.150
Comparison of log effluent HPC (cfu/mL) for PDS01–PDS09 and PDS13–PDS14

Group	N	Missing	Mean	Standard deviation
PDS01–PDS09	468	6	3.15	0.59
PDS13–PDS14	104	0	2.73	0.52

BIOLOGICAL CHARACTERIZATION

This section was organized by major subsections for clarity. Biological characterization or stability was assessed by bulk water measures of bioactivity and biofilm measures of bioactivity.

Bulk Biostability

Bulk biostability was assessed by HPC changes as generated by inhibitor dose and type as well as by comparison to the pH controls.

Bulk Effluent HPC

HPC is the most common measure of bioactivity in drinking water distribution systems and is described in the following text using statistical analyses to assess change in response to variations in operation conditions.

Statistical Comparison of Effluent HPC Between PDSs. The following sections present a statistical comparison of effluent HPC among PDSs.

Statistical Comparison Between Control PDSs and Inhibitor PDSs. The experimental facilities allowed for side-by-side testing of 14 separate PDS hybrid lines (each with four identical pipe materials) that received the same finished water blend. The individual hybrid PDSs received a different corrosion control strategy, consisting of one of four inhibitors at a range of doses, and, in two lines, pH control without addition of an inhibitor.

Given that PDS13–PDS14 were the control pipes that received no inhibitor, PDS01–PDS09 (phosphate-based inhibitors) and PDS10–PDS12 (Si-based inhibitors) are compared with them. The statistics test by SigmaPlot confirmed that the effluent HPC data of PDS01–PDS09 and PDS13–PDS14 were normally distributed (normality test passed, $p = 0.198$), and they also passed the equal variance test (p-value was 0.301).

A t-test was used to compare the means of the two data sets, and it showed that the difference in the mean values of the two groups is greater than would be expected by chance; there is a statistically significant difference between these two groups ($p < 0.001$) (Table 5.150). PDSs with phosphate inhibitor had 0.42 log (163%) greater mean values for effluent HPC than the control PDSs. This means that P-based inhibitor addition resulted in an increase in effluent HPC by more than a factor of 2.

Statistical tests by SigmaPlot confirmed that the HPC data of PDS10–PDS12 and PDS13–PDS14 evaluated as a whole were not normally distributed (the p-value was less than 0.050). A Mann–Whitney rank sum test (Table 5.151) was used to compare the data, and it showed that the difference in the median values between the two groups was greater than would be expected by chance; there was a statistically significant difference (p-value ≤ 0.001) between effluent HPC from the Si-based inhibitor and the control PDSs.

Table 5.151
Comparison of log effluent HPC (cfu/mL) for PDS10–PDS12 with PDS13–PDS14

Group	N	Missing	Median	25%	75%
PDS10–PDS12	156	3	3.12	2.69	3.42
PDS13–PDS14	104	0	2.82	2.35	3.15

Table 5.152
Comparison of log effluent HPC (cfu/mL) for PDS01–PDS09 and PDS10–PDS12

Group	N	Missing	Mean	Standard deviation
PDS01–PDS09	468	6	3.15	0.59
PDS10–PDS12	156	3	3.13	0.64

PDSs with silica inhibitor had 0.30 log (99%) greater median values than the control PDSs. The distribution also showed that the lowest 25% of the values in PDS10–PDS12 were 0.34 log higher than the control PDSs, and the lowest 75% of the values were 0.28 log higher. Again, the addition of inhibitors corresponded to a significantly greater effluent HPC, and the increase with Si-based inhibitors was at about the same level as was observed with phosphate-based inhibitors. A possible mechanism to explain the increase in HPC for the P-based inhibitors would be that phosphorus is a macronutrient that is known to be one of the major limiting nutrients in some full-scale distribution systems. However, the observation that the same type of increase occurred with the Si-based inhibitors brings that explanation into question. It has also been observed that sometimes chemical feed streams have biodegradable organics as trace impurities; it is possible that this was a factor for both inhibitors.

The data was also examined for a possible difference between phosphate-based inhibitor PDSs and Si-based inhibitor PDSs. The statistics test by SigmaPlot confirmed that the effluent HPC data of PDS01–PDS09 and PDS10–PDS12 (Table 5.152) were normally distributed (normality test passed, p-value = 0.065). An F-test showed that the two sets of data had equal variances (p-value = 0.331).

A t-test was used to compare the means of the two data sets, and it showed that the difference in the mean values of the two groups was not great enough to reject the possibility that the difference is due to random sampling variability. There was not a statistically significant difference between these two groups (p-value = 0.065) at a 95% confidence level. However, the p-value is quite low, and at a 93.5% confidence level the difference was significant. This makes it very possible that the P-based inhibitors resulted in a greater effluent HPC value than the Si inhibitors. However, the difference is only 0.02 log, which has little practical significance.

Statistical Comparison Between Phosphate-based Inhibitor PDSs. PDS01–PDS09 (phosphate-based inhibitors) received three corrosion control chemicals each at three different doses. PDS01–PDS03 received BOP inhibitor, PDS04–PDS06 received OP, and PDS07–PDS09 received ZOP. The differences between these phosphate-based inhibitors were evaluated using statistical analysis. Statistical tests by SigmaPlot confirmed that the HPC data of PDS01–PDS03, PDS04–PDS06, and PDS07–PDS09 were normally distributed (i.e., normality test passed, p-value = 0.017) and also they had equal variances (equal variance test passed, p-value = 0.611).

Table 5.153
Log effluent HPC (cfu/mL) for phosphate-based inhibitors, PDS01–PDS09

Group	N	Missing	Mean
BOP	156	2	3.25
OP	156	1	3.14
ZOP	156	3	3.06

Table 5.154
All pairwise multiple comparisons of log effluent HPC (cfu/mL) for different phosphate-based inhibitors (Holm–Sidak method)

Comparison	Difference of means	t	Unadjusted *p*-value	Final critical level	Significant?
BOP vs. ZOP	0.19	2.76	0.006	0.017	Yes
BOP vs. OP	0.12	1.72	0.087	0.025	No
OP vs. ZOP	0.07	1.06	0.292	0.050	No

The Holm–Sidak method was used to compare all three sets of data. The statistical values are shown in Table 5.153, and the results of the comparison are shown in Table 5.154.

The differences in the mean values among the treatment groups are greater than would be expected by chance; there is a statistically significant difference (p-value = 0.021). When performing the test, the p-values of all comparisons are computed and ordered from smallest to largest. Each p-value is then compared to a critical level that depends on the significance level of the test, the rank of the p-value, and the total number of comparisons made. A p-value less than the critical level indicates there is a significant difference between the corresponding two groups. From Table 5.154, the mean effluent HPCs of BOP and ZOP were significantly different, with BOP being higher by 0.19 log. The difference between BOP and OP was 0.12 log, but this difference could not be shown to be significantly different (the p-value of 0.087 was greater than the critical p-value level at 0.025). In a similar fashion, the difference between the OP and ZOP PDSs was not significantly different with a p-value equal to 0.292, compared to a critical value of 0.05.

Effluent HPC as a Function of Different Doses of Inhibitors. The averages of effluent HPC in four phases are shown in Figure 5.164 (BOP inhibitor), Figure 5.165 (OP inhibitor), Figure 5.166 (ZOP inhibitor), and Figure 5.167 (Si inhibitor). The graphs do not show any consistent upward or downward trends corresponding to dosage for any of the inhibitors.

HPC Modeling—Data Summary. An empirical model was developed using the entire data set (all phases and all PDSs). The objective of the model was to quantify the impact of water quality on the effluent HPC in the distribution system. Dummy variables for each inhibitor (BOP, OP, ZOP, Si, and pH control) and control lines were incorporated into the model. The use of dummy variables allowed estimation of a single parameter that is associated with each of the four corrosion inhibitors. The bulk HPC model was set up in the same way as the metal release models. Initial model development segregated the data based on inhibitor type. All of the data using phosphate-based inhibitors (PDS01 to PDS09) were combined for analysis. Similarly, the

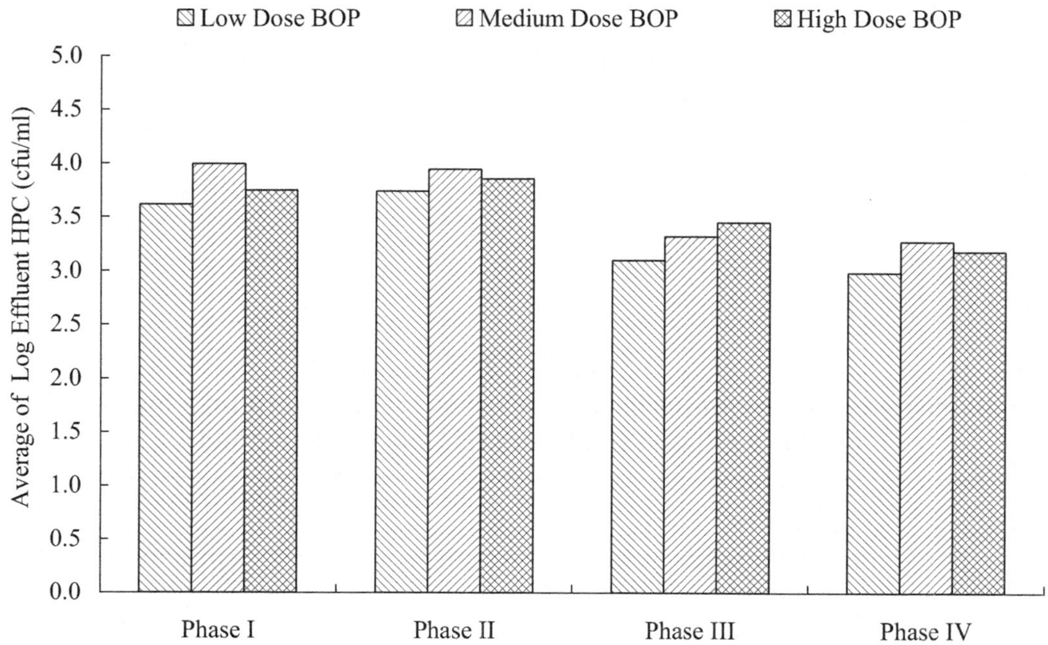

Figure 5.164 Average log effluent HPC for four phases using different doses of BOP inhibitor

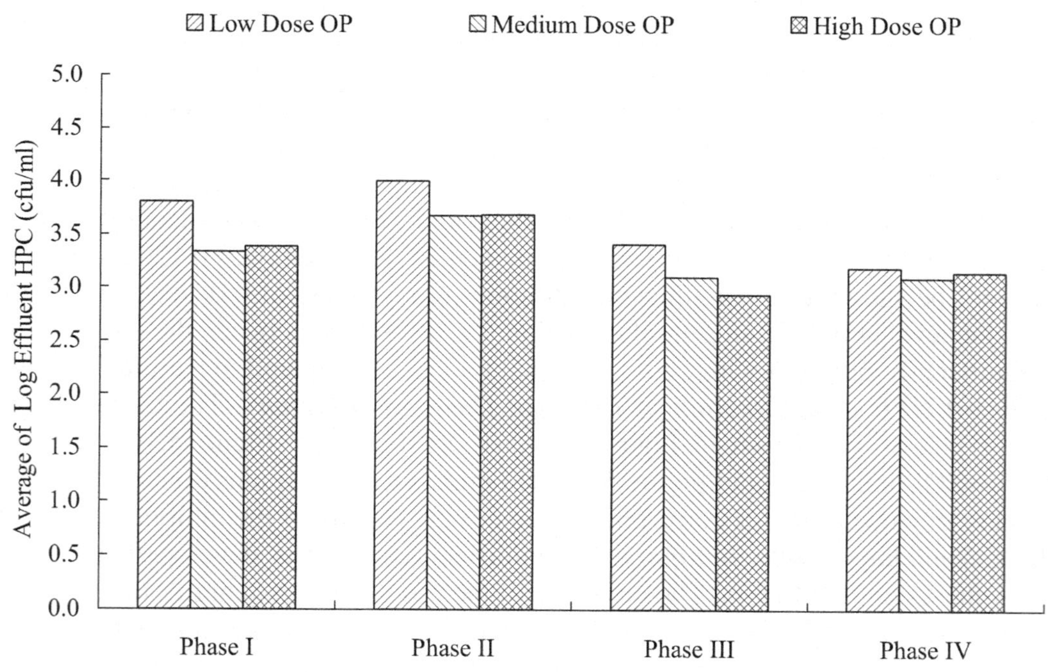

Figure 5.165 Average log effluent HPC for four phases using different doses of OP inhibitor

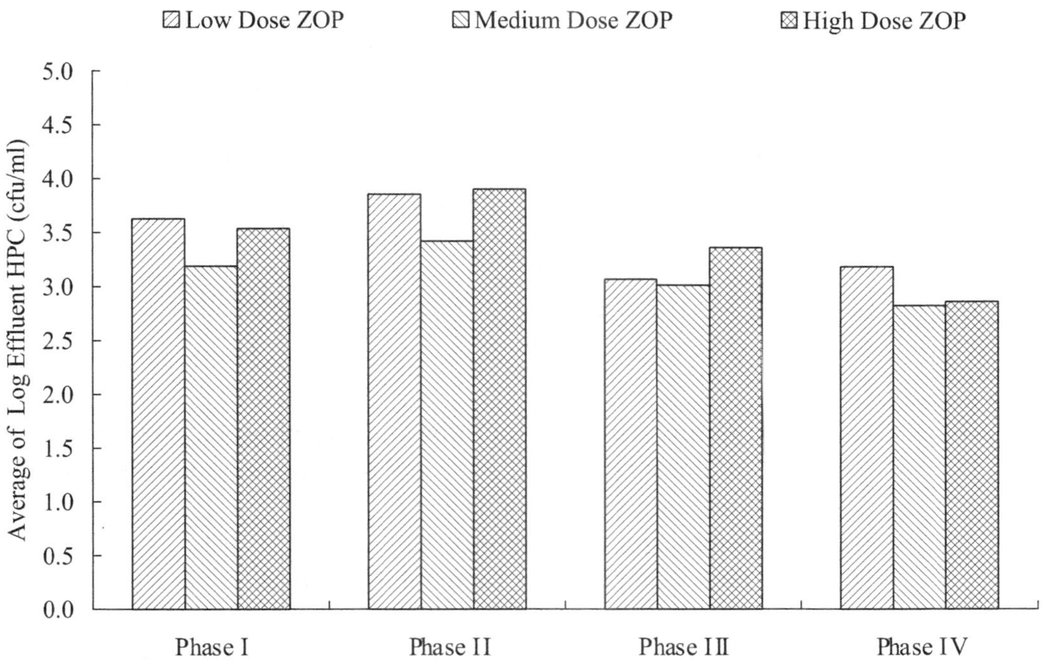

Figure 5.166 Average log effluent HPC for four phases using different doses of ZOP inhibitor

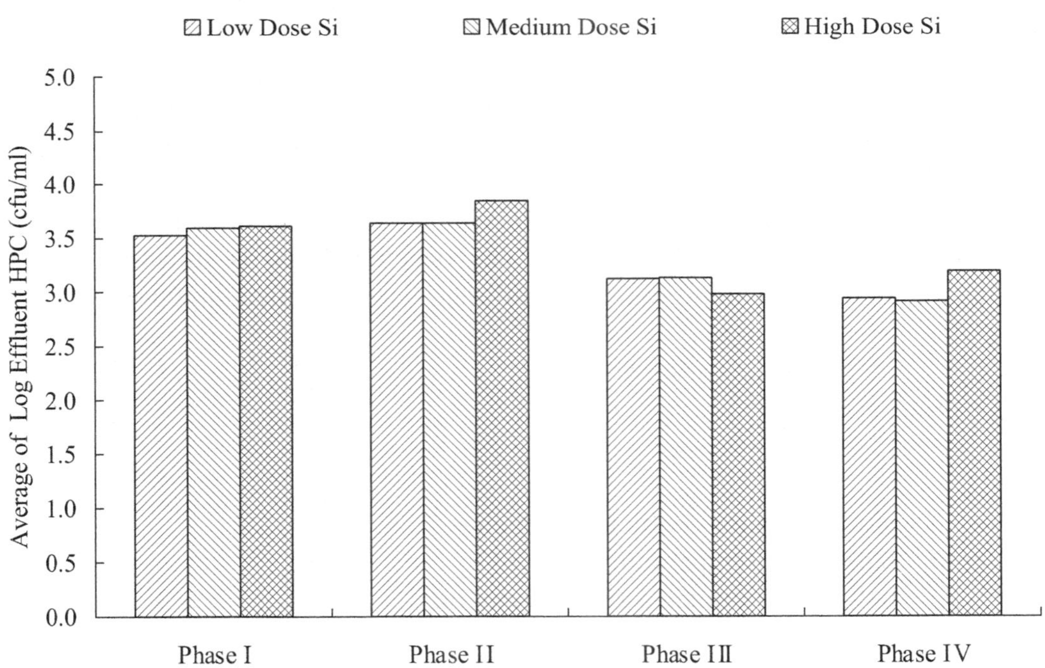

Figure 5.167 Average log effluent HPC for four phases using different doses of Si inhibitor

Si data (PDS10 to PDS12) were evaluated as a group. The pH control data (PDS13 and PDS14) were also evaluated as a separate data set.

Bulk Effluent HPC Model for PDS01–PDS09. An empirical model for log HPC with PDS01–PDS09 data from all four phases was developed and is shown in Equation 5.85. PDS01–PDS09 were fed with blended water and different phosphate inhibitors (PDS01–PDS03 BOP lines, PDS04–PDS06 OP lines, and PDS07–PDS09 ZOP lines).

$$\log \mathrm{HPC}_{\mathrm{eff}} = (3.50 \times \mathrm{BOP} + 3.40 \times \mathrm{OP} + 3.27 \times \mathrm{ZOP}) \times \frac{1.020^{(\mathrm{temp} - 20)}}{\mathrm{TCl}_{2-\mathrm{eff}}^{\ 0.195}} \tag{5.85}$$

where BOP, OP, ZOP = dummy variables for three kinds of phosphate-based inhibitors (blended ortho-polyphosphate, orthophosphate, and zinc orthophosphate)
temp = bulk water temperature, °C
$\mathrm{TCl}_{2\text{-eff}}$ = bulk water effluent total chlorine, mg/L

Model effluent HPC increased as temperature increased and decreased as effluent residual increased. From the dummy variables coefficients, effluent HPC differs from one inhibitor to another slightly. The coefficients show the same results as the statistical comparison. BOP had the highest coefficient and the highest mean of effluent HPC. The ZOP coefficient was the lowest with the ranking of the means BOP > OP > ZOP. Influent total phosphate dropped out of the model, although PDS01–PDS09 were fed by phosphate inhibitor (p-value = 0.99). Predicted versus actual log effluent HPC is shown in Figure 5.168 using Equation 5.85.

The predicted log effluent HPC is from 2.2 to 4.3 log, whereas actual project log HPC is from 0.5 to 4.5 log. All of the models with dummy variables for inhibitor overpredict low observed values and underpredict high values because of the dominance of the dummy variable term in the models.

Bulk Effluent HPC Model for PDS10–PDS12. An empirical model for log effluent HPC for all four phases using PDS10–PDS12 data was developed and is shown in Equation 5.86. PDS10–PDS12 were fed with blend water and different doses of silica inhibitor.

$$\log \mathrm{HPC}_{\mathrm{eff}} = a \times \frac{1.023^{(\mathrm{temp} - 20)}}{\mathrm{TCl}_{2-\mathrm{eff}}^{\ 0.378}} \tag{5.86}$$

where a = silica inhibitor coefficient, which is 3.8872
temp = bulk water temperature, °C
$\mathrm{TCl}_{2\text{-eff}}$ = bulk water effluent total chlorine, mg/L

Model effluent HPC increased as temperature increased and decreased as effluent residual increased in a manner similar to the P-based PDSs. Influent silica dropped out of the model with a p-value of 0.41. Predicted versus actual log effluent HPC is shown in Figure 5.169 using Equation 5.86. The predicted log effluent HPC is from 2.1 to 3.8 log, whereas actual project log HPC is from 1.0 to 4.2 log. All of the models with dummy variables for inhibitor overpredicted

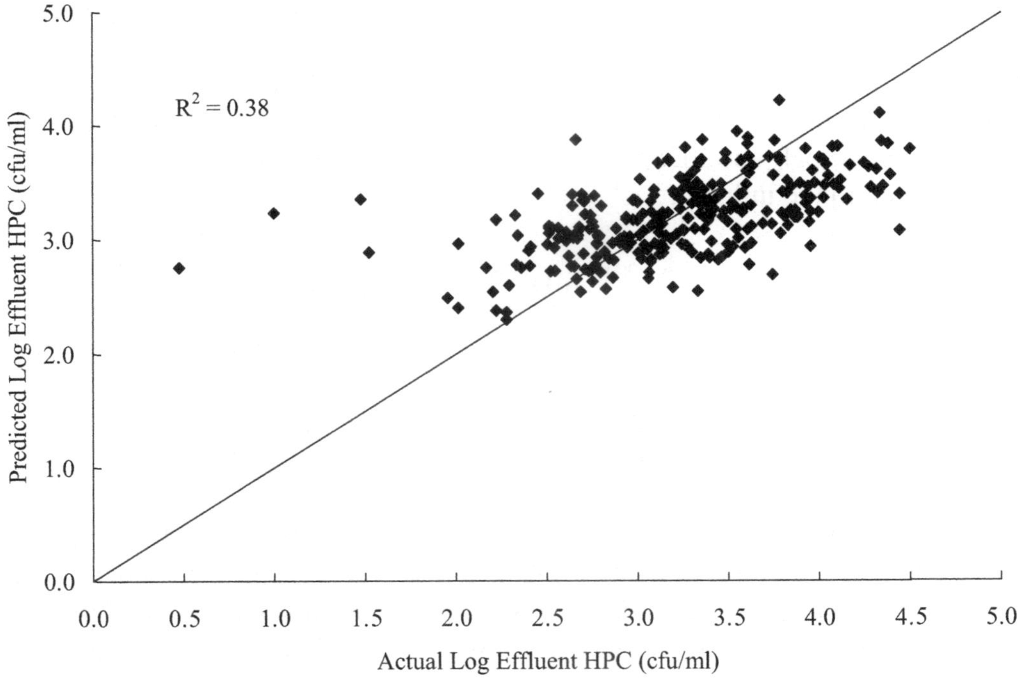

Figure 5.168 Predicted log effluent HPC versus actual project log effluent HPC for PDS01–PDS09 model

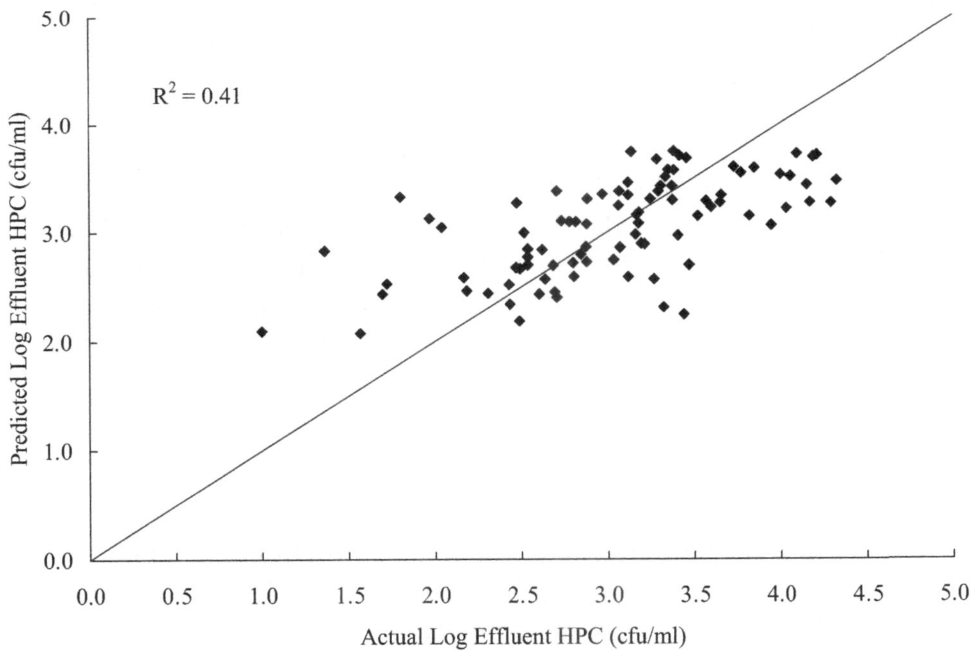

Figure 5.169 Predicted log effluent HPC versus actual project log effluent HPC for PDS10–PDS12 model

low values and underpredicted high values because of the dominance of the dummy variable term in the models.

Bulk Effluent HPC Model for PDS13–PDS14. An empirical model for log effluent HPC for all four phases using PDS13–PDS14 data was developed and is shown in Equation 5.87. PDS13 was operated at pH_s and PDS14 at pH_s+0.3. They were both control lines with no corrosion inhibitors added to the influent.

$$\log HPC_{eff} = a \times \frac{1.021^{(temp-20)} \times pH^{0.995}}{TCl_{2-eff}^{0.0858}} \tag{5.87}$$

where a = pH_s control lines' coefficient, which is 2.5128
temp = bulk water temperature, °C

Model effluent HPC increased as temperature and pH increased and decreased as effluent residual increased. The pH and effluent residual terms were kept in the model at a *p*-value of 0.134 and 0.054, respectively. Even though these *p*-values did not satisfy the general criteria for inclusion of variables in an empirical model ($p < 0.05$), the pH and effluent residual terms were retained in the model. The pH term was included to establish sensitivity to the corrosion control strategy (pH elevation). The chlorine residual term was included to maintain a model form that is consistent with the models for the inhibitor PDSs. In both cases, inclusion of the terms is consistent with the conventional mechanistic understanding of the impact of pH and disinfectant residual on planktonic bacteria. Predicted versus actual log effluent HPC is shown in Figure 5.170 using Equation 5.87. The predicted log effluent HPC was from 2.1 to 3.3 log, whereas the actual project log HPC was from 1.0 to 3.7 log. However, as with all of the models, low values were overpredicted and high values were underpredicted even though the overall behavior was acceptable, most likely because of the large influence of the dummy variables in the models. However, all the models fit the data better using dummy variables.

Bulk Effluent HPC Model for All PDSs. An empirical model for log HPC with data from all four phases for all 14 PDSs was developed and is shown in Equation 5.88. The coefficient matrix is listed in Table 5.155. In comparison with the models presented previously (Equations 5.85 to 5.87), this composite model (Equation 5.88) permitted estimation of unique coefficients for each inhibitor or pH control. This composite model is subsequently referred to as the "all-PDSs model."

$$\log HPC_{eff} = a \times temp^{b} \times TCl_{2-eff}^{c} \tag{5.88}$$

where a = coefficients for different kinds of inhibitors (BOP, OP, ZOP, and silica) and pH control lines
b = coefficients for bulk water temperature
c = coefficients for bulk water influent total chlorine
temp = bulk water temperature, °C
TCl_{2-eff} = bulk water influent total chlorine, mg/L

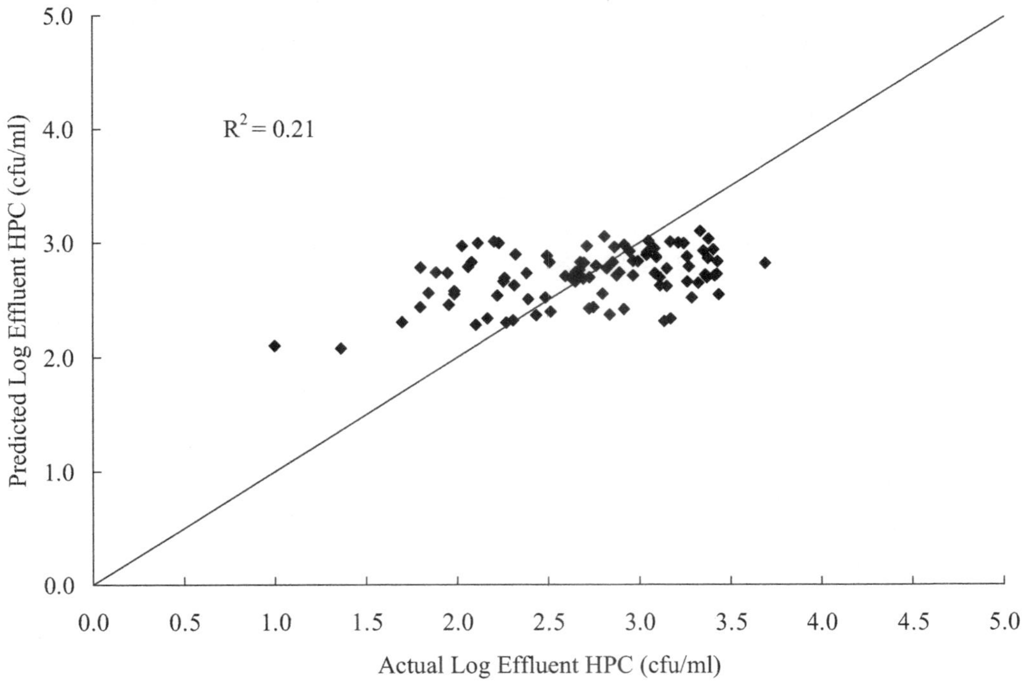

Figure 5.170 Predicted log effluent HPC versus actual project log effluent HPC for the PDS13–PDS14 model

Table 5.155
Coefficient matrix for all-PDSs model

Parameter	a	b	c
BOP	0.65	0.56	–0.19
OP	1.63	0.26	–0.21
ZOP	0.74	0.50	–0.21
Si	0.90	0.48	–0.35
pH	0.66	0.47	–0.071

The dummy variable coefficients ("a") listed in Table 5.155 correspond to each of the inhibitors and to the pH control lines. The values were similar except for the OP lines. As a result of the greater dummy variable value for the OP lines, the corresponding coefficients for temperature and residual were lower than for the other PDSs. Because the value of the dummy variable and the coefficients both affect the predicted HPC value, the coefficients cannot be used for a direct comparison of the impact of the inhibitors relative to one another. For example, according to the statistical comparison discussed previously, the effluent HPC of BOP lines were not significantly different from those of the OP lines. However, the dummy variable coefficient of BOP is approximately 1 unit less than that of OP.

Table 5.156 shows the sensitivity test for the all-PDSs bulk HPC model. The user values were selected to mimic project operating conditions during phase IV of the project. These user

Table 5.156
Sensitivity test for all-PDSs log effluent HPC (cfu/mL) model

Parameter	Project values		User values			Model predicted values			Range (Maximum–Minimum)
	Minimum	Maximum	Average	Maximum	Minimum	Average	Maximum	Minimum	
BOP									
Temp (°C)	10.5	29.1	21.3	23.8	16.8	2.97	3.16	2.60	0.56
TCl_{eff} (mg/L)	0.60	5.21	2.62	4.3	1.1	2.97	2.70	3.49	–0.79
OP									
Temp (°C)	10.4	29	21.1	23.6	16.7	2.88	2.96	2.71	0.25
TCl_{eff} (mg/L)	0.97	3.54	2.92	4.66	1.13	2.88	2.61	3.50	–0.89
ZOP									
Temp (°C)	11.9	29.3	21.3	24.3	16.4	2.82	3.01	2.47	0.53
TCl_{eff} (mg/L)	0.57	4.77	2.36	4.6	1.1	2.82	2.46	3.30	–0.84
Si									
Temp (°C)	11.9	29.3	21.3	24.3	16.4	2.90	3.09	2.56	0.53
TCl_{eff} (mg/L)	0.57	4.87	2.36	4.6	1.1	2.90	2.29	3.80	–1.50
pH									
Temp (°C)	11.9	29.3	21.3	24.3	16.4	2.57	2.73	2.28	0.46
TCl_{eff} (mg/L)	0.57	5.07	2.36	4.6	1.1	2.57	2.45	2.72	–0.26

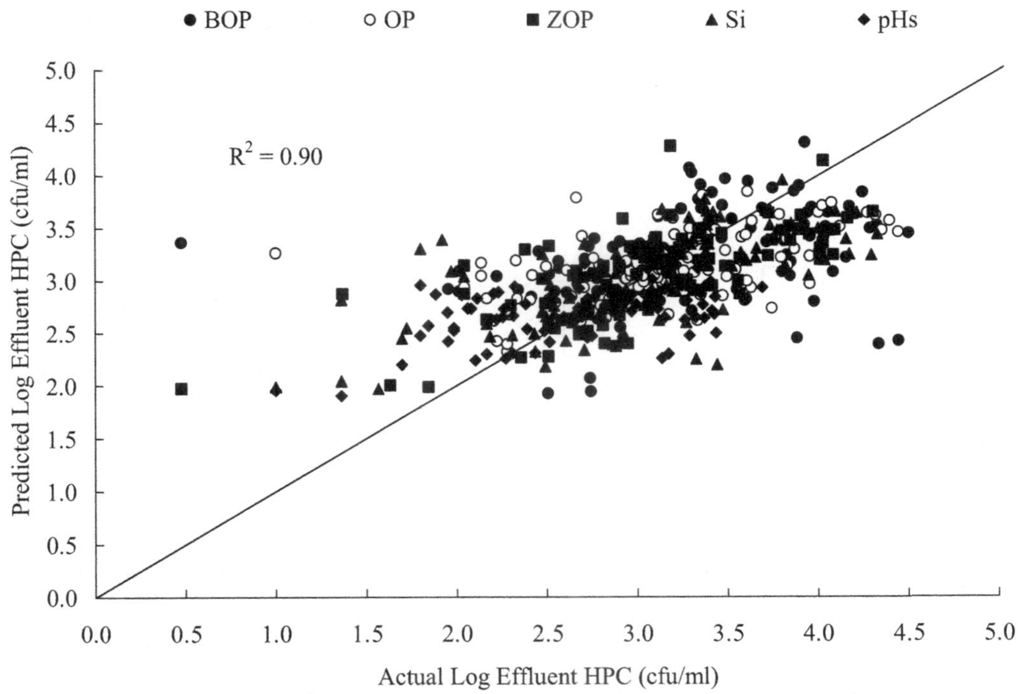

Figure 5.171 Predicted log effluent HPC versus actual project log effluent HPC for all-PDSs bulk effluent HPC model

values were specific to each different set of PDSs (BOP user values from PDS1–PDS3, OP from PDS04–PDS06, and ZOP from PDS07–PDS09). The differences in the calculated ranges were not only from the model coefficients but also from the user values.

Everything except the pH control PDSs showed a greater sensitivity to residual than to temperature. OP showed the lowest sensitivity to temperature, whereas the Si PDSs showed the greatest sensitivity to residual. The pH PDSs showed the lowest sensitivity to residual of all the PDSs.

The main advantage of the all-PDSs model is that it achieves an improvement in the prediction accuracy for all the conditions studied, as shown in Figure 5.171. The coefficient of determination R^2 quantifies the strength of the association of the predicted values to the actual values observed. The R^2 was 0.90, which is an excellent value for microbial enumeration techniques covering a broad range of materials and water quality conditions. The predicted log effluent HPC range was from approximately 1.90 to 4.30 log, whereas the actual HPC values ranged from 0.49 to 4.49 log. The visual trend of the predicted values versus the actual values was good, and the all-PDSs model seemed to work acceptably well as a tool for describing the data. However, there was still an overprediction of the lowest values and an underprediction of the highest values, but the fraction of data affected by this discrepancy was lower than for the individual models discussed previously.

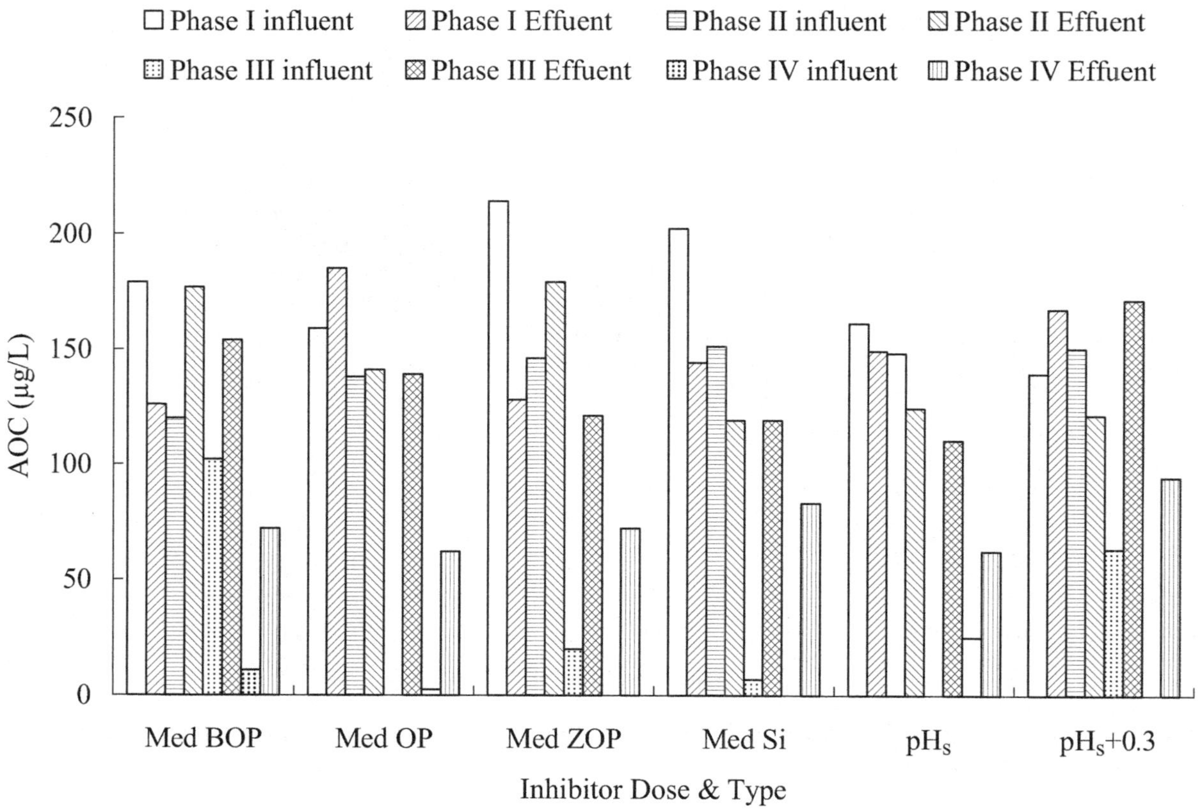

Figure 5.172 Phases I, II, and III AOC in medium-dose inhibitor lines and control lines

Assimilable Organic Carbon

Influent and Effluent AOC. Figure 5.172 shows the influent and effluent AOC values for all four phases. The influent and effluent AOC for phase I and phase II averaged 159 and 147 µg/L as acetate carbon, respectively. These values are nearly equal and easily within the variability of the AOC test. Influent AOC values were abnormally low during phases III and IV, even though acetate standards behaved normally. It may be that something was inhibitory in the influent at the time of sampling given that the corresponding effluent samples in phase III behaved normally. However, by phase IV, even effluent values were becoming low. The apparent inhibitory effect was common to all the PDSs and thus was not related to the inhibitors.

Interpretation of the AOC data is focused on phases I and II as a result (Table 5.157). The data indicate that there was no significant change in AOC levels through the PDSs. The observed change in AOC from the PDS input to the PDS output varied from –57 to +86 µg/L for the phase I and phase II data. The average change in AOC was –37 µg/L with a standard deviation of 62 µg/L. Routine AOC analyses of duplicate samples were completed to assess the precision of the AOC measurement. The average range of the duplicate samples was also 37 µg/L, with a standard deviation of the range equal to 33 µg/L. The analytical precision, evaluated as range, is comparable in magnitude to the observed difference between the PDS input and output AOC concentrations. Thus, it can be concluded that the change in AOC values through the PDSs would not likely be significant in comparison to the precision of the measurement, and a very large replication

Table 5.157
AOC concentration (μg/L) for each PDS in four phases

Phase	PDS	Influent AOC	Effluent AOC	Influent–Effluent AOC
I	02	179	126	53
	05	159	185	–26
	08	214	128	86
	11	202	144	58
	13	161	149	12
	14	139	167	–28
II	02	120	177	–57
	05	138	141	–3
	08	146	179	–33
	11	151	119	32
	13	148	124	24
	14	150	121	29
III	02	102	154	–52
	05	0	139	–139
	08	20	121	–101
	11	7	119	–112
	13	0	110	–110
	14	63	171	–108
IV	02	11	72	–61
	05	2.5	62	–59.5
	08	0	72	–72
	11	0	83	–83
	13	25	62	–37
	14	0	94	–94
Average		93	127	–37
Standard deviation		77	37	62

program would be necessary to correctly establish the true significance of any measured difference between PDS influent and effluent AOC concentrations.

Phases III and IV had lower influent and effluent AOC than phases I and II. Growth and yield controls showed that as much as 67% of the influent values were inhibitory for the AOC assay in phase IV, compared to 33% in phase I. Tables 5.158 and 5.159 present results from supplemental experiments to investigate the reported increase in bulk HPC associated with addition of the inhibitors, relative to the pH control PDSs. The first supplemental experiment was completed early in phase IV, and it was determined that all the inhibitors caused an increase in growth of the P17 and NOX (expressed as an increase in AOC). The addition of a phosphate salt (K_2HPO_4) demonstrated that addition of phosphorus alone (with no commercial inhibitor product) could also greatly stimulate growth. The second experiment, which occurred later in phase IV, exhibited inhibition of growth, which was consistent with the growth and yield control data.

Table 5.158
Corrosion inhibitor spiked blend water identifications

Identification	Description
BOP	Blend water spiked with BOP inhibitor to make 2 mg/L-P in the sample
OP	Blend water spiked with OP inhibitor to make 2 mg/L-P in the sample
ZOP	Blend water spiked with ZOP inhibitor to make 2 mg/L-P in the sample
Silica	Blend water spiked with Si inhibitor to make 12 mg/L-SiO_2 in the sample
K_2HPO_4	Blend water spiked with K_2HPO_4 to make 2 mg/L-P in the sample

Table 5.159
AOC of spiked blend tank samples

	Spike sample identification	Unspiked sample AOC (μg/L)	Spiked sample AOC (μg/L)	Increase in AOC (μg/L)
First experiment	BOP	12	116	104
	OP	12	215	203
	ZOP	12	145	133
	Silica	12	329	317
	K_2HPO_4	12	354	342
Second experiment	BOP	6	138	132
	OP	6	9	3
	ZOP	6	0	–6
	Silica	6	25	19
	K_2HPO_4	6	0	–6

Internal Port HPC

Internal Port HPC Data Summary. To monitor the effects of pipe material, bulk HPCs were monitored through the PDS internal ports of the medium-dose PDSs and the pH_s PDS (PDS02, PDS05, PDS08, PDS11, and PDS13) in phases II, III, and IV. This allowed assessment of the effect of PVC, UCI, LCI, and GS materials on bulk water HPCs.

The actual HPC average values from influent, PVC pipe, LCI pipe, UCI pipe, and GS pipe (effluent) are shown in Figures 5.173–5.177. The mean values are indicated by bars. Maximum and minimum observed values are indicated with whiskers.

Segment Influent HPC and Effluent HPC Statistical Comparison. The segment influent and effluent HPC for each of the four materials for medium-dose inhibitor PDSs and pH control line (PDS13) were compared by paired t-tests. Table 5.160 is a statistical summary for the data sets. From Table 5.160, only the PVC material pipe segment showed a statistically significant increase during the study. Analysis of the changes in bulk HPC levels did not seem to mirror biofilm densities for the material the water was passing through. This is consistent with the bulk

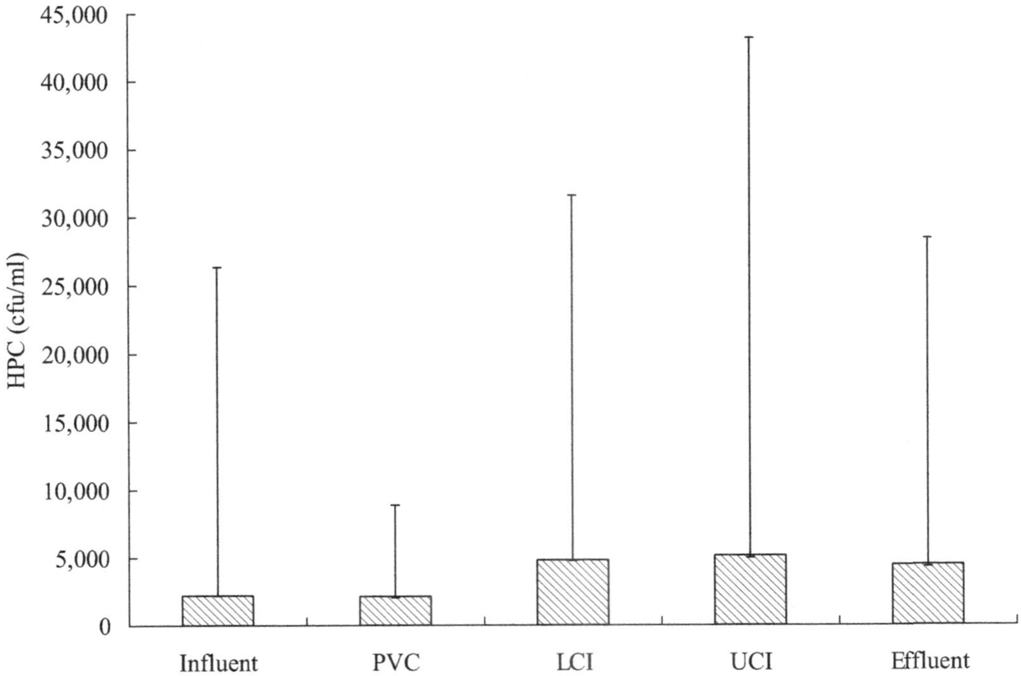

Figure 5.173 Average HPC exiting a particular pipe material for medium-dose BOP PDS (PDS02)

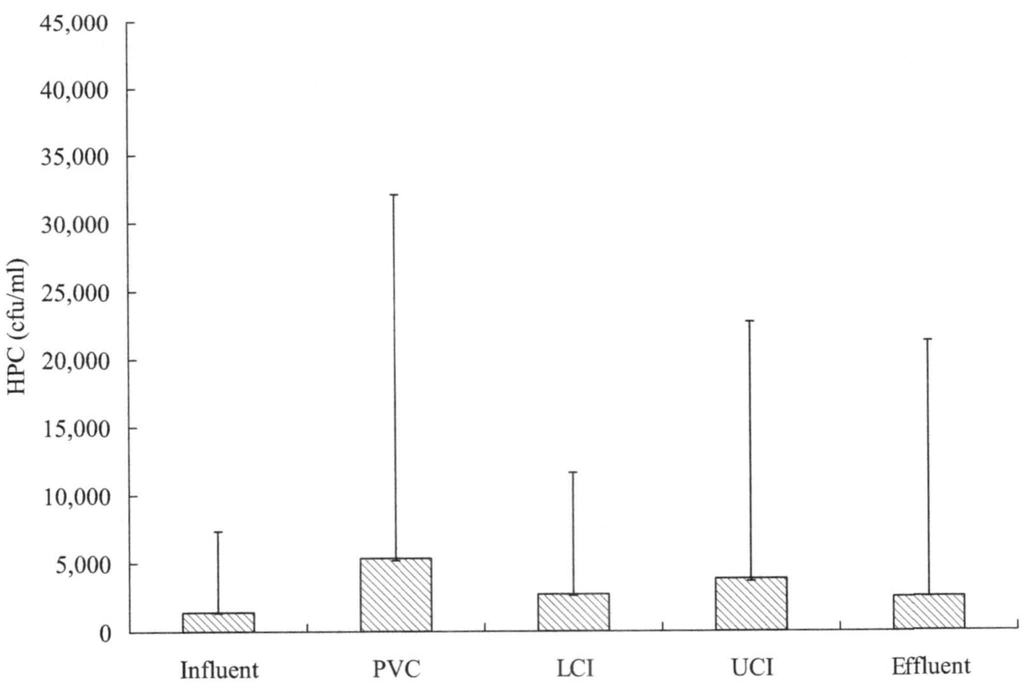

Figure 5.174 Average HPC exiting a particular pipe material for medium-dose OP PDS (PDS05)

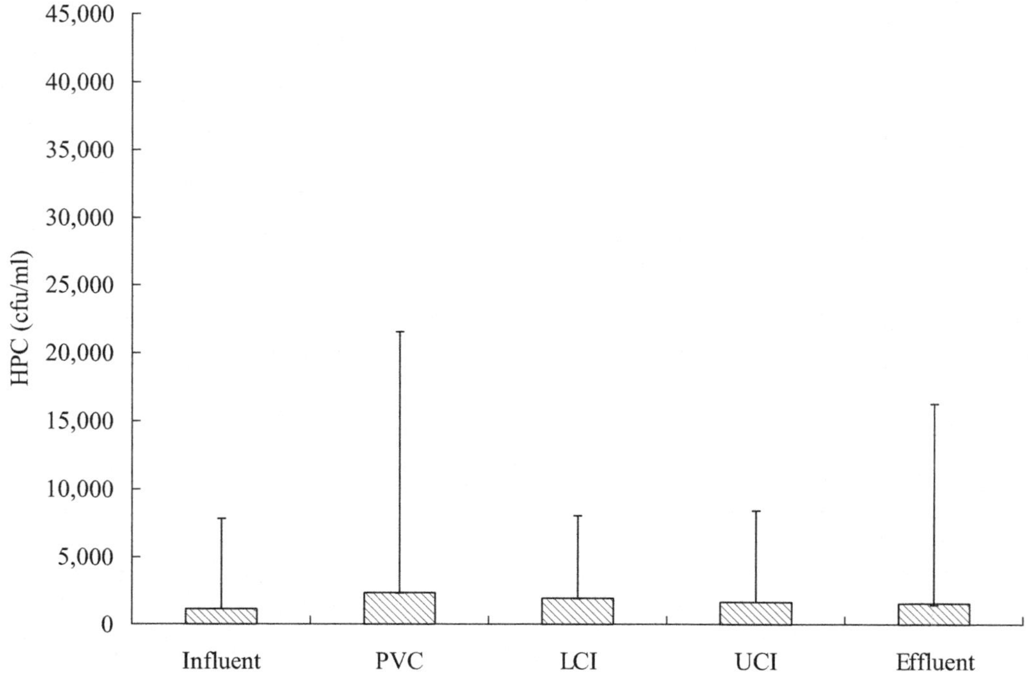

Figure 5.175 Average HPC exiting a particular pipe material for medium-dose ZOP PDS (PDS08)

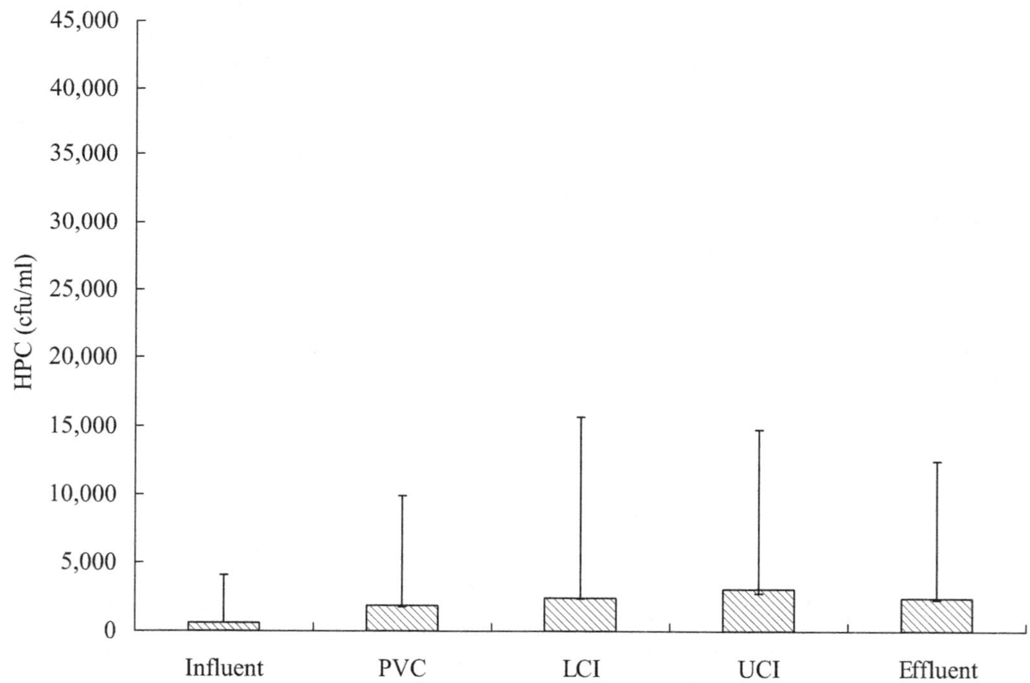

Figure 5.176 Average HPC exiting a particular pipe material for medium-dose Si PDS (PDS11)

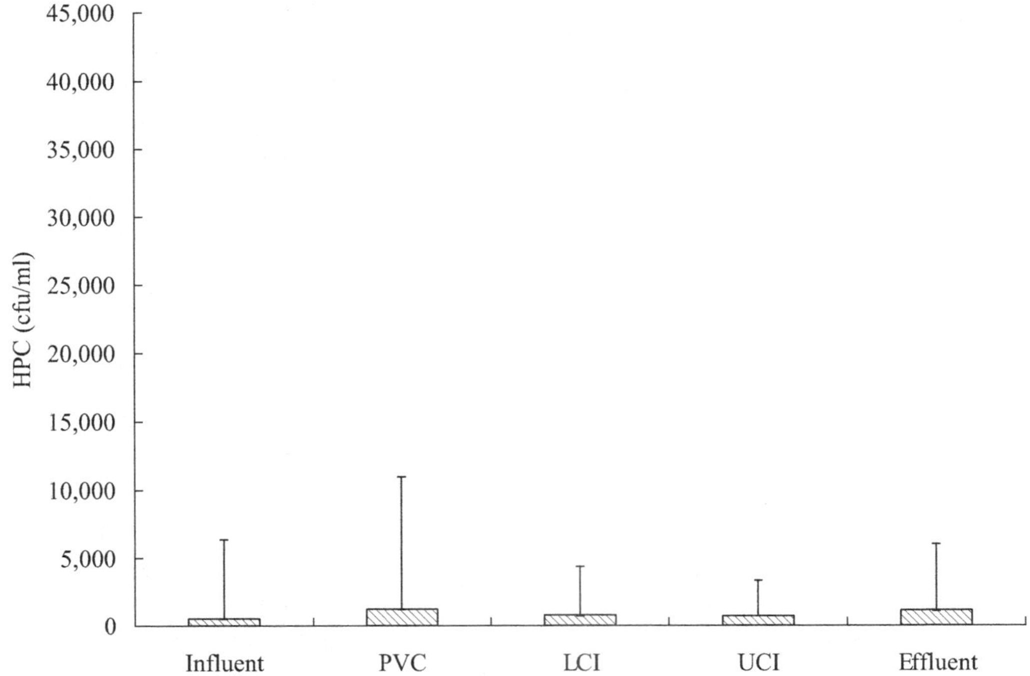

Figure 5.177 Average HPC exiting a particular pipe material for control PDS (PDS13)

Table 5.160
Paired t-test for pipe material segment log HPC (cfu/mL) summary

	Influent	PVC	LCI	UCI	GS
Mean	2.62	3.03	3.06	3.11	3.10
Standard deviation	0.64	0.62	0.60	0.58	0.52
Observations	135	135	135	135	135
p-value		1.5E–07	0.72	0.46	0.84
Significant difference?		Yes	No	No	No

NOTE: The t-test was done on the influent and effluent of PVC, LCI, UCI, and GS material pipes.

HPC model in that residual and temperature were consistently shown to be significant while biofilm density dropped out of the regression and was not significant. However, the increases in the PVC section HPC occurred even when there was very significant residual. It may be that the large increase in HPC through the PVC segment was largely because it was the first segment, and thus the first place where the influent water came in contact with the biofilm. Although bulk HPCs did not correlate with biofilms, the more than double average HPC density increase from influent to the end of the PVC segment is almost certainly not due to growth in the bulk phase.

Taking all the statistical comparisons in context of each other, it is unlikely that the internal port HPC data really quantifies material effects on bulk HPC. Even though the quantity of biofilm biomass sloughing off into the bulk liquid could be related to the biofilm density, the bulk HPC measurements may not reflect the flux of bacteria entering the bulk from the biofilm. The bulk HPC measurements will be significantly affected by the bulk phase disinfectant, resulting in inactivation of bulk phase HPC, particularly if the organisms are not part of a fairly large aggregation of cells. This explanation is consistent with the regression analysis for bulk HPC in which biofilm density was not demonstrated to be significant, but both residual and temperature were significant. The empirical model displayed the greatest sensitivity to bulk residual levels.

Biofilm Biostability

Water quality deterioration due to formation of biofilms in distribution systems is a practical problem. Biofilms are always present, can never practically be removed, and represent reservoirs for microorganisms in drinking water distribution systems. Drinking water utilities have to find a way to co-exist with biofilm growth. Distribution system biofilms can, in a well-managed system, be looked on as being in a pseudo-equilibrium steady state with a stable residual.

In developing an empirical model, the BFHPC data presented a significant challenge because the model would ideally address both pipe material and corrosion inhibitor effects. However, it was not appropriate to use two sets of dummy variables, as a total of 20 such dummy variables would have been required (four pipe materials and five inhibitor treatments). Segregation of the data into 20 subsets would produce undesirably small data sets to support parameter estimation procedures. The adopted models have dummy variables for material, but water quality parameters (i.e., silica, TP, Zn, pH) were evaluated for inclusion in the model to provide sensitivity to the inhibitors instead of dummy variables. In most cases, however, these water quality parameters were not significant and fell out of the model.

Biofilm HPC

BFHPCs were used as a general indicator of bacterial activity in biofilms. The following sections offer statistical analyses of their significance.

Statistical Comparison Between Different PDSs. Statistical analyses were conducted to determine the effects of inhibitors on biofilm activity. The following sections discuss this activity relative to controls and inhibitor type.

Statistical Comparison Between Control and Inhibitor PDSs. The experimental facilities allowed for side-by-side testing of 14 separate PDS hybrid lines (four pipe materials) that received the same finished water blend. The individual hybrid lines received a different corrosion control strategy consisting of an inhibitor at variant doses or pH control without addition of an inhibitor.

Given that PDS13–PDS14 were the control pipes, PDS01–PDS09 (phosphate-based inhibitors) and PDS10–PDS12 (Si-based inhibitors) are compared with them. The normality test failed in both comparisons and the Mann–Whitney rank sum test was used instead of a t-test. The results are shown in Tables 5.161 and 5.162.

The difference in the median values between the two groups in Table 5.161 is greater than would be expected by chance; there is a statistically significant difference ($p = 0.015$). PDSs with phosphate inhibitor had 0.34 log (119%) greater median values for BFHPC than the control PDSs. The distribution also showed that the lowest 25% of the values in PDS01–PDS09 were 0.44 log

Table 5.161
Comparison of log BFHPC (cfu/cm²) for PDS01–PDS09 with PDS13–PDS14

Group	N	Missing	Median	25%	75%
PDS01–PDS09	144	0	5.26	4.65	5.59
PDS13–PDS14	32	0	4.92	4.21	5.21

NOTE: "Missing" denotes the number of missing values for that column or group.

Table 5.162
Comparison of log BFHPC (cfu/cm²) for PDS10–PDS12 with PDS13–PDS14

Group	N	Missing	Median	25%	75%
PDS10–PDS12	48	1	5.28	4.55	5.76
PDS13–PDS14	32	0	4.92	4.21	5.21

NOTE: "Missing" denotes the number of missing values for that column or group.

higher than the same value in the control PDSs. The lowest 75% of the observed values in PDS01–PDS09 were 0.38 log higher than the control PDSs. Thus, through the entire distribution of the data, the phosphate-based inhibitors corresponded to a statistically significant increase in biofilm density.

The difference in the median values between the two groups in Table 5.162 is greater than would be expected by chance; there is a statistically significant difference ($p = 0.041$). PDSs with silica inhibitor had 0.36 log (129%) greater median values than the control PDSs. The distribution also showed that the lowest 25% of the values in PDS01–PDS09 were 0.34 log higher than the control PDSs, and the lowest 75% of the values were 0.55 log higher. Again, the addition of inhibitors corresponded to a significantly greater biofilm density, and the increase with Si-based inhibitors was at about the same level as was observed with phosphate-based inhibitors.

The data was also examined for a possible difference between phosphate-based inhibitor PDSs and Si-based inhibitor PDSs. The results are shown in Table 5.163. From the table, the difference in the median values between the two groups is not great enough to exclude the possibility that the difference is due to random sampling variability; there was not a statistically significant difference ($p = 0.579$). The median value of PDS01–PDS09 was only 0.02 log lower than PDS10–PDS12. At 25% level, the difference in the median value was 0.10 log, and at the 75% level, it was 0.17 log difference. The biofilm density with phosphate-based inhibitors was about the same level as Si-based inhibitors, and there was no statistically significant difference between the biofilm densities observed.

Statistical Comparison Between Phosphate-based Inhibitor PDSs. PDS01–PDS09 (phosphate-based inhibitors) received three different corrosion control chemicals consisting of an inhibitor at various doses. PDS01–PDS03 received BOP inhibitor, PDS04–PDS06 received OP, and PDS07–PDS09 used ZOP. The differences between these phosphate-based inhibitors were evaluated using statistical analysis. The results are shown in Table 5.164.

The differences in the median values among the treatment groups were not great enough to exclude the possibility that the differences were due to random sampling variability; there was not

Table 5.163
Comparison of log BFHPC (cfu/cm²) for PDS01–PDS09 with PDS10–PDS12

Group	N	Missing	Median	25%	75%
PDS01–PDS09	144	0	5.26	4.65	5.59
PDS10–PDS12	48	1	5.28	4.55	5.76

Table 5.164
Friedman repeated-measures ANOVA on ranks of log BFHPC density (cfu/cm²) for different inhibitors

Group	N	Missing	Median	25%	75%
BOP	48	0	5.25	4.37	5.56
OP	48	0	5.34	4.71	5.63
ZOP	48	0	5.24	4.85	5.59

a statistically significant difference ($p = 0.621$). The biofilm density was not shown to vary significantly among the different phosphate-based inhibitors.

Biofilm Density Growth on Different Doses of Inhibitors. The averages of biofilm density in four phases for different materials are shown in Figures 5.178–5.181 (BOP, OP, ZOP, and Si inhibitors, respectively). The graphs do not show any upward or downward slope with dosage for any of the inhibitors except for the silica inhibitor. For the Si inhibitor, biofilm density decreased with the inhibitor dose increases on PVC material.

Biofilm Modeling. *Data Summary.* An empirical model was developed using the entire data set (all phases and all PDSs). The objective of the model was to quantify the impact of water quality on the biofilm density in the distribution system. Dummy variables (PVC, LCI, UCI, and GS) for each pipe material were incorporated into the model. Water quality parameters, including influent dissolved zinc, total phosphate, and silica, were used to evaluate the effect of inhibitors on BFHPC. The use of dummy variables allowed estimation of a single parameter that is associated with each of the four materials.

Model Development. The BFHPC model was developed using a methodology that differed from that used to obtain the metal and bulk HPC models given that replicate experiments (i.e., duplicate coupons incubated in parallel) were available for biofilm samples. This allowed direct estimation of the experimental error (also referred to as the error mean square) for use in the ANOVA.

Four kinds of materials (PVC, UCI, LCI, and GS) were used as the dummy variables instead of the inhibitors and pH control lines (BOP, OP, ZOP, Si, and pH control) as was done with bulk HPC and other water quality parameters. The empirical models presented in this report are intended to provide information regarding the benefit of inhibitor addition for control of biofilms. Inhibitor factors are presented in the biofilm model as water quality parameters because dummy variables are required for the pipe materials and there are problems with having multiple sets of dummy variables. The full model was set up first with all the dummy variables and all the independent water quality parameters. The water quality parameter that failed the hypothesis test (F statistic lower than the calculated or tabular F value) by the greatest amount was eliminated

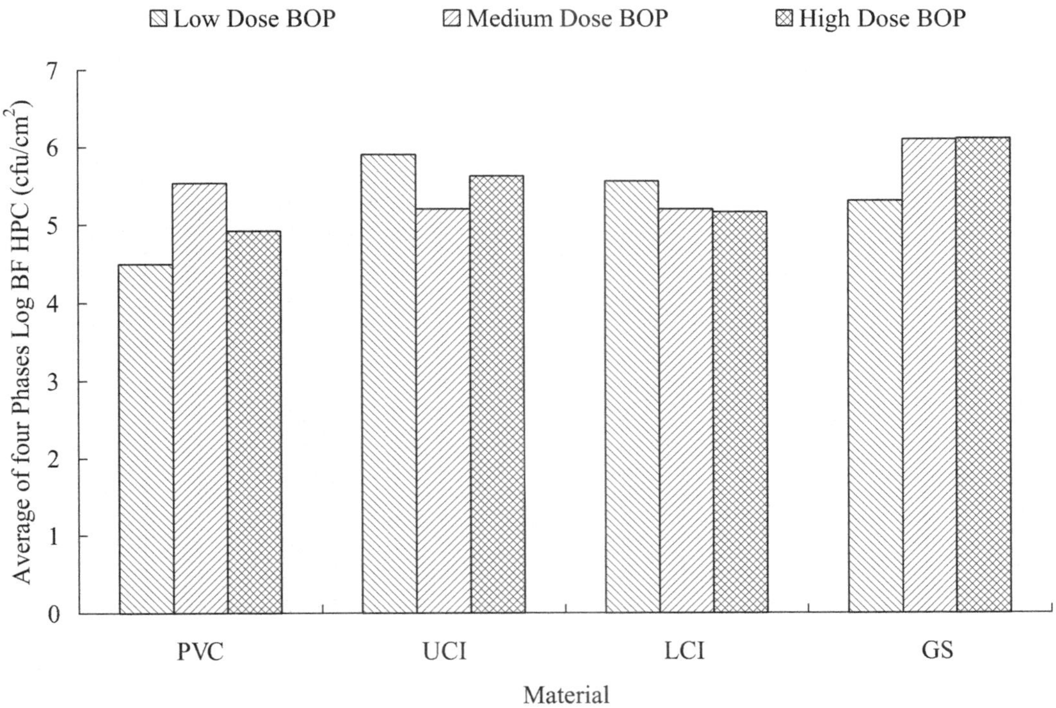

Figure 5.178 Biofilm density of four kinds of materials on different doses of BOP inhibitor

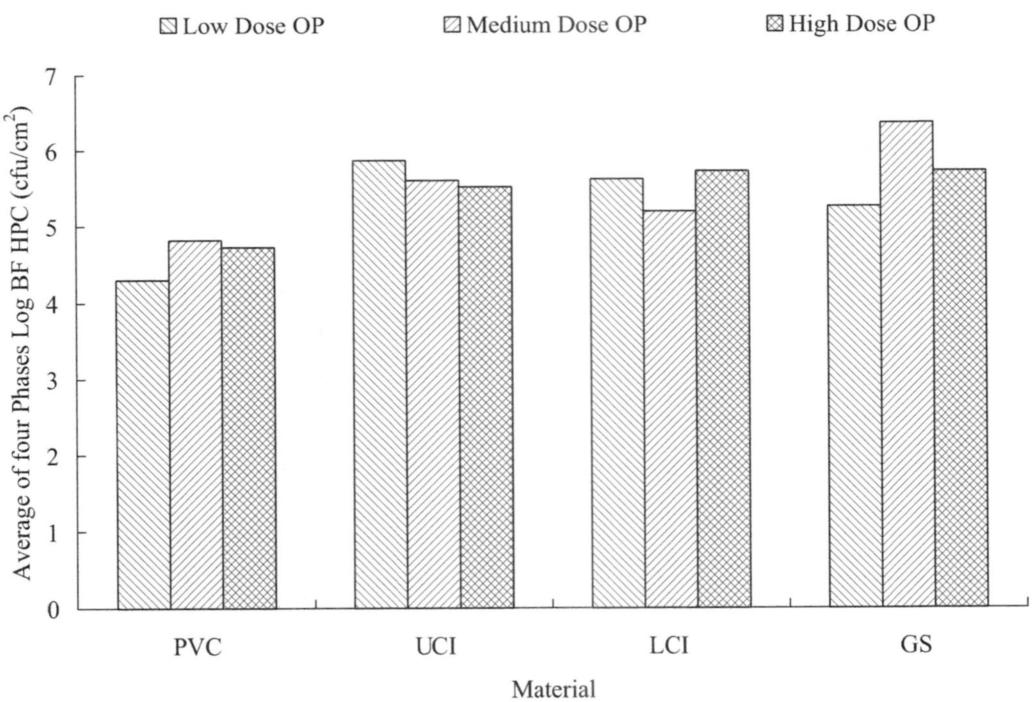

Figure 5.179 Biofilm density of four kinds of materials on different doses of OP inhibitor

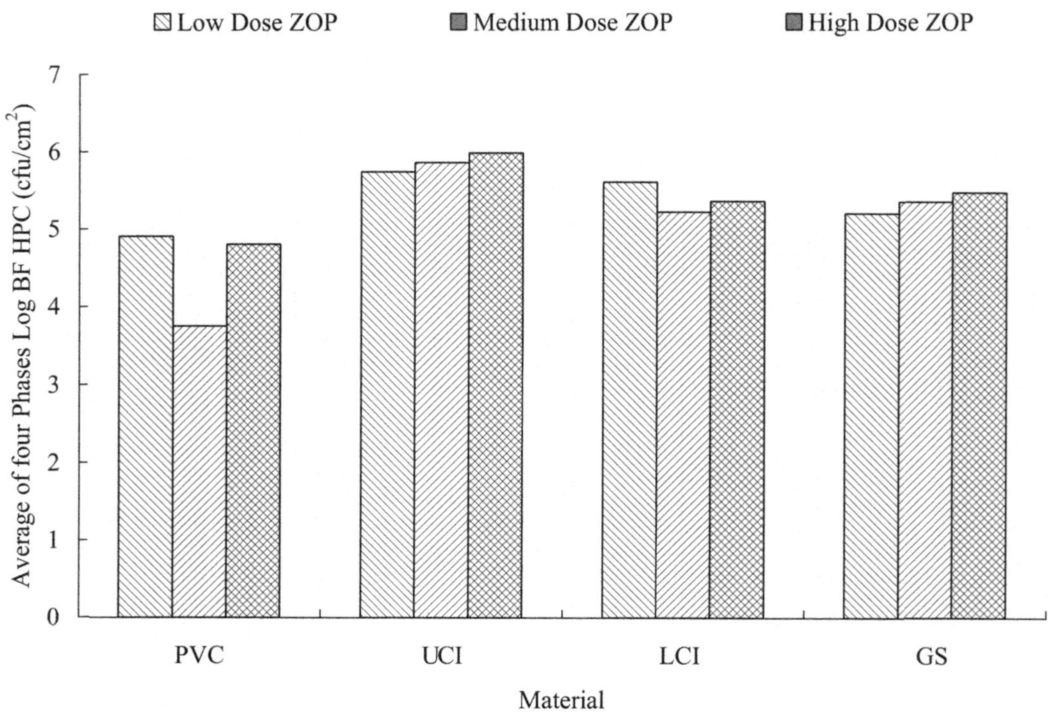

Figure 5.180 Biofilm density of four kinds of materials on different doses of ZOP inhibitor

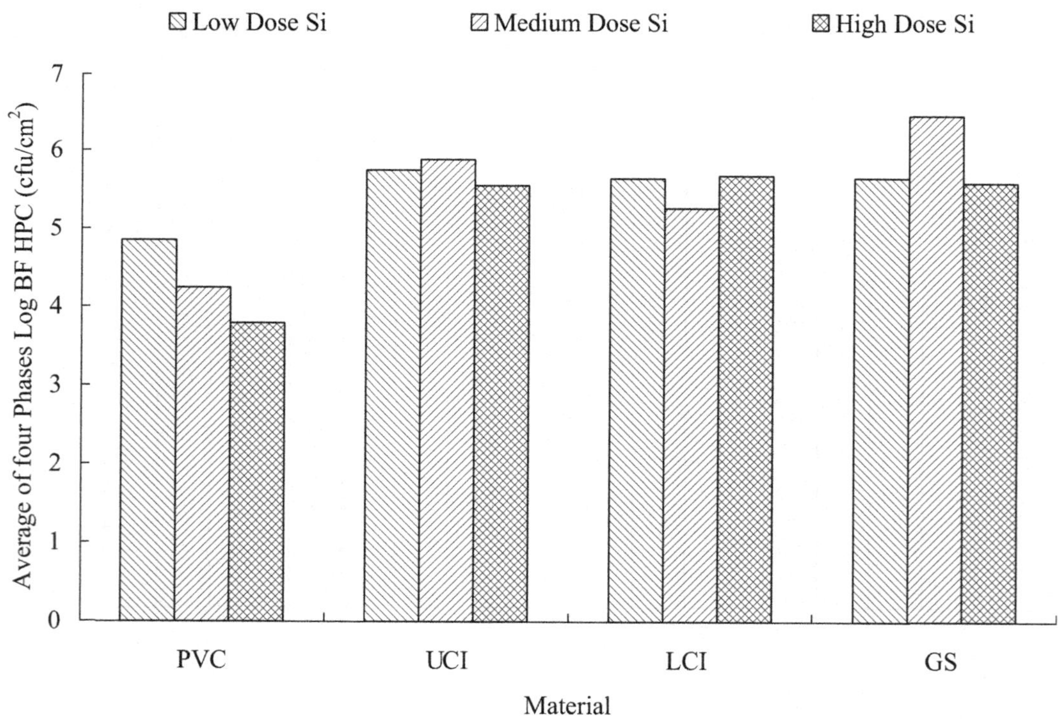

Figure 5.181 Biofilm density of four kinds of materials on different doses of Si inhibitor

from the model. The iteration continued until all the parameters remaining in the model passed the hypothesis test. From the previous discussion of hypotheses tests, phosphate inhibitor PDSs and silica inhibitor PDSs were significantly different from pH control PDSs. Two different assemblies of data were used for model development: segregated PDS data sets (PDS01–PDS09, PDS10–PDS12, and PDS13–PDS14) and a composite data set with data from all the PDSs. These groupings are subsequently described as segregated and pooled data. All remaining independent variables (i.e., material, temperature, residual, and water quality parameters) were statistically significant at a 95% level of confidence in the reported models.

BFHPC Model for PDS01–PDS09. An empirical model for log BFHPC with PDS01–PDS09 data from all phases was developed and is shown in Equation 5.89. PDS01–PDS09 were fed with blended water and different phosphate inhibitors (PDS01–PDS03 BOP lines, PDS04–PDS06 OP lines, and PDS07–PDS09 ZOP lines).

$$\log \text{BFHPC} = (10.19 \times \text{PVC} + 13.37 \times \text{UCI} + 12.33 \times \text{LCI} + 13.21 \times \text{GS}) \times \frac{1.034^{(\text{temp}-20)}}{\text{TCl}_{2-\text{inf}}^{0.640}} \tag{5.89}$$

where PVC, UCI, LCI and GS = dummy variables for four kinds of materials (polyvinyl chloride unlined cast iron, lined cast iron, and galvanized steel)
temp = average bulk water temperature during incubation, °C
$\text{TCl}_{2\text{-inf}}$ = average bulk water influent total chlorine during incubation, mg/L

Model BFHPC increased as temperature increased and decreased as residual increased. Based on the dummy variable coefficients, BFHPC densities differ significantly from one material to another. The highest densities were associated with the unlined metals GS and UCI. PVC had a much lower density than the three other materials (UCI > GS > LCI >> PVC). Total phosphate dropped off from the model, although PDS01–PDS09 were fed by phosphate inhibitor. This implied that there was probably not a consistent trend among the phosphorus inhibitors of BFHPC density varying significantly with dosage, which was shown visually in Figures 5.178–5.181.

Predicted versus actual log BFHPC is shown in Figure 5.182 using Equation 5.89. The predicted log BFHPC is from 3.5 to 6 log, whereas actual project log BFHPC is from 1 to 7 log. The lowest data set, 74.3% (26/35), came from PVC material and 22.9% (8/35) came from LCI; only one came from the GS. The six outliers in the circle shown in the graph were from phases I and IV, which had the incubation time during November to March with lower temperature. All of the models with dummy variables for material inherently underpredict high values and overpredict low values because of the dominance of the dummy variable term in the models. This mathematical artifact was avoided in the pooled data set analysis by doing the model analysis for each material separately; but with the smaller data sets in the segregated models, this was not attempted.

Model for PDS10–PDS12. An empirical model for log BFHPC with PDS10–PDS12 data for all phases was developed and is shown in Equation 5.90. PDS10–PDS12 was fed with blend water and different dosages of silica inhibitor.

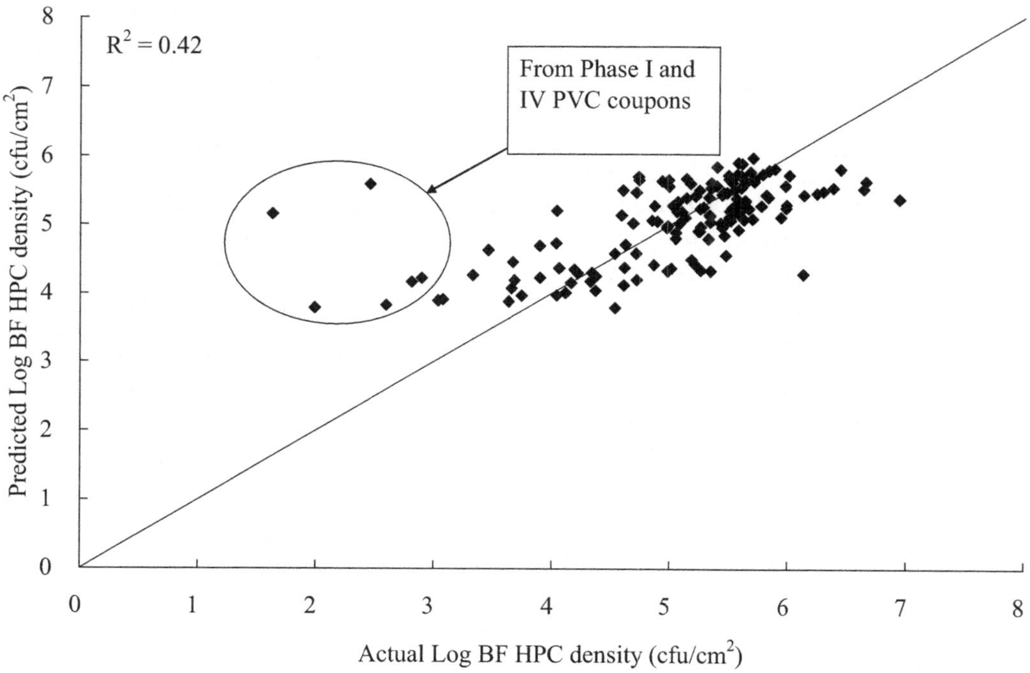

Figure 5.182 Predicted log BFHPC versus actual project log BFHPC for PDS01–PDS09 model

$$\log \text{BFHPC} = (16.70 \times \text{PVC} + 22.70 \times \text{UCI} + 22.08 \times \text{LCI} + 23.57 \times \text{GS}) \times \frac{1.039^{(\text{temp}-20)}}{\text{TCl}_{2-\text{inf}}^{\;0.828} \times \text{Si}_{\text{inf}}^{\;0.077}} \tag{5.90}$$

where PVC, UCI, LCI, and GS = dummy variables for four kinds of materials (polyvinyl chloride, unlined cast iron, lined cast iron, and galvanized steel)
temp = average bulk water temperature during incubation, °C
TCl_{2-inf} = average bulk water influent total chlorine concentration during incubation, mg/L
Si_{inf} = average bulk water influent silica concentration during incubation, mg/L

BFHPC increased as temperature increased and decreased as residual increased also, just as the model for PDS01–PDS09 did. However, silica (the inhibitor) was significant (unlike phosphorus in PDS01–PDS09), implying that the dose of Si inhibitor may have had a significant impact on BFHPC density, with increasing dosage decreasing BFHPC values. It was shown in Figure 5.181 that biofilm density decreased with silica inhibitor dose increase for PVC material, but the other materials did not show an obvious trend. BFHPC densities were higher on unlined metals, again, with a hierarchical sequence of GS > UCI > LCI > PVC.

Predicted versus actual log BFHPC is shown in Figure 5.183 using Equation 5.90. The predicted log BFHPC was from 2.99 to 7.03 log, whereas actual project log BFHPC was from 3.58 log to 6.00 log.

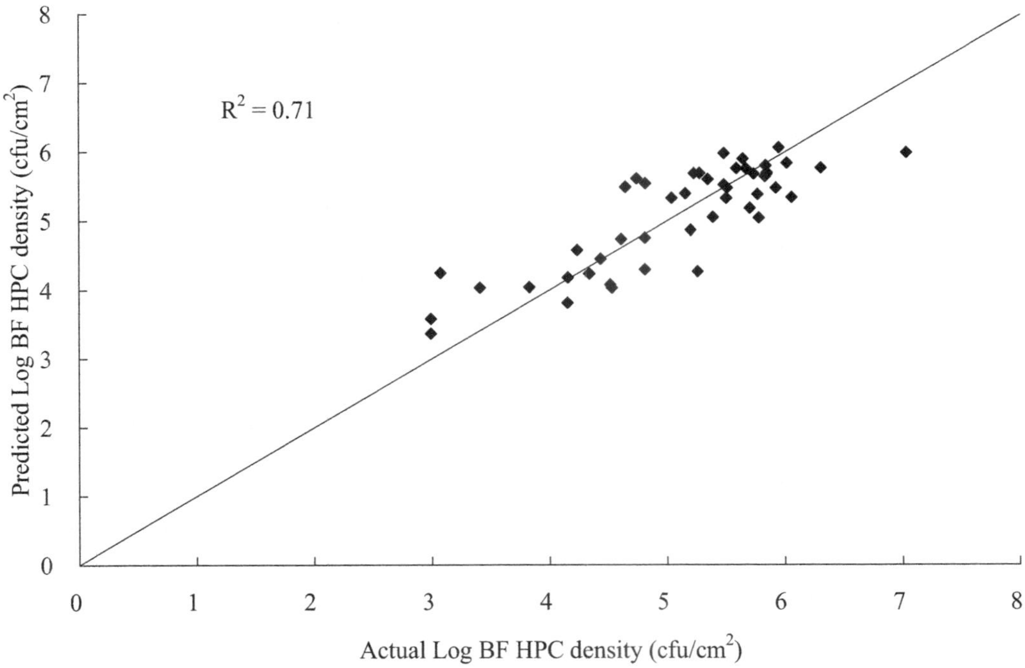

Figure 5.183 Predicted log BFHPC versus actual project log BFHPC for PDS10–PDS12 model

Model for PDS13–PDS14. An empirical model for log BFHPC with PDS13–PDS14 data for all phases was developed and is shown in Equation 5.91. PDS13 was operated at pH_s and PDS14 at pH_s+0.3. They were both control lines, and no corrosion inhibitors were added to the influent.

$$\log \text{BFHPC} = (3.76 \times \text{PVC} + 4.66 \times \text{UCI} + 4.78 \times \text{LCI} + 5.13 \times \text{GS}) \times \frac{\text{TCl}_{2-\text{inf}}^{\ 0.208}}{\text{Si}_{\text{inf}}^{\ 0.140}} \qquad (5.91)$$

where PVC, UCI, LCI and GS = dummy variables for four kinds of materials (polyvinyl chloride, unlined cast iron, lined cast iron, and galvanized steel)
Temp = average bulk water temperature during incubation, °C
$TCl_{2\text{-inf}}$ = average bulk water influent total chlorine concentration during incubation, mg/L

The nonlinear least-squares parameter estimation for an initial model that included both temperature and residual as independent variables did not converge. As a result, temperature was eliminated during development of the model. Influent residual in the model had the opposite effect compared to all the other models, with a counterintuitive increase in biofilm density as residual increased. The possible reason is that temperature and influent residual were confounded (the correlation coefficient between temperature and residual in the data set for PDS13 and PDS14 was 0.926). Because temperature was excluded from the model, the residual term may represent both

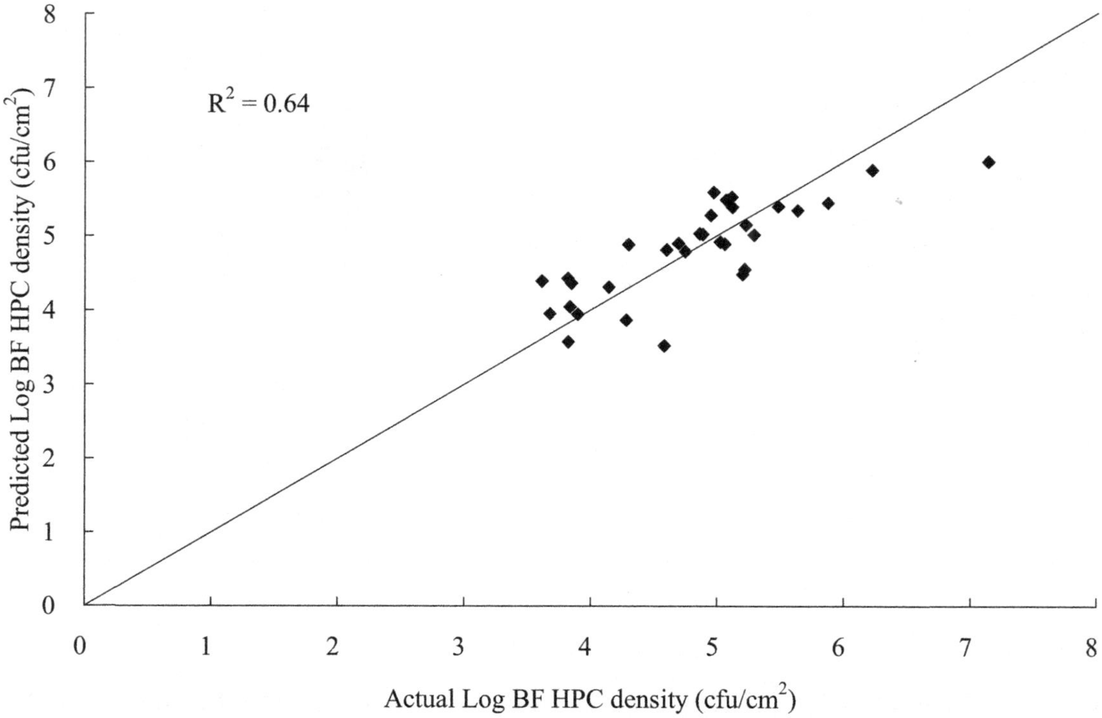

Figure 5.184 Predicted log BFHPC versus actual project log BFHPC for PDS13–PDS14 model

the effect of residual and temperature. Also, the data set of PDS13–PDS14 only included 32 data points, which may have not been sufficient to develop a meaningful model. Based on the dummy variable coefficients, PVC again has the lowest biofilm density, but LCI was actually slightly higher than UCI, which differs from the PDS01–PDS09 and PDS10–PDS12 models (the sequence was GS > LCI > UCI > PVC). Also, for reasons that are not clearly understood, influent silica stayed in the model even though no silica inhibitor was added to these lines.

Predicted versus actual log BFHPC is shown in Figure 5.184 using Equation 5.91. The predicted log BFHPC was from 4.38 log to 5.99 log, whereas actual project log BFHPC was from 3.61 log to 7.13 log.

Model for All PDSs. An empirical model for log BFHPC with data for all PDSs for all study phases was developed and is shown in Equation 5.92. The coefficient matrix is listed in Table 5.165. As previously mentioned, this model is subsequently referred to as the "all-PDSs model." Although sensitivity of BFHPC to pipe material is provided by association of separate parameter estimates (see Table 5.165) for each pipe material, there is no sensitivity to inhibitor in the all-PDSs model. A single value of the coefficient "a" is recommended for all corrosion control strategies.

$$\log \text{BFHPC} = a \times \text{temp}^b \times \text{TCl}_{2\text{-inf}}{}^c \qquad (5.92)$$

where a = coefficients for different inhibitors (BOP, OP, ZOP, silica, and pH control)
b = coefficients for average bulk water temperature during incubation

Table 5.165
Coefficient matrix for all-PDSs log BFHPC model

Material	a	b	c
PVC	0.30	1.31	–0.94
UCI	2.68	0.25	–0.056
LCI	1.46	0.71	–0.62
GS	2.95	0.20	0.0004

c = coefficients for average bulk water influent total chlorine during incubation
temp = average bulk water temperature during incubation, °C
TCl_{2-inf} = average bulk water influent total chlorine during incubation, mg/L

The much lower PVC dummy variable coefficient (1.2 to 2.6 log lower) indicates the importance of material in determining biofilm density. From the dummy variable coefficients, unlined metals resulted in higher biofilm density though PVC resulted in the lowest biofilm density (GS > UCI > LCI > PVC). BFHPC increased as temperature increased and decreased as residual increased. For the GS material, the residual coefficient was a positive number, but it was very low, 0.0004, which is, for all practical purposes, equal to zero. It indicated that the residual had less effect on the biofilm density for galvanized steel, whereas the material impact on density was dominant. The same thing can be said for the other unlined metal, UCI.

PVC and LCI showed sensitivity to both temperature and residual in the range of 1.5 to 2.5 logs (Table 5.166). In contrast, the unlined metals had approximately 0.5 logs of variation in temperature and very little sensitivity to residual. This was because most of the predicted BFHPC value came from the much larger coefficient "a" (see Table 5.165) that the unlined metals had relative to PVC and LCI. The coefficient is a fitting parameter that is largely material specific. However, in the segregated models (PDS01–PDS09, PDS10–PDS12, PDS13–PDS14), the unlined metals showed slightly greater sensitivity to temperature and residual than PVC and LCI (Table 5.167). This seems to imply that it may be misleading to assign too much physical significance to the coefficient values from the regression models as they are influenced by the mathematical form of the equations in addition to the effect of the relevant independent parameters. In particular, the all-PDSs models used a different temperature term from the segregated models and cannot be directly compared.

Table 5.168 shows comparisons of model coefficients (not of sensitivity ranges, as shown in Table 5.167). For the segregated models, the coefficients were identical and suitable for relative comparisons. However, the all-PDSs model showed a superior fit using a different mathematical term for temperature effects; and, thus, the coefficient, even when it seems comparable to other model coefficients, is not. The main observation is that regardless of the model used, the constant coefficient for PVC was always lower (for the same model) than the corresponding coefficients for the other three materials.

The main advantage of the all-PDSs model is that it gives a good predicted versus actual log BFHPC for all the conditions studied, as shown in Figure 5.185 using Equation 5.92. The coefficient of determination R^2 quantifies the strength of the association of the predicted values to the actual values observed. The R^2 was 0.53, which is lower than ideal but within a reasonable range for microbial enumeration techniques covering a broad range of materials and water quality

Table 5.166
Sensitivity test for all PDSs for log BFHPC density (cfu/cm^2)

Parameter	Project values		User values			Model predicted values			Range (Maximum – minimum)
	Minimum	Maximum	Average	Maximum	Minimum	Average	Maximum	Minimum	
PVC									
Temp (°C)	10.4	29.7	21.18	24.3	15.6	3.74	4.48	2.51	1.97
TCl_2 (mg/L)	1.65	7.00	4.72	6.34	3.19	3.74	2.83	5.42	–2.59
UCI									
Temp (°C)	10.4	29.7	21.18	24.3	15.6	5.28	5.46	4.89	0.57
TCl_2 (mg/L)	1.65	7.00	4.72	6.34	3.19	5.28	5.19	5.40	–0.20
LCI									
Temp (°C)	10.4	29.7	21.18	24.3	15.6	4.89	5.40	3.93	1.46
TCl_2 (mg/L)	1.65	7.00	4.72	6.34	3.19	4.89	4.07	6.24	–2.17
GS									
Temp (°C)	10.4	29.7	21.18	24.3	15.6	5.36	5.50	5.04	0.46
TCl_2 (mg/L)	1.65	7.00	4.72	6.34	3.19	5.36	5.36	5.36	<0.01

Table 5.167
Summary of BFHPC density (cfu/cm²) range from sensitivity tests

Material	Model	Temperature range BFHPC	TCl_2 range BFHPC
PVC	All	1.97	–2.59
UCI	All	0.57	–0.20
LCI	All	1.46	–2.17
GS	All	0.46	0.002

Table 5.168
Coefficients summary for all log BFHPC density (cfu/cm²) models

Material	Model	Constant coefficient	Temperature	Residual	Si
PVC	All PDSs	0.30	1.31	–0.94	
	PDS01–PDS09	10.19	1.03	–0.64	
	PDS10–PDS12	16.70	1.04	–0.83	–0.077
	PDS13–PDS14	3.76	N/A	0.21	–0.14
UCI	All PDSs	2.68	0.25	–0.056	
	PDS01–PDS09	13.37	1.03	–0.64	
	PDS10–PDS12	22.70	1.04	–0.83	–0.077
	PDS13–PDS14	4.66	N/A	0.21	–0.14
LCI	All PDSs	1.46	0.71	–0.62	
	PDS01–PDS09	12.33	1.03	–0.64	
	PDS10–PDS12	22.08	1.04	–0.83	–0.077
	PDS13–PDS14	4.78	N/A	0.21	–0.14
GS	All PDSs	2.95	0.20	0.0004	
	PDS01–PDS09	13.21	1.03	–0.64	
	PDS10–PDS12	23.57	1.04	–0.83	–0.077
	PDS13–PDS14	5.13	N/A	0.21	–0.14

NOTE: N/A = When influent temperature data were included in the initial model, the model did not converge.

conditions. The predicted log BFHPC range was from approximately 2.86 to 6.85 log, whereas the actual BFHPC values ranged from 1.5 to 7.5 log. The visual trend of the predicted values versus the actual values was good, and the all-PDSs model seemed to work acceptably well as a tool for describing the data.

*Biofilm TC/*E. Coli

Biofilm density was quantified almost exclusively by BFHPC during the study, as biofilm coliform were only observed in phase III and phase IV samples, and a positive *E. coli* was never

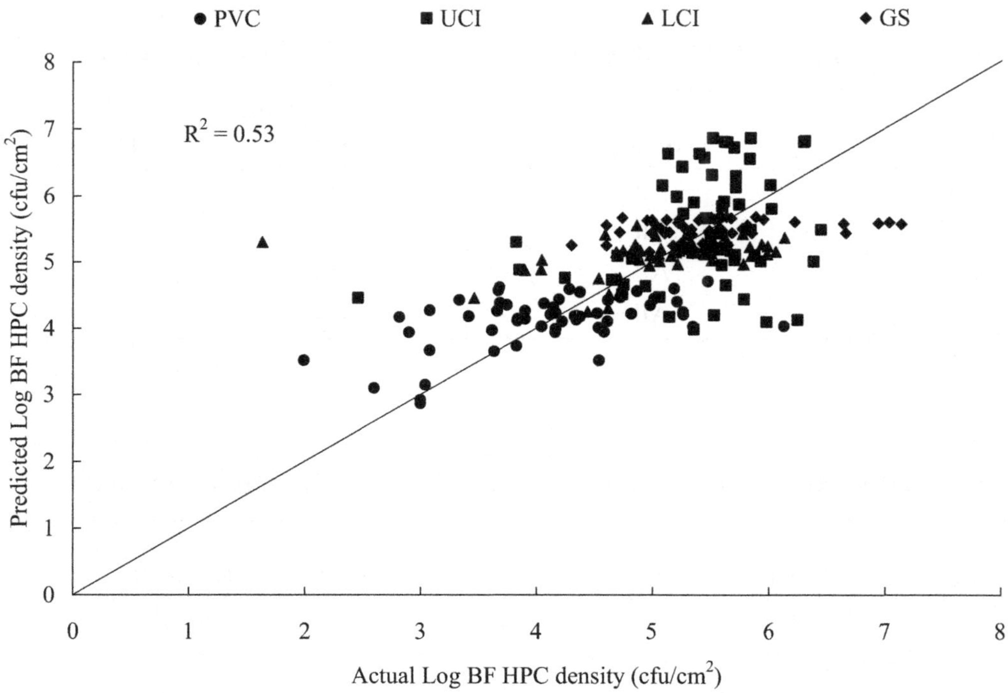

Figure 5.185 Predicted log BFHPC versus actual project log BFHPC for all-PDSs model

observed. The coupon material and the density of the coliforms observed in phases III and IV are listed in Table 5.169. The highest coliform density observed was 153 MPN/cm^2.

From Table 5.169, PDS04 had coliform growth on all four types of material coupons. In all cases, the density of coliforms was a very small fraction of the overall biofilm, being far less than 1% in all cases.

Relationship of Bulk HPC and Biofilm HPC

In the TBW II study, the average density of suspended cells in the PDS bulk water in all distribution systems was about 2.3 orders of magnitude less (comparing 1 mL to 1 cm^2) than the average density of biofilm cells. The average biofilm density was 5.67 log ($10^{5.67}$) cfu/cm^2, and the average effluent bulk HPC was 3.44 log cfu/mL. Empirical models were developed separately for characterization of the bulk and biofilm HPC, which evaluates the significance of BFHPC for the bulk HPC model and evaluates the significance of bulk HPC to the BFHPC model. The results of the regression analysis suggested that biofilm HPCs and bulk HPCs were independent, which is apparent in Figure 5.186 for a composite of all data, and in Figure 5.187 for PVC, Figure 5.188 for UCI, Figure 5.189 for LCI, and Figure 5.190 for GS. The highest R^2 (0.21) was from GS material and the lowest value (0.01) from UCI. Hence, there was poor correlation for BFHPC and bulk HPC for all materials.

Figure 5.190 shows the sensitivity of BFHPC and bulk HPC to temperature and residual terms in the corresponding regression models. There is significant variability associated with temperature or residual for both the biofilm and bulk (i.e., suspended or planktonic) HPC. BFHPC was typically more sensitive to temperature, whereas bulk HPC was more sensitive to residual.

Table 5.169
Observed coliform in TBW II phases III and IV

	PDS	Material	Most probable density on coupon (MPN/cm2)	BFHPC density (MPN/cm2)	Percentage
Phase III	04	PVC (only on duplicate coupon)	0.20	4.25E+03	0.0047
		LCI	0.17	2.44E+03	0.0070
		UCI	0.2	2.16E+05	0.0001
		GS	1.03	1.36E+05	0.0008
	06	GS	0.22	1.38E+05	0.0002
Phase IV	11	UCI	153	4.27E+05	0.0360
	14	GS	1.1	1.29E+05	0.0009

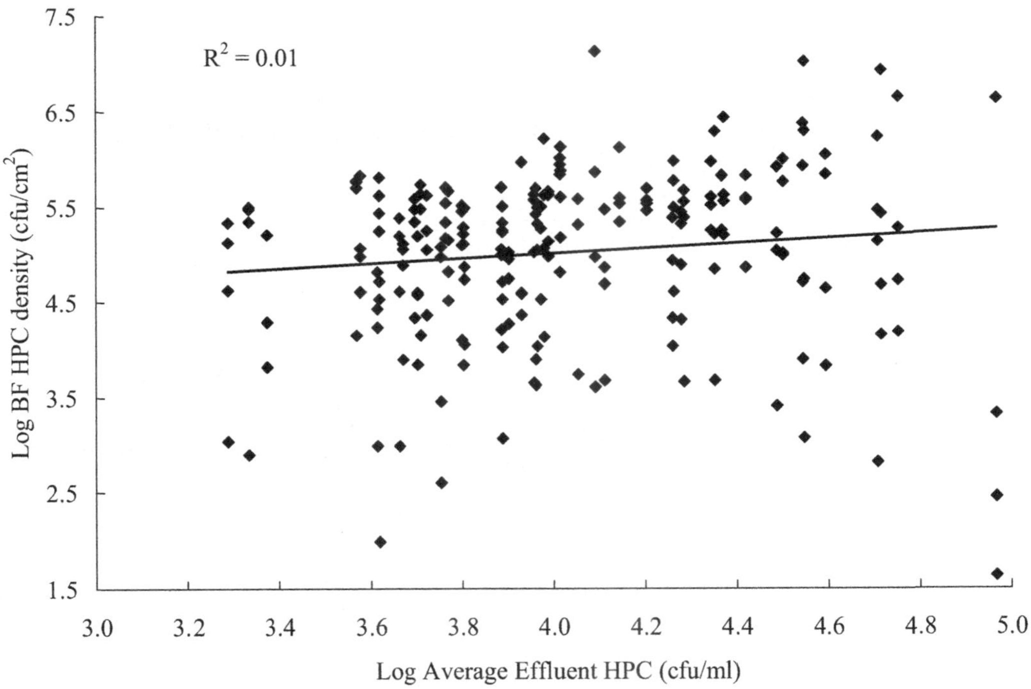

Figure 5.186 Scatter plot of log BFHPC and log average effluent HPC for all materials

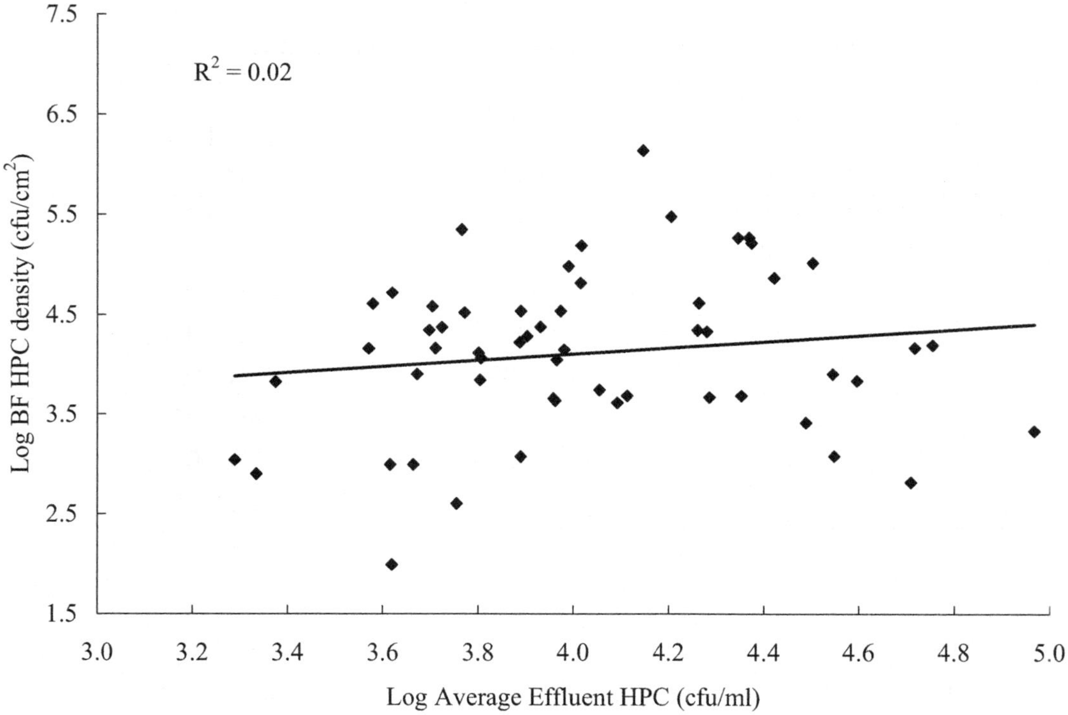

Figure 5.187 Scatter plot of log BFHPC and log average effluent HPC for PVC

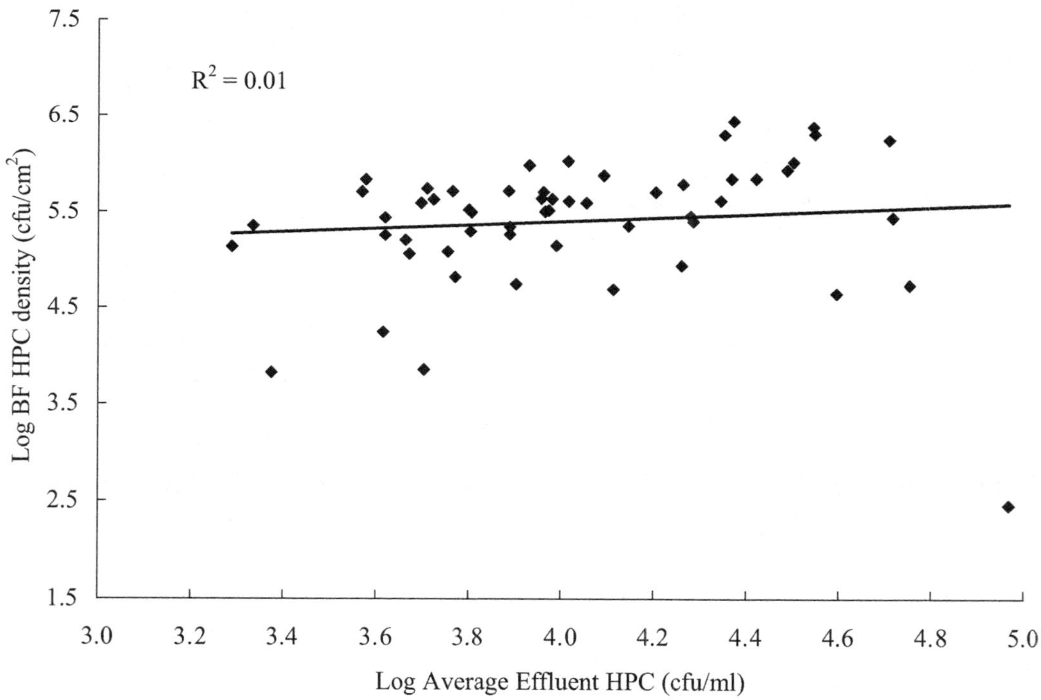

Figure 5.188 Scatter plot of log BFHPC and log average effluent HPC for UCI

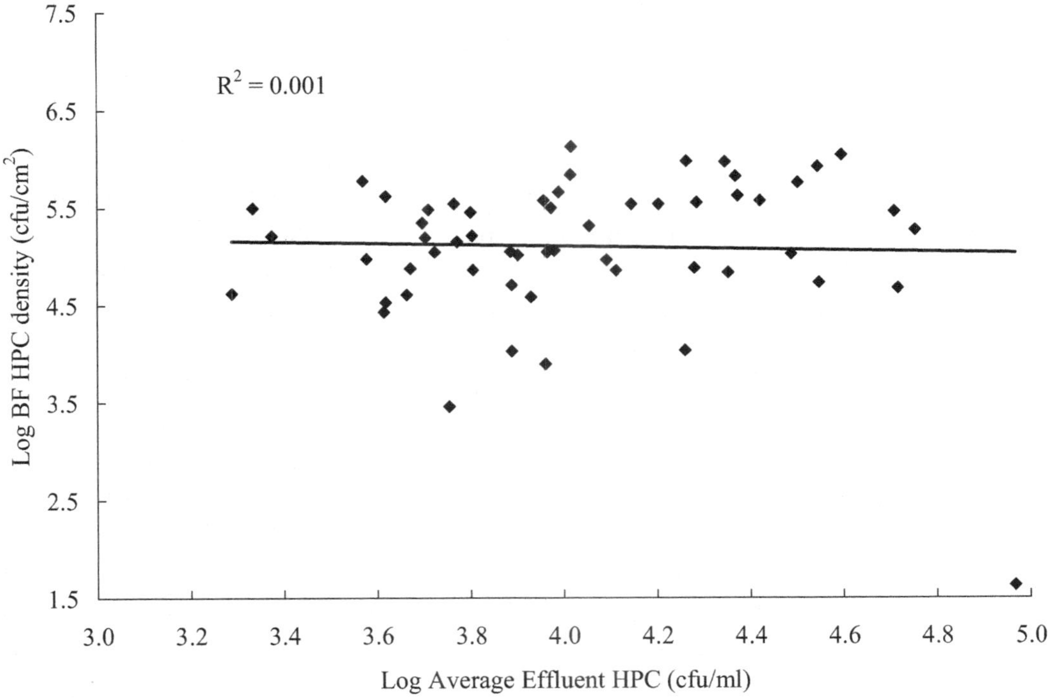

Figure 5.189 Scatter plot of log BFHPC and log average effluent HPC for LCI

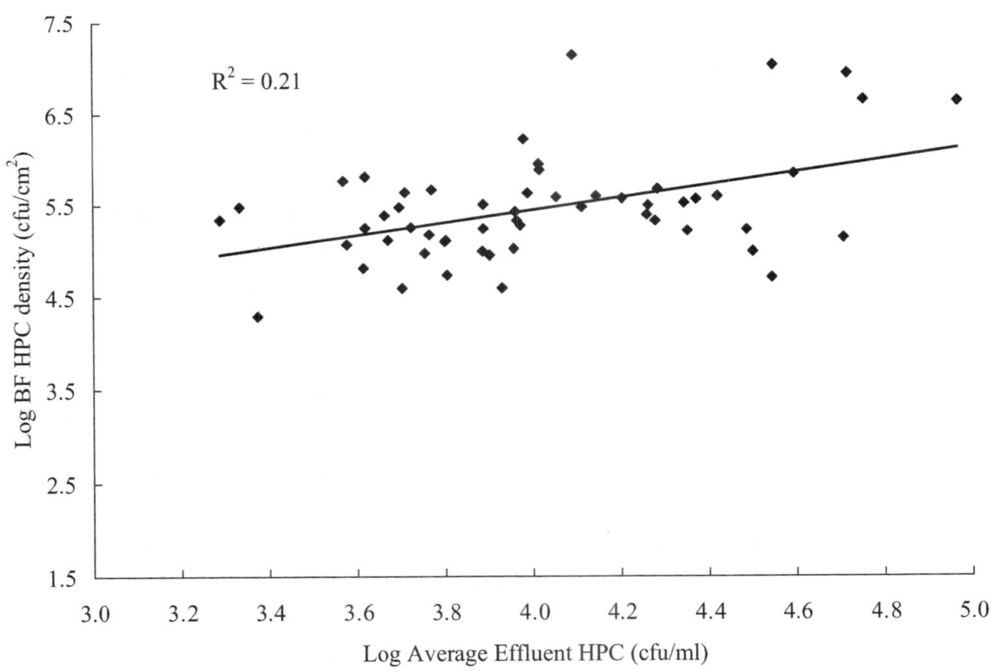

Figure 5.190 Scatter plot of log BFHPC and log average effluent HPC for GS

This is probably because many of the bacteria in the biofilms will be protected from residual because of diffusion limitations as the biofilm gets deeper.

Summary and Conclusions

All inhibitors increased bulk HPC and BFHPC significantly relative to the PDS controls (pH control with no inhibitor added). Phosphate-based inhibitors of all types resulted in a bulk effluent HPC value 0.42 log greater than the control PDSs (i.e., well over a factor of 2, 0.30 log being equal to 2 times the control HPC value), and this difference was statistically significant. Addition of the silica inhibitor resulted in a statistically significant 0.31 log greater median HPC relative to the control PDSs. The lower bulk HPC level observed with silica inhibitors versus phosphate-based inhibitors was significant (at a 93.5% confidence level). The addition of phosphate-based or Si inhibitors increased BFHPC values, and this increase was statistically significant. Similar to the effect on bulk HPC, biofilm density also increased by a factor of 2 or more (0.34 log for P-based inhibitors and 0.36 log for Si inhibitor).

Addition of inhibitors more than doubled the biofilm density and the concentration of HPC in the bulk liquid. The increase in bulk HPC could cause utilities to exceed the maximum allowable HPC of 500 cfu/mL, which is accepted in lieu of the minimum required residual (0.2 mg/L Cl_2 or 0.6 mg/L NH_2Cl as Cl_2) for surface waters. For example in Table 5.150, PDS13 and PDS14 had a mean bulk HPC equal to 2.73 log or 537 cfu/mL. Adding P-based inhibitors more than doubled the bulk HPC (3.15 log or 1,413 cfu/mL). In cases where the HPC value is below the 500 cfu/mL limit prior to inhibitor use, the results suggest that bulk HPC might more than double with addition of inhibitor and thus exceed 500 cfu/mL. This would be true of dead ends and other vulnerable areas, assuming that the distribution system of concern had similar limiting nutrients as the PDS used in this study. This means the impact would be system specific in all likelihood.

Bulk HPC was affected consistently by influent residual and temperature, and was most sensitive to residual. Figure 5.191 shows the predicted bulk HPC using the all-PDSs model with temperature assumed to be constant at the average value for the study. BOP is shown because it had the highest predicted bulk HPC, whereas the pH_s PDS had the lowest. In Figure 5.191, A is 2.8 log cfu/mL HPC at 0.75 mg/L residual, which is a typical operational condition prior to inhibitor addition. The increase in HPC following the addition of 0.5 to 2.0 mg/L P-BOP at the same residual concentration is represented by B, which corresponds to 3.8 log, or a 1.0 log HPC increase. Instead, it is more desirable to increase the residual from 0.75 to 3.8 mg/L in order to maintain the original HPC level. This determination is shown in Figure 5.191 by the horizontal line from A to C and then by tracing a vertical line down from C to the x-axis.

BFHPC was affected by pipe material, effluent residual, and temperature in addition to the relatively small increase due to inhibitor addition. It was most affected by material type, with PVC biofilm density consistently much lower than the other materials. Unlined metals consistently had the highest biofilm densities. After material, temperature was the dominant parameter affecting biofilm densities for the unlined metals (UCI and GS). In contrast, for PVC and LCI, BFHPC had a similar dependence on residual and temperature, with residual being slightly more important than temperature. The lack of residual impact observed for the unlined metals was probably because of diffusion limitations of the chloramine residual when penetrating the denser UCI and GS biofilms.

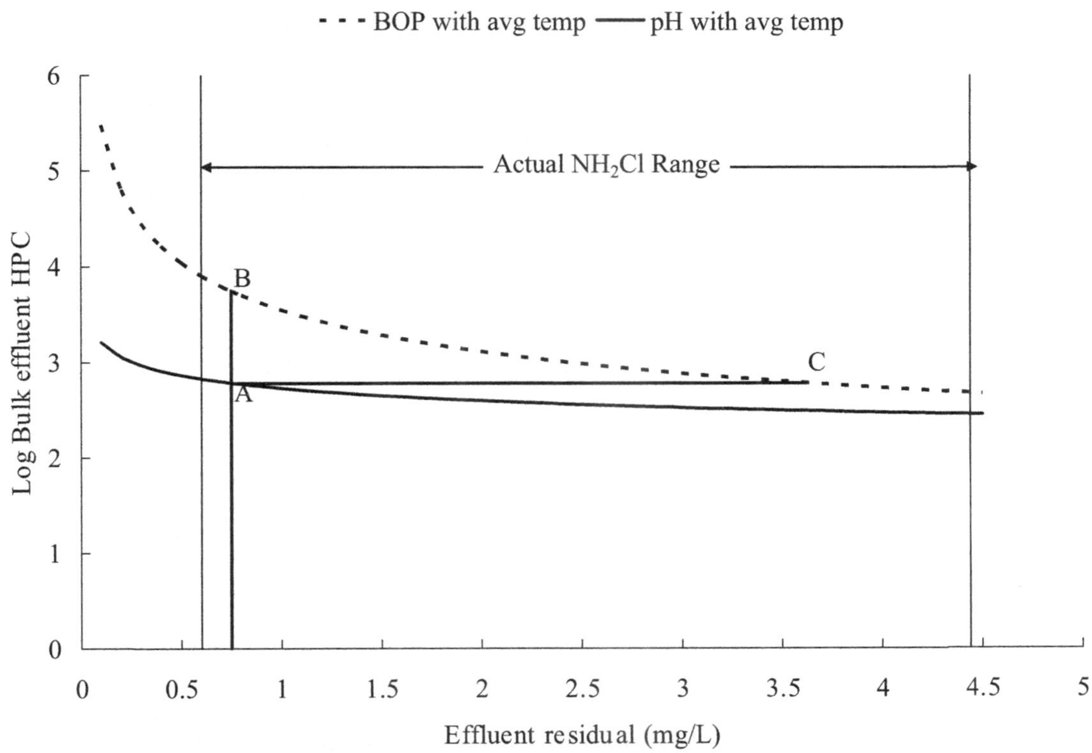

Figure 5.191 Log bulk effluent HPC versus effluent chloramine residual

For BFHPC, the model implied that the effect of material, temperature, and residual were more significant than the increase observed due to inhibitor addition. Unlined metals showed a small but significant dependence on temperature, but little or no effect due to residual. This difference in behavior of the unlined metals versus PVC and LCI is illustrated in Figure 5.192, which shows the sensitivity of the PVC and UCI biofilm to residual. The data implies that the less dense PVC and LCI biofilms were sensitive to residual though the denser UCI and GS biofilms were not. If this was due to diffusion limitations because of the thickness or density of the biofilm, then flushing the biofilms should make them more sensitive to residual, as the shear forces should remove some of the upper biofilm. In summary, residual should be a viable control strategy for lined pipe and PVC. For unlined metals, flushing, pigging, and other strategies to make the biofilms less thick or dense should be used in conjunction with maintaining a significant residual.

Phosphate- or silica-based inhibitors affected both bulk HPC and BFHPC. Supplemental experiments were conducted to determine why inhibitor addition affected HPC levels. The experiments measured the growth response of P17 and NOX (the same organisms used in the AOC method) when spiked with the inhibitors or with a phosphate salt. The results showed that, at least part of the time, phosphorus, all phosphate-based inhibitors, and the Si inhibitor could stimulate very significant increases in growth. These experiments, though not conducted for a large number of samples, confirmed that phosphorus addition could sometimes stimulate very significant growth, and occasionally an unknown nutrient in the Si inhibitor also stimulated biological growth. The observed higher HPCs in inhibitor lines, plus the growth response observed in some

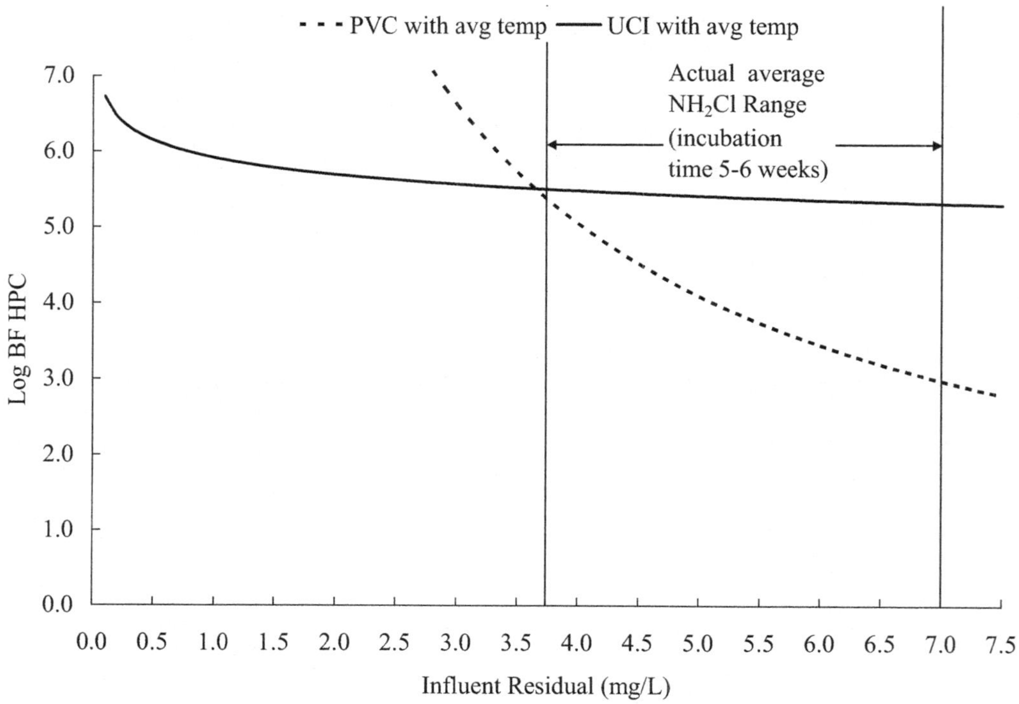

Figure 5.192 Details of predicted PDS log BFHPCs versus chloramine residual at average (23.6°C) temperature

of the supplemental experiments, imply that phosphorus was a limiting nutrient for the water blends studied, and that there may have been other trace nutrients (organic or inorganic) in the Si PDSs that were not specifically identified.

Coliforms were detected in the biofilm with very low frequency (7 samples of a total of 252 samples, 2.8%) and were less than 0.03% of the biofilm in the highest observation of the study. AOC did not change significantly as the bulk water passed through the PDSs.

CHAPTER 6
CONCLUSIONS AND RECOMMENDATIONS

CONCLUSIONS

Conclusions are organized by areas of investigation in the report.

Metal Release

Conclusions within areas of investigation are further subdivided by specific metals.

Control of Iron

Satisfactory control of iron release from UCI and GS pipe materials will require a combination of management of hydraulic conditions and adjustment of water chemistry by alkalinity supplementation, pH elevation, and/or inhibitor addition. Supplementation of alkalinity as currently practiced for SW and RO sources was confirmed to have benefit to reduce iron concentrations. Although the corrosion control strategies that were investigated (pH elevation and inhibitor addition) were also demonstrated to have merit, the mitigation of iron release was modest. The hybrid PDS operation (2-day HRT) experienced a statistically significant increase in iron concentration for all blends and all corrosion control strategies. Elimination of distribution system dead zones and associated extreme HRTs would be necessary to prevent red water episodes.

It is recommended to study the effect of longer incubation time of coupons with addition of inhibitors (more than 3 months) to examine whether a longer treatment period causes variability of scale uniformity or composition. It is also recommended to consider wider ranges of water quality (i.e., alkalinity and pH) than was maintained in this study to examine how scale formation is affected and how consistent the scale's ability is to continually retard iron release.

Control of Copper

The experimental results for copper concentration demonstrate pronounced benefits associated with all of the corrosion control strategies that were investigated. Operation at pH_s in the absence of inhibitor addition (PDS13) resulted in consistent violation of the copper standard. Although operation at an elevated pH would be successful for compliance with the copper standard, the extreme pH conditions that would be required (pH > 8.0) may produce unacceptable scaling conditions. All of the inhibitors were shown to produce compliance with the relevant copper standard for all water quality blends. All of the phosphate-based inhibitors (BOP, OP, and ZOP) produced similar results for a comparable inhibitor total phosphorus dose. Even though the performance with the minimum dose of the Si inhibitor (3 mg/L SiO_2) was inferior to performance with the minimum dose of the phosphate-based inhibitors (0.5 mg/L TP), an increase in the Si dose could be practiced to achieve comparable performance. These results are significant, as the performance with inhibitors indicates that it would be possible to achieve compliance with copper standards for elevated alkalinity conditions without reliance on an elevated pH.

Control of Lead

Elevation of pH achieved compliance with lead standards, but operation at a large positive LSI may produce unacceptable calcium carbonate scale formation. The use of all of the inhibitors achieved control of lead concentration for all of the water quality blends employed in the study for pH conditions that would not promote deposition of calcium carbonate scale. All of the inhibitors were effective, although the ZOP and BOP inhibitors provided somewhat better results at the minimum inhibitor dose than either the OP or Si products. Increases in the dose of OP or Si would improve results to comparable conditions obtained with ZOP or BOP.

Electronoise

Models for the release of total and dissolved iron, copper, and lead were developed using EN parameters and were as accurate as models developed from water quality data for the prediction of total or dissolved iron, copper, and lead release. EN monitoring is faster and less labor intensive than water quality monitoring and represents a significant advance for controlling metal release in drinking water distribution systems.

Iron, copper, or lead release was most accurately described by metal release models using ECNCR, HMCR, LPRCR, or PF. Both general (uniform) and localized (pitting) mechanisms affected the release of iron, copper, and lead.

The EN models are comparable to water quality models for metal release, and because EN is a real-time technique, it offers a tremendous advantage over traditional water quality sampling techniques. Remote access of EN monitoring equipment is possible, and the system requires little to no maintenance, with the exception of a power supply or battery. The rapid turnaround of corrosion rates from EN can be used to proactively estimate metal release in drinking water, and mitigating measures can be implemented before the full adverse impacts are realized.

Identification and Modeling of Corrosion Scale Products

Conclusions pertaining to the identification and modeling of corrosion scale products are organized by metal to aid understanding.

Iron

Surface composition of iron scale on iron coupons exposed to varying inhibitors was analyzed using survey and high-resolution XPS spectra scans. Ferric oxide (Fe_2O_3), magnetite (Fe_3O_4), and hydrated ferric oxide (FeOOH) were the major iron corrosion products detected in the surface composition of the iron scale, which was independent of phosphate- or silica-based inhibition. However, ferric phosphate ($FePO_4$) was detected on the majority of coupons exposed to phosphate inhibition, which was attributed to addition of the inhibitors. This indicates that the formation of solid phosphate scale ($FePO_4$) was a mechanism of reduced iron release in the presence of phosphate-based inhibitors. Silicate scale was detected on all coupons but to a greater extent on all coupons exposed to Si inhibitor; hence, the formation of silica scale was a mechanism for reduced iron release in the presence of silica-based inhibitors. No zinc scale products were identified in the scale of any coupon that indicated zinc in the ZOP inhibitor contributed to inhibition of iron or zinc release. The variation in water quality (alkalinity, sulfates, chlorides, pH, and temperature) by phase did not affect the composition of the surface scales on iron coupons; however, inhibitor type did affect the scale composition.

Thermodynamic models based on water, carbonate, phosphate, and silica species for siderite ($FeCO_3$), $FePO_4$, and $FeSiO_3$ were developed to represent pH only and the addition of phosphate- or silica-based inhibition relative to the description of dissolved iron release. The $FePO_4$ model grossly (10,000 times) underpredicted actual dissolved iron release for phosphate inhibition. The $FeSiO_3$ model also significantly underpredicted actual dissolved iron release. The $FeCO_3$ model slightly overpredicted actual iron release in the pH-only PDSs and slightly overpredicted actual iron release in the presence of silica and phosphate inhibition. It was the most descriptive model developed, which indicates iron release is controlled by $FeCO_3$ whether inhibitors are present or not, but it is significantly inhibited by the formation of phosphate- or silicate-based scales in the presence of phosphate- or silicate-based inhibitor.

Galvanized Steel

Surface characterization of GS coupons found that ferric oxide (Fe_2O_3), magnetite (Fe_3O_4), and hydrated ferric oxide (FeOOH) are the major corrosion products in the surface scale of the GS coupons. There was no evidence of correlation between Fe_2O_3, Fe_3O_4, and FeOOH surface compounds and iron release in this work; however, iron oxides were by far the greatest solid form of iron in the GS films and are undoubtedly involved in the release of iron from galvanized steel. Ferric phosphate ($FePO_4$) was detected on the majority of coupons exposed to phosphate inhibition; in contrast, none of the other coupons exposed to nonphosphate treatments had $FePO_4$. Therefore, addition of corrosion inhibitor changed the chemical composition of corrosion products and resulted in the formation of solid phosphate scale ($FePO_4$) which is a barrier to metal release from galvanized steel.

Variation of phase water quality (alkalinity of 84 to 175 mg/L as $CaCO_3$, chlorides of 35.4 to 122.9 mg/L, and total hardness of 136 to 252 mg/L as $CaCO_3$) did not affect the distribution of solid iron forms in the GS films, which indicates that water quality variations in this work would not adversely affect iron release from galvanized steel.

Scale on coupons exposed to Si and pH treatments did not contain any form of phosphorus. Silicate scale was detected on all coupons but was most abundant in the presence of silica inhibition and significantly inhibited iron release relative to all other methods of inhibition in this work.

Zinc deposits were present in the GS scale for most observations. A majority of zinc existed as zinc oxide (ZnO). Zinc hydroxide ($Zn(OH)_2$) was identified in less than half of all GS coupons. The lack of zinc phosphate ($Zn_3(PO_4)_2$) scale in the presence of ZOP indicates that zinc phosphate precipitation was not a factor in inhibiting iron release.

The same thermodynamic models that were developed to describe iron release from UCI or iron coupons are applicable to describe iron release from GS coupons. Different thermodynamic models were developed to describe dissolved zinc release from GS pipe surfaces assuming zinc carbonate ($ZnCO_3$) was the controlling solid phase in the presence of pH-only, phosphate-based, or silica-based inhibition. Models based on zinc, water, and carbonate equilibrium overpredicted actual zinc concentrations. Incorporation of zinc–phosphate complexes into the zinc, water, carbonate, and phosphate model was inconsequential. Models based on zinc, water, carbonate, and silica equilibrium underpredicted dissolved zinc release but reflected reduced zinc release as observed in the data.

Copper

Thermodynamic models were developed describing copper release assuming different solids that might be expected to form in copper pipes receiving no inhibitor, phosphate inhibitor, or silicate inhibitor. The thermodynamic models were developed using thermodynamic data for each solid and relating the predicted dissolved concentration of cupric ion, Cu^{2+}, to the cupric ion complexing ligands present in the water. The primary ligands responsible for increase in total copper release were all related to alkalinity, including $CuHCO_3^-$, $CuCO_3^0$, $Cu(CO_3)_2^{2-}$, $Cu(OH)CO_3^-$, and $Cu(OH)_2CO_3^{2-}$. When combined, these complexes accounted for approximately 90% of all soluble copper species. Thermodynamic models were also developed assuming $Cu_3(PO_4)_2{\cdot}2H_2O$ and $Cu_3(PO_4)_2$ as controlling solid phases.

For all silicate inhibitors, the cupric hydroxide ($Cu(OH)_2$) and cupric silicate ($CuSiO_3$) models provided the best prediction of copper release, with each model overpredicting by no more than about 0.3 mg/L Cu.

Survey and high-resolution XPS spectra were collected to identify the corrosion products of copper and determine other chemical compounds present. Cuprous oxide (Cu_2O), tenorite (CuO), $Cu(OH)_2$, and Cu(II) salts were detected as the predominant corrosion compounds in the scale formed on copper coupons in the presence of inhibitors. Cupric hydroxide was the most predominant curve for all inhibitor types. The addition of corrosion inhibitors did not affect the relative distribution of copper corrosion products as measured by the four copper compound curves.

Phosphate inhibitor addition was the primary source of phosphorus compounds on the copper surface. Phosphate was detected on many coupons that did not have any calcium phosphate ($Ca_3(PO_4)_2$) scales. The thermodynamic model for cupric phosphate dihydride, $Cu_3(PO_4)_2{\cdot}2H_2O$, predicted dissolved copper release similar to the reduced levels of copper release observed in the phosphate lines.

XPS analysis indicated the presence of a polymeric silica film on copper coupons in the presence of silica inhibition, which is consistent with the presence of $CuSiO_3$ in the surface scale. The thermodynamic model for cupric silicate hydride, $CuSiO_3{\cdot}H_2O$, predicted dissolved copper release similar to the reduced levels of copper release observed in the silicate lines.

Lead

Thermodynamic models were developed considering hydrocerussite ($Pb_3(OH)_2(CO_3)_2$), cerussite ($PbCO_3$), lead(II) oxide (PbO), and lead(IV) oxide (PbO_2) as controlling solid phases in the presence of pH, phosphate, and silica inhibition. Analysis of all the lead/tin coupons suggests that the predominant corrosion products for lead were $Pb_3(OH)_2(CO_3)_2$, $PbCO_3$, and PbO. PbO_2 was detected in more than half of the lead coupons; however, the fraction of lead(IV) species was less than 10% of the lead corrosion products.

Hydrocerussite and cerussite models provided similar predictions of lead release. Lead release was diffusion limited because of the lack of lead surface area in the copper corrosion loops, as evidenced by the overprediction of lead release by thermodynamically developed models.

XPS detected various forms of phosphorus in all lead/tin coupons exposed to phosphate-based inhibition. Most of the phosphorus was found as phosphate; however, polyphosphates were found in the lead scales formed in the presence of BOP and ZOP. Phosphorus was not present in any lead/tin scales formed in the presence of Si, pH_s, and pH_s+0.3, which demonstrated the necessity of phosphate inhibitor addition and formation of phosphate-based scales for inhibition of lead release. The thermodynamic model developed assuming that hydroxypyromorphite ($Pb_5(PO_4)_3OH$), a solid that has been known to persist in water distribution systems (Schock 1989), best predicted lead result, which also indicated that formation of a lead phosphate corrosion scale was necessary to inhibit lead release using phosphate-based inhibitors.

All XPS scans of lead/tin coupons exposed to ZOP indicated the presence of zinc predominantly in the form of ZnO. There was no observed benefit to applying ZOP over OP during the study.

The addition of Si demonstrated an improvement in lead release relative to pH control. This improvement was due to deposition of amorphous silica, which provided a passivating barrier that inhibited lead release.

Surface Roughness

The change in surface roughness of all materials exposed to pH-only, phosphate-based, or silica-based inhibition was independent of inhibitor type or dose but was dependent on the initial surface roughness. The change in roughness of UCI and GS coupons was correlated to alkalinity and temperature; however, roughness change was generally independent of water quality. The order of surface roughness by material was UCI > GS ≈ LCI > PVC for the used material and Cu > Pb for the new materials.

Although there was significant variability in roughness measurements, measurement of mean roughness of replicate coupons after inhibition exposure showed that experimental precision was within 20%. Replicate analysis of surface roughness showed that traditional measurements of precision or accuracy were not suited for assessment of experimental precision or accuracy relative to roughness measurements.

Residual

Chloramine residual in the PDS was not affected by the use of corrosion inhibitors. The chloramine demand in the PDS did not change by inhibitor type or dose. Only the pH control had an effect on chloramine residual decay with the rate of decay decreasing with the increase in water pH.

A first-order model for chloramine dissipation was developed as a function of pipe material, pipe geometry, initial chlorine dose, water quality (temperature, pH, and UV_{254}), and time. The model for chloramine was adequate to predict chloramine residual in the PDS at the 95% confidence level using independent data that was not used to develop the models. Hence, the model for chloramine residual was verified.

The decay rate of chloramine residual in the PDS was dependent on pipe material, temperature, and water quality. Lower rates of chloramine decay occurred in PVC and LCI pipes, whereas chloramine decayed more rapidly in unlined iron-based pipes (UCI and GS). Decline of chloramine residual increased in PDSs irrespective of material as time, UV_{254} (organic content), and temperature increased. The rate of decay decreased as water pH increased.

The wall decay rate for UCI and GS increased with flow velocity. PVC and LCI did not show variation in wall decay rate with Reynolds number. UCI and GS showed a region of rapid increase of wall decay rate under laminar conditions followed by a more gradual increase under transitional and fully turbulent flow. The empirical model was modified to include terms that account for the changes in wall decay rate with the Reynolds number.

Biological Characterization

The biological data showed that there was a statistically significant increase in both bulk water and biofilm bacterial densities (measured as HPC) with the addition of any of the four inhibitors.

Regression analysis of the biofilm data yielded a model showing that PVC and LCI biofilms were less dense than for unlined metals and could be managed by increasing residual maintenance with or without the presence of inhibitors. Biofilm density and bulk HPC levels increased with temperature, which is uncontrollable in distribution systems.

There was no direct correlation between BFHPC and bulk HPC for the combined chlorine residuals that were maintained in this study. The absence of any correlation may result in part by inactivation of sloughed bacteria by the bulk residual.

Inhibitor addition increased bulk HPC from 99% to 163% relative to the pH inhibition PDSs. The increase was about a factor of two times the control bulk HPC or slightly higher. This increase was observed under low-flow velocities and in relatively small-diameter pipe corresponding to areas of the distribution system most likely to have problems, such as dead ends, storage facilities, or any distribution system location with long HRTs.

This study, like the one at Cypress Creek that preceded it (Taylor et al. 2005), have consistently demonstrated that adequate residual maintenance is a very important factor, if not the most important factor, in managing bacterial water quality in nearly all distribution systems.

Nutrients other than carbon may limit biological growth. Acetate did stimulate growth, but so did phosphate salts, phosphorus-based inhibitors, and silicon-based inhibitors. Phosphorus was identified as one of the nutrients stimulating growth, and there was also an unidentified nutrient in the silicon-based inhibitor.

RECOMMENDATIONS

This section is organized by areas of investigation.

Metal Release

Recommendations pertaining to metal release are organized by specific metal to aid reader understanding.

Control of Iron

The experimental system was configured to permit evaluation of iron release from UCI and GS pipe materials for extreme distribution system conditions with low velocities and excessive HRTs. The design did not permit testing of iron release from the PDS pipe materials for hydraulic conditions representative of average distribution system conditions. The pipe sizes were also limited to a single diameter for each pipe material, so that experimental evaluation of the impact of pipe diameter could not be completed. The earlier study included development of a procedure to adjust the iron flux for alternate flow conditions (Reynolds number), pipe length, and pipe diameter. Experimental verification of that procedure with data from full-scale distribution system operation with varying hydraulic conditions and alternate pipe diameters would be desired. Full-scale verification would be appropriate for all of the alternate corrosion control strategies that were investigated: BOP, OP, ZOP, Si, and pH elevation. All of the simulation results suggest that the worst-case expectations for red water episodes would be experienced during laminar flow conditions as provided for the current experimental program.

The evaluation of the kinetics for reversion of the polyphosphate species to orthophosphate in the BOP inhibitor was similarly handicapped by the limited range of hydraulic conditions and pipe diameter. Resolution of the contribution of a bulk reaction and a wall reaction component to the overall first-order reversion rate constant would require data for different pipe diameters for each material. Because the wall reaction term is subject to variation with different mass transfer conditions, experiments must be conducted at varying Reynolds numbers to fully characterize the polyphosphate reversion rate. In addition, examination of variation in water quality on the reversion rates should be studied using a robust range of water quality.

Control of Copper and Lead

Federal regulations for copper and lead establish compliance standards for the 90th percentile concentration. The development of models in the current study is based on description of mean events (approximately 50th percentile events). An attempt was made using the pilot-scale data to define a probability distribution to relate the mean and 90th percentile concentrations. The pilot-scale data indicated that the ratio of 90th percentile to mean concentration was 1.3 and 2.5 for copper and lead, respectively. Examination of data from full-scale operating systems to verify the probability distribution would be relevant. It is quite likely, especially for lead, that greater sources of variation in concentration would be experienced in a prototype system that was present in the PDS because of differences in plumbing materials.

Electronoise

Standardization of EN measurement parameters is recommended for the water industry as other standards of measurement have been developed for the water community. Whereas this work has indicated that the EN models were as descriptive of metal release as water quality models, no EN information transfer could be effective unless standardized methods of measuring EN parameters are developed much like the methods and procedures defined in *Standard Methods*.

Correlation of EN parameters to metal release should be done in other waters throughout the United States and the world. There are fundamental similarities between the measurement of fundamental water quality parameters and EN parameters. As water quality monitoring is vital to the water community, EN parameters appear to have similar potential. Development of EN models that predict distribution system water quality need to be developed by the water community and verified in actual drinking water distribution systems, which would include user-friendly programs and tables for proactive control of metal release.

Identification and Modeling of Corrosion Scale Products

Continued work on identification of controlling scales and thermodynamic modeling is recommended using longer incubation time, more metals, more coupons, and larger ranges of water quality (i.e., alkalinity and pH) than was maintained in this study. Currently, very few sources of thermodynamic data for cupric phosphate and cupric silicate species are available in the literature. A better understanding of the solubility equilibrium constants for these solids would help determine whether these solids, in fact, are responsible for reductions in copper release with inhibitor addition. Further research into the aquatic chemistry of cupric silicates is recommended to determine how silicate species, and hence cupric silicate solubility, vary with pH. However, no binding energies for cupric phosphate and very few binding energies for cupric silicate solids were found in the literature. Additional binding energy data for these solids would help to support positive identifications of their presence in the copper scale. Expanded XPS libraries and thermodynamic databases would significantly enhance insight into formation of corrosion scales.

Work should be initiated on the incorporation of inhibiting materials such as phosphates and silica directly into the interior of pipe walls using nanoparticles. Development of such materials to augment copper pipe and lead-free solders could significantly reduce copper and lead release in drinking water, reduce pipe failure, and reduce cost.

Surface Roughness

Optical profilometry is a valid technique for surface roughness measurement of several materials common in water distribution systems, including GS, LCI, PVC, UCI, copper, and lead/tin.

The development of standard methods for the surface roughness measurement of corrosion coupons by optical profilometry would help standardize the collection and interpretation of results. Studies of surface-related water quality issues such as biofilm density, chlorine dissipation, and polyphosphate reversion could be conducted with the use of an optical profiler.

Investigations of relationships between biofilm density and surface roughness using an optical profiler and several corrosion coupons should be instigated. Within a given material, several coupons of varying roughness would be studied, ranging from new, to moderately rough, to very rough. An optical profiler would measure the surface roughness of each coupon before incubation in a pilot system. After incubation, each coupon would be scraped and cultured to test the biofilm density. In this way, a quantitative relationship relating biofilm density to pipe surface roughness could be developed.

A similar study could be conducted to test the effect of pipe surface roughness on chlorine decay. Within a given material, several coupons of varying roughness would be studied, ranging from new, to moderately rough, to very rough. Each coupon would be placed in a relatively inert glass reactor, and kinetic study of chlorine decay would be conducted. In this way, a quantitative relationship between chlorine decay and surface roughness could be developed. A similar reactor study could be conducted to determine the effect of surface roughness on reversion of polyphosphate to orthophosphate.

Although, in this study, models did not indicate a statistically significant effect of inhibitors on surface roughness over 3 months, similar studies with longer incubation times for corrosion coupons, such as 6 months to 1 year, may demonstrate some effect.

Head loss tests within the distribution system may be more sensitive to the practical effect of corrosion inhibitors on surface roughness. Whereas this study was conducted in a PDS simulating dead-end conditions within a distribution system, studies conducted at the distribution system level would be insightful. A utility could collect Hazen–Williams coefficients for the hydrodynamic model of its distribution system and compare head loss data for multiple locations within the distribution system before and after introduction of inhibitors to directly measure the practical effect of inhibitors on pipe surface roughness. If corrosion inhibitors had a beneficial effect on surface roughness in the distribution system, then one would expect to see decreases in energy consumption for the high-service pumps and booster pumps throughout the system.

Residual

UCI and GS pipes should be replaced as soon as possible because of the high rates of chloramine residual dissipation in these pipes and the possible associated complete loss of residual in the distribution systems. Monitoring of chloramine residuals in the distribution system should include increased monitoring based on water age, season, and pipe material. Dead-end or HRT locations, GS or UCI pipe use locations, and high-temperature seasons warrant increased monitoring of chloramine residual.

The model for chloramine residual could be used to predict disinfectant residual concentrations at different points within the distribution system and can determine the need for locations where rechlorination, flushing, or other means of residual maintenance should be implemented.

The chloramine residual model should be calibrated and verified in actual distribution systems. Calibration of the chloramine residual model in actual distribution systems for varying seasons, pipe materials, water quality conditions, hydraulic conditions, and pipe geometries is recommended.

The model for chloramine residual decay was developed mostly under laminar regime conditions, which might limit its use in full-scale distribution systems. Furthermore, the limited length of pipe utilized in the development of the residual models constrained the velocity of flow that could be achieved while maintaining the required 2-day HRT. The HRT of the experiments under turbulent flow conditions were short, which could have influenced the observed chloramine decay rates. Chloramine dissipation studies in full-scale distributions systems are required to verify the models, especially under turbulent flow regimes.

The pipe materials were restricted to a single diameter used in the PDS construction. Verification of the effect of pipe diameter on chloramine dissipation rates would require a greater range of pipe diameters for each material.

Biological Characterization

Because corrosion inhibitors stimulated growth, it becomes important to know how they affect specific populations that have possible regulatory or health impacts. Research is needed to determine how coliform and opportunistic pathogens are affected by the additional nutrients in corrosion inhibitors (or any other inorganic chemical leaving the plant in the finished water). Changes in the pipe surfaces caused by the use of corrosion inhibitors also might favor one set of organisms over another.

It is recommended that growth response for inorganics (whether with P17 and NOX or indigenous organisms) be evaluated by an initial screening of new chemical streams used during treatment for systems that have a history of biostability problems. It would also be important to have known standard solutions of nutrients (particularly phosphorus and nitrogen forms) evaluated in parallel with the chemical feeds being tested. Potential inorganic nutrients that might affect biostability include P, N, and S forms, metal cations needed for enzymatic catalysis, and so forth. Successive screening can be done by injecting a mixture of inorganics and comparing the growth response to a parallel control, which will determine which inorganics would stimulate microbial activity for a specific finished water.

If autotrophic bacteria are included in assessment of biostability, then ammonia and biodegradable organic nitrogen become potential limiting nutrients. This would be most relevant for systems using chloramines or in source waters with significant ammonia or other N species. However, it cannot be assumed that inorganic compounds do not also fuel heterotrophic growth, even in the absence of significant autotrophic activity such as in this study. The number and identities of the limiting nutrient(s) is probably source specific to a given water and plant.

The PDS data and the supplemental AOC experiments imply that for some waters, carbon may not be the only limiting nutrient. It is possible that other nutrients are limiting or that two or more nutrients may be simultaneously limiting biological growth. AOC assessment of biostability should include utilization of an inorganic growth control for any chemicals that are going to be added during treatment (as well as inorganic growth controls spiked with a phosphate salt) in parallel with the acetate carbon growth control specified in *Standard Methods*. It may be appropriate to quantify trace organic impurities in the chemical feed with NPDOC at high levels, or with AOC at low levels.

REFERENCES

APHA, AWWA, and WEF (American Public Health Association, American Water Works Association, and Water Environment Federation). 1998. *Standard Methods for the Examination of Water and Wastewater*. 20th ed. Washington, D.C.: APHA.

ASTM International. 1977. *User Verification of Performance for Precision and Trueness*. EP-15-A2. West Conshohocken, Pa.: ASTM.

Arévalo, J. 2003. Modeling the Effect of Pipe Material, Water Quality and Time on Chlorine Dissipation. Master's thesis. University of Central Florida, Orlando, Fla.

Arévalo, J., J.S. Taylor, J.D. Dietz, and G. Mutoti. 2004. Modeling Chlorine Dissipation in Distribution Systems. In *Proc. of AWWA ACE Conference*. Orlando, Fla.: AWWA.

Armon, R., and Y. Kott. 1995. Distribution Comparison Between Coliphages and Phages of Anaerobic Bacteria (*Bacteroides fragilis*) in Water Sources, and Their Reliability as Fecal Pollution Indicators in Drinking Water. *Water Science and Technology*, 31(5–6):215–222.

Batté, M.B., P. Koudjonou, L. Laurent, J. Mathieu, J. Coallier, and M. Prévost. 2003. Biofilm Responses to Ageing and to a High Phosphate Load in a Bench-scale Drinking Water System. *Water Research*, 37(6):1351–1361.

Beck, G.S. 1995. Deposition and Pipe Corrosion. *Water Environment and Technology*, February: 46–49.

Benjamin, M.M., S.H. Reiber, J.F. Ferguson, E.A. Vanderwerff, and M.W. Miller. 1990. *Chemistry of Corrosion Inhibitors in Potable Water*. Report No. 2P-3C-9056l-3/93-CM. Denver, Colo.: AwwaRF.

Biswas, P., C. Lu, and R.M. Clark. 1993. A Model for Chlorine Concentration Decay in Pipes. *Water Research*, 27(12):1715–1724.

Boffardi, B.P. 1993. The Chemistry of Polyphosphate. *Materials Performance*, 32(8):50–53.

Butterfield, P.W., A.K. Camper, J.A. Biederman, and A.M. Bargmeyer. 2002. Minimizing Biofilm in the Presence of Iron Oxides and Humic Substances. *Water Research*, 36(15):3898–3910.

Camper, A. 1996. *Factors Limiting Microbial Growth in Distribution Systems: Laboratory and Pilot-scale Experiments*. Denver, Colo.: AwwaRF and AWWA.

Characklis, W.G. 1988. *Bacterial Regrowth in Distribution System*. Denver, Colo.: AwwaRF and AWWA.

Characklis, W.G., and K.C. Marshall. 1990. *Biofilms*. New York: John Wiley.

Churchill, D.M., D.S. Mavinic, D.G. Neden, and D.M. MacQuarrie. 2000. The Effect of Zinc Orthophosphate and pH-Alkalinity Adjustment on Metal Levels Leached Into Drinking Water. *Canadian Journal of Civil Engineering*, 27(1):33–34.

Clark, R.M., and R.C. Haught. 2005. Characterizing Pipe Wall Demand: Implications for Water Quality Modeling. *Journal of Water Resources, Planning and Management*, 131(3):208–217.

Cottis, R.A. 2000. Simulation of Electrochemical Noise Due to Metastable Pitting. In *Journal of Corrosion Science and Engineering* [Online]. Available: <http://www.jcse.org/Volume3/Default.html>. [cited December 5, 2007]

Cullen, C.J. 2002. Equilibration of Pilot-scale Distribution Systems. Master's thesis. University of Central Florida, Orlando, Fla.

Davis, C.C., H.W. Chen, and M. Edwards. 2002. Modeling Silica Sorption to Iron Hydroxide. *Environmental Science and Technology*, 36:582–587.

De Beer, D., R. Srinivasan, and P. Stewart. 1994. Direct Measurement of Chlorine Penetration Into Biofilms During Disinfection. *Applied and Environmental Microbiology*, 60(12):4339–4344.

Dodrill, D.M., and M. Edwards. 1995. Corrosion Control on the Basis of Utility Experience. *Journal AWWA*, 87(7):87–99.

Doshi, B., W.M. Grayman, and D. Guastella. 2003. Field Testing the Chlorine Wall Demand in Distribution Mains. Presented at the AWWA Annual Conference, Anaheim, Calif., June 15–19, 2003.

Duirk, S.E., B. Gombert, J. Choi, and R.L. Valentine. 2002. Monochloramine Loss in the Presence of Humic Acid. *Journal of Environmental Monitoring*, 4(1):85–90.

Duirk, S.E., B. Gombert, J.P. Croué, and R.L. Valentine. 2005. Modeling Monochloramine Loss in the Presence of Natural Organic Matter. *Water Research*, 39(14):3418–3431.

Duranceau, S.J., D. Townley, and G.E.C. Bell. 2004. *Optimizing Corrosion Control in Water Distribution Systems*. Denver, Colo.: AwwaRF.

Eden, D.A. 1998. Electrochemical Noise—The First Two Octaves. Paper presented at Corrosion/98 NACE International, Houston, Texas.

Edwards, M., L. Hidmi, and D. Gladwell. 2003. Phosphate Inhibition of Soluble Copper Corrosion By-product Release. *Corrosion Science*, 44:1057–1071.

Edwards, M., S. Jacobs, and D. Dodrill. 1999. Desktop Guidance for Mitigating Pb and Cu Corrosion By-products. *Journal AWWA*, 91(5):66–77.

Edwards, M., L.S. McNeill, and T.R. Holm. 2001. Role of Phosphate Inhibitors in Mitigating Lead and Copper Corrosion. In *Distribution Systems*. Denver, Colo.: AwwaRF.

Escobar, I.C., and A.A. Randall. 2000. Sample Storage Impact on the Assimilable Organic Carbon (AOC) Bioassay. *Water Research*, 34(5):1680–1686.

Feng, Y., W.K. Teo, K.S. Siow, and A.K. Hsieh. 1996a. The Corrosion Behaviour of Copper in Neutral Tap Water. Part II: Determination of Corrosion Rates. *Corrosion Science*, 38(3):387–395.

Feng, Y., W.K. Teo, K.S. Siow, K.I. Tan, and A.K. Hsieh. 1996b. The Corrosion Behaviour of Copper in Neutral Tap Water. Part I: Corrosion Mechanisms. *Corrosion Science*, 38(3):369–385.

Fletcher, M., and K.C. Marshall. 1982. Are Solid Surfaces of Ecological Significance to Aquatic Bacteria? In *Advances in Microbial Ecology*. Edited by K.C. Marshall. New York: Plenum Press.

Frateur, I., C. Deslouis, L. Kiéné, Y. Lévi, and B. Tribollet. 1999. Free Chlorine Consumption Induced by Cast Iron Corrosion in Drinking Water Distribution Systems. *Water Research*, 33(8):1781–1790.

Gotpagar, J., S. Lyuksyutov, R. Cohn, E. Grulke, and D. Bhattacharyya. 1999. Reductive Dehalogenation of Trichloroethylene With Zero-Valent Iron: Surface Profiling Microscopy and Rate Enhancement Studies. *Langmuir*, 15:8412–8420.

Graham, M.J., and M. Cohen. 1976. Analysis of Iron Corrosion Products Using Mossbauer Spectroscopy. *Corrosion*, 32(11):432.

Guay, C., M. Rodriguez, and J. Sérodes. 2005. Using Ozonation and Chloramination to Reduce the Formation of Trihalomethanes and Haloacetic Acids in Drinking Water. *Desalination*, 176(1–3):229–240.

Hahn, R.B., and F.J. Welcher. 1968. *Inorganic Qualitative Analysis: A Short Course for Introductory Chemistry.* 2nd ed. Princeton, N.J.: Van Nostrand.

Hallam, N.B., J.R. West, C.F. Forster, J.C. Powell, and I. Spencer. 2002. The Decay of Chlorine Associated With the Pipe Wall in Water Distribution Systems. *Water Research*, 36(14):3479–3488.

Hatch, G.B., and O. Rice. 1945. Threshold Treatment of Water Systems: Corrosion Control and Scale Prevention With Glassy Phosphate. *Industrial and Engineering Chemistry*, 37(8):710–715.

Health and Welfare Canada. 1996. *Guidelines for Canadian Drinking Water Quality.* Ontario, Canada: Health Canada.

Hidmi, L., and M. Edwards. 1999. Role of Temperature and pH in $Cu(OH)_2$ Solubility. *Environmental Science and Technology*, 33(15):2607–2610.

Hogness, T.R., and W.C. Johnson. 1957. *An Introduction to Qualitative Analysis.* New York: Holt.

Holden, B., M. Greetham, B.T. Croll, and J. Scutt. 1995. The Effect of Changing Inter Process and Final Disinfection Reagents on Corrosion and Biofilm Growth in Distribution Pipes. *Water Science and Technology*, 32(8):213–220.

Holm, T.R., and M.R. Schock. 1991. Potential Effects of Polyphosphate Products on Lead Solubility in Plumbing Systems. *Journal AWWA*, 83(7):76–82.

Hozalski, R.M., E. Esbri-Amador, and C.F. Chen. 2005. Comparison of Stannous Chloride and Phosphate for Lead Corrosion Control. *Journal AWWA*, 97(3):89–102.

Hua, F., J.R. West, R.A. Barker, and C.F. Forster. 1999. Modeling of Chlorine Decay in Municipal Water Supplies. *Water Research*, 33(12):2735–2746.

Hudson, W.D. 1966. Studies of Distribution System Capacity in Seven Cities. *Journal AWWA*, 58(2):157–164.

Imran, S.A., J.D. Dietz, M. Ginasiyo, and W. Xiao. 2006. Optimizing Source Water Blends for Corrosion and Residual Control in Distribution Systems. *Journal AWWA*, 98(5):107–107.

Imran, S.A., J.D. Dietz, G. Mutoti, A.A. Randall, C.D. Cooper, J. S. Taylor, and V. Desai. 2005a. Red Water Release in Drinking Water Distribution Systems. *Journal AWWA*, 97(9):93–100.

Imran, S.A., J.D. Dietz, G. Mutoti, J.S. Taylor, and A.A. Randall. 2005b. Modified Larson Ratio Incorporating Temperature, Water Age and Electroneutrality Effects on Red Water Release. *Journal of Environmental Engineering*, 131(11):1514–1520.

Jadas-Hécart, A., A. El Morer, M. Stitou, P. Bouillot, and B. Legube. 1992. The Chlorine Demand of a Treated Water. *Water Research*, 26(8):1073–1084.

Jovancicevic, V., and B. Bauer. 1989. Calcium Phosphate Deposition on Iron in Oxygen-Containing Neutral Aqueous Solutions: An Electrochemical Approach. *Langmuir*, 5(1):261–267.

Kamrath, M., P. Mrozek, and A. Wieckowski. 1993. Composition Depth Profiles of Potential-Dependent Orthophosphate Film Formation on Iron Using Auger Electron Spectroscopy. *Langmuir*, 9(4):1016–1023.

Kiéné, L., W. Lu, and Y. Lévi. 1998. Relative Importance of the Phenomena Responsible for Chlorine Decay in Drinking Water Distribution Systems. *Water Science and Technology*, 38(6):219–237.

King, E.J. 1959. *Qualitative Analysis and Electrolytic Solutions.* New York: Harcourt Brace.

Klueh, K.G., and R.B. Robinson. 1988. Sequestration of Iron in Groundwater by Polyphosphates. *Journal of Environmental Engineering*, 114(5):1192–1199.

Koudelka, M., J. Sanchez, and J. Augustynski. 1982. On the Nature of Surface Films Formed on Iron in Aggressive and Inhibiting Polyphosphate Solutions. *Journal of the Electrochemical Society*, 129(6):1186–1191.

Larson, T.E. 1957. Evaluation of the Use of Polyphosphate in the Water Industry. *Journal AWWA*, 83(12):1581–1586.

Larson, T.E., and F.W. Sollo. 1967. Loss in Water Main Carrying Capacity. *Journal AWWA*, 59(12):1565–1572.

Latimer, W.M. 1952. The Oxidation States of the Elements and Their Potentials in Aqueous Solutions. Englewood Cliffs, N.J.: Prentice-Hall.

Laurent, P., and P. Servais. 1995. Fixed Bacterial Biomass Estimated by Potential Exoproteolytic Activity. *Canadian Journal of Microbiology*, 41(8):749–752.

LeChevallier, M.W., T. Babcock, and R.G. Lee. 1987. Examination and Characterization of Distribution System Biofilm. *Applied and Environmental Microbiology*, 53(12):2714–2724.

LeChevallier, M.W., N.E. Shaw, L.A. Kaplan, and T.L. Bott. 1993. Development of a Rapid Assimilable Organic Carbon Method for Water. *Applied and Environmental Microbiology*, 59(5):1526–1531.

LePuil, M., Y.C. Chang, A.A. Randall, and J.S. Taylor. 2005. Effect of Distribution System Materials and Water Quality on HPC and Biofilm Proliferation. In *Water Quality in the Distribution System*. Edited by W.C. Lauer. Denver, Colo.: AWWA.

Lippold, S., and J. Podlesny. 1998. *Wyko Surface Profilers Technical Reference Manual*. Tuscon, Ariz.: Veeco Instruments.

MacQuarrie, D.M., D.S. Mavinic, and D.G. Neden. 1997. Greater Vancouver Water District Drinking Water Corrosion Inhibitor Testing. *Canadian Journal of Civil Engineering*, 24:34–52.

Maddison, L.A, G.A. Gagnon, and J.D. Eisnor. 2001. Corrosion Control Strategies for the Halifax Regional Distribution System. *Canadian Journal of Civil Engineering*, 28:305–313.

Malcolm Pirnie, Inc. 1997. *Complaint Assessment for Tucson Water*. White Plains, N.Y.: Malcolm Pirnie.

Margerum, D.W., L.M. Schurter, J. Hobson, and E.E. Moore. 1994. Water Chlorination Chemistry: Nonmetal Redox Kinetics of Chloramine and Nitrite Ion. *Environmental Science and Technology*, 28(2):331–337.

McNeill, L.S., and M. Edwards. 2000. Phosphate Inhibitors and Red Water in Stagnant Iron Pipes. *Journal of Environmental Engineering*, 126(12):1096–1096.

McNeill, L.S., and M. Edwards. 2001. Iron Pipe Corrosion in Distribution Systems. *Journal AWWA*, 93(7):88–100.

Moriarty, B.E. 1990. Surface Studies of Corrosion Inhibitors in Cooling Water Systems. *Materials Performance*, 29(1):45–48.

Morin, P., V. Gauthier, S. Saby, and J.C. Block. 1999. Bacterial Resistance to Chlorine Through Attachment to Particles and Pipes Surfaces in Drinking Water Distribution Systems. In *Biofilms in Aquatic Systems*. Cambridge, U.K.: Royal Society of Chemistry.

Mutoti, G., J.D. Dietz, S.A. Imran, J.S. Taylor, and C.D. Cooper. 2007a. Development of a Novel Iron Release Flux Model for Distribution Systems. *Journal AWWA*, 99(1):102–111.

Mutoti, I., J.D. Dietz, J. Arévalo, and J.S. Taylor. 2007b. Combined Chlorine Dissipation: Pipe Material, Water Quality, and Hydraulic Impacts. *Journal AWWA*, 99(10):97–106.

NACE (National Association of Corrosion Engineers). 1993. Electrochemical Noise as the Basis of Corrosion Monitoring. Paper presented at the 12th International Corrosion Congress, Houston, Texas.

Nagy, L.A., and B.H Olson. 1985. Occurrence and Significance of Bacteria, Fungi and Yeasts Associated With Distribution Pipe Surfaces. In *Technology Conference Proceedings: Advances in Water Analysis and Treatment*. Devner, Colo.: AWWA.

Ndiongue, S., P.M. Huck, and R.M. Slawson. 2005. Effects of Temperature and Biodegradable Organic Matter on Control of Biofilms by Free Chlorine in a Model Drinking Water Distribution System. *Water Research*, 39:953–964.

Nikuradse, J. 1950. Laws of Flow in Rough Pipes. Translated from "Strömungsgesetze in rauhen Rohren" *Forsch. Arb. Ing.-Wes*. No. 361 (1933).

Norton, C.D., and M.W. LeChevallier. 1997. Chloramination: Its Effect on Distribution System Water Quality. *Journal AWWA*, 89(7):66–77.

Ollos, P.J., P.M. Huck, and R.M. Slawson. 2001. Factors Affecting Biofilm Accumulation in Model Distribution Systems. *Journal AWWA*, 95:87–96.

Ollos, P.J., R.M. Slawson, and P.M. Huck. 1998. Bench Scale Investigations of Bacterial Regrowth in Drinking Water Distribution Systems. *Water Science and Technology*, 38(8–9):275–282.

Palit, A., and S.O. Pehkonenb. 2000. Copper Corrosion in Distribution Systems: Evaluation of a Homogeneous Cu_2O Film and a Natural Corrosion Scale as Corrosion Inhibitors. *Corrosion Science*, 42(10):1801–1822.

Pankow, J.F. 1991. *Aquatic Chemistry Concepts*. Chelsea, Mich.: Lewis Publishers.

Persson, P., N. Nilsson, and S. Sjoberg. 1996. Structure and Bonding of Orthophosphate Ions at the Iron Oxide–Aqueous Interface. *Journal of Colloid and Interface Science*, 177(1):263–275.

Pisigan, R.A., and J.E. Singley. 1985. Effects of Water Quality Parameters on the Corrosion of Galvanized Steel. *Journal AWWA*, 77(11):76–82.

Plottu-Pecheux, A., C. Democrate, B. Houssais, D. Gatel, and J. Cavard. 2001. Controlling the Corrosiveness of Blended Waters. *Desalination*, 138(1–3):237–249.

Pontius, F.W. 1991. The New Lead and Copper Rule. *Journal AWWA*, 83(7):12–20.

Powell, J.C., N.B. Hallam, J.R. West, C.F. Forster, and J. Simms. 2000. Factors Which Control Bulk Chlorine Decay Rates. *Water Research*, 34(1):117–126.

PQ Corporation. 2001. *PQ Soluble Silicates: For Protection of Water Systems From Corrosion*. PQ Bulletin 37-3. Valley Forge, Pa.: PQ Corporation.

Pryor, M.J., and M. Cohen. 1951. The Mechanism of the Inhibition of the Corrosion of Iron by Solutions of Sodium Orthophosphate. *Journal of the Electrochemical Society*, 98(7):263.

Reiber, S.H. 1989. Copper Plumbing Surfaces: An Electrochemical Study. *Journal AWWA*, 81(7):114–122.

Rossman, L.A., R.M. Clark, and W.M. Grayman. 1994. Modeling Chlorine Residual in Drinking Water Distribution Systems. *Journal of Environmental Engineering*, 120(4):803–820.

Rushing, J.C., L.S. McNeill, and M. Edwards. 2003. Some Effects of Aqueous Silica on the Corrosion of Iron. *Water Research*, 37:1080–1090.

Sarin, P., V.L. Snoeyink, J. Bebee, W.M. Kriven, and J.A. Clement. 2001. Physico-chemical Characteristics of Corrosion Scales in Old Iron Pipes. *Water Research*, 35(12):2961–2969.

Sarin, P., V.L. Snoeyink, D.A. Lytle, and W.M. Kriven. 2004. Iron Corrosion Scales: Model for Scale Growth, Iron Release, and Colored Water Formation. *Journal of Environmental Engineering–ASCE*, 130(4):364–373.

Schenk, G.H. 1996. *Qualitative Analysis and Ionic Equilibrium*. 3rd ed. Boston: Houghton Mifflin.

Schindler, P., H. Althaus, F. Hofer, and W. Minder. 1965. Solubility Products of Zinc Oxide, Copper Hydroxide, and Copper Oxide as a Function of Particle Size and Molecular Surface Area: A Contribution to the Thermodynamics of Liquid Surface Boundaries. *Helvetica Chimica Acta*, 48(5):1204–1215.

Schindler, P.W. 1967. Heterogeneous Equilibria Involving Oxides, Hydroxides, Carbonates and Hydroxide Carbonates. In *Equilibrium Concepts in Natural Water Systems*. Washington, D.C.: American Chemical Society.

Schock, M.R. 1989. Understanding Corrosion Control Strategies for Lead. *Journal AWWA*, 81(7):88.

Schock, M.R. 1999. Internal Corrosion and Deposition Control. In *Water Quality and Treatment*. New York: McGraw-Hill.

Schock, M.R., and J. C. Fox. 2001. Solving Copper Corrosion Problems While Maintaining Lead Control in a High Alkalinity Water Using Orthophosphate. Presented at the AWWA Annual Conference, Cleveland, Ohio, August 30, 2001.

Schock, M.R., D.A. Lytle, and J.A. Clement. 1995. *Effect of pH, DIC, Orthophosphate and Sulfate on Drinking Water Cuprosolvency*. Risk Management Research Report EPA/600/R-95/085. Cincinnati, Ohio: USEPA.

Schock, M.R., D.A. Lytle, A.M. Sandvig, J. Clement, and S.M. Harmon. 2005. Replacing Polyphosphate With Silicate to Solve Lead, Copper and Source Water Iron Problems. *Journal AWWA*, 97(11):84–93.

Seal, S., and T.L. Barr. 2001. Application of Photoelectron Spectroscopy in Inorganic and Organic Material System. In *Advances in Surface Science*. Edited by H.S. Nalwa. San Diego, Calif.: Academic Press.

Sharp, W.W., and T. Walski. 1988. Predicting Internal Roughness in Water Mains. *Journal AWWA*, 80(11):34–40.

Sillen, L.G., and A.E. Martell. 1964. *Stability Constants of Metal–Ion Complexes*. London: The Chemical Society, Burlington House.

Stewart, P.S., and J.B. Raquepas. 1995. Implications of Reaction-Diffusion Theory for the Disinfection Theory for the Disinfection of Microbial Biofilms by Reactive Antimicrobial Agents. *Chemical Engineering Science*, 50(19):3099–3104.

Tan, Y.J., S. Bailey, and B. Kinsell. 1996. The Monitoring of the Formation and Destruction of Corrosion Inhibitor Films Using Electrochemical Noise Analysis (ENA). *Corrosion Science*, 38(10):1681–1695.

Tang, J. 2003. Effect of Source Water Blending on Iron and Lead Release: Thermodynamic and Statistical Modeling. Ph.D. diss. University of Central Florida, Orlando, Fla.

Tang, Z., S.K. Hong, W. Xiao, and J. Taylor. 2006. Impacts of Blending Ground, Surface and Saline Waters on Lead Release in Drinking Water Distribution System. *Water Research*, 40(5):943–950.

Taylor, J.S., J.D. Dietz, A.A. Randall, S.K. Hong, C.D. Norris, L.A. Mulford, J.M. Arévalo, S. Imran, M. LePuil, S. Liu, I. Mutoti, J. Tang, W. Xiao, C. Cullen, R. Heaviside, A. Mehta, M. Patel, F. Vasquez, and D. Webb. 2005. *Effects of Blending on Distribution System Water Quality*. Denver, Colo.: AwwaRF and Tampa Bay Water.

Taylor, J.S., J.D. Dietz, S.K. Hong, L.A. Mulford, and A.A. Randall. 2001. *Periodic Report 1: Required Treatment and Water Quality Criteria for Distribution System Blending of Treated Surface, Ground and Saline Sources*. Denver, Colo.: AwwaRF.

USEPA (U.S. Environmental Protection Agency). 1991. Drinking Water; Maximum Contaminant Level Goals and National Primary Drinking Water Regulations for Lead and Copper; Final Rule. EPA-812-Z-94-001. Columbus, Ohio: USEPA.

Valentine, R.L., and C.T. Jafvert. 1988. General Acid Catalysis of Monochloramine Disproportionation. *Environmental Science and Technology*, 22(6):691–696.

Van der Wende, E., W.G. Characklis, and D.B. Smith. 1989. Biofilms and Bacterial Drinking Water Quality. *Water Research*, 23(10):1313–1322.

Vasconcelos, J.J., L.A. Rossman, W.M. Grayman, P.F. Boulos, and R.M. Clark. 1997. Kinetics of Chlorine Decay. *Journal AWWA*, 89(7):54–65.

Vikesland, P.J., K. Ozekin, and R.L. Valentine. 1998. Effect of Natural Organic Matter on Monochloramine Decomposition: Pathway Elucidation Through the Use of Mass and Redox Balances. *Environmental Science and Technology*, 32(10):1409–1416.

Vikesland, P.J., K. Ozekin, and R.L. Valentine. 2001. Monochloramine Decay in Model and Distribution System Waters. *Water Research*, 35(7):1766–1776.

Vikesland, P.J., and R.L. Valentine. 2002a. Iron Oxide Surface-catalyzed Oxidation of Ferrous Iron by Monochloramine: Implications of Oxide Type and Carbonate on Reactivity. *Environmental Science and Technology*, 36(3):512–519.

Vikesland, P.J., and R.L. Valentine. 2002b. Modeling the Kinetics of Ferrous Iron Oxidation by Monochloramine. *Environmental Science and Technology*, 36(4):662–668.

Volk, C., E. Dundore, J. Schiermann, and M. LeChevallier. 2000. Practical Evaluation of Iron Corrosion Control in a Drinking Water Distribution System. *Water Research*, 34(6):1967–1974.

Wagner, C.D., A.V. Naumkin, A. Kraut-Vass, J.W. Allison, C.J. Powell, and J.R. Rumble Jr. 2003. NIST X-ray Photoelectron Spectroscopy Database. NIST Standard Reference Database 20, Version 3.5. Washington D.C.: Measurement Services Division of the National Institute of Standards and Technology (NIST) Technology Services.

Walski, T.M., D.V. Chase, D.A. Savic, W. Grayman, S. Beckwith, and E. Koelle. 2003. *Advanced Water Distribution Modeling and Management*. Waterbury, Conn.: Haestad Press.

Xiao, W. 2004. Effect of Source Water Blending on Copper Release in Pipe Distribution System: Thermodynamic and Empirical Models. Ph.D. diss. University of Central Florida, Orlando, Fla.

Yousef, Y.A., D.B. Williams, and J.P. Rozza. 1994. *Evaluation of Corrosion Inhibitors*. Report for Pinellas County Water System, Largo, Fla.

Zacheus, O.M., E.I. Ilvanainen, T.K. Nissinen, M.J. Lethola, and P.J. Martikainen. 2000. Bacterial Biofilm Formation on Polyvinyl Chloride, Polyethylene and Stainless Steel Exposed to Ozonated Water. *Water Research*, 34(1):63–70.

ABBREVIATIONS

AC	alternating current
ANOVA	analysis of variance
AOB	ammonia oxidizing bacteria
AOC	assimilable organic carbon
avg.	average
AWWA	American Water Works Association
AwwaRF	Awwa Research Foundation
BDOC	biodegradable dissolved organic carbon
BFHPC	biofilm heterotrophic plate count
BF-TC/*E. coli*	biofilm total coliform *Escherichia coli*
BOM	biodegradable organic matter
BOP	blended ortho-polyphosphate
°C	degrees Celsius
CET	SmartCET, patented corrosion monitoring electronoise equipment
cfu	colony-forming units
CL	control limit
cm	centimeter
CPDF	cumulative probability density function
cpu	chloroplatinate unit
CR	corrosion rate
DBP	disinfection by-product
DC	direct current
DCu	dissolved copper
DFe	dissolved iron
DO	dissolved oxygen
DPb	dissolved lead
DPD	N,N-diethyl-p-phenylenediamine
DV	dummy variable
ECN	electrochemical noise
ECNCR	electrochemical noise corrosion rate
EDS	energy dispersive X-ray spectroscopy
EDTA	ethylenediaminetetraacetic acid
EN	electronoise
ESCA	electron spectroscopy for chemical analysis
eV	electron volt
ft^2	square feet
ft/sec	feet per second

g	gram
gal	gallon
gpm	gallons per minute
GS	galvanized steel
GW	groundwater
h, hr	hour
HDA	harmonic distortion analysis
HMCR	harmonic distortion analysis corrosion rate
HPC	heterotrophic plate count
HRT	hydraulic residence time
IC	ion chromatograph
ICP	inductively- (or ion-) coupled plasma spectroscopy
IHWW	Indian Hill Water Works
IKRT	kurtosis of the current
in.	inch
in./hr	inches per hour
IRWD	Irvine Ranch Water District
ISKW	skew of the current
kPa	kilopascal
L	liter
LCI	lined cast iron
LCR	Lead and Copper Rule
LPR	linear polarization resistance
LPRCR	linear polarization resistance corrosion rate
LR	Larson's ratio
LRM	modified Larson's ratio
LR1-T	Larson's ratio with temperature
LSI	Langelier saturation index
LVVWD	Las Vegas Valley Water District
m	meter
m^2	square meter
m^3/d	cubic meters per day
meq/L	milliequivalents per liter
mg	milligram
MG	member government
mgd	million gallons per day
mg/L	milligrams per liter
m/hr	meters per hour
mL	milliliter
mm/yr	millimeters per year

MPN	most probable number
mpy	mils per year
mV	millivolt
μg/L	micrograms per liter
μm	micrometer
μmho/cm	micromhos per centimeter (conductivity unit)
μS/cm	microsiemens per centimeter
N	number of observations
N/A	not applicable
NIST	National Institute of Science and Technology
nm	nanometer
NOM	natural organic matter
NOX	*Spirillum* NOX
NPDOC	nonpurgeable dissolved organic carbon
ntu	nephelometric turbidity unit
OP	orthophosphate
P17	*Pseudomonas fluorescens* P17
PCU	Pinellas County Utilities
PDF	probability density function
PDS	pilot distribution system
PEPA	protein exoproteolytic activity
PF	pitting factor
pH_s	saturation pH
PI	phase I
PII	phase II
PIII	phase III
PIV	phase IV
PKRT	kurtosis of the potential
PP	poly-phosphorus
ppb	parts per billion
ppm	parts per million
psi	pounds per square inch
PSKW	skew of the potential
PVC	polyvinyl chloride
QA	quality assurance
QA/QC	quality assurance/quality control
QC	quality control
RO	reverse osmosis
RPD	relative percent difference
RSD	relative standard deviation
RSOL	solution resistance

SEM	scanning electron microscopy
SM	*Standard Methods*
std. dev.	standard deviation
SW	surface water
T, temp.	temperature
TBW	Tampa Bay Water
TBW I	Tampa Bay Water 1 (previous study)
TBW II	Tampa Bay Water 2 (this study)
TC	total coliform
TCE	trichloroethane
TCP	tailored collaboration project
TCu	total copper
TDS	total dissolved solids
TFe	total iron
TKN	total Kjeldahl nitrogen
TOC	total organic carbon
TP	total phosphorus
TPb	total lead
UCF	University of Central Florida
UCI	unlined cast iron
UCL	upper control limit
U.S.	United States
USEPA	United States Environmental Protection Agency
UV	ultraviolet
UV_{254}	ultraviolet absorbance at 254 nm
UWL	upper warning limit
V	volts
vs.	versus
WL	warning limit
XPS	X-ray photoelectron spectroscopy
ZOP	zinc orthophosphate

CONVERSION TABLE

	To convert		
	From customary units	To SI units	Multiply by
Area	mi^2	km^2	2.590
	or	m^2	2.590×10^6
	acre	ha	4.047×10^{-1}
	ft^2	m^2	9.290×10^{-2}
Flow rate	mgd	ML/d	3.785
	acre foot/year	cubic meters/day	3.38137
	acre foot/year	gallons/day	8.9268×10^2
	gph	m^3/h	3.785×10^{-3}
	gpm	m^3/s	6.309×10^{-5}
	or	L/s	6.309×10^{-2}
	ft^3/h	m^3/h	2.832×10^{-2}
	ft^3/min	m^3/s	4.720×10^{-4}
	or	L/s	4.720×10^{-1}
Head loss	ft	kPa	2.989
	or	m	3.048×10^{-1}
Horsepower	hp	metric hp	1.014
	or	kilowatt	7.457×10^{-1}
Length	mile statute (m)	kilometer	1.609
	yd	meter	9.1440×10^{-1}
	ft	m	3.048×10^{-1}
	in	millimeter	2.540×10^{-1}
Pipe cross-section	in^2	mm^2	6.452×10^2
Pressure	psi	kPa	6.895
	or	bar	6.890×10^{-2}
Velocity	ft/s	m/s	3.048×10^{-1}
Volume	Acre foot	US gal	3.258×10^5
	US gal	m^3	3.785×10^{-3}
	or	ML	3.785×10^{-6}
	or	L	3.785×10
	or	Imperial gal	8.330×10^{-1}
	yd^3	m^3	7.645×10^{-1}
	or	*L*	7.645×10^2
	ft^3	m^3	2.832×10^{-2}
	or	ML	3.785×10^{-5}
	in^3	cm^3	1.638×10^1
	or	L	1.638×10^{-2}
Weight	lg ton	metric ton	1.016
	short ton	metric ton	9.070×10^{-1}
	lb	kilogram	4.536×10^{-1}
Temperature	°F = (9/5) (°C + 32)		

IWA Publishing's authorised EU representative for General Product Safety Regulations is Diane D'Arras, 15 rue Duret, 75116 Paris, France, e-mail: safety@iwap.co.uk.

Printed and bound by CPI Group (UK) Ltd, Croydon, CR0 4YY
17/09/2026
02218808-0008